"The earliest roots of urban ecology lie in the study of wild and domestic organisms in cities. Nilon and Aronson show the depth and rigor of this research tradition, illustrate its crucial application to conservation of global biodiversity, and point to the continued need and creative future for urban biodiversity studies."

Steward T.A. Pickett, *Distinguished Senior Scientist, Cary Institute of Ecosystem Studies, Millbrook, New York, USA*

"This is a blockbuster book for the urban century. From the fundamentals, science and theory of urban biodiversity to practical management issues, Nilon and Aronson have produced a must-have that should be on the bookshelf of every urban professional."

Karen C. Seto, *Frederick C. Hixon Professor of Geography & Urbanization Science, Yale University, New Haven, Connecticut, USA*

"Rapid urbanization at global scale exerts serious threats to urban biodiversity which is of great importance for conservation in terms of implementing the Kunming-Montreal Global Biodiversity Framework. Urban development can not be successful unless its biodiversity is sustainably utilized. The book offers readers opportunities to understand concepts and theory for urban biodiversity and learn best practices in using and conserving it."

Keping Ma, *Vice Chair and Secretary General, Biodiversity Committee, Chinese Academy of Sciences, Beijing, China*

"Old concepts of cities and nature being two separate worlds have been overturned. Studies of the close interplay of urban conditions and high biodiversity are yielding a new understanding of the structure of urban nature and its importance to the human experience. This book sets a new standard for explaining the ecological foundations of urban nature, the current state of knowledge, the role of human actions and management on city species, and a path forward for new research and new collaborations. Cities are expanding rapidly across the globe; this Handbook will be a guide to incorporating modern ecological perspectives to the design of healthier, more resilient urban landscapes and improving conservation of the rich urban biodiversity."

Steven N. Handel, *Distinguished Professor of Ecology, Rutgers, The State University of New Jersey, New Brunswick, New Jersey, USA*

ROUTLEDGE HANDBOOK OF URBAN BIODIVERSITY

This handbook provides a state-of-the-art, comprehensive overview of the expanding field of urban biodiversity.

The field of urban biodiversity has emerged from within the broad discipline of urban ecology in the past two decades and is now a significant field in its own right. In view of this, the *Routledge Handbook of Urban Biodiversity* presents a thorough treatment of this field detailing the history of urban biodiversity, theoretical foundations, current state of knowledge, and application of that knowledge. The handbook is split into four parts:

- Part I: Setting the Stage for Urban Biodiversity Research and Practice
- Part II: Foundational Concepts and Theory in Urban Biodiversity Research
- Part III: Population and Community Ecology of Key Urban Taxa
- Part IV: Urban Biodiversity Practice: Management, Planning, and Design for Healthy Communities

This volume contains interdisciplinary and global contributions from established and early career academics as well as professionals and practitioners, addressing two key fields in urban biodiversity: fundamental research focused on answering questions about the mechanisms explaining the distribution of species among and within cities; and applied research and work by practitioners to address concerns about urban biodiversity conservation, restoration, planning, design, and public involvement.

This handbook is essential reading for students, academics, and professionals interested and working in the fields of urban biodiversity, ecology, nature conservation, urban planning, and landscape architecture.

Charles H. Nilon is a professor in the School of Natural Resources at the University of Missouri. His research and teaching focus on urban wildlife conservation, urban ecology, and the human dimensions of wildlife conservation. Dr. Nilon is a co-director of three urban biodiversity synthesis projects, including the "Urban Biodiversity Research Coordination Network" (UrBioNet), and is co-president of the International Urban Biodiversity and Design Network (URBIO).

Myla F.J. Aronson is an assistant professor in the Department of Ecology, Evolution, and Natural Resources, at Rutgers, The State University of New Jersey. Her research and teaching focuses on the patterns, drivers, and management of biodiversity in human dominated landscapes, particularly plant population and community dynamics in urban areas. Dr. Aronson is co-director of three urban biodiversity synthesis projects, including the "Urban Biodiversity Research Coordination Network" (UrBioNet), and serves on the Advisory Board for the International Urban Biodiversity and Design Network (URBIO).

ROUTLEDGE ENVIRONMENT AND SUSTAINABILITY HANDBOOKS

ROUTLEDGE HANDBOOK OF ENERGY COMMUNITIES AND SMART CITIES
Edited by Maciej M. Sokołowski and Anna Visvizi

ROUTLEDGE HANDBOOK OF GLOBAL LAND AND RESOURCE GRABBING
Edited by Andreas Neef, Chanrith Ngin, Tsegaye Moreda Shegro, and Sharlene Mollett

THE ROUTLEDGE HANDBOOK OF ECOMEDIA STUDIES
Edited by Antonio López, Adrian Ivakhiv, Stephen Rust, Miriam Tola, Alenda Y. Chang and Kiu-wai Chu

ROUTLEDGE HANDBOOK OF URBAN BIODIVERSITY
Edited by Charles H. Nilon & Myla F.J. Aronson

ROUTLEDGE HANDBOOK OF URBAN BIODIVERSITY

Edited by Charles H. Nilon and Myla F.J. Aronson

Routledge
Taylor & Francis Group
LONDON AND NEW YORK

earthscan
from Routledge

Designed cover image: © Getty images

First published 2024
by Routledge
4 Park Square, Milton Park, Abingdon, Oxon OX14 4RN

and by Routledge
605 Third Avenue, New York, NY 10158

Routledge is an imprint of the Taylor & Francis Group, an informa business

British Library Cataloguing-in-Publication Data
A catalogue record for this book is available from the British Library

ISBN: 978-0-367-44454-9 (hbk)
ISBN: 978-1-032-49213-1 (pbk)
ISBN: 978-1-003-01612-0 (ebk)

DOI: 10.4324/9781003016120

Typeset in Times New Roman
by Apex CoVantage, LLC

CONTENTS

CONTRIBUTORS

Riley Andrade Department of Wildlife Ecology and Conservation, University of Florida, Gainesville, Florida, USA

Myla F.J. Aronson Department of Ecology, Evolution, and Natural Resources, Rutgers, The State University of New Jersey, New Brunswick, New Jersey, USA

Katherine C.R. Baldock Department of Geography and Environmental Sciences, Northumbria University, Newcastle, upon Tyne, UK

Marc Barra Agence Régionale de la Biodiversité, L'Institut Paris Region, Paris, France

Loren B. Byrne Department of Biology, Marine Biology and Environmental Science, Roger Williams University, Bristol, Rhode Island, USA

Sarel S. Cilliers Unit of Environmental Sciences and Management, North-West University, Potchefstroom, South Africa

Martin Dallimer Sustainability Research Institute, School of Earth and Environment, University of Leeds, Leeds, UK

Zoe G. Davies Durrell Institute of Conservation and Ecology (DICE), School of Anthropology and Conservation, University of Kent, Canterbury, UK

Cynnamon Dobbs Department of Natural Resources and the Environment, College of Agriculture, Health and Natural Resources, University of Connecticut, Storrs, Connecticut, USA

Monika Egerer Department of Life Science Systems, School of Life Sciences, Technical University of Munich, Freising, Germany

Augustin C. Engman Department of Forestry, Wildlife and Fisheries, University of Tennessee, Knoxville, Tennessee

Karl L. Evans Ecology and Evolutionary Biology, School of Biosciences, The University of Sheffield, Sheffield, UK

Jessica C. Fisher Durrell Institute of Conservation and Ecology (DICE), School of Anthropology and Conservation, University of Kent, Canterbury, UK

Lauren J. Frazee Graduate Program in Ecology and Evolution, Rutgers, The State University of New Jersey, New Brunswick, New Jersey, USA

Mirijam Gaertner Centre for Invasion Biology, Department of Botany and Zoology, Stellenbosch University, South Africa; and Nürtingen-Geislingen University of Applied Sciences (HFWU), Nürtingen, Germany

Mark A. Goddard Department of Geography and Environmental Sciences, Northumbria University, Newcastle upon Tyne, UK

Dagmar Haase Institute of Geography, Humboldt Universität zu Berlin, Germany; and Department of Computational Landscape Ecology, Helmholtz Centre for Environmental Research – UFZ, Halle (Saale), Germany

Amy K. Hahs School of Ecosystem and Forest Sciences, The University of Melbourne, Melbourne, Victoria, Australia

Peter Ibsen Department of Botany and Plant Sciences, University of California Riverside, Riverside, California, USA; and United States Geological Survey, Geosciences, and Environmental Change Science Center, Denver, Colorado, USA

Maria Ignatieva School of Design, University of Western Australia, Perth, Western Australia, Australia

Katherine N. Irvine Social, Economic, and Geographic Sciences Department, James Hutton Institute, Craigbuckler, Aberdeen, UK

G. Darrel Jenerette Department of Botany and Plant Sciences, University of California Riverside, Riverside, California, USA

Lea R. Johnson Longwood Gardens, Kennett Square, Pennsylvania, USA

Kirsten Jung Evolutionary Ecology and Conservation Genomics, Ulm University, Ulm, Germany

Dave Kendal Future in Nature, Melbourne, Victoria, Australia

Sonja Knapp Department of Community Ecology, Helmholtz Centre for Environmental Research – UFZ, Halle (Saale), Germany

Ryo Kohsaka The University of Tokyo, Graduate School of Agricultural and Life Sciences, Tokyo, Japan

Ingo Kowarik Technische Universität Berlin, Institute of Ecology, Berlin, Germany; and Berlin-Brandenburg Institute of Advanced Biodiversity Research (BBIB), Berlin, Germany

Dion Kucera Department of Botany and Plant Sciences, University of California Riverside, Riverside, California, USA

Frank A. La Sorte Cornell Lab of Ornithology, Cornell University, Ithaca, New York, USA

Christopher A. Lepczyk College of Forestry, Wildlife, and Environment, Auburn University, Auburn, Alabama, USA

Susannah B. Lerman USDA Forest Service Northern Research Station, Amherst, Massachusetts, USA

Mia K. Lippey Department of Entomology and Nematology, UC Davis, Davis, California, USA

Jeremy Lundholm St Mary's University, Halifax, Nova Scotia, Canada

Ian MacGregor-Fors University of Helsinki, Faculty of Biological and Environmental Sciences, Ecosystems, and Environment Research Programme, Lahti, Finland

J. Scott MacIvor Department of Biological Sciences, University of Toronto Scarborough, Toronto, Ontario, Canada

Emily K. Meineke Department of Entomology and Nematology, UC Davis, Davis, California, USA

Emily Minor Department of Biological Sciences, University of Illinois at Chicago, Chicago, Illinois, USA

Ulla Mörtberg Department for Sustainable Development, Environmental Science and Engineering, KTH Royal Institute of Technology, Stockholm, Sweden

Maximilian Nawrath Sustainability Research Institute, School of Earth and Environment, University of Leeds, Leeds, UK

Charles H. Nilon School of Natural Resources, University of Missouri, Columbia, Missouri, USA

Karen M. O'Neill Human Ecology Department, Rutgers, The State University of New Jersey, New Brunswick, New Jersey, USA

Josep Padullés Cubino Department of Botany and Zoology, Faculty of Science, Masaryk University, Brno, Czech Republic

Kirsten M. Parris School of Ecosystem and Forest Sciences, The University of Melbourne, Melbourne, Victoria, Australia

Stephanie Piper Department of Botany and Plant Sciences, University of California Riverside, Riverside, California, USA

Christine Rega-Brodsky Department of Biology, Pittsburg State University, Pittsburg, Kansas, USA

Debra Roberts Sustainable and Resilient City Initiatives Unit, eThekwini Municipality, and School of Life Sciences, University of KwaZulu-Natal, Durban, South Africa

Allison H. Roy US Geological Survey, Massachusetts Cooperative Fish and Wildlife Research Unit, University of Massachusetts, Amherst, Massachusetts, USA

Charlie M. Shackleton Department of Environmental Science, Rhodes University, Makhanda, South Africa

Emilie K. Stander Department of Science and Engineering, Raritan Valley Community College, Branchburg, New Jersey, USA

Katalin Szlavecz Department of Earth and Planetary Sciences, The Johns Hopkins University, Baltimore, Maryland, USA

Caragh G. Threlfall School of Natural Sciences, Macquarie University, Sydney, New South Wales, Australia

Yuta Uchiyama Graduate School of Human Development and Environment, Kobe University, Kobe, Japan

Timothy Vargo Urban Ecology Center, Milwaukee, Wisconsin, USA

Paige S. Warren Department of Environmental Conservation, University of Massachusetts, Amherst, Massachusetts, USA

Peter Werner Darmstadt, Germany

Diana Wiesner Ceballos, Fundación Cerros de Bogotá, Bogotá D.C., Colombia

Nicholas S.G. Williams School of Ecosystem and Forest Sciences, The University of Melbourne, Melbourne, Victoria, Australia

Jun Yang Ministry of Education Key Laboratory for Earth System Modeling, Department of Earth System Science, Tsinghua University, Beijing, China

Ken P. Yocom Department of Landscape Architecture, University of Washington, Seattle, Washington, USA

Elsa Youngsteadt Department of Applied Ecology, North Carolina State University, Raleigh, North Carolina, USA

Wayne Zipperer USDA Forest Service, Southern Research Station, Gainesville, Florida, USA

ACKNOWLEDGMENTS

The discovery, science, and practice of conserving and enhancing nature in our world's cities and towns requires interdisciplinary and transdisciplinary collaborations and knowledge co-production among environmental, social, and ecological scientists, land managers, urban designers, municipalities, indigenous peoples, and the public. We have been lucky to have our science, practice, and lives enhanced by all these voices.

Developing an edited volume of this scale during a global pandemic relied heavily on our outstanding collaborators. Writing chapters can be a thankless process, but the authors of the chapters in this volume gave their time and energy into producing excellent syntheses of urban biodiversity science and practice. We thank you for your contributions, and also for your patience through this process.

This handbook was built out of our many collaborations supported in part by the National Center for Ecological Analysis and Synthesis (NCEAS), the US National Science Foundation Research Coordination Network UrBioNet (NSF RCN: DEB#1354676/1355151), the National Socio-Environmental Synthesis Center (SESYNC), the Synthesis Centre for Biodiversity Sciences (sDiv) at the German Centre for Integrative Biodiversity Research, and the Helmholtz Centre for Environmental Research (UFZ). We also thank Evan Hogyeum Joo for his careful review and copy-editing of references. We thank the editorial staff at Routledge, including Hannah Ferguson, for suggesting a handbook on urban biodiversity, and Katie Stokes, who saw us through to the end.

Finally, scientifically sound work requires careful peer-review. We thank the following referees who graciously committed their time and expertise in reviewing chapters:

Riley Andrade, Department of Wildlife Ecology and Conservation, University of Florida, Gainesville, Florida, USA

James Aronson, Missouri Botanical Garden, St. Louis, Missouri, and Ecological Health Network, Brookline, Massachusetts, USA

Bruce Clarkson, Environmental Research Institute, University of Waikato, Hamilton, New Zealand

Cynnamon Dobbs, Department of Natural Resources and the Environment, College of Agriculture, Health and Natural Resources, University of Connecticut, Storrs, Connecticut, USA

Diana Dushkova. Helmholtz Centre for Environmental Research – UFZ, Leipzig, Germany

Jenn Engelke, Department of Landscape Architecture, College of Built Environments, University of Washington, Seattle, Washington, USA
Aaron Grade, U.S. Global Change Research Program, National Coordination Office, Washington, D.C., USA
Daijiang Li, Department of Biological Sciences, Louisiana State University, Baton Rouge, Louisiana, USA
Han Li, Department of Biology, University of Nebraska Omaha, Omaha, Nebraska
Basil V. Iannone III, School of Forest, Fisheries, and Geomatics Sciences, University of Florida IFAS, Gainesville, Florida, USA
Maria Ignatieva, School of Design, University of Western Australia, Perth, Western Australia, Australia
Rebecca Jordan, Community Sustainability, College of Agriculture and Natural Resources, Michigan State University, East Lansing, Michigan, USA
Jennifer Adams Krumins, Department of Biology, Montclair State University, Montclair, New Jersey, USA
Dexter Locke, Northern Research Station, USDA Forest Service, Baltimore Field Station, Baltimore, Maryland, USA
Travis Longcore, Institute of the Environment and Sustainability, University of California Los Angeles, Los Angeles, California, USA
Emily Minor, Department of Biological Sciences, University of Illinois at Chicago, Chicago, Illinois, USA
Ana Novoa, Institute of Botany, Czech Academy of Sciences, Pruhonice, Czech Republic
David J. Nowak, Northern Research Station, USDA Forest Service, State University of New York, College of Environmental Science and Forestry, Syracuse, New York, USA
Stephan Pauleit, Technische Universität München, Freising, Germany
Max Piana, Northern Research Station, USDA Forest Service, University of Massachusetts, Amherst, Massachusetts, USA
Clara Pretgitzer, Natural Areas Conservancy, New York, New York, USA
Petr Pyšek, Institute of Botany, Czech Academy of Sciences, Pruhonice, Czech Republic
Donald A. Rakow, School of Integrative Plant Science - Horticulture Section, College of Agriculture and Life Sciences, Cornell University, Ithaca, New York, USA
Freya Rowland, United States Geological Survey, Columbia Environmental Research Center, Columbia, Missouri, USA
Tanja Straka, Institute of Ecology, Technische Universität Berlin, Berlin, Germany
Sarah Weiskopf, National Climate Adaptation Science Center, United States Geological Survey, Reston, Virginia, USA
Ken P. Yocom, Department of Landscape Architecture, University of Washington, Seattle, Washington, USA

1

INTRODUCTION TO THE *ROUTLEDGE HANDBOOK OF URBAN BIODIVERSITY*

Charles H. Nilon and Myla F.J. Aronson

In the early 21st century, humans, for the first time in history, became an urban species. As of 2021, more than 56% of humans live in cities and towns (UN Habitat 2022), making the human experience a decidedly urban one. The nature that humans come into contact with as they live, work, and play is dominated by urban biota, such as the ubiquitous rock dove (i.e. common pigeon, *Columbia livia*), the sewer rat (i.e. Norway/brown rat, *Rattus norvegicus*), and the less noticeable, but certainly not less common, annual meadow grass (*Poa annua*) growing in sidewalk cracks across the world's cities. However, cities also support the plants and animals native to their biogeographic settings, providing opportunities for residents to experience unique aspects of global biodiversity.

Urban systems exemplify human-environment interactions and thus offer unique opportunities to study how humans influence ecological processes and in turn how ecological processes influence human sociocultural and economic practices. Urban ecology has been defined as "the investigation of living organisms in relation to their environment in towns and cities" (Sukopp 2002, p. 373) to understand "the full complexity of the relationship between the biological community and the urban environment due to the interactions between human culture and the natural environment" (Douglas 2011, p. 3) and is an "interdisciplinary research field at the interface of natural sciences, social scientists and humanities, as well as engineering" (Richter and Weiland 2011, p. 3). In these definitions, biological components and communities are at the forefront, pointing to the importance of understanding the natural history of species and their community interactions in cities, hence the rise of urban biodiversity science from within urban ecology.

While naturalists have studied the biodiversity of urban areas for hundreds of years (Sukopp 2002), the biodiversity supported by cities has only recently begun to be recognized as an important component of global biodiversity. The field of urban biodiversity has emerged as a significant subfield from within the broad discipline of urban ecology in the past two decades. Urban biodiversity research builds on the long history of natural studies in London (Fitter 1945), New York City (Kieran 1959), Berlin (Lachmund 2003; Lachmund 2013; Kowarik 2020), Warsaw (Luniak and Pisarski 1994), and other cities. This early natural history work expanded to studies by ecologists between the 1950s and 1980s that was highlighted by research in Chicago (Schmid 1975), Birmingham and the West Midlands (Teagle 1978), Tokyo (Numata 1982), and Berlin (Sukopp 1990). This

DOI: 10.4324/9781003016120-1

work also recognized the relationship between socioeconomic and demographic patterns that shape cities and the distribution of species within cities.

The new focus by mainstream ecologists on urban ecosystems in the 1990s led to funded research initiatives in the US, UK, and Germany, in particular (Nilon et al. 2003), resulting in research on urban biodiversity using contemporary ecological methods which emphasize question-driven research and a focus on understanding drivers of urban processes, detailed ecological responses, and a focus on understanding mechanisms (Pickett et al. 2001; McDonnell and Hahs 2013). This approach has led to work that address questions on biotic homogenization (McKinney 2006), the role of scale and broader landscape patterns (Goddard et al. 2010), population status and reproductive success (e.g. Kowarik and von der Lippe 2018), socioeconomic status in urban neighborhoods and species diversity (e.g. Fan et al. 2019; Kuras et al. 2020), and functional trait diversity (e.g. Nero 2019). Contemporary studies have emphasized the comparative approach to understanding urban biodiversity (McDonnell et al. 2009). For example, the GLOBNET project (Niemelä and Kotze 2009) used carabid beetles in eight cities in North America and Europe to address questions from community ecology. The Urban Biodiversity Research Coordination Network (UrBioNet) synthesized global data on urban biodiversity to address ecological theories on hierarchical filters that shape the distribution of species among and within cities (Aronson et al. 2014; Aronson et al. 2016; La Sorte et al. 2018).

Urban biodiversity as a discipline is on a trajectory to continue with two strong subfields: (1) research focused on answering fundamental questions about the mechanisms explaining the distribution of species among and within cities; and (2) applied research and practice by practitioners to address concerns about urban biodiversity and conservation, monitoring, restoration, planning, design, and public involvement. A recent paper by Rega-Brodsky et al. (2022) reviewed the state of urban biodiversity research since 1990, finding that particular taxa, such as plants and birds, are still the most studied to date, but there has been an increased focus on a variety of taxa, such as mammals and pollinators, over the last ten years. However, there has been little integration of multi-trophic interactions, mechanistic research, and significant geographic biases exist, with developing countries, particularly those in global biodiversity hotspots, being severely understudied.

Applied urban biodiversity research and practice has been influenced by local, regional, and international programs to address global declines in biodiversity (Elmqvist et al. 2013). Studies that incorporate systematic approaches to describing natural and built habitats (Löfvenhaft 2002; Farinha-Marques et al. 2017) and methods for assessing species abundance and composition (Hermy and Cornelis 2000; Sattler et al. 2014) have advanced our understanding of biodiversity in cities, particularly in the most built areas (Zerbe et al. 2003; Snep et al. 2011), and have been crucial to successful conservation and restoration projects (Goode 1989; Luniak 1990; Adams 2005; Stewart et al. 2009; Aronson et al. 2017). Research linking urban biodiversity to ecosystem services (Tratalos et al. 2007), planning and governance (Puppim de Oliveira et al. 2011; Nilon et al. 2017), community science (Anderson et al. 2014), and international initiatives to monitor and conserve urban biodiversity (Kohsaka et al. 2013) increasingly shape how local governments address conservation in their cities.

This edited volume addresses both fundamental research and applied research and practice of urban biodiversity through 28 peer-reviewed chapters detailing the history of urban biodiversity, theoretical foundations, current state of knowledge, and application of that knowledge. We start this volume with four chapters that are fundamental for a deep grounding in understanding the patterns and processes that drive urban biodiversity by examining the history of urban biodiversity research and conservation, and the diversity of cities and their underlying environmental gradients, unique climates, and landscapes (Part 1). We turn next to addressing the current state of knowledge, first with nine chapters that synthesize the large body of literature establishing key concepts, theories, and knowledge gaps on the patterns and drivers of urban biodiversity (Part 2). Then we turn to the state of knowledge and knowledge gaps of key urban taxa – plants, insects, birds, bats, amphibians, and freshwater invertebrates and fishes (Part 3). Finally, we end this volume with nine chapters addressing applied research and practice, including monitoring of biodiversity, planning, design and management of green spaces, and the social justice aspects of urban biodiversity (Part 4).

Here, we have brought together a diverse group of researchers and practitioners from around the world to comprehensively cover the science and practice of urban biodiversity. Authors include established and early career researchers, designers, city planners, and conservation professionals. Within each chapter, the authors synthesize the literature and case studies with a focus on management and conservation, methods, and practice. Each chapter ends with key companion papers for exploring the topics in more detail. In this book, we have targeted researchers, university faculty, graduate students, and advanced undergraduates who will value a book that provides in-depth overviews of the broad range in the field of urban biodiversity, and professionals working in nature conservation, landscape architecture, urban planning, and nature conservation outreach and education who value a book that covers the scientific aspects of biodiversity and their real-world applications.

References

Adams, L.W. 2005. Urban wildlife ecology and conservation: A brief history of the discipline. *Urban Ecosystems* 8:139–156.

Anderson, P.M.L., G. Avlonitis, H. Ernstson. 2014. Ecological outcomes of civic and expert-led urban greening projects using indigenous plant species in Cape Town, South Africa. *Landscape and Urban Planning* 127:104–113.

Aronson, M.F.J., F.A. La Sorte, C.H. Nilon, M. Katti, M.A. Goddard, C.A. Lepczyk, P.S. Warren, N.S.G. Williams, S. Cilliers, B. Clarkson, C. Dobbs, R. Dolan, M. Hedblom, S. Klotz, J.L. Kooijmans, I. Kühn, I. MacGregor-Fors, M. McDonnell, U. Mörtberg, P. Pyšek, S. Siebert, J. Sushinsky, P. Werner, M. Winter. 2014. A global analysis of the impacts of urbanization on bird and plant diversity reveals key anthropogenic drivers. *Proceedings of the Royal Society B: Biological Sciences* 281:20133330.

Aronson, M.F.J., C.A. Lepczyk, K.L. Evans, M.A. Goddard, S.B. Lerman, J.S. MacIvor, C.H. Nilon, T. Vargo. 2017. Biodiversity in the city: Key challenges for urban green space management. *Frontiers in Ecology and the Environment* 15:189–196.

Aronson, M.F.J., C.H. Nilon, C.A. Lepczyk, T.S. Parker, P.S. Warren, S.S. Cilliers, M.A. Goddard, A.K. Hahs, C. Herzog, M. Katti. 2016. Hierarchical filters determine community assembly of urban species pools. *Ecology* 97:2952–2963.

Douglas, I. 2011. Introduction. Pages 3–6 in I. Douglas, D. Goode, M.C. Houck, R. Wang, editors. *The Routledge Handbook of Urban Ecology*, 1st edition. Routledge, London.

Elmqvist, T., M. Fragkias, J. Goodness, B. Güneralp, P.J. Marcotullio, R.I. McDonald, S. Parnell, M. Schewenius, M. Sendstad, K.C. Seto, C. Wilkinson, editors. 2013. *Urbanization, Biodiversity and Ecosystem Services: Challenges and Opportunities*. Springer Netherlands, Dordrecht, Netherlands.

Fan, C., M. Johnston, L. Darling, L. Scott, F.H. Liao. 2019. Land use and socio-economic determinants of urban forest structure and diversity. *Landscape and Urban Planning* 181:10–21.

Farinha-Marques, P., C. Fernandes, F. Guilherme, J.M. Lameiras, P. Alves, R.G.H. Bunce. 2017. Urban Habitats Biodiversity Assessment (UrHBA): A standardized procedure for recording biodiversity and its spatial distribution in urban environments. *Landscape Ecology* 32:1753–1770.

Fitter, R.S.R. 1945. *London's Natural History*. Collins, London.

Goddard, M.A., A.J. Dougill, T.G. Benton. 2010. Scaling up from gardens: Biodiversity conservation in urban environments. *Trends in Ecology and Evolution* 25:90–98.

Goode, D.A. 1989. Urban nature conservation in Britain. *Journal of Applied Ecology* 26:859–873.

Hermy, M., J. Cornelis. 2000. Towards a monitoring method and a number of multifaceted and hierarchical biodiversity indicators for urban and suburban parks. *Landscape and Urban Planning* 49:149–162.

Kieran, J. 1959. *A Natural History of New York City*. Houghton Mifflin, New York.
Kohsaka, R., H.M. Pereira, T. Elmqvist, L. Chan, R. Moreno-Peñaranda, Y. Morimoto, T. Inoue, M. Iwata, M. Nishi, M. da L. Mathias, C.S. Cruz, M. Cabral, M. Brunfeldt, A. Parkkinen, J. Niemelä, Y. Kulkarni-Kawli, G. Pearsell. 2013. Indicators for management of urban biodiversity and ecosystem services: City Biodiversity Index. Pages 699–718 in T. Elmqvist, M. Fragkias, J. Goodness, B. Güneralp, P.J. Marcotullio, R.I. McDonald, S. Parnell, M. Schewenius, M. Sendstad, K.C. Seto, C. Wilkinson, editors. *Urbanization, Biodiversity and Ecosystem Services: Challenges and Opportunities*. Springer Netherlands, Dordrecht, Netherlands.
Kowarik, I. 2020. Herbert Sukopp – An inspiring pioneer in the field of urban ecology. *Urban Ecosystems* 23:445–455.
Kowarik, I., M. von der Lippe. 2018. Plant population success across urban ecosystems: A framework to inform biodiversity conservation in cities. *Journal of Applied Ecology* 55:2354–2361.
Kuras, E.R., P.S. Warren, J.A. Zinda, M.F.J. Aronson, S. Cilliers, M.A. Goddard, C.H. Nilon, R. Winkler. 2020. Urban socioeconomic inequality and biodiversity often converge, but not always: A global meta-analysis. *Landscape and Urban Planning* 198:103799.
Lachmund, J. 2003. Exploring the city of Rubble: Botanical fieldwork in bombed cities in Germany after World War II. *OSIRIS* 18:234–254.
Lachmund, J. 2013. *Greening Berlin: The Co-Production of Science, Politics, and Urban Nature*. MIT Press, Cambridge, MA.
La Sorte, F.A., C.A. Lepczyk, M.F.J. Aronson, M.A. Goddard, M. Hedblom, M. Katti, I. MacGregor-Fors, U. Mörtberg, C.H. Nilon, P.S. Warren, N.S.G. Williams, J. Yang. 2018. The phylogenetic and functional diversity of regional breeding bird assemblages is reduced and constricted through urbanization. *Diversity and Distributions* 24:928–938.
Löfvenhaft, K., C. Björn, M. Ihse. 2002. Biotope patterns in urban areas: A conceptual model integrating biodiversity issues in spatial planning. *Landscape and Urban Planning* 58:223–240.
Luniak, M., editor. 1990. *Urban Ecological Studies in Central and Eastern Europe*. Polish Academy of Sciences Institute of Zoology, Warsaw, Poland,
Luniak, M., B. Pisarski. 1994. State of research into the fauna of Warsaw (up to 1990). *Memorabilia Zoologica* 49:155–165.
McDonnell, M.J., A.K. Hahs. 2013. The future of urban biodiversity research: Moving beyond the 'low-hanging fruit'. *Urban Ecosystems*:1–13.
McDonnell, M.J., A.K. Hahs, J.H. Breuste. 2009. *Ecology of Cities and Towns: A Comparative Approach*. Cambridge University Press, Cambridge.
McKinney, M.L. 2006. Urbanization as a major cause of biotic homogenization. *Biological Conservation* 127:247–260.
Nero, B.F. 2019. Woody species and trait diversity-functional relations of green spaces in Kumasi, Ghana. *Urban Ecosystems* 22:593–607.
Niemelä, J., D.J. Kotze. 2009. Carabid beetle assemblages along urban to rural gradients: Areview. *Landscape and Urban Planning* 92:65–71.
Nilon, C.H., M.F.J. Aronson, S.S. Cilliers, C. Dobbs, L.J. Frazee, M.A. Goddard, K.M. O'Neill, D. Roberts, E.K. Stander, P. Werner, M. Winter, K.P. Yocom. 2017. Planning for the future of urban biodiversity: A global review of city-scale initiatives. *BioScience* 67:332–342.
Nilon, C.H., A.R. Berkowitz, K.S. Hollweg. 2003. Introduction: Ecosystem understanding is a key to understanding cities. Pages 1–13 in A.R. Berkowitz, C.H. Nilon, K.S. Hollweg, editors. *Understanding Urban Ecosystems: A New Frontier for Science and Education*. Springer, New York.
Numata, M. 1982. The development of the Chiba Bayshore cities and related ecological problems. *Integrated Ecological Studies in Bay-Coast Cities* IV:1–17.
Pickett, S.T.A., M.L. Cadenasso, J.M. Grove, C.H. Nilon, R.V. Pouyat, W.C. Zipperer, R. Costanza. 2001. Urban ecological systems: Linking terrestrial ecological, physical, and socioeconomic components of metropolitan areas. *Annual Review of Ecology and Systematics* 32:127–157.
Puppim de Oliveira, J.A., O. Balaban, C.N.H. Doll, R. Moreno-Peñaranda, A. Gasparatos, D. Iossifova, A. Suwa. 2011. Cities and biodiversity: Perspectives and governance challenges for implementing the convention on biological diversity (CBD) at the city level. *Biological Conservation* 144:1302–1313.
Rega-Brodsky, C.C., M.F.J. Aronson, M.R. Piana, E.-S. Carpenter, A.K. Hahs, A. Herrera-Montes, S. Knapp, D.J. Kotze, C.A. Lepczyk, M. Moretti, A.B. Salisbury, N.S.G. Williams, K. Jung, M. Katti, I. MacGregor-Fors, J.S. MacIvor, F.A. La Sorte, V. Sheel, C.G. Threfall, C.H. Nilon. 2022. Urban biodiversity: State of the science and future directions. *Urban Ecosystems* 25:1083–1096.

Richter, M., U. Weiland. 2011. *Applied Urban Ecology: A Global Framework*. Wiley-Blackwell, Oxford.

Sattler, T., G.B. Pezzatti, M.P. Nobis, M.K. Obrist, T. Roth, M. Moretti. 2014. Selection of multiple umbrella species for functional and taxonomic diversity to represent urban biodiversity. *Conservation Biology* 28:414–426.

Schmid, J.A. 1975. *Urban Vegetation: A Review and Chicago Case Study*. Department of Geography Research Paper No. 161. University of Chicago, Chicago, IL.

Snep, R.P., M.F. Wallis DeVries, P. Opdam. 2011. Conservation where people work: A role for business districts and industrial areas in enhancing endangered butterfly populations? *Landscape and Urban Planning* 103:94–101.

Stewart, G.H., M.E. Ignatieva, C.D. Meurk, H. Buckley, B. Horne, T. Braddick. 2009. URban Biotopes of Aotearoa New Zealand (URBANZ) (I): Composition and diversity of temperate urban lawns in Christchurch. *Urban Ecosystems* 12:233–248.

Sukopp, H. 1990. Urban ecology and its application in Europe. Pages 1–22 in H. Sukopp, S. Henjy, editors. *Urban Ecology: Plants and Plant Communities in Urban Environments*. SPB Academic, Amsterdam, Netherlands.

Sukopp, H. 2002. On the early history of urban ecology in Europe. *Preslia* 74:373–393.

Teagle, W.G. 1978. *The Endless Village: The wildlife of Birmingham, Dudley, Sandwell, Walsall and Wolverhampton*. Nature Conservancy Council, West Midland Region, Shrewsbury.

Tratalos, J., R.A. Fuller, P.H. Warren, R.G. Davies, K.J. Gaston. 2007. Urban form, biodiversity potential and ecosystem services. *Landscape and Urban Planning* 83:308–317.

UN Habitat. 2022. *World Cities Report: Envisaging the Future of Cities*. United Nations Human Settlements Programme (UN-Habitat), Nairobi, Kenya.

Zerbe, S., U. Maurer, S. Schmitz, H. Sukopp. 2003. Biodiversity in Berlin and its potential for nature conservation. *Landscape and Urban Planning* 62:139–148.

PART 1

Setting the stage for urban biodiversity research and practice

2

HISTORY OF URBAN BIODIVERSITY RESEARCH AND PRACTICE

Charles H. Nilon

Introduction: urban biodiversity is about studying the nature that occurs in cities

In this chapter, I explore the history of basic and applied research in urban biodiversity up to the 1990s, when, as several authors in this book discuss, urban ecology emerged as a unified sub-discipline within ecology. This chapter will investigate the role of natural history and nature study in developing an understanding of urban biodiversity, the role of ecology in defining and studying biodiversity in cities, and describe the importance of applied ecologists in developing the field of urban biodiversity.

Urban biodiversity research and practice is about people's interest in the nature around them and how they understand the occurrence and distribution of the species that share cities with them. Urban biodiversity research and practice involve how people study nature in cities and how they use that information. Savard et al. (2000) noted that biodiversity has been an important component of urban ecology studies and is incorporated into the following concepts: hierarchy of scales from individual lots in a city to regional landscapes; species identity; habitat fragmentation and habitat quality; role of local and landscape attributes; and the role of planning, design, and management.

A key part of understanding the history of urban biodiversity research and practice is that it takes place in urban spaces. Lachmund (2013), in discussing urban ecology studies in Berlin, noted that urban spaces are not just locations where scientists make observations and collect data but also places where ecologists and residents have assigned meanings and attributes that influence research and practice.

Natural history and cities

The history of research in and practice of urban biodiversity is intertwined with the development of urban ecology and the associated sub-disciplines of urban forestry, urban wildlife, and urban nature conservation. However, the discipline has its origin in studies of the natural history of cities – the observations that urban residents make about the nature around them. These observations connect spiritual, cultural, and scientific interest in nature.

Cultural and spiritual interests in biodiversity have a long history in cities. For hundreds of years, people have observed and valued the species associated with mountains, rivers and streams,

DOI: 10.4324/9781003016120-3

sacred groves, and cemeteries that are found in and around cities (Ormsby 2021). De Lacy and Shackleton (2017) studied the history of species and vegetation associated with Christian churches, Muslim mosques, and Hindu/Buddhist shrines and temples, noting the old age of many of the sites and their scattered distribution throughout many of the world's cities. These observations are supported by studies from very different cities. Dickson (2001) found that the earliest recorded list of plants in Glasgow, Scotland, was written by a Christian pastor in the mid 1500s. The species list was attached to a published description of medicinal plants. Barau et al.'s (2013) study of the Emir's Palace in Kano, Nigeria, documents the history of gardens dating from the 1600s and the oral history documenting the medicinal, spiritual, and cultural values associated with different tree species in the palace gardens.

There is also a long history of broader interest in documenting the occurrence of species in cities. Sukopp (2002) described reports on the flora of settled places and different habitats in those places from as early as 1561, and Fitter (1945) described a report of a "botanical ramble" in London in 1629. Dickson (2001) referenced a published flora of Glasgow, Scotland, from 1793, Luniak and Pisarski (1994) documented collections of insects and other invertebrates in Warsaw from the 18th century, and Douglas and Goode (2010) noted published lists of the flora and/or fauna of New York City, London, and Paris from the mid to late 1700s.

More organized efforts to document species in cities were driven by natural history societies that engaged middle-class residents of cities in the 1800s (Phillips 2003). Natural history societies continued into the 20th century and remained important in documenting the flora and fauna of cities. Studies from these natural history societies provided important information on the distribution and occurrence of different taxonomic groups such as the herpetofauna of Baltimore, Maryland, USA (Mansueti 1941), and the flora of Syracuse, New York, USA (Goodrich 1912). Some early ecologists worked in and around cities during this period. Henry Cowles' (1901) publication on plant communities of the Chicago region, Illinois, USA, included the results of his research as well as locations of the sites he studied and directions on reaching the sites using public transportation. Publications on the natural history of London (Fitter 1945) and New York City (Kieran 1959) summarized the type of information on the biodiversity that was compiled by naturalists and ecologists during the first half of the 20th century.

The emphasis on natural history also saw the development of the Nature Study Movement in the United States and the Nature Movement in the United Kingdom, which emphasized programs for schoolchildren in urban and rural areas to connect them with nature through in-depth study of plants and animals in and around where they lived (Jenkins and Swinnerton 1996; Armitage 2007). Nature study programs were important in developing an interest in more detailed and ecological approaches to natural history and conservation. For example, McAllister and Marshall (2000) related their lifelong commitment to studying birds from their exposure to their school's nature study curriculum.

Post–World War II ecologists: urban areas as unique places for research

The period between the end of World War II and 1990 saw the development of research focused on biodiversity in cities. Sukopp's (2002) paper on the history of urban ecology in Europe described research by ecologists and others in London, Paris, New York, Vienna, and Brussels during this period. Research on plant and animal communities in urban areas emerged as a research topic among ecologists starting their careers in the 1950s. The work done by these researchers was important to the development of urban biodiversity. These studies are relevant to contemporary research because of the breadth of topics covered and the focus on similar habitats and topics that are the focus of current urban biodiversity research.

The work of four ecologists who worked between the 1950s and 1980s, Herbert Sukopp, Makato Numata, Maciej Luniak, and Forest Stearns, is helpful in understanding the development of urban biodiversity as a discipline. Their work builds on an earlier tradition of natural history studies in cities but also reflects their training as ecologists who were developing their careers during this postwar period.

Herbert Sukopp was trained as a plant ecologist in the early 1950s (Lachmund 2003; Lachmund 2013; Kowarik 2020; Kowarik 2022). His research on urban biodiversity initially focused on vegetation of large wastelands sites in Berlin that were created as a result of bombing during World War II and were not redeveloped due to the unique political situation in the city. His work evolved from documenting the unique assemblage of plants on these sites, to studies of the mechanisms driving plant species composition in cities, and ultimately led to development of broader theories about urban biodiversity and urban ecology (Lachmund 2003; Kowarik 2022). Sukopp's work was particularly relevant because of his leadership of the Institute of Ecology at the Technical University of Berlin (Kowarik 2022), where he supervised a number of studies on the vegetation of wasteland sites that developed theories on the mechanisms shaping the vegetation of urban areas. Sukopp and his colleagues organized and published the results of many symposia on plants and plant communities in urban environments that summarized much of the work done in Europe and Asia before the 1990s (Sukopp et al. 1990).

Numata's (1991) urban biodiversity research on Tokyo and the Chiba Bay region of Japan started with his involvement with the UNESCO Man and the Biosphere Program's (MAB) international work on human settlements. A plant ecologist, his research on urban biodiversity expanded from an initial study of changing species composition in a remnant forest in the center of Tokyo, to consider the role of plant, bird, mammal, and microorganism biodiversity on urban ecosystem structure and function (Numata 1991). He led a research team studying biodiversity and ecosystem processes in the urban region surrounding Chiba Bay (Numata 1981).

Maciej Luniak was an ornithologist with the Institute of Zoology of the Polish Academy of Sciences. In 1954, he initiated a study of bird species composition and abundance in Warsaw that covered a range of habitat types and land uses. This research continued through the 1980s, allowing a comparison of change over time in breeding and winter communities (Luniak 1996). Under Luniak's direction, the Institute of Zoology carried out three main research projects between 1974 and 1990 (Luniak 1990a): (1) a long-term (1961–2000) study of bird populations and communities in Warsaw (Luniak 2001); (2) a long-term study of bird communities of new housing developments (Luniak 1994); and (3) a study of terrestrial invertebrates, synanthropic and parasitic fauna, and mammals (Luniak and Pisarski 1994). The research group also studied changes in vegetation, birds, and small mammals in a suburban area near Warsaw as part of a larger interdisciplinary study of the suburban development of an agricultural area. The project studied changes in biodiversity that were associated changes in environmental and socioeconomic change in the region (Mackin-Rogalska et al. 1988). Luniak was instrumental in connecting European researchers studying urban biodiversity by organizing workshops and symposia (Luniak 1990b) and comparative studies of birds among 27 cities in Central and Eastern Europe Warsaw (Luniak 1990c).

Forest Stearns, a forest ecologist at the University of Wisconsin–Milwaukee, was an advocate of urban ecology research, noting that it was primarily the study of plants and animals in cities (Stearns 1970). Stearns and his graduate students developed an urban ecology research program focused on Milwaukee and its suburbs, where they studied wildlife and wildlife habitats, plant communities, effects of urbanization on plant growth, primary productivity in residential areas, spontaneous vegetation, weed communities, and soils and seed bank (Stearns 1984). His research group studied these topics ranging from plant and small mammal communities in remnant forest patches in Milwaukee and its suburbs (Levenson 1981; Matthiae and Stearns 1981) to the vegetation of residential lots in a Milwaukee suburb (Dorney et al. 1984).

International collaborations among ecologists studying biodiversity in cities became more common in in the 1970s and 1980s. Researchers in socialist countries in Eastern Europe collaborated on a joint project to define plant and animal communities associated with different types of buildings and managed habitats in the region's cities (Trojan 1990). The MAB influenced the development of urban ecology as a discipline with its consideration of the ecological and social aspects of urban ecosystems (McDonnell and Niemelä 2011). The Rome MAB project, led by the botanist Valerio Giacomini, incorporated studies of mammals, arthropods, and vegetation. Researchers in Rome considered biodiversity at regional and local scales and linked their work to studies of residents' perceptions of the local environment (Celesti-Grapow and Fanelli 1991; Bonnes 2004). National MAB committees also addressed biodiversity in cities. The German MAB program organized a conference on soil organisms (German MAB Committee 1989) and the UK MAB provided initial support to the UK MAB Urban Forum, which brought together researchers and practitioners working on urban biodiversity and nature conservation (Goode 1999; Adams 2005). The MAB urban projects are examples of collaborative research across countries on biodiversity in cities that would become more common in the future.

Urban biodiversity studies before 1990 addressed many of conceptual questions that will be covered in this book. These include biodiversity in socioecological systems, novel habitats, and residential landscapes. Emlen's (1974) study of birds in remnant habitats and in residential areas of Tucson, Arizona, USA, was instrumental for understanding the influence of urban development on species composition and feeding and nesting guild structure. Vessel and Wong (1987) and Mühlenbach (1979) studied the synanthropic flora of rail yards and railroad right-of-ways in St. Louis, Missouri, USA, replicating a study he had conducted in Riga, Latvia, 40 years before, while also documenting the contemporary research on the topic in other cities. Researchers in the 1980s explicitly studied urbanization gradients that were influencing species composition of insects (Klausnitzer and Richter 1983) and small mammals (Nilon and VanDruff 1987). Schmid's (1975) study of vegetation study of front-yard vegetation in Chicago neighborhoods addressed questions about human facilitation of vegetation through planting and the role of socioeconomic and demographic factors that shape differences in vegetation among neighborhoods. Owen's (Owen and Owen 1975; Owen 1991) long-term study on biodiversity in a suburban Leicester, UK, garden documented changes in plant and invertebrates composition and abundance.

Researchers expanded efforts to describe species composition and distribution across the range of land uses in cities. Three publications on birds covering 16 cities (Kelcey and Rheinwald 2005), vegetation covering 16 cities (Kelcey and Müller 2011), and mammals and invertebrates covering 9 cities (Kelcey 2015) summarized many of the postwar studies and documented species diversity patterns over time and how they are connected to land use and demographic changes within cites. Others expanded the use of city floras (Clemants and Moore 2003) and ornithological atlases (Luniak 2017) as tools for assessing species composition, distribution, and abundance across all land uses in a city. Floras incorporated the natural history tradition of documenting plant species in a region with contemporary efforts to understand plant species distributions, the occurrence of native and non-native species, and changes in species composition and distribution over time (Dickson et al. 2001). Trepl (1994) and Pyšek (1989, 1993) contributed important work synthesizing data from floras and research studies to develop theory on explaining patterns of plant species composition in cities. Bird atlases used a standard protocol developed in the 1970s for documenting species occurrence and evidence of nesting and reproduction and were used extensively in European cities (Luniak 2017; Luniak 2001).

Role of applied research: management, conservation, design, and planning

Applied research on urban biodiversity in the 1970s and 1980s was conducted in support of management and conservation efforts. This work was often funded by government agencies resulting in a number of studies published as reports, conference proceedings, and theses or dissertations. A short review from the United Kingdom, Germany, and the United States illustrates the scope of this research.

Studies to support conservation efforts by local governments were an important part of applied research on urban biodiversity in the United Kingdom. The focus on the biodiversity of green areas in cities started in 1974 when England's nature conservation agency, the Nature Conservancy Council, commissioned the West Midlands Wildlife Survey (Teagle 1978). The survey, in the second largest metropolitan area in the UK, documented species in a mosaic of habitats shaped the industrial legacy of the region, remnant nature areas, and built spaces (Teagle 1978; Goode 1989; Goode 2018).

The West Midlands Wildlife Trust built on Teagle's (1978) and worked with local governments in the West Midlands to develop criteria and methods for assessing biodiversity in the region and across a range of urban habitat types. The findings of their assessment were incorporated into designing a local network of protected areas called Sites of Importance for Nature Conservation (Box et al. 1994; Box 1999). Box (1999) noted the uniqueness of habitats associated with the postindustrial landscape and stressed the importance of identifying and conserving the biodiversity found in these spaces. The Greater London Ecology Unit also developed protocols for vegetation surveys of green spaces in each of the 33 local authorities in the region that were incorporated into planning a network of Sites of Importance for Nature Conservation (Goode 1989). The champion for applied urban biodiversity research in the UK was led by George Barker, who organized and funded projects in his role as coordinator for urban nature conservation for the Nature Conservancy Council, and as a global advocate through international collaborations in the US, Germany, Poland, and South Africa, and as editor of a newsletter, *Urban Wildlife News*, that was read in 38 countries (Goode 2018; Marren 2018).

Applied work in Germany was influenced by the work of Herbert Sukopp and his colleagues at the Technical University of Berlin and centered around the activities of the Biotope Mapping in Urban Areas Working Group (Sukopp and Weiler 1988; Werner 1999). The working group was organized to assist cities and towns in developing standardized approaches for assessing plant and animal diversity and recommending mapping methods for identifying distinct biotope types and recording data on the flora and fauna (Sukopp and Weiler 1988). Biotope mapping often focused on identifying the species and habitats associated with "four types of nature" that define urban green spaces in cities: (1) remnant natural areas, (2) agricultural areas, (3) parks and gardens, and (4) ruderal sites (Kowarik 1992; Werner 1999). In Berlin and other cities, the sites that were studied using the biotope mapping protocols were incorporated into efforts to protect unique wasteland habitats (Werner 1999; Kowarik 2022).

Applied urban biodiversity research in the United States emerged in the 1970s and connected to the disciplines of urban wildlife conservation and urban forestry (Noyes and Progulske 1974; Kielbaso 1978). Work by Alered Geis was instrumental to much of the initial urban wildlife research (Adams 2005). Geis studied the bird communities of two planned residential developments in Columbia, Maryland, USA (Geis 1974; Geis 1976), and Coldspring Town in Baltimore, Maryland (Geis 1980), and framed his research around the need for data that could be used to conserve bird communities during and after development projects. Other studies in residential neighborhoods followed, focusing on habitat features that were predictors of bird species composition and abundance (Williamson and DeGraaf 1981; DeGraaf and Wentworth 1986; Johnsen and VanDruff 1987). Similar studies of other vertebrate taxa occurred during this period (VanDruff and Rowse 1986; Adams 2005).

Symposia organized by the National Institute for Urban Wildlife (Adams and Leedy 1987; Adams and Leedy 1991; Adams 2005) brought together researchers and practitioners to address urban wildlife conservation topics, reflecting the diversity of topics that incorporated urban biodiversity, with sections on ecology of urban wildlife, planning and design, management issues and successes, and public participation and education. Applied research was influenced by planning and design professionals who addressed biodiversity and conservation in their practice (Spirn 1984; DeGraaf 1987). Biodiversity surveys similar to those previously described for the West Midlands and London were carried out in the Portland, Oregon – Vancouver, Washington Metropolitan Area (Lev and Sharp 1991), Tucson, Arizona (Shaw and Supplee 1987), and other US cities to support planning initiatives to protect unique habitats.

Research in urban forestry included studies of the urban forest that emphasized the ecological aspects of street and park trees as part of urban vegetation (McBride and Jacobs 1976; Talarchek 1990; Nowak 1991; Welch 1994), the role of surrounding land uses on remnant forest patches (Airola and Buchholz 1984; Rudnicky and McDonnell 1989), and the importance of residential yards as a component of urban vegetation (Richards et al. 1984). Research in these areas was broadly defined as studies on the ecology of the urban forest and is summarized in three journal special issues edited by Rowen Rowntree, the leader of the US Forest Service urban forest ecology project in the 1980s (Rowntree 1984; Rowntree 1988).

Conclusions

Urban biodiversity research and practice can be viewed through the lens of how people come to understand the nature around them. The topics covered in this chapter provide a brief overview of an extensive literature that is not always acknowledged by contemporary urban biodiversity researchers. The remaining chapters in this book view research and practice through contemporary and emerging theories and practices; however, the core questions addressed in earlier studies remain interesting and relevant. Understanding how earlier descriptive natural history studies, pioneering ecological studies of novel habitats, and applied research to address conservation, management, and planning issues have addressed these core questions will enhance future efforts to understand urban biodiversity.

Key companion papers

Adams, L.W. 2005. Urban wildlife ecology and conservation: A brief history of the discipline. *Urban Ecosystems* 8:139–156.

Gilbert, O. 1989. *The Ecology of Urban Habitats*. Chapman and Hall, London.

McKinney, M.L. 2002. Urbanization, biodiversity, and conservation. *BioScience* 52:883–890.

Sukopp, H. 2002. On the early history of urban ecology in Europe. *Preslia* 74:373–393.

Weiland, U., M. Richter. 2009. Lines of tradition and recent approaches to urban ecology, focusing on Germany and the USA. *GAIA – Ecological Perspectives for Science and Society* 18:49–57.

References

Adams, L.W. 2005. Urban wildlife ecology and conservation: A brief history of the discipline. *Urban Ecosystems* 8:139–156.

Adams, L.W., D.L. Leedy, editors. 1987. *Integrating Man and Nature in the Metropolitan Environment: Proceedings of a National Symposium on Urban Wildlife*. National Institute for Urban Wildlife, Columbia, MD.

Adams, L.W., D.L. Leedy, editors. 1991. *Wildlife Conservation in Metropolitan Environments*. National Institute for Urban Wildlife, Columbia, MD.

Airola, T.M., K. Buchholz. 1984. Species structure and soil characteristics of five urban forest sites along the New Jersey Palisades. *Urban Ecology* 8:149–164.

Armitage, K.C. 2007. 'The child is born a naturalist': Nature study, woodcraft Indians, and the theory of recapitulation. *The Journal of the Gilded Age and Progressive Era* 6:43–70.

Barau, A.S., A.N.M. Ludin, I. Said. 2013. Socio-ecological systems and biodiversity conservation in African city: Insights from Kano Emir's Palace gardens. *Urban Ecosystems* 16:783–800.

Bonnes, M. 2004. Inhabitants' environmental perceptions in the city of Rome within the framework for urban biosphere reserves of the UNESCO programme on man and biosphere. *Annals of the New York Academy of Sciences* 1023:175–186.

Box, J. 1999. Nature conservation and post-industrial landscapes. *Industrial Archaeology Review* 21:137–146.

Box, J., A. Douse, T. Kohler. 1994. Non-statutory sites of importance for nature conservation in the West Midlands. *Journal of Environmental Planning and Management* 37:361–367.

Celesti-Grapow, L., G. Fanelli. 1991. Ecological study on flora and vegetation of the Aurelio area in Rome. Pages 104–124 in M. Bonnes, editor. *Urban Ecology Applied to the City of Rome*. Instituto di Psicologia, Consiglio Nazionale delle Richerche, Rome.

Clemants, S.E., G. Moore. 2003. Patterns of species diversity in eight Northeastern United States cities. *Urban Habitats* 1:4–16.

Cowles, H.C. 1901. *The Plant Societies of Chicago and Vicinity*. The Geographic Society of Chicago, Chicago, IL.

DeGraaf, R.M. 1987. Urban wildlife habitat research – Application to landscape design. Pages 107–111 in L.W. Adams, D.L. Leedy, editors. *Integrating Man and Nature in the Metropolitan Environment*. National Institute for Urban Wildlife, Columbia, MD.

DeGraaf, R.M., J.M. Wentworth. 1986. Avian guild structure and habitat associations in suburban bird communities. *Urban Ecology* 9:399–412.

De Lacy, P., C. Shackleton. 2017. Aesthetic and spiritual ecosystem services provided by urban sacred sites. *Sustainability* 9:1628.

Dickson, J.H. 2001. *The Changing Flora of Glasgow: Urban and Rural through the Centuries*. Edinburgh University Press, Edinburgh.

Dorney, J.R., G.R. Guntenspergen, J.R. Keough, F. Stearns. 1984. Composition and structure of an urban woody plant community. *Urban Ecology* 8:69–90.

Douglas, I., D. Goode. 2010. Urban natural histories to urban ecologies: The growth of the study of urban nature. Pages 74–83 in I. Douglas, D. Goode, M.C. Houck, R.Wang, editors. *The Routledge Handbook of Urban Ecology*. Routledge, London.

Emlen, J.T. 1974. An urban bird community in Tucson, Arizona: Derivation, structure, regulation. *The Condor* 76:184–197.

Fitter, R.S.R. 1945. *London's Natural History*. Collins, London.

Geis, A.D. 1974. Effects of urbanization and type of urban development on bird populations. Pages 97–105 in J.H. Noyes, D.R. Progulske, editors. *Wildlife in an Urbanizing Environment*. University of Massachusetts Cooperative Extension Service, Amherst, MA.

Geis, A.D. 1976. Bird populations in a new town. *Atlantic Naturalist* 31:141–146.

Geis, A.D. 1980. Breeding and wintering bird populations at Cylburn Park and vicinity before the construction of Coldspring Town. *Atlantic Naturalist* 33:5–8.

German National Committee of the UNESCO Programme Man and the Biosphere, editors (German MAB Committee). 1989. *Report on the MAB Workshop International Scientific Workshop on Soils and Soil Zoology in Urban Ecosystems as a Basis for Management and Use of Green/Open Spaces*. Duetsches MAB-Nationalkomitee, Bonn, Germany.

Goode, D.A. 1989. Urban nature conservation in Britain. *Journal of Applied Ecology*:859–873.

Goode, D.A. 1999. Habitat survey and evaluation for nature conservation in London. *Deinsea* 5:27–39.
Goode, D.A. 2018. *George Barker 1940–2018: A Tribute*. www.thenatureofcities.com/2018/05/22/george-barker-1940-2018-tribute/
Goodrich, L.L.H. 1912. *Flora of Onondaga County as Collected by the Members of the Syracuse Botanical Club*. The McDonnell Co., Syracuse, NY.
Jenkins, E.W., B.J. Swinnerton. 1996. The school nature study union 1903–94. *History of Education* 25:181–198.
Johnsen, A.M., L.W. VanDruff. 1987. Summer and winter distribution of introduced bird species and native bird species richness within a complex urban environment. Pages 123–128 in L.W. Adams, D.L. Leedy, editors. *Integrating Man and Nature in the Metropolitan Environment*. National Institute for Urban Wildlife, Columbia, MD.
Kelcey, J.G., editor. 2015. *Vertebrates and Invertebrates of European Cities: Selected Non-Avian Fauna*. Springer Science+Business Media, New York.
Kelcey, J.G., N. Müller, editors. 2011. *Plants and Habitats of European Cities*. Springer, New York.
Kelcey, J.G., G. Rheinwald, editors. 2005. *Buchholz*. Ginster Verlag, Katharinen, Germany.
Kielbaso, J.J. 1978. The urban forestry working group of the society of American foresters. *Journal of Arboriculture* 4:43–45.
Kieran, J. 1959. *A Natural History of New York City*. Houghton Mifflin, New York.
Klausnitzer, B., K. Richter. 1983. Presence of an urban gradient demonstrated for carabid associations. *Oecologia* 59:79–82.
Kowarik, I. 1992. Das Besondere der städtischen Flora und Vegetation Schriftenr. *Deutsch. Rat Landespflege* 61:33–47.
Kowarik, I. 2020. Herbert Sukopp – An inspiring pioneer in the field of urban ecology. *Urban Ecosystems* 23:445–455.
Kowarik, I. 2022. The Berlin school of urban ecology and the emergence of wasteland ecology. Pages 139–152 in P. Boschiero, T. Folkerts, L. Latini, editors. *Natur Park Schöneberger Südgelände and Berlin's Urban Nature*. Fondazione Benetton Studi RiercheAntiga, Treviso, Italy.
Lachmund, J. 2003. Exploring the city of rubble: Botanical fieldwork in bombed cities in Germany after World War II. *Osiris*:234–254.
Lachmund, J. 2013. *Greening Berlin: The Co-Production of Science, Politics, and Urban Nature*. MIT Press, Cambridge, MA.
Lev, E., L. Sharp. 1991. The Portland-vancouver natural areas inventory: Field surveys and preliminary wildlife data. Pages 237–242 in L.W. Adams, D.L. Leedy, editors. *Wildlife Conservation in Metropolitan Environments*. National Institute for Urban Wildlife, Columbia, MD.
Levenson, J.B. 1981. Woodlots as biogeographic islands in southeastern Wisconsin. Pages13–39 in R.L. Burgess, D.M. Sharpe, editors. *Forest Island Dynamics in Man-Dominated Landscapes*. Springer-Verlag, New York.
Luniak, M. 1990a. Investigations on urban ecology by the Institute of Zoology, Polish Academy of Sciences in Warsaw (1974–1990). Pages 19–25 in M. Luniak, editor. *Urban Ecological Studies in Central and Eastern Europe*. Polish Academy of Sciences Institute of Zoology, Warsaw, Poland.
Luniak, M., editor. 1990b. *Urban Ecological Studies in Central and Eastern Europe*. Polish Academy of Sciences, Warsaw.
Luniak, M. 1990c. Avifauna of cities in Central and Eastern Europe – Results of the international inquiry. Pages 131–149 in M. Luniak, editor. *Urban Ecological Studies in Central and Eastern Europe*. Polish Academy of Sciences Institute of Zoology, Warsaw, Poland.
Luniak, M. 1994. The development of bird communities in new housing estates in Warsaw. *Memorabilia Zoologica* 49:257–267.
Luniak, M. 1996. Inventory of the avifauna of Warsaw – Species composition, abundance, and habitat distribution. *Acta Ornithologica* 31:67–80.
Luniak, M. 2001. *Ptaki Warszawy: 1962–2000*. Atlas Warszawy. Polish Academy of Sciences Stanislaw Leszczycki Institute of Geography and Spatial Planning, Warsaw, Poland.
Luniak, M. 2017. Urban ornithological atlases in Europe: A review. Pages 209–223 in E. Murgui, M. Hedblom, eds. *Ecology and Conservation of Birds in Urban Environments*. Springer, Cham, Switzerland.
Luniak, M., B. Pisarski. 1994. State of research into the fauna of Warsaw (up to 1990). *Memorabilia Zoologica* 49:155–165.

Mackin-Rogalska, R., J. Pinowski, J. Solon, Z. Wojcik. 1988. Changes in vegetation, avifauna, and small mammals in a suburban habitat. *Polish Ecological Studies* 14:293–330.

Mansueti, R.J. 1941. *A Descriptive Catalogue of the Amphibians and Reptiles Found in and Around Baltimore City, Maryland: Within a Radius of Twenty Miles*. The Natural History Society of Maryland, Baltimore, MD.

Marren, P. 2018, June 11. George Barker obituary. *The Guardian*. www.theguardian.com/environment/2018/jun/11/george-barker-obituary

Matthiae, P.E., F. Stearns. 1981. Mammals in forest islands in southeastern Wisconsin. Pages 55–66 in R.L. Burgess, D.M. Sharpe, editors. *Forest Island Dynamics in Man-Dominated Landscapes*. Springer-Verlag, New York.

McAllister, T., D. Marshall. 2000. Hometown. Pages 6–10 in M.C. Houck, M.J. Cody, editors. *Wild in the City: A Guide to Portland's Natural Areas*. Oregon Historical Society Press, Portland, OR.

McBride, J., D. Jacobs. 1976. Urban forest development: A case study, Menlo Park, California. *Urban Ecology* 2:1–14.

McDonnell, M.J., J. Niemelä. 2011. The history of urban ecology. *Urban Ecology* 9:34–49.

Mühlenbach, V. 1979. Contributions to the synanthropic (adventive) flora of the railroads in St. Louis, Missouri, USA. *Annals of the Missouri Botanical Garden*:1–108.

Nilon, C., L.W. VanDruff. 1987. Analysis of small mammal community data and applications to management of urban greenspaces. Pages 53–59 in L.W. Adams, D.L. Leedy, editors. *Integrating Man and Nature into the Metropolitan Environment*. National Institute for Urban Wildlife, Columbia, MD.

Nowak, D.J. 1991. *Urban Forest Development and Structure: Analysis of Oakland, California*. Ph.D. Dissertation, University of California, Berkeley.

Noyes, J.H., D.R. Progulske. 1974. *A Symposium on Wildlife in an Urbanizing Environment, November 27–29, 1973, Springfield, Massachusetts*. Cooperative Extension Service, University of Massachusetts, Amherst.

Numata, M., editor. 1981. *Chiba Bay-Coast Cities Project: Integrated Ecological Studies in Bay-Coast Cities III*. Chiba University, Chiba.

Numata, M. 1991. Urban ecosystem studies in the Tokyo Metropolitan Area. *Shukutoku University Bulletin* 25:41–53.

Ormsby, A.A. 2021. Diverse values and benefits of urban sacred natural sites. *Trees, Forests and People* 6:100136.

Owen, J. 1991. *The Ecology of a Garden: The First Fifteen Years*. Cambridge University Press, Cambridge.

Owen, J., D.F. Owen. 1975. Suburban gardens: England's most important nature reserve? *Environmental Conservation* 2:53–59.

Phillips, D. 2003. Friends of nature: Urban sociability and regional natural history in Dresden, 1800–1850. *Osiris* 18:43–59.

Pyšek, P. 1989. On the richness of central European urban flora. *Preslia* 61:329–334.

Pyšek, P. 1993. Factors affecting the diversity of flora and vegetation in Central European settlements. *Vegetatio* 106:89–100.

Richards, N.A., J.R. Mallette, R.J. Simpson, E.A. Macie. 1984. Residential greenspace and vegetation in a mature city: Syracuse, New York. *Urban Ecology* 8:99–125.

Rowntree, R.A. 1984. Ecology of the urban forest: Introduction to part I. *Urban Ecology* 8:1–11.

Rowntree, R.A. 1988. Ecology of the urban forest: Introduction to part III. *Landscape and Urban Planning* 15:1–10.

Rudnicky, J.L., M.J. McDonnell. 1989. Forty-eight years of canopy change in a hardwood-hemlock forest in New York City. *Bulletin of the Torrey Botanical Club* 116:52–64.

Savard, J-P.L., P. Clergeau, G. Mennechez. 2000. Biodiversity concepts and urban ecosystems. *Landscape and Urban Planning* 48:131–142.

Schmid, J.A. 1975. *Urban Vegetation. A Review and Chicago Case Study*. Department of Geography, University of Chicago, Chicago, IL.

Shaw, W.W., V. Supplee. 1987. Wildlife conservation in rapidly expanding metropolitan areas: Information, institutional, and economic constraints and solutions. Pages 191–197 in L.W. Adams, D.L. Leedy, editors. *Integrating Man and Nature into the Metropolitan Environment*. National Institute for Urban Wildlife, Columbia, MD.

Spirn, A.W. 1984. *The Granite Garden: Urban Nature and Human Design*. Basic Books, New York.

Stearns, F. 1970. Urban ecology today. *Science* 170:1006–1007.

Stearns, F. 1984. Research in plant ecology at UWM – 1965 to 1984. *Field Station Bulletin* 17:18–30.

Sukopp, H. 2002. On the early history of urban ecology in Europe. *Preslia* 74:373–393.

Sukopp, H., S. Hejný, I. Kowarik. 1990. *Urban Ecology: Plants and Plant Communities in Urban Environments*. SPB Academic Pub, The Hague, Netherlands.

Sukopp, H., S. Weiler. 1988. Biotope mapping and nature conservation strategies in urban areas of the Federal Republic of Germany. *Landscape and Urban Planning* 15:39–58.

Talarchek, G.M. 1990. The urban forest of New Orleans: An exploratory analysis of relationships. *Urban Geography* 11:65–86.

Teagle, W.G. 1978. *The Endless Village. Nature Conservancy Council*. West Midlands Region, Shrewsbury.

Trepl, L. 1994. Towards a theory of urban biocoenoses: Some hypotheses and research questions. *Memorabilia Zoologica* 49:15–19.

Trojan, P. 1990. Introduction. Pages 5–7 in M. Luniak, editor. *Urban Ecological Studies in Central and Eastern Europe*. Polish Academy of Sciences Institute of Zoology, Warsaw, Poland.

VanDruff, L.W., R.N. Rowse. 1986. Habitat association of mammals in Syracuse, New York. *Urban Ecology* 9:413–434.

Vessel, M.F., H.H. Wong. 1987. *Natural History of Vacant Lots*. University of California Press, Berkeley, CA.

Welch, J.M. 1994. Street and park trees of Boston: A comparison of urban forest structure. *Landscape and Urban Planning* 29:131–143.

Werner, P. 1999. Why biotope mapping in populated areas? *Deinsea* 5:9–26.

Williamson, R., R.M. DeGraaf. 1981. Habitat associations of ten bird species in Washington, DC. *Urban Ecology* 5:125–136.

3

COMPARATIVE URBAN ECOLOGY AND THE GRADIENT APPROACH

Extending knowledge by acknowledging the diversity of cities around the world

Amy K. Hahs

Introduction

With the exception of Antarctica, cities and towns occur across all of the world's continents. These urban landscapes come in many different forms and sizes; from small city-states such as the Vatican, where approximately 800 people live in an area of 0.17 square miles, through to Tokyo, Japan, the world's largest megacity with 37 million people in 13,500 square kilometres (Peach 2018, 2019).

This chapter provides an overview of the diversity of city contexts, before examining how this diversity offers an exciting opportunity to use a comparative approach to better understand how the environmental, social, and cultural dimensions of urbanisation shapes the responses of urban biodiversity.

Diversity of city contexts

Let's start this section with a short exercise, as you are probably already more aware of differences within and across cities than you may realise.

Think about the city or town where you currently live. If you were going to write a letter to a relative or someone who was planning to visit your city in the near future (or if you were going to visit them in their city or town), what are some of the things it would be helpful to know? If you really took time to think about it deeply, some of the points you might share are as follows:

What are the geographic coordinates? Many cities share the same name but are in different countries. For example, Melbourne is a city of 5 million people in south-eastern Australia and is also a city of 85,677 in Florida, USA (www.melbourneflorida.org/about). Since each city has a unique set of geographic coordinates, providing the coordinates is a simple way of ensuring people know the exact location of your city.

What is the weather like throughout the year? Cities are located across many different climatic zones, and this will influence the biodiversity present and potentially their response to urbanisation (Richards et al. 2019). For example, Helsinki, Finland, is located at 60°N latitude, and the plants and animals that survive there must have the capacity to survive long periods of snow each

DOI: 10.4324/9781003016120-4

year. In contrast, the city of Singapore is located at 1°N latitude and does not experience any days of snow, or even a "winter" season. The plants and animals living in these two environments are quite different and will tend to follow well-established trends associated with latitudinal gradients, such as increases in mammalian body size (Rodríguez et al. 2006). This will have implications for the direction and mechanisms of the biodiversity response and the types of questions that are relevant. For example, tree growth may be more rapid in Helsinki compared to surrounding areas due to longer growing season associated with urban heat island effect, while in Singapore tree growth may be slower in urban compared to less urbanised landscapes due to spatial constraints; changes in the dates associated with the onset of spring may not be relevant in Singapore.

What type of environment is your city located within? Historically, cities have tended to be located in highly productive landscapes with access to fresh water, timber, grazing lands, and other resources that helped support higher densities of people. Subsequent developments in engineering and transport have enabled cities to become established in less productive environments, such as deserts. Cities also vary in their altitude, soil types, proximity to the coast, and other factors which influence the vegetation communities present. This has implications for the animals, insects, and microorganisms that occupy the landscape. The response of biodiversity to urbanisation will be moderated by the identity and characteristics of the regional flora and fauna and how urbanisation has altered disturbance regimes and environmental conditions in the area. For example, cities in temperate regions with highly efficient stormwater drainage results in much drier conditions in urban areas, while irrigation and supplemental watering in arid landscapes can result in much higher levels of moisture in the landscape than may have been there previously. Therefore, understanding the local environment is important for guiding how to frame your research questions and how to interpret the results.

How large is your city, and what types of urban form are present? Cities come in many different shapes and sizes. Indeed, tourism is based on the premise that each city is unique and has features that are not found everywhere. This variability means that the level of impact of urbanisation will vary depending on the characteristics of the city. For example, the city with the lowest population in the world is Pitcairn Island, with an official population of 50 people in 2022 (Government of Pitcairn Island 2022). In contrast, Tokyo is a city of 36 million people (World Population Review 2022), and the Beijing-Tianjin-Hebei megaregion has a population of 100 million people. Therefore, you would expect the impacts of urbanisation in Pitcairn Island to be substantially smaller than the impacts present in Tokyo or Beijing-Tianjin-Hebei. Population density can also vary within a city, with high densities often associated with medium to high-rise buildings or informal settlements, while suburban landscapes composed of single family homes with gardens will have lower densities of people. The built form will also vary between locations within and across cities in terms of building heights, materials, extent of impermeable surfaces, and vegetation. The degree of artificial light at night, anthropogenic noise, visual presence of people, and change in conditions associated with urban heating and cooling will also vary in response to the built form and the lived experiences of the people who live there. This offers an important opportunity to understand how urbanisation affects biodiversity by comparing how the change of conditions associated with different forms of urbanisation moderate the response of biodiversity in that landscape.

What is the history of your city? Some of the oldest cities in the world have been around for millennia, while the newest cities may have just begun to be constructed. In older cities, the effects of urbanisation have been present for so long it may be more difficult to detect changes in biodiversity compared to more recently built cities where the change is quite recent, and the impacts on flora and fauna can be explicitly documented. Cities also vary depending on their role; for example, port cities or cities associated with industry will have a built form that reflects this purpose. Cities developed under colonisation, segregation, or apartheid will also have a built form that reflects this history, with implications for the responses of biodiversity (du Toit et al. 2021; Schell et al. 2020).

Now that we have articulated some of the key dimensions that can be used to characterise urban areas, let's look at how we can draw upon this diversity to establish research that helps reveal general principles common to all cities, but also investigate the relative importance of local factors as modifying influences. Before we take this closer look, let's establish what we mean when we propose to use "a comparative approach".

What is "comparative urban ecology"?

Comparative studies deliberately set out to create a research framework that allows two or more systems or locations to be examined (or compared) and similarities or differences between them to emerge. This approach has a strong tradition of use in humanities (Griffiths 2017; Pickvance 2001) but has also been widely used within ecology for many decades (Heal and Grime 1991; Martinez 2018).

When we were looking at the diversity of urban forms and contexts in the previous section, many of the examples used were contrasting situations, such as arctic versus tropic or coastal versus montane. Contrasting situations allow for a comparative approach based on specific typology. However, there are usually a wide range of conditions between the *punctuated end- or mid-points*, which allow us to investigate how things change along a *gradient* of conditions. Before we proceed too much further into talking about comparative ecology, it is worth making a quick detour to unpack the concept of *gradients* in the context of ecology.

A brief detour to unpack gradients

Linear and non-linear gradients

The gradient approach in ecology was first articulated by Robert Whittaker, an ecologist who spent a lot of time working in the Great Smoky Mountains, USA. Working in the field, he observed that the plant species showed consistent patterns of change with distance from the edge of the ponds and that these were often strongly related to changes in the soil moisture (Whittaker 1956). This is an example of a direct, linear gradient (Figure 3.1), where a specific condition varies in a predictable ordered way along a physical linear transect.

Linear gradients can be sampled using a transect approach. To help understand the difference between a transect and a gradient, you can think of a long measuring tape laid out perpendicular to the edge of a wetland or river. The transect would be the measuring tape itself, with sampling occurring at specific distances along the tape. The gradient would be the measurements of the actual soil moisture values at those specific distances along the tape measure.

Examples of linear gradients in the real world are relatively uncommon. More often, the conditions change with a non-linear distribution across space, such as a mosaic of soil types or time since fire. In this situation, we can deliberately set out to sample the variation in conditions present in that landscape, and then *a posteriori* order them into a direct gradient of nutrient content or stand age.

Direct and indirect gradients

In the previous section we were looking at examples of direct gradients. They are direct gradients because the environmental condition of interest was deliberately and explicitly measured. However, there are occasions where the environmental condition may be difficult to measure directly, or where there is a complex gradient with multiple interacting conditions changing in parallel. For example, altitudinal gradients are commonly employed in mountain ecology to understand how ecological systems change from the base of the mountain to the mountain peak. In this situation, the change in elevation represents associated changes in temperature, moisture, soil depth, and many other specific environmental conditions. We can use elevation as a proxy for this complex gradient to ask simple questions such as "how does the density of trees change with increasing elevation?" In this case, the use of the proxy allows the question to be addressed more quickly and at less cost than if all of the potential environmental conditions were measured. It also offers a means of representing likely variability in a variable that is hard to measure in the field – for example, how atmospheric moisture changes with elevation. However, it is important to keep in mind that indirect gradients only offer insights into correlative relationships about *what* changes. For a deeper understanding of the mechanisms and drivers of how things change, you would need to structure your study in a way that captures a direct measure of the potential gradient.

Urban environments also represent complex gradients of change where we can use a generalised metric as a proxy for changes in temperature, noise, artificial light at night, and many other impacts associated with urbanisation (see section on *Characteristics of urban areas-commonalities and differences*). For example, increases in the proportion of impermeable surfaces are often accompanied by increases in the Urban Heat Island effect, pollutant loads, and higher levels of artificial light at night. Therefore, we can use proportion of impermeable surfaces as an indirect proxy to understand changes in stream health across an urban landscape. However, we would need to use direct measures of stormwater temperatures, pollutant loads, or artificial light at night levels to determine the conditions that are responsible for the observed changes in stream health.

Direct and indirect gradients are very useful in comparative ecology research. For example, in montane ecology, the tree line represents the point on the mountain at which the growing conditions are no longer suitable for trees to persist. However, the elevation and conditions at which the tree line occurs differs between tree species, and the amount of solar radiation, which changes with aspect, and depending on whether the mountain is located at the equator or at higher latitudes in the northern and southern hemispheres (Körner and Paulsen 2004). In this example, the direct gradient used to identify the conditions that determine the tree line can be placed into a broader global context through the indirect gradient of elevation. The value and role of direct and indirect gradients is similar within the discipline of urban ecology. The direct gradients are critical for understanding specific mechanisms through which biodiversity is impacted by urbanisation; however, the indirect gradients that are common to all urban areas allow us to investigate how consistent a response is in a broader global context.

Punctuated and continuous gradients

Much of the early research in urban ecology set out to examine how an ecological response varied across a landscape depending on the level of urbanisation it was exposed to. This was the premise behind the urban-rural gradient approach proposed by McDonnell and Pickett (1990), and which has continued to be a foundational concept in urban ecology research (McDonnell and Hahs 2008;

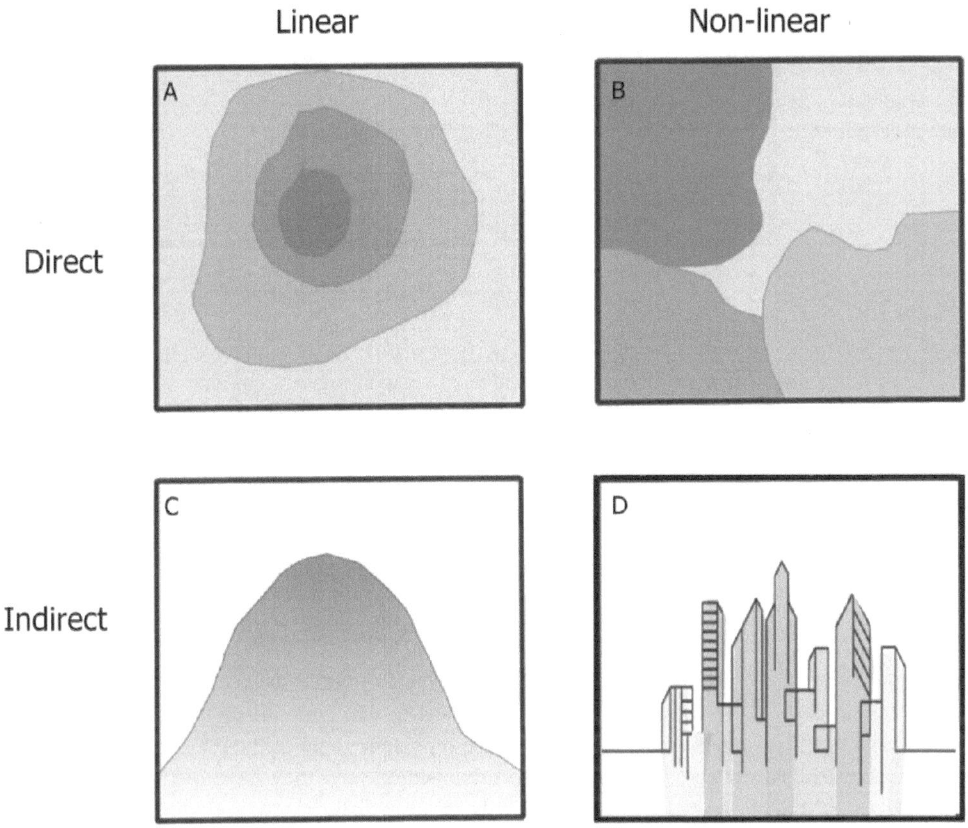

Figure 3.1 Examples of linear and non-linear gradients representing direct and indirect measurements of environmental conditions. The grey shading indicates different levels of condition within the image. Line art in D is City Line Vectors by Vecteezy. www.vecteezy.com/free-vector/city-line

Rega-Brodsky et al. 2022). Within this large body of research, the gradient has been represented both as *punctuated* categories (e.g. urban-suburban-rural contrasts), as well as along *continuous* linear and non-linear gradients of direct and indirect environmental change (see Figure 3.2). The way that the gradient is conceptualised will determine the type of inferences that can be drawn from the studies (Caryl et al. 2014). Punctuated gradients where there are comparisons between categories will reveal the magnitude and directionality of differences in the ecological response at specific conditions (e.g. parks vs residential landscapes). Continuous gradients will also reveal the magnitude and direction of change but offer additional insights into the relationship between the urban condition and the biodiversity response (e.g. are there intermediate responses or thresholds that can be used to inform future design or management decisions).

While the relevance and value of the urbanisation gradient concept has been challenged (e.g. Ramalho and Hobbs 2012), and the relationship with the "continuum of urbanity" needs to be examined further (Pickett and Zhou 2015), there is no denying the formative influence it has had in urban ecology (e.g. McDonnell and Hahs 2008; Rega-Brodsky et al. 2022). Since the gradient

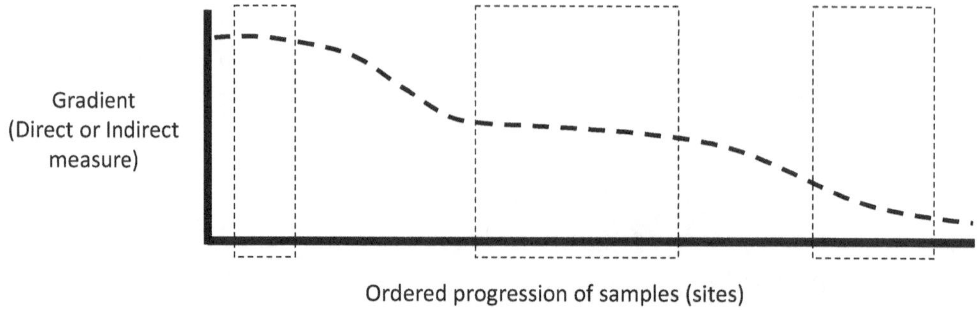

Figure 3.2 Examples of relationship between continuous variables representing environmental conditions (solid line) and punctuated or categorical representations (dashed boxes).

approach is also embedded within comparative ecology, it was important to take this small detour to clarify some of the important terminology and distinctions that sit within the concept. This will become clearer in the section *Approaches to comparative studies in urban ecology*, when we being to examine different framings for comparative ecology research.

Characteristics of urban areas: commonalities and differences

Despite the diversity of urban and city forms, histories and contexts articulated at the start of this chapter, the process of urbanisation consistently results in five generalised types of disturbances that are common to all cities (see Table 3.1). These represent some of the gradients in conditions that will affect the response of biodiversity to urbanisation, and they have distinct footprints in time and space (McDonnell and Hahs 2015). Many of these disturbances are interrelated, which is one of the reasons why urbanisation is often represented as an indirect gradient representing more complex, parallel, and interacting changes.

When it comes to studying responses of urban biodiversity, these five impacts offer commonalities that can be used to connect research conducted in urban landscapes around the world, regardless of the size of the city, the physical context, or local histories. Comparing the responses across and between cities allows us to investigate whether there are any general principles (or "laws")

Table 3.1 Five major categories of general disturbances commonly associated with urbanisation

Disturbance type	*Examples*
Changes to the landscape	Loss or transformation of pre-urban habitats
Changes to the physical environment	Changes in local microclimates, soil properties, nutrients, and chemical or sensory pollution
Increased presence and activities associated with larger populations of people	Visual disturbances, culturally informed decision-making, and management practices
Changes to the disturbance regimes	Suppression of fire, change in flooding frequencies
Changes to species assemblages	Species introductions, local extinctions, changes in interaction networks

that emerge consistently, or whether there are important moderating factors. This is equivalent to identifying the conditions that contribute to the tree line in montane ecology (general principle or "law") and understanding how the elevation of the tree line changes depending on the geographic location of the mountain (moderating factor).

By identifying general principles, we can create a basic understanding of the system that can then be used to inform the planning, design, and management of urban landscapes in situations where specific knowledge of the local system is incomplete or not currently available. The discipline of landscape ecology offers an excellent example of how general principles have been successfully used to inform planning, design, and management. The concepts of patch-matrix-corridor, edge effects, and stepping stones are essentially general principles that have been used to shape efforts to conserve species, maintain patch integrity, and increase connectivity for the past 40 years. However, the way they are applied or interpreted may vary depending on the system they are supporting; stepping stones designed for bees may look quite different to stepping stones designed for bears.

While commonalities are helpful for understanding general principles, there are occasions where the specific responses of biodiversity within a system will vary based on the particular characteristics of the city they were captured in (similar to how stepping stones are likely to have different designs based on the taxa they will be supporting). This is where the complementary use of direct and indirect gradients can come into play. The direct measurements are helpful for examining relationships between specific urban impacts and an observed biodiversity response, while the indirect gradients provide a common measure that can be used to situate the study within the broader global context (Martinez 2018).

Ultimately, the goal of comparative approach is to use commonalities and differences to establish a framework where you can investigate different levels of complexity within the system (Table 3.2). The simplest approach are correlative case studies (A) where the patterns of response are characterised along an indirect or generalised gradient. These provide an initial signal of how biodiversity responds to a generalised aspect of the urban environment in that particular context (e.g. butterfly diversity along a gradient of vegetation cover). However, to confirm if this is a consistent pattern across multiple cities and contexts, you would need to replicate that study in multiple cities before you can confirm if it is a broad generalisation (B) that is found across most cities. However, to understand why butterfly diversity is displaying this response to urbanisation, you need to undertake a more mechanistic case study (C), where you directly measure dimensions of the environment, such as temperature and moisture, density of host plants for adult and larval stages, insecticide concentrations, and permeability of the landscape to determine whether the trend is driven by physiological tolerances (temperature and moisture), food availability, suitable conditions for successful reproduction (density of host plants), threatening processes (insecticide exposure), or dispersal limitations (landscape permeability). The most sophisticated levels of study will try to understand these mechanistic relationships and how consistent they are across different cities or systems with the ultimate goal of identifying mechanistic "laws" (D) that can be used to guide actions in locations where ecological knowledge is low. In the next section, we will

Table 3.2 Relationships between gradients and comparative ecology

	Indirect (generalised) gradient	*Direct (specific) gradient*
Comparisons within a system/city	A) Correlative case study	C) Mechanistic case study
Comparisons between systems/cities	B) Confirmed broad generalisation	D) Confirmed mechanistic "law"

take a closer look at how different comparative ecology studies have been set up and the types of questions that can be addressed with each of the different approaches.

Approaches to comparative studies in urban ecology

Within urban ecology, there are many ways to set up a comparative study, and they each reveal a different aspect of the system and how it changes over time, across space, or between contexts. The following are examples of the main forms of comparative research that have been employed in urban ecology research.

Comparisons within cities and between cities

The first critical distinction in comparative ecology research is to establish whether you are working at the city or sub-city scale. Comparisons at the sub-city scale are characterised by distinct "sites" or locations within a city (e.g. the influence of wetland size and landscape context on the composition or calling behaviour of amphibians). These studies reveal more of the subtle details and potential mechanisms through which urban environments filter biodiversity responses. In contrast, studies conducted at the city scale collapse this heterogeneity into a single metric for that city, such as a species list of the plants or birds present. Studies conducted at the city scale are useful for revealing the influence of city scale characteristics (e.g. city age and development history; population size or extent; distribution and connectivity of green spaces) on urban biodiversity responses.

Comparative studies at both scales can use similar metrics to represent the underlying gradients of interest and biodiversity response. However, the inferences and knowledge contributions will differ between the two scales. Comparisons at the city scale will only reveal broad generalisations (A or B in Table 3.2). Understanding the mechanisms behind these patterns will always require more focused studies conducted at the sub-city scale (C in Table 3.2), repeated across multiple cities representing a gradient of city sizes or contexts (D in Table 3.2).

Comparisons across geographical space

The most common comparative approach that can be found in the published literature to date compares ecological responses at different locations across geographical space. It is the approach used to compare responses between different levels of urbanisation, as well as being used when studies seek to compare how an outcome is influenced by land use type, such as vegetation in parks compared to vegetation in residential gardens (e.g. Ortega-Álvarez and MacGregor-Fors 2009; Zhou et al. 2012); or where the land use type is consistent but something specific changes, such as bird assemblages in streets with native versus introduced street trees (e.g. de Castro Pena et al. 2017).

There are many examples in the literature of how an ecological response varies in response to increasing urbanisation. However, they are highly variable in terms of how they represent the gradient (Uchida et al. 2021). While many of the early studies of gradients presented simple comparisons between categories of "urban", "suburban", and "rural", this practice is now discouraged as these categories can be highly variable between cities and even between studies (Moll et al. 2019). This makes it difficult to compare findings across studies and compromises the ability to identify general principles and points of difference. For example, Figure 3.3 presents the potential gradients in human population density between five different cities. All these gradients could be categorised into "urban", "suburban", and "rural" components that make sense for that city. However, when it comes to comparing the response of biodiversity to these urbanisation gradients, the

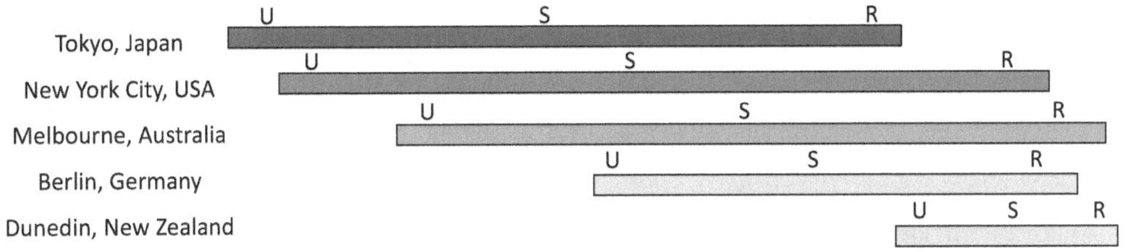

Figure 3.3 Comparisons of urban (U), suburban (S), and rural (R) conditions in five different cities with varying total population sizes. For example, the most urban conditions in Berlin may be more similar to suburban conditions in New York City. NOTE: if the comparison was made using an alternative measure, the relative differences may display a different pattern. For example, if the previous information was changed to reflect impermeable surface cover, the scale would be limited to a range of 0–100%, and some of the differences in the intensity of the land use (e.g. population density) within a specific range of land cover (impermeable surfaces) would be obscured.

categorisation becomes highly problematic as the most "urban" site in Dunedin may be equivalent to the "rural" site for Tokyo. However, if the actual population density (or other measure of the gradient) was provided, the data could quickly and easily be compared between the different cities as there is a common scale acting as a reference frame.

Comparisons across time

Cities and towns are dynamic systems. They are constantly changing through periods of growth and construction, decay and neglect, storms and disturbances, or renewals and disruptions associated with new technologies or philosophies. While there is a common perception that the central area of the city may have been developed for the longest period, this may not necessarily be the case; it is possible for all locations across the city to undergo change over time.

There are many ways to incorporate a temporal dimension into a comparative ecology framework and important reasons why this should be considered (Ramalho and Hobbs 2012; Ossola et al. 2021). The following are an illustrative but non-exhaustive list as a full overview of the possible approaches would require a chapter in itself.

- Age of city/development (length of time a disturbance has been in a landscape)
- Detecting change over time – decadal?
- Detecting response to urban interventions – were urban greening activities "successful"
- Cumulative species loss and cumulative species gain

Comparisons across historical contexts

As we discussed in Section 1, the history of development for individual cities can vary greatly, and this history is captured and reflected in the spatial arrangement of important features (e.g. city walls, Creighton 2007) and the materiality of the buildings, parks, and other components of the built form (Fletcher 2010). This offers the opportunity to compare outcomes between historical contexts, such as comparing modern plant extinction rates between cities that were urbanised prior to 1600 AD, between 1600–1800 AD and after 1800 AD (Hahs et al. 2009). It also offers the opportunity to identify shared features that appear across many climatic and geographic contexts, which can be used to explore variation across cities. For example, cities established and developed through colonisation often feature a common pool of street tree species or a botanic garden that has been designed, constructed, and planted with a shared ideology and purpose (Hunte et al. 2019; Shackleton and Gwedla 2021). These shared features can become a shared foundation for future questions, such as determining the potential impacts of climate change on growth and performance of common street trees (e.g. Limoges et al. 2018), or how we decolonise our understandings of nature and biodiversity in urban landscapes (e.g. Porter et al. 2020; Trisos et al. 2021; Chacko 2022).

Another form of comparison across historical contexts would be to examine the ecological implications of development trajectories: How do ecological responses to greenfield development vary compared to infill development or urban renewal? Does rapid urban expansion have a distinct signal in the impacts on biodiversity compared to locations where urban expansion has been more gradual? What are the implications of a land sparing compared to a land sharing approach to development (e.g. Montoya-Molina 2016)?

Comparisons across cultural contexts

Closely related to the historical context, the cultural context of a city can also offer opportunities for comparative approaches. Historically, the majority of urban ecology research to date has been conducted in countries associated with the Global North. This means that not only are the studies largely associated with temperate biomes and similar functional groups of species, they are also largely based in cities that have similar architectural and built form affinities and closely aligned ideologies that inform how people live in and experience those cities. While some interesting comparisons are possible within the literature from the Global North, there is an enormous potential to really challenge and broaden our understanding of how urbanisation affects urban biodiversity (or humans' understandings of and connections to biodiversity) by considering insights from research conducted in cities across Africa, Latin America, Asia, and other countries within the Global South (e.g. Shackleton et al. 2021; McHale et al. 2013). For example, informal settlements are a more prevalent built form within cities in the Global South, yet they vary considerably in terms of their origins, construction, utilities and services, infrastructure, and status within the local municipal authorities (e.g. Dovey et al. 2020). This will have implications for the relationships these built forms and the people living within them have with the biodiversity and more-than-human elements of the landscape. Similarly, there are differences between cultures in the way that they view and understand the world, which means that the frameworks and concepts that have arisen from research on "Western, educated, industrialized, rich and democratic (WEIRD) societies" (Henrich et al. 2010) do not fit outside of that context. For example, the human-centric framing of Ecosystem Services is problematic for Indigenous and other societies and cultures that are grounded in reciprocity as an ancestral principle, where reciprocal contributions between people and nature offers a more supportive pathway towards sustainability and conservation (Ojeda et al. 2022).

Comparisons across social-technological-economic contexts

In many ways, comparisons across social, technological, or economic gradients could be considered a subset of previous types of comparisons, such as those across geographic space or cultural contexts. However, they are presented here as a distinct category as the complexity within each of these systems are worth considering on their own merits (McPhearson et al. 2016).

The social-political systems prevalent within a city will determine to what extent changes in the urban environment arise from top-down (government decisions and policies) or bottom-up (arising from individuals or social movements) processes, as well as determining the scale and pace of change. For example, Singapore is one of the most highly vegetated large cities in the world, and it has been able to achieve this due to government commitment to developing policies and committing resources to implementing change (Tan et al. 2013). Similarly, the scale of change associated with the Cheonggyecheon River restoration in Seoul, South Korea, was made possible due to government commitments and investment (Shin and Lee 2006). In contrast, the restoration of the Merri Creek in Melbourne, Australia, began due to the commitment of dedicated members of the public who were determined to drive that change (Bush et al. 2003).

Cities contain very complex networks of technology and infrastructure, which can vary along gradients. Infrastructure relates to the physical features and services that support the function of the city (e.g. transport, power, communication, and water) and can include nature-based solutions. Technology relates to the application of scientific knowledge to improve practical outcomes, with examples ranging from smart cities and the internet of things to the range of processes and systems used to treat wastewater. They are presented together here as often the available technologies will shape the infrastructure present. This means that when it comes to setting up comparative studies investigating the impacts of infrastructure, it is also important to account for variations in technologies. For example, street lighting has undergone large shifts in available technologies over the past century, beginning with electrification, and then transitions from low pressure sodium bulbs to LEDs (Gaston et al. 2015), and we are now experiencing innovations to minimise the harmful effects of artificial night at light by changing the spectral signatures of the globes used, as well as creating smart lighting that customises how the light is switched on or off. Comparative studies between locations in the landscape with these different lighting technologies are one of the key sources of evidence that has helped drive the current innovations towards reduced artificial light at night impacts (Haddock et al. 2019). As the density of streetlights is also often related to the density of human populations and urbanisation, the amount of artificial light at night is also used as an indirect gradient of urbanisation both within and between cities. Other examples of comparative studies that investigate impacts of infrastructure and technology on biodiversity are the systems used to manage water distribution across urban areas (Webber et al. 2022). Comparisons between areas with different technologies present allow us to measure the improved biodiversity outcomes from water harvesting and re-use, or the use of nature-based solutions.

As the variations in technology and infrastructure may also vary depending on the context, comparative studies must be very clear in setting up what is the specific gradient they are interested in understanding and set up their study design to allow for additional influences to be accounted for. For example, the Smart Cities and Internet of Things approaches and projects will vary between locations based on the focus of the system (e.g. water/lighting) and how they are configured.

Comparisons within and between taxa (or studies)

Over the past three decades, most of the research on urban biodiversity has focused on a single taxonomic group within a study (Rega-Brodsky et al. 2022). These studies have allowed us to develop a better understanding of changes in (1) species assemblages, (2) functional traits and phylogenetic diversity, (3) physiological responses (e.g. growth rates or stress levels), (4) reproductive success, (5) eco-evolutionary adaptations and many other responses of urban biodiversity along direct and indirect gradients. In some cases, a common methodology was used which allowed for direct comparisons between studies. For example, the GlobeNet project investigated changes in carabid beetle assemblages across gradients in multiple cities (Niemelä et al. 2000). In other cases, the diversity of methodologies employed were so widely varied it was not possible to conduct a meta-analysis, so the emergent generalities had to be extracted through the less precise method of vote-counting (e.g. Williams et al. 2015).

Comparisons between taxa are less common in the literature but offer an opportunity to understand how consistent a particular response might be. For example, within terrestrial and aquatic urban habitats, the community weighted mean body size of some taxa increased with increasing temperatures and decreased for others (Merckx et al. 2018). This highlights the importance of examining biodiversity responses across a large range of taxa, before any broad generalisations can be made.

The methods used to compare across taxa can be very challenging as they may involve different sampling methodologies (e.g. sweep netting for insects compared to placing tiles to sample terrestrial snails and slugs, or recording vocalisations of bats). The responses reported in the literature may also vary – calling activity is commonly reported as an indication of bat activity, while percent cover or relative frequency may be used to represent abundance of different plant species. Additional challenges are that with the exception of body size, there are very few functional traits that can be applied across multiple taxa, and even then, the measures used for the calculation will vary. For example, intertegular distance or the width between the base of the wings is the standard measure of body size for bees (e.g. Raiol et al. 2021), while snout-vent length is a widely used standard measure of body size for reptiles (e.g. Pincheira-Donoso and Meiri 2013). However, it is possible to work around these challenges, and the potential advances in knowledge makes it worth pursuing when the capacity exists.

There are four general approaches to comparing across taxa. The first is to purposively sample multiple taxonomic groups in an individual study. In this case, multiple taxa are sampled simultaneously at a shared set of locations, and then responses to gradients are investigated for each taxonomic group and any similarities or differences are identified (e.g. Threlfall et al. 2017). The advantage of this approach is that the samples were captured from a common set of locations and conditions, so any differences can be more robustly attributed to the response of the species. The disadvantage is that resources, funding, and scientific interests don't always align in a way that makes this possible. An alternative is to compile a data set that contains equivalent information for multiple taxonomic groups, and then use a global analysis to compare responses between taxa (e.g. Aronson et al. 2014). The advantage of this approach is that the methodology used to compile and analyse the data sets were consistent between taxa, so any emergent trends are more likely to reflect taxonomic differences. The disadvantage is that often the types of data sets that can be compiled in this way are highly simplified, so the findings that do emerge reflect broad trends rather than insights into more mechanistic understandings of what may be contributing to those differences. The third option is to conduct a meta-analysis (e.g. Beninde et al. 2015) or systematic review (e.g. Brum et al. 2023); however the challenges experienced when applying these approaches within a taxonomic group will be amplified, and the additional insights gained may not always be as constructive as originally hoped. The final option is to provide a narrative comparison that places the findings of a specific study against the broader literature. However, the main value from this approach is that it would help to identify potential knowledge gaps or fruitful lines of enquiry for future research – rather than offering insights into general principles like the other three approaches.

Comparisons between realms (marine, freshwater, terrestrial)

The final comparison I would like to cover looks across and between realms, such as terrestrial, marine, and freshwater environments, as well as the atmospheric and belowground environments. Research in each of these realms has developed in parallel, but often from different disciplinary traditions (Bugnot et al. 2019). Finding ways to compare research questions, methodologies, and existing knowledge offers opportunities to test the relevance of generalities developed in one realm to further our understanding of biodiversity in another realm (Pinsky et al. 2022). This is still a relatively undeveloped area of enquiry, although with great potential for future research, and anyone interested in this approach is encouraged to look more closely at the references cited previously as a helpful starting point.

Setting up successful comparative ecology research

When setting up a research project that uses a comparative approach framework, there are a few simple steps that can guide you through this process.

1 *What is your question?* One of the first decisions that need to be made is to clearly articulate the knowledge gap or research question you are seeking to address with the study. This chapter has highlighted that there are many different ways to configure a comparative study (direct or indirect gradients represented as punctuated/categorical or continuous variables; across time, space or other contexts; within or across cities; etc). Having a clear question to act as a firm reference point will help determine which approach will offer the greatest insights for the available resources. If you are having trouble addressing your question with your available resources, you can think creatively about how you can adjust your question or methods to fit the available resources, change your question, or work to secure additional resources. Trying to push through with an ambitious project that isn't properly resourced is the path to delays, heartache, or broken dreams – avoid it when you can!
2 *What is the response you will be measuring to address your question?* Which taxa are you working on? Will you be recording physiological measurements or compiling data from existing sources? Does your question require physical measurements in the field (e.g. field surveys, genetic samples, physiological measurements), or can it be compiled using existing data (e.g. spatial datasets or existing literature)? Is the taxonomic group likely to reveal a response within the timeframes of your study? Do you need to collect data at the scale of the city or at multiple locations within a city? Resolving the finer details of your research will help determine if you are conducting a mechanistic study that requires specific environmental factors to be quantified, or if it is a more general study where an indirect gradient can be applied.
3 *Which categories of environmental change are relevant to your study system or research question?* The section on "Characteristics of urban areas: commonalities and differences" outlines some of the important dimensions of change within and between urban areas, which can be used to help establish the important parameters for your comparative study. The relative importance of each dimension will vary between studies. Your question will help determine where the emphasis and the focus of your efforts should be.
4 *Will my proposed research answer my question?* Peer review is a fundamental underpinning in academic research. Seek opportunities to share your work early and often with fellow students, peers, colleagues, and others in your field. They can help point out any gaps or missed opportunities and will give you the greatest chance of getting your papers published smoothly since it has already been through a friendly review process.
5 *Will my proposed research facilitate comparisons for future research?* By paying attention to the tools you use to measure and analyse your data, it is possible to enable more rapid and efficient comparisons with other studies. For example, Moll et al. (2019) have reviewed some of the metrics used to represent urban conditions in wildlife research. This understanding can be helpful for creating research that can be directly compared to other contexts.

Synthesis

Comparative approaches offer important opportunities to contextualise our research and situate our findings against a broader global understanding. This is critical if we are to develop a strong understanding of which patterns, processes, and mechanisms related to urban biodiversity are common across most cities and which are more likely to vary depending on the context.

Key companion papers

Boone, C.G., et al. 2012. A comparative gradient approach as a tool for understanding and managing urban ecosystems. *Urban Ecosystems* 15(4):795–807.

Du Toit, M.J., A.K. Hahs, I. MacGregor-Fors. 2021. The effect of landscape history on the urban environment: Past landscapes, present patterns. Pages 51–78 in C.M. Shackleton, S.S. Cilliers, E. Davoren, M.J. du Toit, editors. *Urban Ecology in the Global South*. Springer, Cham.

Martinez, P. 2018. The comparative method in biology and the essentialist trap. *Frontiers in Ecology and Evolution* 6:130.

McDonnell, M.J., A.K. Hahs. 2013. The future of urban biodiversity research: Moving beyond the 'low-hanging fruit'. *Urban Ecosystems* 16(3):397–409.

Ossola, A., M. Cadenasso, E. Meineke. 2021. Valuing the role of time in urban ecology. *Frontiers in Ecology and Evolution* 9:620620.

Ramalho, C.E., R.J. Hobbs. 2012. Time for a change: Dynamic urban ecology. *Trends in Ecology & Evolution* 27(3):179–188.

Acknowledgements

Much of my learning, thinking, and the development of my understanding on this topic has occurred while I have been living and working on the unceded lands of the Wurundjeri Woi-Wurrung, Bunurong, Boon Wurrung, Wadawurrung, and Dja Dja Wurrung peoples, in the cities and towns that are now commonly known as Melbourne, Ballarat, and Creswick.

It has also been my honour and privilege to work with and learn from the original champions of the comparative approach to urban ecology – Mark McDonnell and Steward Pickett, as well as more recent collaborators such as the members of the UrBioNet Patterns, Drivers, and Traits Working Group, and the Comparative Urban Ecology Research Training Network (CURT). The ideas and ways of thinking presented in this chapter are deeply and fundamentally influenced by conversations I have held with all of these people, and I am sincerely grateful to them for helping to develop, expand, and solidify my understanding of comparative approaches to urban ecology. This chapter was also improved by comments from the reviewers, and I thank them for their constructive feedback.

References

Aronson, M.F.J., F.A. La Sorte, C.H. Nilon, M. Katti, M.A. Goddard, C.A. Lepczyk, P.S. Warren, N.S.G. Williams, S. Cilliers, B. Clarkson, C. Dobbs, R. Dolan, M. Hedblom, S. Klotz, J.L. Kooijmans, I. Kühn, I. MacGregor-Fors, M. McDonnell, U. Mörtberg, P. Pyšek, S. Siebert, J. Sushinsky, P. Werner, M. Winter I. 2014. A global analysis of the impacts of urbanization on bird and plant diversity reveals key anthropogenic drivers. *Proceedings of the Royal Society B: Biological Sciences* 281(1780):20133330.

Beninde, J., M. Veith, A. Hochkirch. 2015. Biodiversity in cities needs space: A meta-analysis of factors determining intra-urban biodiversity variation. *Ecology Letters* 18(6):581–592.

Brum, P.H.R., S.R.A. Gonçalves, C. Strüssmann, A.L. Teixido. 2023. A global assessment of research on urban ecology of reptiles: Patterns, gaps and future directions. *Animal Conservation* 26:1–13.

Bugnot, A.B., G.C. Hose, C. Walsh, O. Floer, K. French, K. Dafforn, J. Hanford, E. Lowe, A.K. Hahs. 2019. Urban impacts across realms: Making the case for inter-realm monitoring and management. *Science of the Total Environment* 648:711–719.

Bush, J., B. Miles, B. Bainbridge. 2003. Merri Creek: Managing an urban waterway for people and nature. *Ecological Management & Restoration* 4(3):170–179.

Caryl, F.M., A.K. Hahs, L.F. Lumsden, R. Van der Ree, C. Wilson, B.A. Wintle. 2014. Continuous predictors of species distributions support categorically stronger inference than ordinal and nominal classes: An example with urban bats. *Landscape Ecology* 29(7):1237–1248.

Chacko, X.S. 2022. Stringing, reconnecting, and breaking the colonial 'Daisy Chain': From botanic garden to seed bank. *Catalyst: Feminism, Theory, Technoscience* 8(1). https://doi.org/10.28968/cftt.v8i1.35133

Creighton, O. 2007. Contested townscapes: The walled city as world heritage. *World Archaeology* 39(3): 339–354.

De Castro Pena, J.C., F. Martello, M.C. Ribeiro, R.A. Armitage, R.J. Young, M. Rodrigues. 2017. Street trees reduce the negative effects of urbanization on birds. *PloS ONE* 12(3):e0174484.

Dovey, K., M. van Oostrum, I. Chatterjee, T. Shafique. 2020. Towards a morphogenesis of informal settlements. *Habitat International* 104:102240.

Du Toit, M.J., A.K. Hahs, I. MacGregor-Fors. 2021. The effect of landscape history on the urban environment: Past landscapes, present patterns. Pages 51–78 in C.M. Shackleton, S.S. Cilliers, E. Davoren, M.J. du Toit, editors. *Urban Ecology in the Global South*. Springer, Cham, Switzerland.

Fletcher, R. 2010. Urban materialities: Meaning, magnitude, friction, and outcomes. Pages 459–483 in D. Hicks, M.C. Beaudry, editors. *The Oxford Handbook of Material Culture Studies*. Oxford University Press, New York.

Gaston, K.J., M.E. Visser, F. Hölker. 2015. The biological impacts of artificial light at night: The research challenge. *Philosophical Transactions of the Royal Society B: Biological Sciences* 370(1667):20140133.

Government of Pitcairn Island. 2022. *Welcome to the Official Web Site of the Government of the Pitcairn Islands*. Accessed 16/06/2022 www.government.pn/

Griffiths, D. 2017. The comparative method and the history of the modern humanities. *History of Humanities* 2(2):473–505.

Haddock, J.K., C.G. Threlfall, B. Law, D.F. Hochuli. 2019. Responses of insectivorous bats and nocturnal insects to local changes in street light technology. *Austral Ecology* 44(6):1052–1064.

Hahs, A.K., M.J. McDonnell, M.A. McCarthy, P.A. Vesk, R.T. Corlett, B.A. Norton, S.E. Clemants, R.P. Duncan, K. Thompson, M.W. Schwartz, N.S.G. Williams. 2009. A global synthesis of plant extinction rates in urban areas. *Ecology Letters* 12:1165–1173.

Heal, O.W., J.P. Grime. 1991. Comparative analysis of ecosystems: Past lessons and future directions. Pages 7–23 in J. Cole, G. Lovett, S. Findlay, editors. *Comparative Analyses of Ecosystems*. Springer, New York.

Henrich, J., S.J. Heine, A. Norenzayan. 2010. Most people are not WEIRD. *Nature* 466:29.

Hunte, N., A. Roopsind, A.A. Ansari, T.T. Caughlin. 2019. Colonial history impacts urban tree species distribution in a tropical city. *Urban Forestry & Urban Greening* 41:313–322.

Körner, C., J. Paulsen. 2004. A world-wide study of high altitude treeline temperatures. *Journal of Biogeography* 31(5):713–732.

Limoges, S., T.T.H. Pham, P. Apparicio. 2018. Growing on the street: Multilevel correlates of street tree growth in Montreal. *Urban Forestry & Urban Greening* 31:15–25.

Martinez, P. 2018. The comparative method in biology and the essentialist trap. *Frontiers in Ecology and Evolution* 6:130.

McDonnell, M.J., A.K. Hahs. 2008. The use of gradient analysis studies in advancing our understanding of the ecology of urbanizing landscapes: Current status and future directions. *Landscape Ecology* 23(10):1143–1155.

McDonnell, M.J., A.K. Hahs. 2015. Adaptation and adaptedness of organisms to urban environments. *Annual Review of Ecology, Evolution, and Systematics* 46:261–280.

McDonnell, M.J., S.T.A. Pickett. 1990. Ecosystem structure and function along urban-rural gradients: An unexploited opportunity for ecology. *Ecology* 71(4):1232–1237.

McHale, M.R., D.N. Bunn, S.T.A. Pickett, W. Twine. 2013. Urban ecology in a developing world: Why advanced socioecological theory needs Africa. *Frontiers in Ecology and the Environment* 11(10):556–564.

McPhearson, T., S.T.A. Pickett, N.B. Grimm, J. Niemelä, M. Alberti, T. Elmqvist, C. Weber, D. Haase, J. Breuste, S. Qureshi. 2016. Advancing urban ecology toward a science of cities. *BioScience* 66(3):198–212.

Merckx, T., C. Souffreau, A. Kaiser, L.F. Baardsen, T. Backeljau, D. Bonte, K.I. Brans, M. Cours, M. Dahirel, N. Debortoli, K. de Wolf, J.M.T. Engelen, D. Fontaneto, A.T. Gianuca, L. Govaert, F. Hendrickx, J. Higuti, L. Lens, K. Martens, H. Matheve, E. Matthysen, E. Piano, R. Sablon, I. Schon, K.V. Doninck,

L. De Meester, H.V. Dyck. 2018. Body-size shifts in aquatic and terrestrial urban communities. *Nature* 558(7708):113–116.

Moll, R.J., J.D. Cepek, P.D. Lorch, P.M. Dennis, E. Tans, T. Robison, J.J. Millspaugh, R.A. Montgomery. 2019. What does urbanization actually mean? A framework for urban metrics in wildlife research. *Journal of Applied Ecology* 56(5):1289–1300.

Montoya-Molina, S., C. Giraldo-Echeverri, J. Montoya-Lerma, J. Chará, F. Escobar, Z. Calle. 2016. Land sharing vs. land sparing in the dry Caribbean lowlands: A dung beetles' perspective. *Applied Soil Ecology* 98:204–212.

Niemelä, J., J. Kotze, A. Ashworth, P. Brandmayr, K. Desender, T. New, L. Penev, M. Samways, J. Spence. 2000. The search for common anthropogenic impacts on biodiversity: A global network. *Journal of Insect Conservation* 4(1):3–9.

Ojeda, J., A.K. Salomon, J.K. Rowe, N.C. Ban. 2022. Reciprocal contributions between people and nature: A conceptual intervention. *BioScience* 72:952–962.

Ortega-Álvarez, R., I. MacGregor-Fors. 2009. Living in the big city: Effects of urban land-use on bird community structure, diversity, and composition. *Landscape and Urban Planning* 90(3–4):189–195.

Ossola, A., M. Cadenasso, E.K. Meineke. 2021. Valuing the role of time in urban ecology. *Frontiers in Ecology and Evolution* 9:620620.

Peach, J. 2018. *The World's 'Smallest' Cities*. thisbigcity. https://thisbigcity.net/microcities-five-of-the-worlds-smallest-cities/

Peach, J. 2019. *The World's 'Biggest' Cities*. thisbigcity. https://thisbigcity.net/megacities-five-of-the-worlds-biggest-cities/

Pickett, S.T.A., W. Zhou. 2015. Global urbanization as a shifting context for applying ecological science toward the sustainable city. *Ecosystem Health and Sustainability* 1(1):1–15.

Pickvance, C.G. 2001. Four varieties of comparative analysis. *Journal of Housing and the Built Environment* 16(1):7–28.

Pincheira-Donoso, D., S. Meiri. 2013. An intercontinental analysis of climate-driven body size clines in reptiles: No support for patterns, no signals of processes. *Evolutionary Biology* 40(4):562–578.

Pinsky, M.L., L. Comte, D.F. Sax. 2022. Unifying climate change biology across realms and taxa. *Trends in Ecology & Evolution* 37(8):672–682.

Porter, L., J. Hurst, T. Grandinetti. 2020. The politics of greening unceded lands in the settler city. *Australian Geographer* 51(2):221–238.

Raiol, R.L., M. Gastauer, A.J. Campbell, R.C. Borges, M. Awade, T.C. Giannini. 2021. Specialist bee species are larger and less phylogenetically distinct than generalists in tropical plant – bee Interaction networks. *Frontiers in Ecology and Evolution* 9:699649.

Ramalho, C.E., R.J. Hobbs. 2012. Time for a change: Dynamic urban ecology. *Trends in Ecology & Evolution* 27(3):179–188.

Rega-Brodsky, C.C., M.F.J. Aronson, M.R. Piana, E.-S. Carpenter, A.K. Hahs, A. Herrera-Montes, S. Kanpp, D.J. Kotze, C.A. Lepczyk, M. Moretti, A.B. Salisbury, N.S.G. Willaims, K. Jung, M. Katti, I. MacGregor-Fors, J.S. Maclvor, F.A. LaSorte, V. Sheel, C.G. Threfall, C.H. Nilon. 2022. Urban biodiversity: State of the science and future directions. *Urban Ecosystems* 25:1083–1096.

Richards, D., M. Masoudi, R.R.Y. Oh, E.S. Yando, J. Zhang, D.A. Friess, A. Grêt-Regamey, P.Y. Tan, P.J. Edwards. 2019. Global variation in climate, human development, and population density has implications for urban ecosystem services. *Sustainability* 11(22):6200.

Rodríguez, M.Á., I.L. López-Sañudo, B.A. Hawkins. 2006. The geographic distribution of mammal body size in Europe. *Global Ecology and Biogeography* 15(2):173–181.

Schell, C.J., K. Dyson, T.L. Fuentes, S. Des Roches, N.C. Harris, D.S. Miller, C.A. Woelfle-Erskine, M.R. Lambert. 2020. The ecological and evolutionary consequences of systemic racism in urban environments. *Science* 369(6510):eaay4497.

Shackleton, C.M., S.S. Cilliers, M.J. du Toit, E. Davoren. 2021. The need for an urban ecology of the Global South. Pages 1–26 in C.M. Shackleton, S.S. Cilliers, E. Davoren, M.J. du Toit, editors. *Urban Ecology in the Global South*. Springer, Cham, Switzerland.

Shackleton, C.M., N. Gwedla. 2021. The legacy effects of colonial and apartheid imprints on urban greening in South Africa: Spaces, species, and suitability. *Frontiers in Ecology and Evolution* 8:579813.

Shin, J.H., I.K. Lee. 2006, November. Cheong Gye Cheon restoration in Seoul, Korea. *Proceedings of the Institution of Civil Engineers-Civil Engineering* 159(4):162–170.

Tan, P.Y., J. Wang, A. Sia. 2013. Perspectives on five decades of the urban greening of Singapore. *Cities* 32:24–32.

Threlfall, C.G., L. Mata, A. Mackie, A.K. Hahs, N.E. Stork, N.S.G. Williams, S.L. Livesley. 2017. Increasing biodiversity in urban green spaces through simple vegetation interventions. *Journal of Applied Ecology* 54(6):1874–1883.

Trisos, C.H., J. Auerbach, M. Katti. 2021. Decoloniality and anti-oppressive practices for a more ethical ecology. *Nature Ecology & Evolution* 5(9):1205–1212.

Uchida, K., R.V. Blakey, J.R. Burger, D.S. Cooper, C.A. Niesner, D.T. Blumstein. 2021. Urban biodiversity and the importance of scale. *Trends in Ecology & Evolution* 36(2):123–131.

Webber, J.L., T. Fletcher, R. Farmani, D. Butler, P. Melville-Shreeve. 2022. Moving to a future of smart stormwater management: A review and framework for terminology, research, and future perspectives. *Water Research* 218:118409.

Whittaker, R.H. 1956. Vegetation of the Great Smoky Mountains. *Ecological Monographs* 26(1):1–80.

Williams, N.S., A.K. Hahs, P.A. Vesk. 2015. Urbanisation, plant traits and the composition of urban floras. *Perspectives in Plant Ecology, Evolution and Systematics* 17(1):78–86.

World Population Review. 2022. *Tokyo Population 2022*. Accessed 16/06/2022 https://worldpopulationreview.com/world-cities/tokyo-population

Zhou, D., T. Fung, L.M. Chu. 2012. Avian community structure of urban parks in developed and new growth areas: A landscape-scale study in Southeast Asia. *Landscape and Urban Planning* 108(2–4):91–102.

4

URBAN CLIMATE AND THE BIOPHYSICAL ENVIRONMENT

Peter Ibsen, Dion Kucera, Stephanie Piper, and G. Darrel Jenerette

Introduction

Since the first settlements circa 8,000 BCE, there have been strong linkages among climate, biodiversity, and society (Sandweiss et al. 1999). Urban form and the life found within cities are molded by the regional and global climate. The urban biophysical environment also has a strong influence on influencing local, regional, and global climate dynamics. These interactions of urban climate and the biophysical environment impact urban residents; from an individual's temporary physical comfort, to determining enduring patterns of biodiversity. In this chapter, we explore the dynamics of the urban climate across spatial scales, from citywide to parcels. We explain how urban climate is defined at these scales, how climate interacts with the biophysical and social environment, and how those interactions drive dynamics of urban biodiversity. We use this framework of scale and connection to examine how a changing climate will interact with urban climate (Figure 4.1). Finally, we make a case for what the goals of urban climate research should be in the coming years.

Urban climates at citywide, neighborhood, and patch scales

Citywide scale

Our knowledge of urban climate is rooted in the energy balance of an urban area (Oke 1988; Grimmond et al. 2011, 2010). An urban area receives short-wave (K↓, irradiance from the sun) and long-wave radiation (L↓, re-radiated energy from the upper-atmosphere and gray infrastructure). Received energy is reflected (K↑) and re-radiated (L↑) back into the atmosphere, and the residual energy that flows in and out of the urban environment is the net radiation flux (Q^*), which is subdivided into stored energy (ΔQ_s) and transformed energy (Q_E and Q_h). Q_E is latent heat, energy that is absorbed or released when liquid or vapor changes states. Q_h is sensible heat, which is energy transferred between surfaces and the atmosphere. Also included in Q^* is any anthropogenic heat flux (Q_f). Anthropogenic heat sources include thermal energy from transportation, the heating and cooling of urban structures, and industrial activities. All the energy inputs must equal the energy outputs. This equality is known as the urban energy balance (Grimmond et al. 2010).

DOI: 10.4324/9781003016120-5

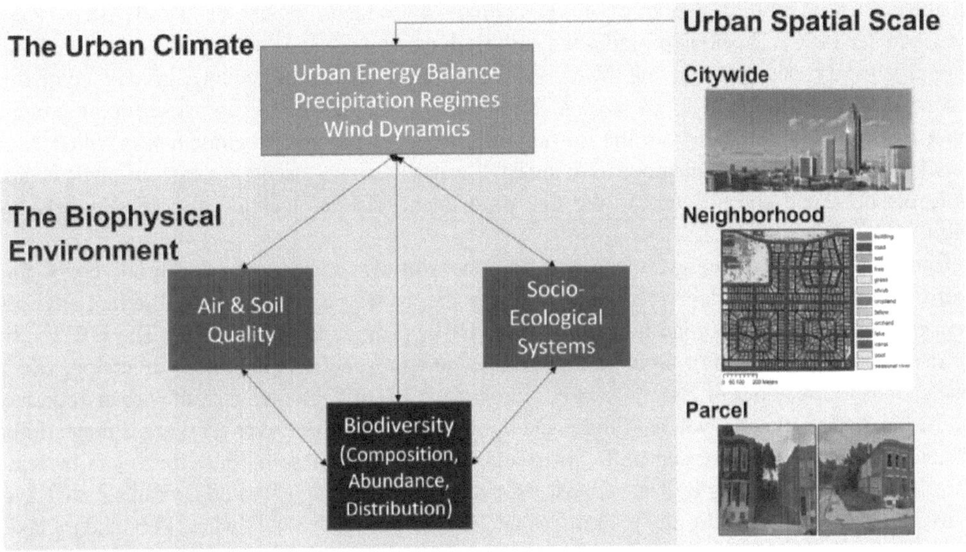

Figure 4.1 A conceptual outline of Chapter 4. The urban climate is composed of energy balances, regional precipitation, and wind regimes. The urban climate interacts with the biophysical environment through air and soil and through human decision-making. These factors determine the biodiversity of the city and, along with the initial climate factors, are influenced by biodiversity. These relationships are modified by the scale at which they occur – the citywide, the neighborhood, and the parcel scale. Images for neighborhood and parcel scale come from Jenerette et al. 2016b and Borowy and Swan 2020, respectively.

A conceptual outline of Chapter 4

Equation 4.1: $K{\downarrow} + L{\downarrow} - K{\uparrow} - L{\uparrow} = Q^*$

$$Q^* = Q_E + Q_h + \Delta Q_s + Q_f$$

A city's geography strongly influences urban climate. Latitude, orographic or maritime effects, and continentality will shape the levels of energy received and emitted from urban areas. These geographic factors drive patterns of wind and precipitation, which in turn influence the energy balance. Moreover, the ratio of latent heat flux to all evaporative heat fluxes (the evaporation fraction, Qe/(Qe + Qh) is a function of wind direction, which is related to local and regional land covers (Templeton et al. 2018). Urban design also affects net radiation, Q^*. A powerful example of this effect is the phenomenon of the Urban Heat Island (UHI). The UHI occurs when an urban area registers warmer air temperatures than its rural counterparts. What drives the urban heat island is the overall difference in Q^* between urban and rural areas. Compared to rural areas, cities have a greater percentage of non-permeable surfaces, such as concrete, asphalt, brick, and stonework. These surfaces have a greater capacity to store heat and release it later as long-wave radiation ($L{\downarrow}$), which results in a larger sensible heat flux, Q_h, during nighttime hours in urban areas. Similarly, that ability to store heat causes urban surfaces to be hotter than rural surfaces during the daytime, which is known as the Surface Urban Heat Island (Moffett et al. 2019).

The UHI has been an active area of research since the early 19th century. *The Climate of London*, published in 1818 by Luke Howard, a meteorologist known as the "father of meteorology", was an exhaustive text on multiple dynamics of urban climate in London, England from 1807–1818 (Parry and Chandler 1966). The book provides the earliest documentation of the UHI, with measurements from urban and rural areas showing lower mean air temperatures in the "country greens" compared to central London. In the methods used, a "thermometer was attached to a post set in the ground, under a Portugal laurel, and from the lowness of this tree the whole instrument was within three feet of the turf" is similar to methods used today. "Each sensor was placed in a custom-made vented white plastic shield and attached to the tree in an inconspicuous location at approximately 2m height" (Shiflett et al. 2017).

Understanding these energy fluxes and how they influence urban climate, biodiversity, and their interactions is scale dependent. At the mesoscale, or whole city scale, the thermal environment is controlled by the urban boundary layer (UBL) (Song and Wang 2016). The UBL is the vertical layer from the top of urban surfaces up to an atmospheric temperature inversion, which corresponds to the upper limits of urban pollution (Oke 1976). Downwind of the urban boundary layer is an "urban plume", where climate and atmospheric dynamics over rural areas may still be affected by the temperature and pollution found in the UBL. These plume effects can increase warming over rural areas as well as increase precipitation. In Chicago, warmed urban air can travel up to 70 km downwind of the city at altitudes of 500–1,000 m above the surface (a boundary layer UHI) (Cosgrove and Berkelhammer 2018).

The same urban form which contributes to the UHI also can drive increases in urban precipitation. Warm city surfaces increase the convective updraft of warm air, which can lead to greater rain in cities compared to rural counterparts (Schmid and Niyogi 2017). Furthermore, pollutants and aerosols originating in urban centers can result in increased cloud formation and rain events downwind from urban centers (Shepherd et al. 2002). This leads to an increase in precipitation over the urban core and downwind of the city, with downwind rainfall increasing with city size and urban cloud condensation nuclei (Sarangi et al. 2018). Thunderstorms are increased in large cities in warm and humid environments due to similar mechanisms (Haberlie et al. 2015). Urban induced rain effects are expected to be larger in coastal cities, where background aerosol concentrations may be low (van den Heever and Cotton 2007). The effect of urban pollution on aerosol formation is high in tropical forests, where urban pollutants can enhance aerosol production up to 400% (Shrivastava et al. 2019). Modeling has shown that high-aerosol loading can increase city-edge precipitation up to 100 mm/hr, while urban land cover can increase urban rainfall by 20 mm/hr (Schmid and Niyogi 2017). However, urban rain is not increased uniformly, rather, the urban landscape clusters the rain into more intense pockets (Paul et al. 2018).

Neighborhood scale

A fundamental characteristic of the urban environment is the spatial heterogeneity of urban forms. The type and density of buildings, vegetation, and infrastructure can all vary over short distances (Cadenasso et al. 2007). The multiple combinations of urban forms regarding structure height, structure density, and material, with vegetated land cover types are called local climate zones (LCZs) (Stewart and Oke 2012). LCZs relate microclimate variation to visible and categorical spatial structures. These visible structures are often obtainable from satellite imagery, which is useful for urban climate studies examining the correlation between changes in microclimate and LCZ over time (Collins and Dronova 2019), as well as in regions which there is little on the ground data of urban microclimate, such as sub-Saharan Africa (Brousse et al. 2019). Using LCZs in conjunction with air-temperature measurements highlights both the correlation between vegetation and cooler microclimates, and how urban form can alter the distribution of microclimates. Whereas increasing vegetation is often hypothesized to be a major driver of heat reduction, in Phoenix, Arizona, compact urban residential areas provided significant mid-afternoon air-cooling compared to more open landscapes, likely due to an increase in surface shading, more so than the vegetation effect (Middel et al. 2014).

The "oasis effect" describes a phenomenon where a city, or a portion thereof, is cooler than the surrounding landscape (Manoli et al. 2020). This difference is most common in seasonally dry climates. Finally, the Park Cool Oasis describes how parks have the capacity to mitigate the UHI by cooling the surrounding urban environment (Cao et al. 2010) and hypothesized to be predominantly driven by leaf area index (amount of vegetation surface area per m^2), plant density, and tree cover (X. Wang et al. 2018). Even with similar urban forms, such as parks, differences in vegetated land covers can result in a diversity of microclimate within the same region. In Oakland, California, surface temperatures were significantly higher in parks with recreational fields compared to those with more woody vegetation, even when there was no difference in remotely sensed vegetation (calculated as the Normalized Difference Vegetation Index, NDVI) (Dronova et al. 2018). Though it may seem that the Dronova et al. (2018) study is contradictory to the results of X. Wang et al. (2018), these studies highlight the drivers of vegetation-derived cooling. Greater tree density and canopy will result in greater shading of surfaces and a reduction in sensible heat flux, while recreation fields are highly irrigated, resulting in an increase in latent heat flux. Water use and shading are the two ends of vegetation-derived urban cooling.

Patch scale

Differences in vegetation type and density have the potential to create microclimates within cities. The density of vegetation as well as the percentage of trees, grass, roads, and built-up areas all correlate with daytime surface temperatures, creating a mosaic of horizontal and vertical microclimates within a single urban block (Jenerette et al. 2016b). Increases in NDVI are strongly correlated with reductions of air temperature across horizontal scales from 10 m to 90 m (Shiflett et al. 2017; Ziter et al. 2019). The daytime temperature above lawns is significantly lower at near-surface levels (0.1 m), compared to a height relevant to the human experience (1.5 m). However, areas covered with short and tall trees all had cooler air temperatures at 1.5 m compared to 0.1 m (Crum and Jenerette 2017). These spatial variations in vegetation-derived cooling are likely driven by latent heat energy transfer.

Across cities, urban microclimates vary depending on their regional mesoclimate. In a study of six cities in the United States, urban microclimates in more mesic regions are both warmer and drier during midsummer mornings when compared to their rural counterparts. However, in more arid cities, early mornings are both cooler and more humid than nearby rural areas. Furthermore, in all the study cities, relative humidity was less variable than in rural areas (Hall et al. 2016). This convergence has been hypothesized to represent a type of urban homogenization, where ecological processes in cities experience less variation than nearby rural lands (Groffman et al. 2017).

At the microscale, urban climate can be highly spatially and temporally heterogeneous, where cooler and warmer pockets of air and surface temperatures shift over short distances or timescales (Pickett et al. 2017). Shifts in climate at a scale smaller than the mesoscale, or mesoclimate (Crum et al. 2017), are generally known as microclimates. Microclimates are often dependent on urban form, that is, the physical layout of built structures (Ossola et al. 2019). For example, when streets are compact and surrounded by high-rises, the urban form both reduces direct shortwave radiation absorbed by the surface and, depending on the direction of the street relative to the prevailing wind, creates wind tunnels (Liu et al. 2017; Middel et al. 2014). However, street canyon airflow can be enhanced by pitched and round roof geometries (Llaguno-Munitxa et al. 2017), with a street's height-to-width aspect ratio being the most important built-environment factor affecting street canyon winds (Karra et al. 2017). Convection is important in providing urban cooling (Aram et al. 2019), as well as in removing pollutants from the city (Karra et al. 2017).

Urban climate effects on air quality

Anthropogenic activities do more than contribute to urban heat flux. Anthropogenic activities, including combustion and the heating and cooling of structures, emit multiple pollutants and greenhouse gasses, which can create urban "domes", where, for example, near surface CO_2 is significantly higher than in rural areas (Kaye et al. 2006). The density of vehicular roadways in urban cores is a major driver of the urban CO_2 dome (Brondfield et al. 2012). Even 20 years ago, in Phoenix, Arizona, weekday mean maximum sampled CO_2 concentrations were 528.8 ppm compared to 378.2 in their rural neighbors (Idso et al. 2001). Elevated urban CO_2 is hypothesized to be a driver of the accelerated growth found in urban trees as well (Smith et al. 2019).

Anthropogenic emissions to the atmosphere affect urban adjacent regions in addition to the urban biophysical environment. Nitrogenous compounds are emitted in urban areas and precipitate as either dry deposition or wet deposition (also known as acid rain). These nitrogen depositions can alter the growth patterns and diversity of urban plant life as well as plant diversity outside the city. Increases in dry nitrogen deposition derived from urban Southern California exacerbate global climate drivers of plant community shifts. Native California shrubs experience plant water use efficiency in a high nitrogen environment, which results in greater mortality during severe droughts, and a resultant change of local biodiversity dominated by native shrubs to one dominated by invasive grasses (Valliere et al. 2017). Other nitrogenous emissions (NOx specifically) interact with biological volatile organic compounds (BVOCs), such as the terpene isoprene, which are emitted by some urban plants to form ozone, O_3. In the stratosphere, ozone acts as an effective barrier against UV radiation; however, in cities, ozone is a major contributor to smog. Furthermore, ozone formation is accelerated by higher temperatures, creating a potential feedback effect in conjunction with the urban heat islands (Y. Wang et al. 2018).

Aerosols also are direct and indirect causes of long-term negative health effects. Recently, urban vegetation has been examined as a potential mitigation tool for atmospheric aerosols. The leaf tissue of urban tree species can capture particulate matter in the trichomes (fine hairs), or waxy layers found on the leaf surface (Grote et al. 2016). However, the efficiency of particulate matter removal depends both on the size of the aerosol, as well as on the species of tree (Wang et al. 2019). A 10-to-20-fold difference in particulate deposition on the leaf can occur depending on the leaf area index of the species (Janhäll 2015). While certain urban tree species may reduce harmful aerosols, there is a cost. The *Tipuana tipu*, a common urban tree species found in South America and the Southwestern United States, was found to have reduced growth correlated to accumulation of PM_{10} (Locosselli et al. 2019). These interactions of urban atmosphere and trees are all strongly influenced by urban microclimates, where the built structures of a city can have significant effects on the functioning of urban biodiversity.

Urban climate affects soil biophysical properties

Urban soils are not identical to soils in undisturbed environments, and climate is a driver of some of those differences. Urban heat island effects extend to the soil, and urban soils tend to be warmer when compared to suburban and rural soils. This difference is due to re-radiation of $L\downarrow$ to the ground, since cities have increased haze from pollutants compared to rural areas, as well as an increase of radiation reflected downwards by urban buildings (Landsberg 1981; Stewart and Oke 2012). Increased temperature of urban soils affects rates of litter decomposition and biogeochemical cycling. Pollution slows microbial activity, and therefore decomposition, but heat islands speed it up. In NYC, leaf litter decomposition is about twice as fast as in rural areas (Pouyat et al. 1997).

In colder cities, salt pollution from deicing practices is an urban concern, particularly in colder climates during winter months. The amount of salt applied has increased over time and will likely continue to do so with increased occurrence of large blizzards due to global change (Kelly et al. 2019). Salts can create a nutrient imbalance by altering the ions used during nutrient cycling. They can also reduce soil water potential, therefore hurting plant health. Finally, salts also aid in the release of heavy metals, building on an existing soil pollution concern (Schuler and Relyea 2018).

Atmospheric nitrogenous urban pollutants such as nitric oxides (collectively NO_X) and ammonia (NH_3) are deposited both on plant surfaces and directly to the soil or ground surface. When deposited on plants and soils, nitrogenous deposition can be taken up and used by plants and soil microbes. However, when pollution lands on impervious surfaces, it can remain there, or it can travel far from the source during rain, flooding, or irrigation events. Polluted waters traveling long distances over impervious surfaces can result in hotspots of nitrogen buildup. Interestingly, endpoints of nitrogen-enriched waters result in fertile "accidental" communities. In arid cities, low-lying urban areas with poor drainage and management can form into unexpected wetlands, which provide habitat for plants and wildlife (Palta et al. 2017). Accidental wetlands in the arid urban city of Phoenix, Arizona, had a greater richness of plant species compared to actively restored wetlands within the same watershed, with some wetland species, such as *Samolus parviflorus* and *Stemodia durantifolia*, only being recorded at the accidental sites (Bateman et al. 2015).

Urban climate and socio-ecological systems

The built environment, which alters urban climates, is reflective of socioeconomics. In 2007, Ostrom (2007) proposed a framework for studying socio-ecological systems (SES); this framework has formed the scaffolding for subsequent research on the topic (McGinnis and Ostrom 2014; McPhearson et al. 2016). In this framework, social actors and governance systems interact to utilize a natural commons. However, social systems do not necessarily abuse the commons, rather, social systems can self-organize to allow for a socially desirable use of natural resources (Partelow 2018).

Income inequality has a strong relationship with urban climate and biodiversity. In arid cities, where liberal access to water is a luxury, higher income areas see a corresponding increase in local vegetative biodiversity, a phenomenon known as the "luxury effect" (Hope et al. 2008). However, the regional climate of cities may moderate luxury effects. The luxury effect has been shown to hold for numerous aspects of the urban ecosystem, including arthropod and bird diversity (Leong et al. 2018), lizard diversity (Ackley et al. 2015), urban trees (Clarke et al. 2013), and pollinator abundance (Baldock et al. 2019) in cities around the world. Income has been shown to be an important factor in determining the richness of various taxa, raising valid concerns of environmental injustice (Chamberlain et al. 2020). However, the effect of income on urban biodiversity is complex. A study of six US cities found no effect when examining household yard flora, which implies that the influence of socioeconomic factors varies among regions, spatial, and temporal scales (Padullés Cubino et al. 2018). At global scales, the income-diversity relationship is stronger in more arid cities, though multiple other untested factors also mediate this relationship, including land use history, age of city structures, type of biodiversity, and social history (Kuras et al. 2020). Across the United States, urban ecosystems converge to be more similar among one another than to the natural ecosystems they replaced. The large amount of heterogeneity found in cities could suggest that a large amount of niche differentiation would occur (Katayama et al. 2014). However, the unique combination of natural and anthropogenic filters found in cities generally result in a convergence of ecological structures and functions (Groffman et al. 2014; Alberti et al. 2020). Furthermore, urban areas can buffer the temporal and spatial disturbances found in rural systems. Arid cities maintain irrigated vegetation, which acts as a yearlong source of habitat and diversity.

Convergence has been shown to hold for numerous aspects of cities, including the structure and composition of urban yards (Pearse et al. 2018), soil organic matter (Herrmann et al. 2018), carbon budgets, and soil moisture (Swan et al. 2011). Similarly, western black widows (*Latrodectus hesperus*) are more genetically similar between Las Vegas and Phoenix than between the populations just outside each of those cities (Miles et al. 2018). Convergence has also been found at the continental scale, such as for total nitrogen and organic matter of urban soils (Trammell et al. 2020). However, a paucity of urban ecology research from the Global South limits our understanding of convergence of urban ecosystems at the global scale (Ziter 2016).

Urban climate and biodiversity

Many cities across the globe incorporate similar urban forms and combinations of built structures with a variety of urban vegetation. Nevertheless, differences in urban vegetation diversity are strongly based on regional climate. While many urban centers boast tree-lined streets, a city's regional climate has a strong influence on what species of tree are found there. Cities that share similar climates are more likely to share a similar taxonomic diversity of urban tree species (Jenerette et al. 2016a). Globally, both the mean and minimum annual temperatures are good predictors of what tree species can be found within an urban region (Kendal et al. 2018). Similar patterns were found when comparing vegetation types within the same urban structure. In recreational parks, woody and turf species both had lower taxonomic diversity in hotter cities (Ibsen et al. 2020). However, the urban microclimate can be altered by the types of vegetation found in cities (Gillner et al. 2015). Urban climate both affects the arrangement and function of urban biodiversity and is also reciprocally modified by biodiversity.

Urban climate drives changes in vegetative and animal diversity, influencing present function and life cycles. As the UHI warms cities, ectothermic animals (those whose body temperatures fluctuate according to its surroundings) could potentially experience large effects due to urban climate. A sampling of multiple common bee species found in residential yards and parks of varying UHI intensity found populations with the lowest thermal tolerances, those that dwell in underground cavities, declined the most with exposure to greater UHI intensity (Hamblin et al. 2017). Similarly, when comparing the same species of a common lizard (*Anolis cristellus*) between urban and rural habitats in Puerto Rico, the urban dwelling species displayed a greater tolerance to ambient heat, and urban individuals showed genetic variation in thermal acclimation genes, indicating a potential urban-climate-induced evolutionary divergence (Campbell-Staton et al. 2020). The effect on life cycle and development is also found in some spider species. Black widow spiders (*Latrodectus hersperus*) exposed to temperatures found in urban Phoenix, Arizona (33°C), had slower development and lower body mass than those exposed to ambient desert temperatures (27°C) (Johnson et al. 2020). For urban trees, warmer planting environments, open growth conditions, and elevated CO_2 can all induce faster growth in younger saplings (Smith et al. 2019). Persistent interactions between urban climate and ecologically relevant traits can become a driver of urban-induced evolution (Alberti et al. 2020).

While urban climates, macro-, meso-, and microscales, have strong influences on biodiversity, the reverse can also be true, where urban vegetation diversity can influence urban climate. We have discussed how urban vegetation can reduce land and air temperatures within a city, shrinking the UHI. When we break down the term "vegetation" into specific species and functional groups, there can be large differences in climate mitigation potential. The surface temperatures of the urban forest vary by species. In a study of approximately 200 different tree species in the city of Basel, Switzerland, trees with larger leaves had higher crown temperatures compared to smaller-leaved species (Leuzinger et al. 2010), though certain species with large leaves and high stomatal conductance (e.g. *Tilia sp.*) resisted that trend, indicating the importance of leaf conductance in surface cooling. The relationship between leaf conductance and microclimate cooling is an important issue for urban planners when deciding on what urban tree species to plant and for what benefits. Not only do different species of urban tree have a broad spectrum of transpiration rates (Pataki et al. 2011), but those rates also have species-specific responses to increases in aridity (Litvak et al. 2012), which would be an important consideration in arid and semi-arid cities when estimating vegetation-derived cooling benefits.

Positive benefits humans receive from natural systems and biodiversity are known as ecosystem services (Bolund and Hunhammar 1999), and urban climate indirectly affects those relationships. Increases in regional aridity can increase nocturnal cooling benefits provided by urban vegetation (Ibsen et al. 2021). However, achieving vegetation-derived cooling benefits in arid and semi-arid cities requires a trade-off of a social disservice of excessive water use to maintain that vegetation (Roman et al. 2020). In mesic cities, urban tree canopy has a non-linear relationship with vacant lots, where both low and high vacancy blocks in Toledo had the largest percentages of tree canopy in the city (Berland et al. 2020). Depending on the regional climate, vegetation can be, as Berland et al. (2020) state, a "beauty or blight". In complex socio-ecological systems, individual, local, and broad cultural norms can determine the value of urban nature (Keeler et al. 2019). These cultural norms are strongly driven by climate, where the aesthetic pallet of planted urban vegetation will vary in hot and arid cities compared to milder and mesic cities, where more abundant water resources allows for more natural regeneration (Avolio et al. 2018; Nowak and Greenfield 2018; Roman et al. 2018; Andrade et al. 2019).

Urban climate futures

In our current era of rapid urbanization and anthropogenic climate change, understanding the linkages between climate, biodiversity, and society are more important than ever. Since the preindustrial era, the global temperature has increased 0.76°C (DaMatta et al. 2010). Moreover, the intensity of UHI and urban heatwaves are increasing globally (Ramamurthy and Bou-Zeid 2017), with the UHI acting as a multiplier of global climate change in local urban areas. These shifts in the urban climate will have significant effects on urban biotic communities. Urban ecological researchers, local to federal government, and regional stakeholders have been making plans on how to conserve urban biodiversity in a changing climate. As of 2020, the US Forest Service has developed a 20-year evaluation of 12 potential "Climate-Ready Trees" for California, which could potentially survive and thrive in a hotter and drier environment (McPherson et al. 2018). However, while many cities have projects focused on conserving either biodiversity or ecosystem services, few accomplish both (Nilon et al. 2017).

The changes to global climate could prove disastrous for the many species of life found in cities, and many urban areas have developed climate resiliency plans in response. However, urban areas unintentionally serve as "arks" for many taxa and can be very biodiverse. For instance, 30% of all Australian endangered species are found in Australian cities (Ives et al. 2016). In highly urban Southern California, 33 species of feral parrot were identified in the mid-1990s, despite no parrot being native to the region (Uehling et al. 2019). Finally, Los Angeles has one of the most diverse urban forests in the United States, home to over 500 tree species (Jenerette et al. 2016a). This compares favorably with the tree diversity of the United States and Canada, which is home to 679 species in all (Brockman 2002). These hotspots of biodiversity were accidentally created; imagine how cities would look if cities were managed to include and protect this biodiversity. Accidental urban hotspots of biodiversity may be negatively impacted by climate change, but cities can mitigate these effects by actively managing urban green spaces to court this novel biodiversity, while taking care to not introduce problematic invasive species.

The next stage of research connecting urban climate and urban biodiversity

Urban climate strongly influences the biophysical environment across spatial, temporal, and social scales, and studies in a variety of regions and unique cities have explored the multitude of relationships between climate and biodiversity. With this library of information going back over 100 years, we must ask, what is the next stage of research exploring urban climate's interactions with urban biodiversity? The next era of research needs to fill large outstanding gaps in the field. Three major gaps are as follows:

1 Partitioning the interactions of climate and urban biodiversity by synthesizing studies using multiple cities in varying regions, political climates, and historical legacies.
2 Including local stakeholders' knowledge in the process of research studies and direct application of results. Use transdisciplinary approaches to develop novel case studies of urban climate effects on urban biodiversity.
3 Directing research and application to address the legacy and present-day issues regarding linkages between biodiversity and environmental justice.

None of these gaps have easy solutions, and many could be considered "wicked problems". However, initial steps are being made to resolve these gaps.

Urban climate and biodiversity are tightly bound, however, often that relationship is obfuscated. Partitioning out the relative effects of vegetation type on climate at the city scale is still an active area of research, at the microscale and especially among multiple cities. Multicity studies are not new but have been coming more into the foreground. Wider availability of remote sensing data and the funding of large collaborative urban research networks have allowed for many comparisons of climate's influence across a variety of cities (Kendal et al. 2018; Krayenhoff et al. 2018). Furthermore, in the United States, Long-Term Ecological Research (LTER) sites, funded by the National Science Foundation, have been developed in arid (the Phoenix Cap-LTER) and mesic (previously the Baltimore Ecosystem Study and now the Minneapolis-St. Paul Metropolitan Area) cities. These LTERs provide the time to test urban climate hypotheses locally, before scaling the hypothesis up to multiple cities (Pickett et al. 2020). In addition to long-term studies, recent studies are looking to discern the influences of local biophysical and social legacies (Roman et al. 2018) and urban management practices (Cavender-Bares et al. 2020; Larson et al. 2020) in relation to urban climate. A deeper understanding of urban biodiversity drivers' effect sizes across climates will fill a needed gap in the current literature.

Developing local studies which incorporate local stakeholder knowledge into research planning, and conducting research directly applicable by stakeholders, is a necessary corollary to macrosystem studies. Relationships between urban climate and the biophysical environment permeate urban society across ecological, cultural, political, and economical scales. Individuals across fields and communities have been enacting on-the-ground changes to the urban environment to enhance biodiversity. Cities in multiple regions have municipal programs of mass tree plantings (McPhearson et al. 2010; Pincetl 2010; Carmichael and Mcdonough 2019), industries have developed methods of implementing green infrastructure to moderate urban climate impact, and individuals have acted as "lone rangers" to bring more biodiversity to their local neighborhoods (Schifman et al. 2017). A transdisciplinary approach is needed to release the emergent properties of this broad and deep brain trust of urban climate actors.

Global climate change has the potential to unequally affect urban climates. Urban climate researchers need to be filling the gap of studies examining both how these climate shifts will change the distribution and diversity of urban life, but also what methods can be implemented to address environmental justice and inequities in urban areas. Access to urban biodiversity is not equal among racial groups or income brackets in the US; however there are city specific exceptions to those trends. Similarly, legacies of injustice such as discriminatory housing practices have exposed more black Americans to extreme urban heat (Shandas et al. 2019). Yet there are still regional differences in the inequity of urban heat risk; Midwest cities display the lowest inequities, while Southeast and Western cities display the greatest (Hoffman et al. 2020). To meaningfully address climate-driven inequities, research must both understand the nuances of city-specific biodiversity distributions and promote transdisciplinary approaches, based in historical and cultural truths, to reduce inequity (Schwarz et al. 2018).

These three goals act to link the historical social legacies of urban ecosystems to local and macro scale processes. In a changing climate, research cannot ignore the small results to only focus on the large. Cities can be compared globally, but each urban area has its own unique character and, in many cases, climate. Moreover, there are large gaps in urban ecological data coming out of the global south (Ziter 2016). By identifying the patterns of climate and biodiversity that exist across cities and filling the gaps of knowledge from unstudied urban systems, researchers can provide lasting theoretical frameworks to help us understand these complex dynamics that have been entwined with human existence for millennia.

Key companion papers

Cadenasso, M.L., S.T.A. Pickett, K. Schwarz. 2007. Spatial heterogeneity in urban ecosystems: Reconceptualizing land cover and a framework for classification. *Frontiers in Ecology and the Environment* 5:80–88.

Groffman, P.M., et al. 2014. Ecological homogenization of urban USA. *Frontiers in Ecology and the Environment* 12:74–81.

Jenerette, G.D., et al. 2016. Climate tolerances and trait choices shape continental patterns of urban tree biodiversity. *Global Ecology and Biogeography* 25:1367–1376.

Oke, T.R. 1988. The urban energy balance. *Progress in Physical Geography: Earth and Environment* 12:471–508.

Stewart, I.D., T.R. Oke. 2012. Local climate zones for urban temperature studies. *Bulletin of the American Meteorological Society* 93:1879–1900.

References

Ackley, J.W., J. Wu, M.J. Angilletta, S.W. Myint, B. Sullivan. 2015. Rich lizards: How affluence and land cover influence the diversity and abundance of desert reptiles persisting in an urban landscape. *Biological Conservation* 182:87–92.

Alberti, M., E.P. Palkovacs, S. Des Roches, L. De Meester, K.I. Brans, L. Govaert, N.B. Grimm, N.C. Harris, A.P. Hendry, C.J. Schell, M. Szulkin, J. Munshi-South, M.C. Urban, B.C. Verrelli. 2020. The complexity of urban eco-evolutionary dynamics. *BioScience* 70:772–793.

Andrade, R., K.L. Larson, D.M. Hondula, J. Franklin. 2019. Social – Spatial analyses of attitudes toward the desert in a Southwestern U.S. city. *Annals of the American Association of Geographers*:1–20.

Aram, F., E.H. García, E. Solgi, S. Mansournia. 2019. Urban green space cooling effect in cities. *Heliyon* 5:e01339.

Avolio, M.L., D.E. Pataki, T.L.E. Trammell, J. Endter-Wada. 2018. Biodiverse cities: The nursery industry, homeowners, and neighborhood differences drive urban tree composition. *Ecological Monographs* 88(2):259–276.

Baldock, K.C.R., M.A. Goddard, D.M. Hicks, W.E. Kunin, N. Mitschunas, H. Morse, L.M. Osgathorpe, S.G. Potts, K.M. Robertson, A.V. Scott, P.P.A. Staniczenko, G.N. Stone, I.P. Vaughan, J. Memmott. 2019. A systems approach reveals urban pollinator hotspots and conservation opportunities. *Nature Ecology and Evolution* 3(3):363–373.

Bateman, H.L., J.C. Stromberg, M.J. Banville, E. Makings, B.D. Scott, A. Suchy, D. Wolkis. 2015. Novel water sources restore plant and animal communities along an urban river. *Ecohydrology* 8:792–811.

Berland, A., D.H. Locke, D.L. Herrmann, K. Schwarz. 2020. Beauty or blight? Abundant vegetation in the presence of disinvestment across residential parcels and neighborhoods in Toledo, OH. *Frontiers in Ecology and Evolution* 8:566759.

Bolund, P., S. Hunhammar. 1999. Ecosystem services in urban areas. *Ecological Economics* 29(2):293–301.

Borowy, D., C.M. Swan. 2020. A multi-trait comparison of an urban plant species pool reveals the importance of intraspecific trait variation and its influence on distinct functional responses to soil quality. *Frontiers in Ecology and Evolution* 8:1–20.

Brockman, C.F. 2002. *Trees of North America*. St. Martin's Press, New York.

Brondfield, M.N., L.R. Hutyra, C.K. Gately, S.M. Raciti, S.A. Peterson. 2012. Modeling and validation of on-road CO2 emissions inventories at the urban regional scale. *Environmental Pollution* 170:113–123.

Brousse, O., S. Georganos, M. Demuzere, S. Vanhuysse, H. Wouters, E. Wolff, C. Linard, N.P.M. van Lipzig, S. Dujardin. 2019. Using local climate zones in sub-Saharan Africa to tackle urban health issues. *Urban Climate* 27:227–242.

Cadenasso, M.L., S.T.A. Pickett, K. Schwarz. 2007. Spatial heterogeneity in urban ecosystems: Reconceptualizing land cover and a framework for classification. *Frontiers in Ecology and the Environment* 5(2):80–88.

Campbell-Staton, S.C., K.M. Winchell, N.C. Rochette, J. Fredette, I. Maayan, R.M. Schweizer, J. Catchen. 2020. Parallel selection on thermal physiology facilitates repeated adaptation of city lizards to urban heat islands. *Nature Ecology and Evolution* 4:652–658.

Cao, X., A. Onishi, J. Chen, H. Imura. 2010. Quantifying the cool island intensity of urban parks using ASTER and IKONOS data. *Landscape and Urban Planning* 96:224–231.

Carmichael, C.E., M.H. Mcdonough. 2019. Community stories: Explaining resistance to street tree-planting programs in Detroit, Michigan, USA. *Society & Natural Resources* 32:588–605.

Cavender-Bares, J., J. Padullés Cubino, W.D. Pearse, S.E. Hobbie, A.J. Lange, S. Knapp, K.C. Nelson. 2020. Horticultural availability and homeowner preferences drive plant diversity and composition in urban yards. *Ecological Applications* 30(4):e02082.

Chamberlain, D., C. Reynolds, A. Amar, D. Henry, E. Caprio, P. Batáry. 2020. Wealth, water and wildlife: Landscape aridity intensifies the urban luxury effect. *Global Ecology and Biogeography* 29:1595–1605.

Clarke, L.W., G.D. Jenerette, A. Davila. 2013. The luxury of vegetation and the legacy of tree biodiversity in Los Angeles, CA. *Landscape and Urban Planning* 116:48–59.

Collins, J., I. Dronova. 2019. Urban landscape change analysis using local climate zones and object-based classification in the Salt Lake metro region, Utah, USA. *Remote Sensing* 11(13):1615.

Cosgrove, A., M. Berkelhammer. 2018. Downwind footprint of an urban heat island on air and lake temperatures. *NPJ Climate and Atmospheric Science* 1(46):1–10.

Crum, S.M., G.D. Jenerette. 2017. Microclimate variation among urban land covers: The importance of vertical and horizontal structure in air and land surface temperature relationships. *Journal of Applied Meteorology and Climatology* 56:2531–2543.

Crum, S.M., S.A. Shiflett, G.D. Jenerette. 2017. The influence of vegetation, mesoclimate and meteorology on urban atmospheric microclimates across a coastal to desert climate gradient. *Journal of Environmental Management* 200:295–303.

DaMatta, F.M., A. Grandis, B.C. Arenque, M.S. Buckeridge. 2010. Impacts of climate changes on crop physiology and food quality. *Food Research International* 43:1814–1823.

Dronova, I., M. Friedman, I. McRae, F. Kong, H. Yin. 2018. Spatio-temporal non-uniformity of urban park greenness and thermal characteristics in a semi-arid region. *Urban Forestry & Urban Greening* 34:44–54.

Gillner, S., J. Vogt, A. Tharang, S. Dettmann, A. Roloff. 2015. Role of street trees in mitigating effects of heat and drought at highly sealed urban sites. *Landscape and Urban Planning* 143:33–42.

Grimmond, C.S.B., M. Blackett, M.J. Best, J.J. Baik, S.E. Belcher, J. Beringer, S.I. Bohnenstengel, I. Calmet, F. Chen, A. Coutts, A. Dandou, K. Fortuniak, M.L. Gouvea, R. Hamdi, M. Hendry, M. Kanda, T. Kawai, Y. Kawamoto, H. Kondo, E.S. Krayenhoff, S.H. Lee, T. Loridan, A. Martilli, V. Masson, S. Miao, K. Oleson, R. Ooka, G. Pigeon, A. Porson, Y.H. Ryu, F. Salamanca, G.J. Steeneveld, M. Tombrou, J.A. Voogt, D.T. Young, N. Zhang. 2011. Initial results from phase 2 of the international urban energy balance model comparison. *International Journal of Climatology* 31:244–272.

Grimmond, C.S.B., M. Blackett, M.J. Best, J. Barlow, J.J. Baik, S.E. Belcher, S.I. Bohnenstengel, I. Calmet, F. Chen, A. Dandou, K. Fortuniak, M.L. Gouvea, R. Hamdi, M. Hendry, T. Kawai, Y. Kawamoto, H. Kondo, E.S. Krayenhoff, S.H. Lee, T. Loridan, A. Martilli, V. Masson, S. Miao, K. Oleson, G. Pigeon, A. Porson, Y.H. Ryu, F. Salamanca, L. Shashua-Bar, G.J. Steeneveld, M. Tombrou, J. Voogt, D. Young, N. Zhang. 2010. The international urban energy balance models comparison project: First results from phase 1. *Journal of Applied Meteorology and Climatology* 49:1268–1292.

Groffman, P.M., M. Avolio, J. Cavender-Bares, N.D. Bettez, J.M. Grove, S.J. Hall, S.E. Hobbie, K.L. Larson, S.B. Lerman, D.H. Locke, J.B. Heffernan, J.L. Morse, C. Neill, K.C. Nelson, J. O'Neil-Dunne, D.E. Pataki, C. Polsky, R.R. Chowdhury, T.L.E. Trammell. 2017. Ecological homogenization of residential macrosystems. *Nature Ecology & Evolution* 1:0191.

Groffman, P.M., J. Cavender-Bares, N.D. Bettez, J.M. Grove, S.J. Hall, J.B. Heffernan, S.E. Hobbie, K.L. Larson, J.L. Morse, C. Neill, K. Nelson, J. O'Neil-Dunne, L. Ogden, D.E. Pataki, C. Polsky, R.R. Chowdhury, M.K. Steele. 2014. Ecological homogenization of urban USA. *Frontiers in Ecology and the Environment* 12:74–81.

Grote, R., R. Samson, R. Alonso, J.H. Amorim, P. Cariñanos, G. Churkina, S. Fares, D. Le Thiec, Ü. Niinemets, T.N. Mikkelsen, E. Paoletti, A. Tiwary, C. Calfapietra. 2016. Functional traits of urban trees: Air pollution mitigation potential. *Frontiers in Ecology and the Environment* 14:543–550.

Haberlie, A.M., W.S. Ashley, T.J. Pingel. 2015. The effect of urbanisation on the climatology of thunderstorm initiation. *Quarterly Journal of the Royal Meteorological Society* 141:663–675.

Hall, S.J., J. Learned, B. Ruddell, K.L. Larson, J. Cavender-Bares, N. Bettez, P.M. Groffman, J.M. Grove, J.B. Heffernan, S.E. Hobbie, J.L. Morse, C. Neill, K.C. Nelson, J.P.M. O'Neil-Dunne, L. Ogden, D.E. Pataki, W.D. Pearse, C. Polsky, R.R. Chowdhury, M.K. Steele, T.L.E. Trammell. 2016. Convergence of microclimate in residential landscapes across diverse cities in the United States. *Landscape Ecology* 31:101–117.

Hamblin, A.L., E. Youngsteadt, M.M. López-Uribe, S.D. Frank. 2017. Physiological thermal limits predict differential responses of bees to urban heat-island effects. *Biology Letters* 13:20170125.

Herrmann, D.L., L.A. Schifman, W.D. Shuster. 2018. Widespread loss of intermediate soil horizons in urban landscapes. *Proceedings of the National Academy of Sciences* 115(26):6751–6755.

Hoffman, J., V. Shandas, N. Pendleton. 2020. The effects of historical housing policies on resident exposure to intra-urban heat: A study of 108 US urban areas. *Climate* 8:12.

Hope, D., C. Gries, W. Zhu, W.F. Fagan, C.L. Redman, N.B. Grimm, A.L. Nelson, C. Martin, A. Kinzig. 2008. Socioeconomics drive urban plant diversity. *Proceedings of the National Academy of Sciences of the United States of America* 100(15):8788–8792.

Ibsen, P.C., D. Borowy, T. Dell, H. Greydanus, N. Gupta, D.M. Hondula, T. Meixner, M.V. Santelmann, S.A. Shiflett, M.C. Sukop, C.M. Swan, M.L. Talal, M. Valencia, M.K. Wright, G.D. Jenerette. 2021. Greater aridity increases the magnitude of urban nighttime vegetation-derived air cooling. *Environmental Research Letters* 16:034011.

Ibsen, P.C., D. Borowy, M. Rochford, C.M. Swan, G.D. Jenerette. 2020. Influence of climate and management on patterns of taxonomic and functional diversity of recreational park vegetation. *Frontiers in Ecology and Evolution* 8:1–13.

Idso, C.D., S.B. Idso, R.C. Balling. 2001. An intensive two-week study of an urban CO2 dome in Phoenix, Arizona, USA. *Atmospheric Environment* 35:995–1000.

Ives, C.D., P.E. Lentini, C.G. Threlfall, K. Ikin, D.F. Shanahan, G.E. Garrard, S.A. Bekessy, R.A. Fuller, L. Mumaw, L. Rayner, R. Rowe, L.E. Valentine, D. Kendal. 2016. Cities are hotspots for threatened species. *Global Ecology and Biogeography* 25:117–126.

Janhäll, S. 2015. Review on urban vegetation and particle air pollution – Deposition and dispersion. *Atmospheric Environment* 105:130–137.
Jenerette, G.D., L.W. Clarke, M.L. Avolio, D.E. Pataki, T.W. Gillespie, S. Pincetl, D.J. Nowak, L.R. Hutyra, M. McHale, J.P. McFadden, M. Alonzo. 2016a. Climate tolerances and trait choices shape continental patterns of urban tree biodiversity. *Global Ecology and Biogeography* 25:1367–1376.
Jenerette, G.D., S.L. Harlan, A. Buyantuev, W.L. Stefanov, J. Declet-Barreto, B.L. Ruddell, S.W. Myint, S. Kaplan, X. Li. 2016b. Micro-scale urban surface temperatures are related to land-cover features and residential heat related health impacts in Phoenix, AZ USA. *Landscape Ecology* 31:745–760.
Johnson, J.C., E. Garver, T. Martin. 2020. Black widows on an urban heat island: Extreme heat affects spider development and behaviour from egg to adulthood. *Animal Behaviour* 167:77–84.
Karra, S., L. Malki-Epshtein, M.K.A. Neophytou. 2017. Air flow and pollution in a real, heterogeneous urban street canyon: A field and laboratory study. *Atmospheric Environment* 165:E370–384.
Katayama, N., T. Amano, S. Naoe, T. Yamakita, I. Komatsu, S. Takagawa, N. Sato, M. Ueta, T. Miyashita. 2014. Landscape heterogeneity-biodiversity relationship: Effect of range size. *PLoS ONE* 9:1–8.
Kaye, J.P., P.M. Groffman, N.B. Grimm, L.A. Baker, R.V. Pouyat. 2006. A distinct urban biogeochemistry? *Trends in Ecology and Evolution* 21(4):192–199.
Keeler, B.L., P. Hamel, T. McPhearson, M.H. Hamann, M.L. Donahue, K.A. Meza Prado, K.K. Arkema, G.N. Bratman, K.A. Brauman, J.C. Finlay, A.D. Guerry, S.E. Hobbie, J.A. Johnson, G.K. MacDonald, R.I. McDonald, N. Neverisky, S.A. Wood. 2019. Social-ecological and technological factors moderate the value of urban nature. *Nature Sustainability* 2:29–38.
Kelly, V.R., S.E.G. Findlay, K.C. Weathers. 2019. *Road Salt: The Problem, the Solution, and How to Get There*. Cary Institute of Ecosystem Studies, Millbrook, NY.
Kendal, D., C. Dobbs, R.V. Gallagher, L.J. Beaumont, J. Baumann, N.S.G. Williams, S.J. Livesley. 2018. A global comparison of the climatic niches of urban and native tree populations. *Global Ecology and Biogeography* 27:629–637.
Krayenhoff, E.S., M. Moustaoui, A.M. Broadbent, V. Gupta, M. Georgescu. 2018. Diurnal interaction between urban expansion, climate change and adaptation in US cities. *Nature Climate Change* 8:1097–1103.
Kuras, E.R., P.S. Warren, J.A. Zinda, M.F.J. Aronson, S. Cilliers, M.A. Goddard, C.H. Nilon, R. Winkler. 2020. Urban socioeconomic inequality and biodiversity often converge, but not always: A global meta-analysis. *Landscape and Urban Planning* 198:103799.
Landsberg, H.E. 1981. *The Urban Climate*. Academic Press, New York.
Larson, K.L., R. Andrade, K.C. Nelson, M.M. Wheeler, J.M. Engebreston, S.J. Hall, M.L. Avolio, P.M. Groffman, M. Grove, J.B. Heffernan, S.E. Hobbie, S.B. Lerman, D.H. Locke, C. Neill, R.R. Chowdhury, T.L.E. Trammell. 2020. Municipal regulation of residential landscapes across US cities: Patterns and implications for landscape sustainability. *Journal of Environmental Management* 275:111132.
Leong, M., R.R. Dunn, M.D. Trautwein. 2018. Biodiversity and socioeconomics in the city: A review of the luxury effect. *Biology Letters* 14.
Leuzinger, S., R. Vogt, C. Körner. 2010. Tree surface temperature in an urban environment. *Agricultural and Forest Meteorology* 150:56–62.
Litvak, E., H.R. McCarthy, D.E. Pataki. 2012. Transpiration sensitivity of urban trees in a semi-arid climate is constrained by xylem vulnerability to cavitation. *Tree Physiology* 32:373–388.
Liu, F., H. Qian, X. Zheng, L. Zhang, W. Liang. 2017. Numerical study on the urban ventilation in regulating microclimate and pollutant dispersion in urban street canyon: A case study of Nanjing New Region, China. *Atmosphere* 8:164.
Llaguno-Munitxa, M., E. Bou-Zeid, M. Hultmark. 2017. The influence of building geometry on street canyon air flow: Validation of large eddy simulations against wind tunnel experiments. *Journal of Wind Engineering and Industrial Aerodynamics* 165:115–130.
Locosselli, G.M., E.P. de Camargo, T.C.L. Moreira, E. Todesco, M. Fátima Andrade, C.D.S. de André, P.A. de André, J.M. Singer, L.S. Ferreira, P.H.N. Saldiva, M.S. Buckeridge. 2019. The role of air pollution and climate on the growth of urban trees. *Science of the Total Environment* 666:652–661.
Manoli, G., S. Fatichi, E. Bou-Zeid, G.G. Katul. 2020. Seasonal hysteresis of surface urban heat islands. *Proceedings of the National Academy of Sciences of the United States of America* 117:7082–7089.
McGinnis, M.D., E. Ostrom. 2014. Social-ecological system framework: Initial changes and continuing challenges. *Ecology and Society* 19(2):30.
McPhearson, P.T., M. Feller, A. Felson, R. Karty, J.W.T. Lu, M.I. Palmer, T. Wenskus. 2010. Assessing the effects of the urban forest restoration effort of MillionTreesNYC on the structure and functioning of New York City ecosystems. *Cities and the Environment (CATE)* 3:1–21.

McPhearson, P.T., S.T.A. Pickett, N.B. Grimm, J. Niemelä, M. Alberti, T. Elmqvist, C. Weber, D. Haase, J. Breuste, S. Qureshi. 2016. Advancing urban ecology toward a science of cities. *BioScience* 66:198–212.

McPherson, E.G., A.M. Berry, N.S. van Doorn. 2018. Performance testing to identify climate-ready trees. *Urban Forestry & Urban Greening* 29:28–39.

Middel, A., K. Häb, A.J. Brazel, C.A. Martin, S. Guhathakurta. 2014. Impact of urban form and design on mid-afternoon microclimate in Phoenix local climate zones. *Landscape and Urban Planning* 122:16–28.

Miles, L.S., R.J. Dyer, B.C. Verrelli. 2018. Urban hubs of connectivity: Contrasting patterns of gene flow within and among cities in the western black widow spider. *Proceedings of the Royal Society B: Biological Sciences* 285:20181224.

Moffett, K.B., Y. Makido, V. Shandas. 2019. Urban-rural surface temperature deviation and intra-urban variations contained by an urban growth boundary. *Remote Sensing* 11(22):2683.

Nilon, C.H., M.F.J. Aronson, S.S. Cilliers, C. Dobbs, L.J. Frazee, M.A. Goddard, K.M. O'Neill, D. Roberts, E.K. Stander, P. Werner, M. Winter, K.P. Yocom. 2017. Planning for the future of urban biodiversity: A global review of city-scale initiatives. *BioScience* 67:332–342.

Nowak, D.J., E.J. Greenfield. 2018. US urban forest statistics, values, and projections. *Journal of Forestry* 116(2):164–177.

Oke, T.R. 1976. The distinction between canopy and boundary-layer urban heat islands. *Atmosphere* 14:268–277.

Oke, T.R. 1988. The urban energy balance. *Progress in Physical Geography: Earth and Environment* 12:471–508.

Ossola, A., D. Locke, B. Lin, E. Minor. 2019. Greening in style: Urban form, architecture and the structure of front and backyard vegetation. *Landscape and Urban Planning* 185:141–157.

Ostrom, E. 2007. A diagnostic approach for going beyond panaceas. *Proceedings of the National Academy of Sciences of the United States of America* 104(39):15181–15187.

Padullés Cubino, J., J. Cavender-Bares, S.E. Hobbie, D.E. Pataki, M.L. Avolio, L.E. Darling, K.L. Larson, S.J. Hall, P.M. Groffman, T.L.E. Trammell, M.K. Steele, J.M. Grove, C. Neill. 2018. Drivers of plant species richness and phylogenetic composition in urban yards at the continental scale. *Landscape Ecology* 34:63–77.

Palta, M.M., N.B. Grimm, P.M. Groffman. 2017. 'Accidental' urban wetlands: Ecosystem functions in unexpected places. *Frontiers in Ecology and the Environment* 15:248–256.

Parry, M., T.J. Chandler. 1966. The climate of London. *The Geographical Journal* 132:84.

Partelow, S. 2018. A review of the social-ecological systems framework: Applications, methods, modifications, and challenges. *Ecology and Society* 23(4):36.

Pataki, D.E., H.R. McCarthy, E. Litvak, S. Pincetl. 2011. Transpiration of urban forests in the Los Angeles metropolitan area. *Ecological Applications* 21:661–677.

Paul, S., S. Ghosh, M. Mathew, A. Devanand, S. Karmakar, D. Niyogi. 2018. Increased spatial variability and intensification of extreme monsoon rainfall due to urbanization. *Scientific Reports* 8:1–10.

Pearse, W.D., J. Cavender-Bares, S.E. Hobbie, M.L. Avolio, N. Bettez, R.R. Chowdhury, L.E. Darling, P.M. Groffman, J.M. Grove, S.J. Hall, J.B. Heffernan, J. Learned, C. Neill, K.C. Nelson, D.E. Pataki, B.L. Ruddell, M.K. Steele, T.L.E. Trammell. 2018. Homogenization of plant diversity, composition, and structure in North American urban yards. *Ecosphere* 9:e02105.

Pickett, S.T.A., M.L. Cadenasso, M.E. Baker, L.E. Band, C.G. Boone, G.L. Buckley, P.M. Groffman, J.M. Grove, E.G. Irwin, S.S. Kaushal, S.L. Ladeau, A.J. Miller, C.H. Nilon, M. Romolini, E.J. Rosi, C.M. Swan, K. Szlavecz. 2020. Theoretical perspectives of the Baltimore Ecosystem Study: Conceptual evolution in a social-ecological research project. *BioScience* 70:297–314.

Pickett, S.T.A., M.L. Cadenasso, E.J. Rosi-Marshall, K.T. Belt, P.M. Groffman, J.M. Grove, E.G. Irwin, S.S. Kaushal, S.L. LaDeau, C.H. Nilon, C.M. Swan, P.S. Warren. 2017. Dynamic heterogeneity: A framework to promote ecological integration and hypothesis generation in urban systems. *Urban Ecosystems* 20:1–14.

Pincetl, S. 2010. Implementing municipal tree planting: Los Angeles million-tree initiative. *Environmental Management* 45:227–238.

Pouyat, R.V., M.J. McDonnell, S.T.A. Pickett. 1997. Litter decomposition and nitrogen mineralization in oak stands along an urban-rural land use gradient. *Urban Ecosystems* 1:117–131.

Ramamurthy, P., E. Bou-Zeid. 2017. Heatwaves and urban heat islands: A comparative analysis of multiple cities. *Journal of Geophysical Research: Atmospheres* 122:168–178.

Roman, L.A., T.M. Conway, T.S. Eisenman, A.K. Koeser, C. Ordóñez Barona, D.H. Locke, G.D. Jenerette, J. Östberg, J. Vogt. 2020. Beyond 'trees are good': Disservices, management costs, and tradeoffs in urban forestry. *Ambio* 50:615–630.

Roman, L.A., H. Pearsall, T.S. Eisenman, T.M. Conway, R.T. Fahey, S. Landry, J. Vogt, N.S. van Doorn, J.M. Grove, D.H. Locke, A.C. Bardekjian, J.J. Battles, M.L. Cadenasso, C.C.K. van den Bosch, M. Avolio, A. Berland, G.D. Jenerette, S.K. Mincey, D.E. Pataki, C. Staudhammer. 2018. Human and biophysical legacies shape contemporary urban forests: A literature synthesis. *Urban Forestry & Urban Greening* 31:157–168.

Sandweiss, D.H., K.A. Maasch, D.G. Anderson. 1999. Transitions in the mid-Holocene. *Science* 283:499–500.

Sarangi, C., S.N. Tripathi, Y. Qian, S. Kumar, L.R. Leung. 2018. Aerosol and urban land use effect on rainfall around cities in Indo-Gangetic basin from observations and cloud resolving model simulations. *Journal of Geophysical Research: Atmospheres* 123:3645–3667.

Schifman, L.A., D.L. Herrmann, W.D. Shuster, A. Ossola, A. Garmestani, M.E. Hopton. 2017. Situating green infrastructure in context: A framework for adaptive socio-hydrology in cities. *Water Resources Research* 53:10139–10154.

Schmid, P.E., D. Niyogi. 2017. Modeling urban precipitation modification by spatially heterogeneous aerosols. *Journal of Applied Meteorology and Climatology* 56:2141–2153.

Schuler, M.S., R.A. Relyea. 2018. A review of the combined threats of road salts and heavy metals to freshwater systems. *BioScience* 68:327–335.

Schwarz, K., A. Berland, D.L. Herrmann. 2018. Green, but not just? Rethinking environmental justice indicators in shrinking cities. *Sustainable Cities and Society* 41:816–821.

Shandas, V., J. Voelkel, J. Williams, J. Hoffman. 2019. Integrating satellite and ground measurements for predicting locations of extreme urban heat. *Climate* 7:1–13.

Shepherd, M.J., H. Pierce, A.J. Negri. 2002. Rainfall modification by major urban areas: Observations from spaceborne rain radar on the TRMM satellite. *Journal of Applied Meteorology* 41:689–701.

Shiflett, S.A., L.L. Liang, S.M. Crum, G.L. Feyisa, J. Wang, G.D. Jenerette. 2017. Variation in the urban vegetation, surface temperature, air temperature nexus. *Science of the Total Environment* 579:495–505.

Shrivastava, M., M.O. Andreae, P. Artaxo, H.M.J. Barbosa, L.K. Berg, J. Brito, J. Ching, R.C. Easter, J. Fan, J.D. Fast, Z. Feng, J.D. Fuentes, M. Glasius, A.H. Goldstein, E.G. Alves, H. Gomes, D. Gu, A. Guenther, S.H. Jathar, S. Kim, Y. Liu, S. Lou, S.T. Martin, V.F. McNeill, A. Medeiros, S.S. de Sá, J.E. Shilling, S.R. Springston, R.A.F. Souza, J.A. Thornton, G. Isaacman-VanWertz, L.D. Yee, R. Ynoue, R.A. Zaveri, A. Zelenyuk, C. Zhao. 2019. Urban pollution greatly enhances formation of natural aerosols over the Amazon rainforest. *Nature Communications* 10:1046.

Smith, I.A., V.K. Dearborn, L.R. Hutyra. 2019. Live fast, die young: Accelerated growth, mortality, and turnover in street trees. *PLoS ONE* 14:1–17.

Song, J., Z. Wang. 2016. Diurnal changes in urban boundary layer environment induced by urban greening. *Environmental Research Letters* 11:114018.

Stewart, I.D., T.R. Oke. 2012. Local climate zones for urban temperature studies. *Bulletin of the American Meteorological Society* 93:1879–1900.

Swan, C.M., S.T.A. Pickett, K. Szlavecz, P. Warren, K.T. Willey. 2011. Biodiversity and community composition in urban ecosystems: Coupled human, spatial, and metacommunity processes. Pages 179–186 in J. Niemelä, editor. *Urban Ecology: Patterns, Processes, and Applications*. Oxford University Press, Oxford.

Templeton, N.P., E.R. Vivoni, Z.-H. Wang, A.P. Schreiner-McGraw. 2018. Quantifying water and energy fluxes over different urban land covers in Phoenix, Arizona. *Journal of Geophysical Research: Atmospheres* 123:2111–2128.

Trammell, T.L.E., D.E. Pataki, R.V. Pouyat, P.M. Groffman, C. Rosier, N. Bettez, J. Cavender-Bares, M.J. Grove, S.J. Hall, J. Heffernan, S.E. Hobbie, J.L. Morse, C. Neill, M. Steele. 2020. Urban soil carbon and nitrogen converge at a continental scale. *Ecological Monographs* 90:e01401.

Uehling, J.J., J. Tallant, S. Pruett-Jones. 2019. Status of naturalized parrots in the United States. *Journal of Ornithology* 160:907–921.

Valliere, J.M., I.C. Irvine, L. Santiago, E.B. Allen. 2017. High N, dry: Experimental nitrogen deposition exacerbates native shrub loss and nonnative plant invasion during extreme drought. *Global Change Biology* 23:4333–4345.

van den Heever, S.C., W.R. Cotton. 2007. Urban aerosol impacts on downwind convective storms. *Journal of Applied Meteorology and Climatology* 46:828–850.

Wang, H., B.A. Maher, I.A. Ahmed, B. Davison. 2019. Efficient removal of ultrafine particles from diesel exhaust by selected tree species: Implications for roadside planting for improving the quality of urban air. *Environmental Science & Technology* 53:6906–6916.

Wang, X., H. Cheng, J. Xi, G. Yang, Y. Zhao. 2018. Relationship between park composition, vegetation characteristics and cool island effect. *Sustainability* 10:10–15.

Wang, Y., H. Du, Y. Xu, D. Lu, X. Wang, Z. Guo. 2018. Temporal and spatial variation relationship and influence factors on surface urban heat island and ozone pollution in the Yangtze River. *Science of the Total Environment* 631–632:921–933.

Ziter, C.D. 2016. The biodiversity-ecosystem service relationship in urban areas: A quantitative review. *Oikos* 125:761–768.

Ziter, C.D., E.J. Pedersen, C.J. Kucharik, M.G. Turner. 2019. Scale-dependent interactions between tree canopy cover and impervious surfaces reduce daytime urban heat during summer. *Proceedings of the National Academy of Sciences of the United States of America* 116:7575–7580.

5

UNDERSTANDING LAND USE, LAND COVER, AND LANDSCAPE PATTERNS OF THE WORLD'S CITIES FOR SUSTAINABLE BIODIVERSITY PLANNING

Cynnamon Dobbs, Dagmar Haase, and Ulla Mörtberg

Introduction

Our globe is rapidly urbanizing, and the land surface covered by cities is increasingly growing (Elmqvist et al. 2018). More than 50% of the global population today lives in cities and peri-urban areas and this number is only expected to increase. Within the coming decade, the urban population is expected to grow by an additional billion people (Haase et al. 2018). During the same period, cities are expected to double in population size and also in area. This means that the world is increasingly urban, interconnected, and constantly changing (Keys et al. 2019). If current trends continue, by 2050 the global urban population is estimated to be >6 billion. In terms of the Earth's surface, this also means that more than 60% of the area projected to be urban in 2030 has yet to be built (Elmqvist et al. 2018). Most of this urban growth is assumed to happen in small and medium-sized cities, not in large or megacities (Güneralp et al. 2020).

Along with ongoing urbanization, built land cover in the Global South – holding most of the biodiversity hotspots around the globe – is expected to increase by more than 200% between 2000 and 2030 (Elmqvist et al. 2018). By 2030, at the end of the coming decade, urban areas will determine quality of life and health for 5 billion people, most of all in Africa, Asia, and Latin America (Huang et al. 2019). By 2050, up to 75% of the world's population is expected to become urban (Angel et al. 2011a, 2011b). Future urban growth will primarily occur in areas with limited financial and infrastructural resources and thus pose significant challenges to meet several basic needs, such as food and freshwater provisioning, and to support as fair as possible access to those resources, as well as preserving remaining biodiversity (McDonald et al. 2019).

Urban areas are expanding faster than urban populations (Elmqvist et al. 2018). This urban expansion heavily relies and depends on natural resources, including soil and water, so on a global scale, urban expansion consumes virgin land for agriculture and pastures (Martin et al. 2019), with knock-on effects on biodiversity and ecosystem services elsewhere (Aronson et al. 2014; McDonald et al. 2019). Urban expansion is accelerating in areas adjacent to biodiversity hotspots

DOI: 10.4324/9781003016120-6

Figure 5.1 Urban sprawling in Sao Paulo, Brazil, towards the Atlantic Forest biodiversity hotspot. Picture credit: Cynnamon Dobbs.

and even faster in low-elevation, biodiversity-rich coastal zones than in other areas (McDonald et al. 2019; Figure 5.1).

However, urban growth also holds multiple opportunities to vastly improve global sustainability. This requires not only an acknowledgment of the dependence by individual cities on regional ecosystems and natural resources but also on other regions around the world (Haase 2019). When looking at cities today, one of the core questions worldwide is how to keep biodiversity alive in and around urban areas and available for future generations (Haase 2013). Over the past decades, urban planning has been focused on how to minimize negative effects of settlement, transport, and industrial growth and consequent land consumption, to slow the loss and degradation of natural and semi-natural habitats using land zoning and conservation stewardship schemes (Andersson et al. 2014; McDonald et al. 2019). Doing so, cities play a central role in achieving key biodiversity and sustainable development goals.

Urban land use – history, patterns, and drivers of change

Urbanization is one of the most important global change processes. As the share of people in, and the footprint of, urban areas, continue to grow globally and locally, understanding urbanization processes and resulting land use – both their spatial patterns and use intensity – is increasingly important with respect to natural resources, aging, justice, health, and global change (Seto and Reenberg 2014; Haase et al. 2018). Rapid urbanization is a "child" of the early Anthropocene (Haraway 2016) having started with and having been initiated the industrial revolution in the 19th century and having accelerated during the Great Acceleration after WWII (Steffen et al. 2015). In 1800, only 3% of the world's population lived in cities; this figure rose to 50% by the end of the 20th century. In 1950, there were 83 cities with populations that exceeded 1 million; by 2010, this number had risen to >460 (Haase et al. 2018). Today, as in the past, the majority of the world's cities have been growing annually up to >5% (Haase and Schwarz 2016; Haase et al. 2018). However, there have simultaneously been conurbations exhibiting negative net growth rates, most of them are to be found in the early industrialized world, namely, Europe, US, and Japan (Haase 2014). Most of future urban growth will take place in Asia and the West African urban belt, with annual urban growth rates of 3–5% (UN 2019). The UN predicts that by 2050, 65% of the populations in developing countries and nearly 90% of the populations in developed countries will live in urban areas (UN 2019). Very recent migration from cities due to the 2020 COVID-19 pandemic might alter these predictions, but so far, no resilient data is available.

Urban expansion over the last three decades has been greatest along coastlines and low-lying coastal zones (Seto et al. 2011). Current urban hotspots are situated on the coastlines of South Asia, Southeast Asia, Southeast China, the United States' East Coast, Western Europe, Japan, West Africa, and the Atlantic coast of Latin America, including the ten largest megacities, which are in developing countries such as Guangzhou, Kolkata, Shanghai, Mumbai, and Bangkok (Nicholls et al. 2008). A recent study found that at the turn of the 21st century, 11% of all urban land (over 70,000 km^2) was located within low-elevation coastal zones (Güneralp et al. 2015; Figure 5.2), defined as "the contiguous area along the coast that is less than 10 meters asl" (McGranahan et al. 2007).

There is wide variability in terms of the spatial configuration of urban areas across the world. Although a variety of socioeconomic and biophysical factors influence the spatial growth of cities and their relative influence varies between regions (Seto et al. 2012), it is claimed that globalization leads to a proliferation of similar urban forms across different continents (Leichenko and Solecki 2005). At least one study found that income, in interaction with city size, appears to have a pronounced effect on urban growth, particularly in smaller cities (Jenerette and Potere 2010).

Particularly in the developed world, post–WWII motorization, poor planning, and market failures led to urban sprawl, which is defined broadly as "excessive spatial growth of cities" (Brueckner 2000), more specifically, as spatial growth of cities that creates forms of suburban development that lack accessibility and open space. As a spatial-temporal process, urban sprawl can be seen as a low-density expansion or "leapfrog" development of large urban areas into the surrounding rural landscape (Nilsson et al. 2014). For example, from 1990 to 2006, urban land and associated infrastructure across Europe grew at an annual rate of about 1,000 km^2, which is equivalent to the area of the German capital of Berlin. Nevertheless, the most prominent case of this kind of urban growth has been the expansion of the cities in the United States after WWII (Batty et al. 1999). The development of large, suburbanized peripheries around historically compact European cities (Haase and Nuissl 2010) came to be known as the "in-between cities" or *in German "Zwischenstadt*" – a settlement form in between the urban and the rural, which is mainly composed of detached houses and industrial, commercial, and retail sites that dominate the urban-to-rural interface (Nilsson et al. 2014). Conversely, rapidly growing urban areas in Asia and Africa contain many rural features in their peri-urban spaces, including various forms of gardening and farming and thus are distinctly different than suburbanization seen in the Global North, producing a full

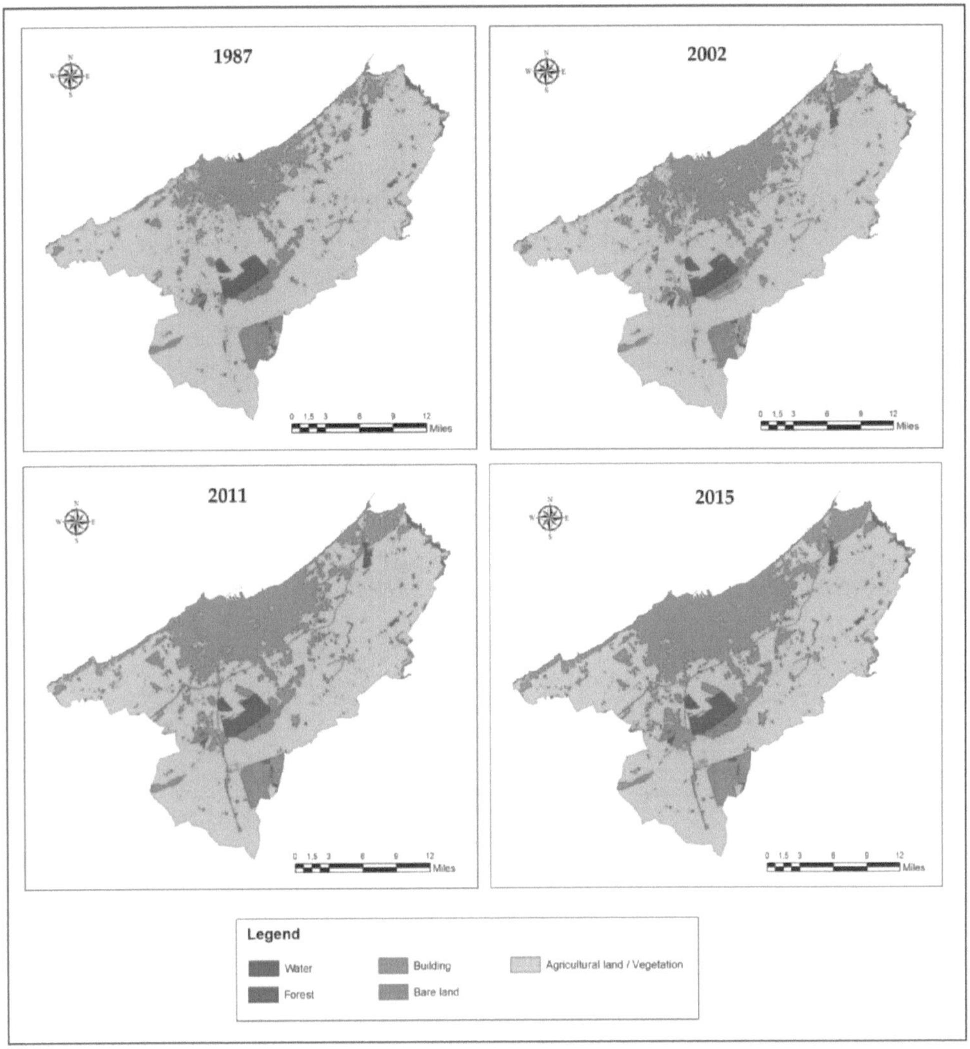

Figure 5.2 A typical feature of coastal sprawling growth dynamics of urban land in Morocco between 1987 and 2015 (Bahi et al. 2017).

range of ecosystem diversity responses (Haase et al. 2018; Figure 5.3). South America presents similar trends with sprawling occurring with low densities creating what is called as new peripheries (Inostroza et al. 2013).

The process of urbanization can be clearly observed along the rural-urban gradient – that is, the ideal typical transect that links the urban (built, populated) and the rural (open, vegetated), which displays a typical configuration of population density, coverage of built-up area, respective impervious cover, and demographic structure, including lifestyles and travel behaviour (Haase and Nuissl 2010). Along the rural-urban gradient, an increasing amount of land consumption occurs – namely the transformation of green spaces to built-up areas, described as landscape urbanization,

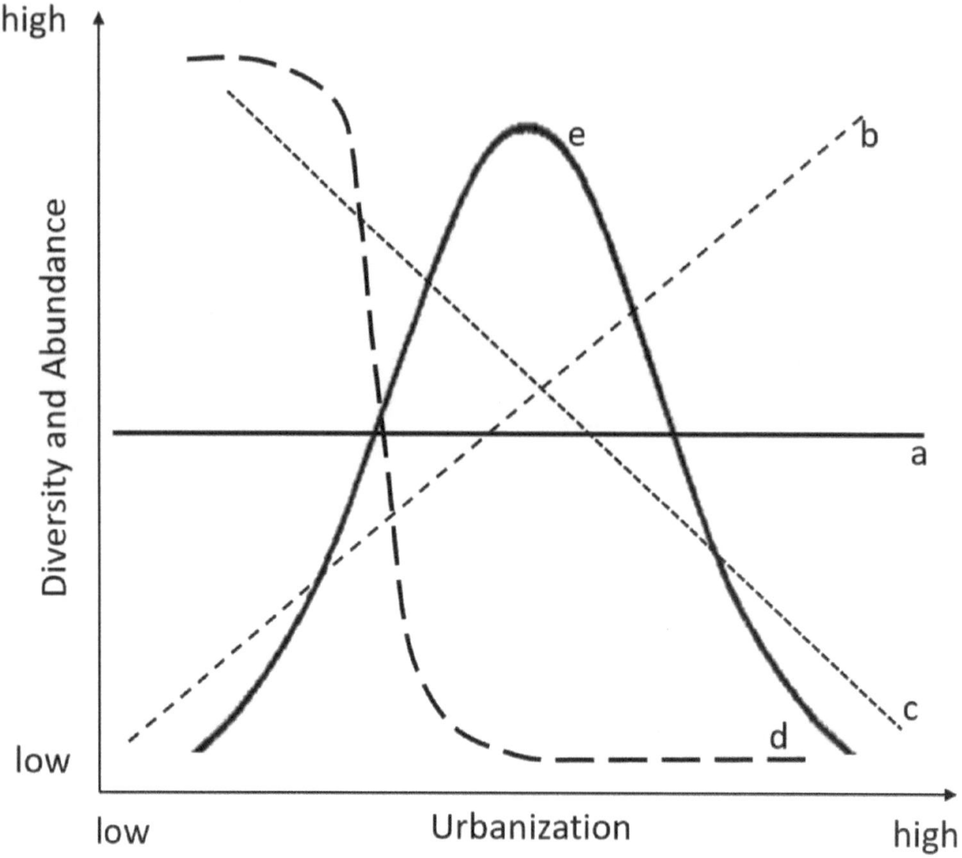

Figure 5.3 Stylized responses of organisms on and along the rural-urban or urbanization gradient: (a) no response, (b) positive response, (c) negative response, (d) tipping point response, and (e) intermediate response (according to Wang et al. 2020, adapted).

in contrast to demographic urbanization (Bai et al. 2011). What is more, recently we see in many parts of the urban world a trend that peri-urban landscapes as complementary providers of urban ecosystem services get "filled" by former urban land uses (services, health care, leisure, sports, shopping, gardening, etc.) when open or low-density areas in cities are reduced by densification (Hedblom et al. 2017).

Urban expansion in the Global South, including cities from Latin America, Asia, and Africa, show different patterns than regions from the Global North. Latin America has more than 80% of its population living in cities and has rapid rates of urbanization and sprawling, leading to land use change affecting the ecosystem structure, impacting biodiversity loss within and beyond the city limits (Grimm et al. 2008; Dobbs et al. 2019). The drivers of urban growth in this region include growing population and economic development (Inostroza et al. 2013), besides rural to urban migration, which has led to large socioeconomic inequity and ecological legacy (Isendahl and Smith 2013).

Urban expansion shows different rates around the planet: West Asian and Middle Eastern countries have a higher level of urbanization, related to the expansion of urban area, while Southeast

Asian countries and China showed a high-speed of urbanization, related to the rate of urban growth (Chen et al. 2019). Africa is recognized as one of the fastest-growing regions in the world, adding seven megacities over 10 million people, but most of the increase in population is occurring in small and medium-size cities. There is also a large transformation of rural towns into urban areas (UN Habitat 2011; Lwasa 2014). The underlying processes of urbanization in Africa are different from other continents with some similarities to Latin America, distinguishing the role of unplanned and unregulated growth, legacy of colonialism, structural adjustment, and neoliberalism that leads to weak planning institutions on the dynamics of urbanization (Pieterse 2011). There is a recognition of complex settlement-governance arrangements with weak local authorities and poor land use capacity that goes in detriment of biodiversity, derived from unregulated urban expansion, increased food demand, fragmentation from transportation development, bushmeat trade, and environmental degradation. In addition, ethnic conflicts and civil wars also influence urbanization by increasing informal settlements in peri-urban areas through displaced people (McGranahan et al. 2009; Güneralp et al. 2018). Thus, countries with weak governance show the fastest rates of urbanization, enhancing biodiversity loss from urban expansion.

Addressing the effects of urban expansion on habitat loss is highly dependent on the governance, and it can or cannot effectively shape urban growth. Higher impact on biodiversity loss tends to occur largely in countries with low levels of political stability and weak land governance, which influence how law, regulations, and institutions decide on the management of land resources (Huang et al. 2018). Effective policy and implementation of land use planning instead can decrease threats to biodiversity by enhancing features that promote biodiversity conservation (Halleux et al. 2012).

Urban biodiversity and landscape ecology

Most measures of biodiversity, such as richness and abundance, depend on landscape patterns resulting from land use/land cover distribution, quantity, and quality. Landscape ecology provides methods and tools for assessing at multiple scales, measures of biodiversity for supporting decision-making in urban areas. Landscape ecology is usually used for informing land planning for conservation (Karimi and Hockings 2018) and adaptive management (Chacón-Moreno et al. 2020), but it also provides a great framework to understand patterns of biodiversity in urban areas. Land cover and vegetation cover have been identified as explanatory variables of diversity of plant and birds in urban areas worldwide (Aronson et al. 2014).

Landscape metrics such as habitat type and the surrounding matrix were used in early studies of urban biodiversity (Ricketts 2001), while it has been distinguished that landscape factors such as permeability affects species dispersal. Patch area follows the species-area relationship (Arrhenius 1921), where larger patches support greater species richness. A meta-analysis by Beninde et al. (2015) recognized that the strongest effect on urban biodiversity relates to habitat patch area and structures supporting connectivity, "corridors". Beninde et al. (2015) also found that corridors more than stepping stones have a larger effect on species richness.

Urban biodiversity conservation needs spatially explicit data on vegetation patterns that can be identified as habitat and allow assessing species' abundance and distribution. Mapping vegetation patterns can help identify areas depleted of biodiversity or prioritizing areas for improvement and can aid in prioritizing where to enhance connectivity of vegetation patches. Not only vegetation needs to be mapped for assessing urban biodiversity, but also a mix between mapping land cover and land use becomes highly relevant. Biodiversity can be supported by water elements and green spaces and also by different vegetated areas such as cemeteries, vacant lots, community gardens, and even by some grey infrastructure (i.e. buildings; Jarvis and Young 2005; Dobbs et al. 2018). Therefore, mapping land use/land cover needs to focus on identifying patterns of these elements useful for assessing biodiversity and not only for mapping vegetation patches. Analyses should explore horizontal and vertical heterogeneity of vegetation in and around cities, to quantify the amount, connectivity, complexity and heterogeneity, and structural complexity of urban vegetation and water elements.

Data on connectivity, complexity, and heterogeneity can be derived from different remote sensing data, ranging from mid resolution data such as MODIS to high resolution data such as WorldView or LIDAR (Lausch et al. 2016a, 2016b and 2018). Remote sensing data can be processed to obtain vegetation indices, phenology metrics, landscape metrics, and topography metrics (Dobbs et al. 2018; Castillo-Cabrera et al. 2021; Figure 5.4). The use of high-resolution imagery has the advantage of detecting small objects, which allows the identification of vegetation of residential gardens, small green spaces, green roofs, small ponds or creeks, and can even allow the discrimination among species through spectral analysis. Mid-resolution imagery allows understanding the dynamics involved in land use/land cover change, including sprawling, in-filling, and abandonment of urban spaces. Images from MODIS can aid in understanding present and past phenology metrics that can provide information of patterns of species plantations in relation to evergreen and

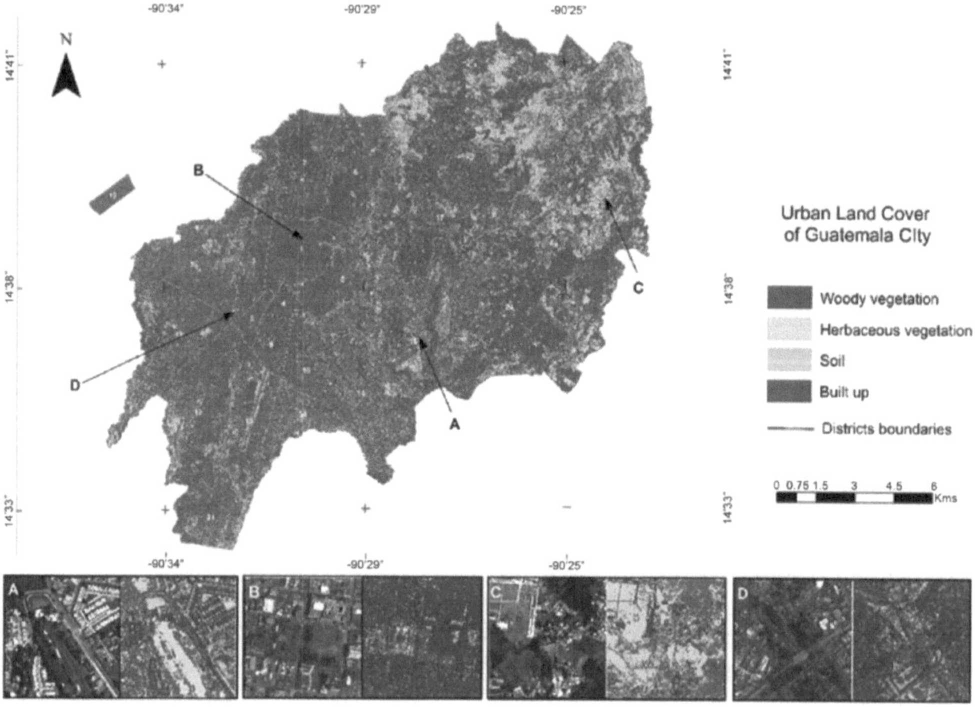

Figure 5.4 Urban Land Cover for Guatemala City using the WorldView-2 sensor. Left images are true colour composition of the WorldView-2 image and right images show the RF classification result. Images in (A) show a detail from zone 16 with low built-up area and new housing near a golf course. (B) Detail from zone 1, in this case "Plaza de la Constitución" (Main square). (C) Rural area with more soil cover of agricultural fields. (D) Vegetation in a dense built-up area known as "El Trébol". (WorldView-2 images courtesy of Digital Globe Foundation) (Castillo-Cabrera et al. 2021).

deciduous vegetation and inform on differences in phenology due to urban heat island effects. Mid-resolution data can inform on habitat loss and fragmentation due to land use and land cover change (Hedblom and Mörtberg 2011); however it would not be informative on vegetation structure, highly relevant for biodiversity assessment.

Data on structural complexity as surrogate of habitat delimitation can be obtained from LIDAR data. This data can give us information on potential habitat for different birds by detecting buildings that can be used as perch, for example, or changes in vertical structure affecting fauna habitat. Vertical in combination with horizontal structure can be used to model dispersion, especially in relation to invasive species.

Remote sensing data can be linked to species distribution modelling, which can be assessed in areas subjected to urbanization for understanding the impacts of sprawling or detecting where vegetation types are located within the urban area (Martinez-Harms et al. 2016; Hedblom and Mörtberg 2011; Morales and Fernández 2020). These data can also be used in models assessing habitat network size and their structural and functional connectivity integrating networks models such as Conefor Sensinode, MatrixGreen, or GUIDOS software (Saura and Torné 2009; Zetterberg et al. 2010; Dobbs et al. 2018; Furberg et al. 2020).

Therefore, mapping for biodiversity needs to include distribution of green spaces, both public and private, including residential areas, and can be derived from the analysis of vegetation indices or the digitization of land cover. Vegetation type, at a gross classification such as evergreen and deciduous, can be derived from phenotypic analysis (Li et al. 2017) from satellite imagery or from species distribution modelling (Fernández and Morales 2019) and also from LIDAR that can discriminate among vegetation height, informing on structural complexity also. The possibility of performing temporal analysis from satellite imagery will inform on habitat loss dynamics and changes in vegetation management trends. Finally, land use/land cover analysis through satellite imagery can inform on spatiotemporal connectivity dynamics (Hedblom and Söderström 2008; Hepcan 2013; Dobbs et al. 2018; Furberg et al. 2020). Trait analysis can also be done through land use/land cover analysis. Wellmann et al. (2018) and Haase et al. (2019) both propose the use of remote sensing and the application of spectral libraries for subpixel vegetation fractions for assessing a spatiotemporal analysis of spectral traits (diversities), which allows to discriminate land uses and link it to functional diversity of urban areas (see also a perspectives paper by Andersson et al. 2021).

Another thing to be considered is habitat type, history, and scale. Threatened species can be associated with urbanized areas at the national scale (Lenzen et al. 2009; Ives et al. 2016), but studies at large spatial scales cannot unravel the role of different urban land use types for hosting self-sustaining populations of endangered species. Scaling down to 1-km^2 grids, for example, revealed a negative influence of urbanization on an endangered plant species in a citywide survey (Schmidt et al. 2014). It is therefore necessary to consider the relative contribution to species conservation of natural historical remnants versus urbanized land use types.

Vegetation patterns and its relation to urban biodiversity

Urban vegetation is a composite of remnant natural habitats, regenerating vegetation, vacant land, managed parks, gardens, and streetscapes (Chong et al. 2014). The ability of green patches for providing habitat for biodiversity depend on how these are distributed in space, how well connected they are, and structure and composition in the vertical and horizontal planes (Aronson et al. 2017; Brunbjerg et al. 2018). Certain species forage on the ground, while others are arboreal, and certain species require more space to survive, therefore vegetation patterns directly affect urban biodiversity (Chong et al. 2014).

The scale of analysis of vegetation patterns is also relevant to assess effects on urban biodiversity. Brunbjerg et al. (2018) found that vegetation cover at small scales (100–250 m radius) affects bees and hoverflies, intermediate scales (250–500 m) influences bat activity, while birds respond differently to vegetation characteristics from small to large scales (1000 m). However, they also found confounding effects related to the structural complexity of vegetated patches, where variability on canopy height can also serve as an explanatory variable for species richness at a landscape scale.

Dynamics in rapidly urbanizing areas create highly fragmented features, leading to landscapes that are composed by heterogeneous, complex, and discontinuous vegetation patches (Liu et al. 2014). The heterogeneous nature of vegetation management in cities leads to a highly variable combination of small decisions, which affects the survival of species needing larger habitat (Aronson et al. 2017). This has consequences for ecosystem function and the ecosystem services provided by vegetation patches exacerbating ecological and environmental problems and biodiversity loss (Grimm et al. 2008; Wu 2010; Dobbs et al. 2019).

An analysis made at a landscape scale for 100 cities around the globe found that the best descriptors for vegetation were the fragmentation of the vegetation patches, their amount and configuration. Another emerging pattern showed that more populated cities tended to have more fragmented vegetation patches, and this was more profound in cities located in warmer climates and with larger inequalities. Cities that are less populated and from countries with reduced inequalities have more and better distributed vegetation (Dobbs et al. 2018). Vegetation cover is thus highly influenced by the level of socioeconomic inequality, showing an influence in the form of fragmentation and unequal distribution of green patches. This also suggests that unequal societies might be leading to biotic impoverishment. Vegetation patterns in cities around the world vary, and there is a need not only to look for wide patterns for urban biodiversity but to address how local vegetation patterns affect the different taxa of urban biodiversity.

Brunbjerg et al. (2018) found that species richness for multiple taxa in Birmingham (UK) was related to high vegetation cover, while higher bird richness and bat activity was related to the diversity of tree canopy height, confirming that the structural complexity in the vertical plane is very relevant for biodiversity conservation in urban areas. However, also simple measures such as vegetation cover in a city have been useful for assessing patterns of urban biodiversity. Chong et al. (2014) found differences for bird and butterfly diversity in Singapore when comparing natural and cultivated greenery derived from remote sensing analysis. They also suggest that the presence of vegetation always increases species richness at a landscape scale. The authors also found that road density, causing fragmentation of habitat, decreases species richness for birds and butterflies.

Haase and Gläser (2009) showed for the city of Leipzig, Germany, that despite urbanization, high-biodiversity areas like wetlands and floodplains did not simply disappear, but lost most of their floodplain properties: flood loam is still present in most places, whereas surface sealing and the depression of the groundwater table have resulted in a distinct change of water storage capacity and, subsequently, tree and non-tree species composition. The study further showed that management in the form of state forestry was able to maintain the medieval forest trees species composition until today. Thus, despite urbanization, the current forest area is still able to serve for both recreation and core habitat of *Quercus robur* as a fundamental niche for many organisms. Hence, small parts of the floodplain forest should be used as a coppice-with-standards forest since this management system can be conducive to a higher occurrence of *Quercus robur* and, above all, a high biodiversity in the floodplain forests under today's urbanization patterns. When it comes to urban green space and merging cultural ecosystem services and biodiversity, another example comes from Stockholm, Sweden, where a larger amount of mature and old deciduous trees has historically been favoured in and near the city, leaving relatively large and old stands of native tree species supporting high biodiversity (Mörtberg and Wallentinus 2000).

Another study by Ignatieva et al. (2020) impressively shows that urban lawns – the most common part of open green spaces and urban green infrastructures – can serve as nature-based solutions in finding resilient sustainable alternatives for urban green spaces in the time of growing urbanization, increased urban land use competition, various user demands, and related societal challenges of the urban environment. The authors hypothesize that various local native plant communities rich in species and able to withstand harsh conditions, such as heavy trampling and droughts, could form a solid ground for both recreational space of urban residents while being host for bees, bumblebees, butterflies, and other insects.

Planning for sustainable urban development

To support biodiversity and ecosystem services (ES) in cities, different planning concepts have been launched over the years, such as green infrastructure and nature-based solutions, as well as more overarching urban planning strategies for better balancing urban services and ecosystem services, in order to reach a more sustainable urban development. Green infrastructure (GI) is a multifunctional network of green and blue spaces and the processes that govern its development (Ahern 2007; Pauleit et al. 2019). From a planning perspective, it can be defined as a strategically planned network of natural and semi-natural areas, designed, and managed to deliver ecosystem services as well as enhancing biodiversity (EC 2013, 2019). In an urban context, GI is a global effort to improve multiple dimensions of urban sustainability (Kabisch et al. 2017), and several similar approaches have been developed, such as greenbelts, green wedges, and greenways (Pauleit et al. 2019). Among the wide range of ecosystem services that could be supported along with biodiversity are water purification, air quality, recreation, and climate mitigation and adaptation.

The EU strategy on biodiversity for 2020 (EC 2011) included a commitment to develop a GI strategy, entitled The EU Strategy on Green Infrastructure (EC 2013), which targets both nature and people and supports national and regional decision-makers and planners in their work with green infrastructure. According to EC (2013), GI can be a reservoir of biodiversity that can be drawn upon to repopulate and revitalize degraded environments and seeks to reverse some of the loss and fragmentation of ecosystems. Green infrastructure, as defined by the EU, includes biodiversity-rich nature areas such as forest, natural grasslands, wetlands, and water bodies, and in the EU, the Natura 2000 network of protected nature areas constitutes the backbone of the green infrastructure. Furthermore, GI also includes semi-natural spaces such as parks, agricultural land, allotments, community and private gardens, green spaces connected to grey infrastructure, and covers artificial features built to enhance ecosystem services or assist wildlife movement, such as green roofs and walls or eco-bridges and fish ladders (EC 2013; Pauleit et al. 2019).

From a biodiversity perspective, GI as a planning tool can be seen as consisting of outlined areas of high biodiversity value (protected or not), functioning as nodes or core areas in a network, where the links would be zones or structures supporting connectivity among these nodes (similar to definitions by Opdam et al. 2006; Carlier and Moran 2019; EC 2019). Landscape metrics representing habitat quality, quantity, and connectivity and related modelling tools have been used to shape and evaluate GI for planning purposes (see review by Bolliger and Silbernagel 2020). In particular, connectivity has been modelled using different GIS-based approaches, aiming to find zones that enable movements and dispersal between nodes or habitat patches (Nor et al. 2017; Perkl et al. 2016; Zhang et al. 2019; Bolliger and Silbernagel 2020). However, there is yet no consensus on how to design GI for promoting urban biodiversity, not at least in the view of simultaneously promoting urban ecosystem services (Hostetler et al. 2011; Badiu et al. 2016; Chatzimentor et al. 2020). Still, major efforts on developing GI are currently undertaken and methodological frameworks suggested (Gavrilidis et al. 2019), to accelerate the much-needed integration of these issues into the urban planning process.

The GI-related concept of nature-based solutions (NBS) has been defined as "solutions that are inspired and supported by nature, which are cost-effective, simultaneously provide environmental, social and economic benefits and help build resilience" (EC 2015) and are seen as a fundamental component of the transition towards sustainable development. The concept of NBS can be seen as integrating existing approaches, such as green infrastructure, with assessments of the social and economic benefits of resource-efficient and systemic solutions that combine technical, governance, regulatory, and social innovation (Raymond et al. 2017).

The recent EU Biodiversity Strategy calls for NBS in urban areas and NBS in planning, specifically requiring Member States to become more active in terms of greening cities (EC 2020). The promotion of healthy ecosystems, GI, and nature-based solutions should be systematically integrated into urban planning, including public spaces, infrastructure, and design of buildings and their surroundings. In line with this, European cities are required to develop Urban Greening Plans, which should include measures to create biodiverse and accessible urban forests, parks and gardens, green roofs and walls, tree-lined streets, and urban farms, meadows, and hedges. They should also help improve connections among green spaces, eliminate the use of pesticides, and limit excessive mowing of urban green spaces and other biodiversity harmful practices. Such plans could mobilize policy, regulatory, and financial tools.

Urban green planning has generally focused on conservation, creating vegetation patches and connectivity zones, overlooking the role of other land uses for biodiversity conservation. Lowering the impact of other land uses, and even encouraging traditional ways these are developed, will enhance their role in contributing to urban biodiversity (Hostetler et al. 2011). Urban planning for biodiversity and ES needs to involve different planning levels, sectors, and spatial scales, from regional or even national, to municipal, city districts and projects, down to private gardens where information and voluntary input can be mobilized (Van der Jagt et al. 2018). With this comes inevitably some challenges and problems, such as limited cooperation between stakeholders and planning and governance boundaries (Mell 2014). Challenges with urban planning for biodiversity and ES have also been recognized in urban areas with large informal settlements, such as in developing countries.

In response to some of these challenges, the approach of integrating green and grey infrastructure has been proposed by the World Bank and World Resources Institute (Browder et al. 2019). For example, in Nairobi, Kenya, a project combining blue-green and grey infrastructure showed that involvement in co-development of small-scale GI changed people's perception, valuation, and stewardship of NBS and ES (Mulligan et al. 2020). According to Browder et al. (2019), integrating GI into traditional infrastructure planning and into mainstream project appraisal processes and investments can produce lower cost, more resilient services, including water security, and climate resilience and help in achieving development goals.

Urban planning strategies

Recently, spatial strategies for urban development have been developed on an overarching urban planning level, in order to reach more sustainable urban development (e.g. EC 2015; UN Habitat 2016). These strategies can be seen as a response to main problems with overall trends in urban development, attributed to both sprawling cities and uninformed compaction of cities. Thus, a major challenge is to simultaneously develop compact, energy-efficient, and yet green, biodiverse, and livable cities (Grimm et al. 2008; EEA 2009; OECD 2012; Elmqvist et al. 2013; Energy Sector Management Assistance Program 2014).

Urban sprawl has been characterized by incremental development with little comprehensive public planning (Ewing 2008; Botequilha-Leitão 2012), while the main impacts of urban sprawl come from the inefficient and segregated land use with high travel demands and low energy and land use efficiency and from loss of open and green spaces, with economic, social, and environmental consequences (Real Estate Research Corporation 1974; Travisi et al. 2010; OECD 2012; Wilson and Chakraborty 2013). On the other hand, a compact city with high density when it comes to only few criteria (e.g. number of residents per area unit) may lead to unwanted effects such as increased traffic congestion and a lack of vegetation and open space in cities (OECD 2012). For a compact city to be sustainable, mixed land use and multifunctionality is seen as equally important. According to the OECD (2012) and the Energy Sector Management Assistance Program (2014), a sustainable compact city is therefore characterized by shorter intra-urban travel distances, less car dependence, and better access to public transport, a diversity of local jobs and public services, as well as ecosystem services and public open space and implies an optimum use of land resources.

A spatial strategy that has been promoted as more flexible in this context is the polycentric city, where functions such as institutions, companies, and workplaces are disseminated from the urban centre to other parts of the urban region (EC 1999; Faludi 2005; UN 2011). In European cities, this strategy has been broadly implemented as a basis of EU spatial development policy (EC 1999). It is intermediate between the compact monocentric city and the sprawling city (Botequilha-Leitão 2012) and permits clustering of land uses to several urban centres to reduce transportation, as well as protection of green spaces (Mörtberg et al. 2013, 2017).

In view of this, a set of spatial strategies have been agreed upon at a global level in the New Urban Agenda, adopted in 2016, including compactness, polycentrism, mixed use, and prioritization of urban renewal (EC 2015; UN 2015). Since spatial planning encompasses different sectoral policies, strategies may respond to multiple objectives, including biodiversity conservation but also protection of cultural and natural heritage, social inclusion, reduction of air and water pollution, resilience to natural hazards, and climate change mitigation and adaptation (EC 2011; UN Habitat 2016). Cortinovis et al. (2019) analyzed a set of these spatial strategies in European cities using spatial-quantitative indicators during the years 2006–2012. Among the findings were that during this time-frame, most cities achieved a more compact urban form and functional mix, while, on the other hand, green spaces increased only in 20% of the cities, and very few halted land take. This shows both great potential and difficulties in urban planning for biodiversity and ES, but on the other hand, the uptake of these issues and approaches in the planning system is currently ongoing (Khoshkar et al. 2020).

Urban planning support systems that take into account a wide spectrum of sustainability goals, integrating energy and transport, as well as urban services and ecosystem services, are emerging (e.g. Deal et al. 2012). In such models, urban-regional systems can be explored through scenarios of future development, including polycentricity or other concepts, and can be evaluated in a systematic way. Furthermore, these can support planning in collaboration with stakeholders, integrating scientific as well as practitioner's knowledge in a collaborative geodesign process (Batty 2012; Foster 2020; Page et al. 2020).

In urban contexts, there is a need to better understand and evaluate people's diverse interactions with GI as well as enhance stewardship and thus move beyond policymaking, narrowly focusing on biodiversity alone. It also provides support for a context-sensitive and site-specific approach to planning and management (Elands et al. 2018) and nongovernmental engagement of civil society in the form of nature stewardship (Andersson et al. 2014).

Conclusions

Cities have a role to play in conserving biodiversity. The understanding of land use/land cover patterns is highly relevant for assessing patterns for urban biodiversity, and it requires a multiple-scale

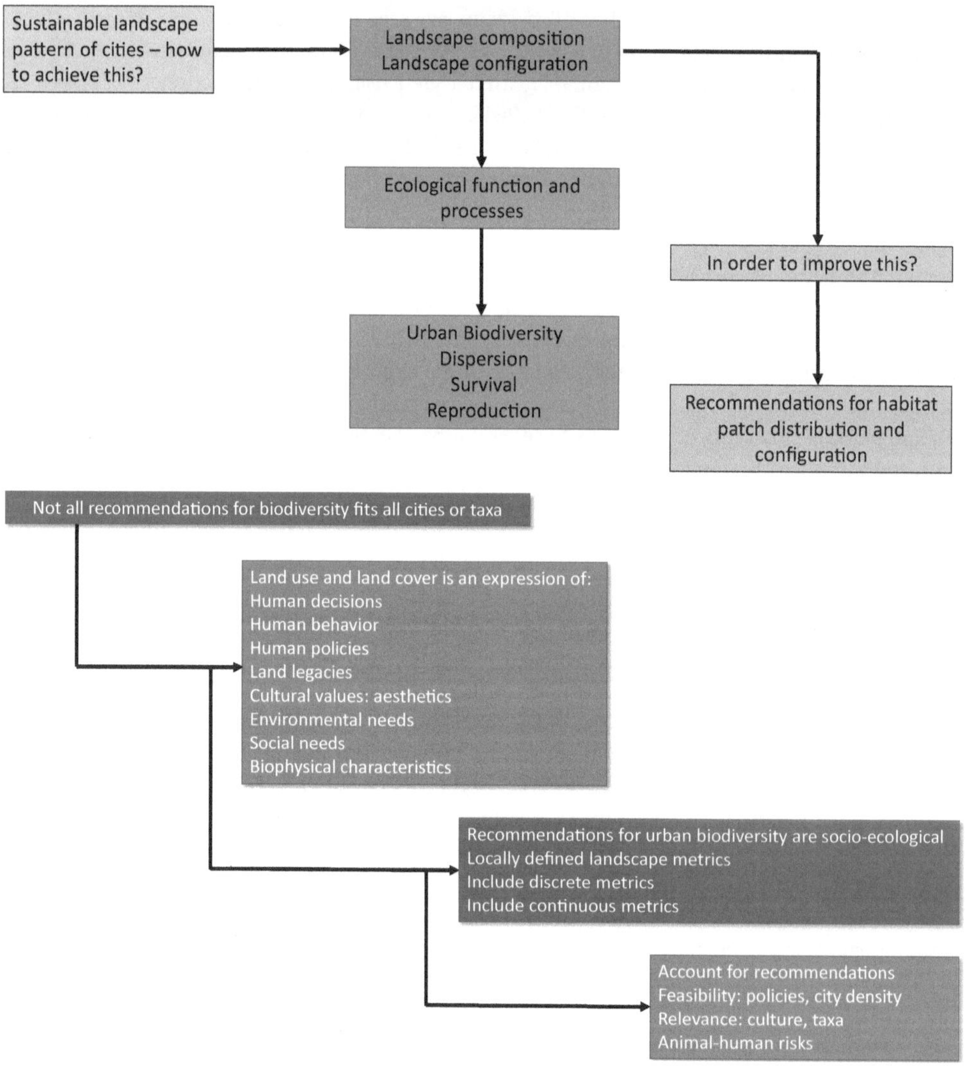

Figure 5.5 How to achieve sustainable land cover and land use in cities to preserve biodiversity? An iterative approach resulting from the synthesis provided in this chapter.

approach to be relevant for different taxa. However, consideration needs to be taken when generalizing the effects of land use/land cover change on urban biodiversity given different trends in urbanization that might lead to differentiated patterns according to the variety of climatic, social, and cultural contexts of cities.

The search for sustainability and resilience in the development of cities might change the already studied effect of urbanization on biodiversity, by shifts in current urban landscape patterns. This raises an opportunity for incorporating more profound concepts of landscape ecology, animal behaviour, and population ecology into planning of future cities.

It is also important to understand the local context to determine which actions need to be taken to promote planning towards enhancing biodiversity and sustainability. We need to remember that land use/land cover is an expression of human decisions (i.e. where we decide to construct a new development), behaviour (i.e. how we decide tree cover will be placed in this new development), policies (i.e. how much green a new development requires according to the number of houses), and cultural values (i.e. are there any remnants of indigenous forest where the new development is happening?), among others (Figure 5.5). Therefore, recommendations for conservation of enhancing urban biodiversity needs to be locally pertinent following a socio-ecological approach. Recommendations should include discrete and continuous landscape metrics that are locally defined, assessing its relevance and feasibility.

Finally, given current times, we need to include in the decision-making towards planning land use/land cover both the benefits and the risks involved in increasing the relationships between animals and humans. Increasing biodiversity in cities can provide health benefits by exposure to nature that is better distributed by applying a landscape approach in urban planning, highly relevant during times of quarantine, such as the COVID-19 pandemic (Ramirez-Rubio et al. 2019; Azevedo et al. 2020). However, increasing biodiversity in cities could also increase the frequency of unfavourable animal encounters (i.e. mosquitos, rats). Having a larger variety of species in urban areas, producing areas of buffers or barriers, and understanding in a more robust way the patterns of dispersion can reduce the risks for future pandemic mobility (Azevedo et al. 2020; Zhao et al. 2020).

Key companion papers

Beninde, J., M. Veith, A. Hochkirch. 2015. Biodiversity in cities needs space: A meta-analysis of factors determining intra-urban biodiversity variation. *Ecology Letters* 18:581–592.

Bolliger, J., J. Sibernagel. 2020. Contribution of connectivity assessments to green infrastructure (GI). *International Journal of Geo-Information* 9:212.

Elmqvist, T., et al. 2018. *The Urban Planet: Patterns and Pathways to the Cities We Want.* Cambridge University Press, Cambridge.

Seto, K.C., M. Fragkias, B. Güneralp, M.K. Reilly. 2011. A meta-analysis of global urban land expansion. *PLoS ONE* 6(8):e23777.

References

Ahern, J. 2007. Planning and design for sustainable and resilient cities: Theories, strategies and best practice for green infrastructure. Pages 135–176 in V. Novotny, J. Ahern, P. Brown, editors. *Water-Centric Sustainable Communities.* Wiley-Blackwell, Hoboken, NJ.

Andersson, E., S. Barthel, S. Borgström, J. Colding, T. Elmqvist, C. Folke, Å. Gren. 2014. Reconnecting cities to the biosphere: Stewardship of green infrastructure and urban ecosystem services. *Ambio* 43:445–453.

Andersson, E., D. Haase, P. Anderson, C. Cortinovis, J. Goodness, D. Kendal, A. Lausch, T. McPhearson, D. Sikorska, T. Wellmann. 2021. What are the traits of a social-ecological system? Towards a framework in support of urban sustainability. *NJP Urban Sustainability* 1:14.

Angel, S., J. Parent, D.L. Civco, A.M. Blei. 2011b. *Making Room for a Planet of Cities*. Policy Focus Report/ Code PF027. Lincoln Institute of Land Policy, Cambridge, MA.

Angel, S., J. Parent, D.L. Civco, A.M. Blei, D. Potere. 2011a. The dimensions of global urban expansion: Estimates and projections for all countries, 2000–2050. *Progress in Planning* 75:53–107.

Aronson, M.F.J., F.A. La Sorte, C.H. Nilon, M. Katti, M.A. Goddard, C.A. Lepczyk, P.S. Warren, W.P.S. Williams, S. Cilliers, B. Clarkson, C. Dobbs, R. Dolan, M. Hedblom, S. Klotz, J. Louwe Kooijmans, I. Kühn, I. MacGregor-Fors, M. McDonnell, U. Mörtberg, P. Pysek, S. Sieberg, J. Sushinsky, P. Werner, M. Winter. 2014. A global analysis of the impacts of urbanization on bird and plant diversity reveals key anthropogenic drivers. *Proceedings of the Royal Society B* 281:20133330.

Aronson, M.F.J., C.A. Lepczyk, K.L. Evans, M.A. Goddard, S.B. Lerman, J.S. MacIvor, C.H. Nilon, T. Vargo. 2017. Biodiversity in the city: Key challenges for urban green space management. *Frontiers in Ecology and the Environment* 15(4):189–196.

Arrhenius, O. 1921. Species and area. *Journal of Ecology* 9:95–99.

Azevedo, J.C., S. Luque, C. Dobbs, G. Sanesi, T.C.H. Sunderland. 2020. The ethics of isolation, the spread of pandemics and landscape ecology. *Landscape Ecology* 35:2133–2140.

Badiu, D.L., C.I. Iojă, M. Pătroescu, J. Breuste, M. Artmann, M.R. Niță, S.R. Grădinaru, C.A. Hossu, D.A. Onose. 2016. Is urban green space per capita a valuable target to achieve cities' sustainability goals? Romania as a case study. *Ecological Indicators* 70:53–66.

Bahi, H., H. Rhinane, A. Bensalmia, U. Fehrenbach, D. Scherer. 2017. Effects of urbanization and seasonal cycle on the surface urban heat island patterns in the coastal growing cities: A case study of Casablanca, Morocco. *Remote Sensing* 9:91.

Bai, X., J. Chen, P. Shi. 2011. Landscape urbanization and economic growth in China: Positive feedbacks and sustainability dilemmas. *Environmental Science & Technology* 46:132–139.

Batty, M. 2012. Building a science of cities. *Cities* 29(1):S9–S16.

Batty, M., Y. Xie, Z. Sun. 1999. The dynamics of urban sprawl. *CASA Working Paper Series*, Paper 15, Centre for Advanced Spatial Analysis (CASA). University College London, London.

Beninde, J., M. Veith, A. Hochkirch. 2015. Biodiversity in cities needs space: A meta-analysis of factors determining intra-urban biodiversity variation. *Ecology Letters* 18:581–592.

Bolliger, J., J. Silbernagel. 2020. Contribution of connectivity assessments to green infrastructure (GI). *International Journal of Geo-Information* 9(4):212.

Botequilha-Leitão, A. 2012. Eco-polycentric urban systems: An ecological region perspective for network cities. *Challenges* 3:1–42.

Browder, G., S. Ozment, I. Rehberger Bescos, T. Gartner, G.-M. Lange. 2019. *Integrating Green and Grey – Creating Next Generation Infrastructure*. World Bank and World Resources Institute, Washington, DC.

Brueckner, J.K. 2000. Urban sprawl: Diagnosis and remedies. *International Regional Science Review* 23(2):160–171.

Brunbjerg, A.K., J.D. Hale, A.J. Bates, R.E. Fowler, E.J. Rosefeld, J.P. Sadler. 2018. Can patterns of urban biodiversity be predicted using simple measures of green infrastructure? *Urban Forestry & Urban Greening* 32:143–153.

Carlier, J., J. Moran. 2019. Landscape typology and ecological connectivity assessment to inform greenway design. *Science of the Total Environment* 651:3241–3252.

Castillo-Cabrera, F., T. Wellmann, D. Haase. 2021. Urban green fabric analysis promoting sustainable planning in Guatemala City. *Land* 10:18.

Chacón-Moreno, E., I. Olivares, G. Navarro, A.J. Albarrán, Y. Paredes, C.I. Aranguren, G.J. Nagy. 2020. Landscape ecology and conservation for building resilience and adaptation to global change in Venezuela. Pages 147–160 in W.L. Filho, G.J. Nagy, M. Borga, D. Chávez Muñoz, A. Magnuszewski, editors. *Climate Change, Hazards and Adaptation Options*. Springer Nature Switzerland, Cham, Switzerland.

Chatzimentor, A., E. Apostolopoulou, A.D. Mazaris. 2020. A review of green infrastructure research in Europe: Challenges and opportunities. *Landscape and Urban Planning* 198:103775.

Chen, M., Y. Sui, W. Liu, H. Liu, Y. Huang. 2019. Urbanization patterns and poverty reduction: A new perspective to explore the countries along the Belt and Road. *Habitat International* 84:1–14.

Chong, K.Y., S. Teo, B. Kurukulasuriya, Y.F. Chung, S. Rajathurai, H.T.W. Tan. 2014. Not all green is good: Different effects of the natural and cultivated components of urban vegetation on bird and butterfly diversity. *Biological Conservation* 171:299–309.

Cortinovis, C., D. Haase, B. Zanon, D. Geneletti. 2019. Is urban spatial development on the right track? Comparing strategies and trends in the European Union. *Landscape and Urban Planning* 181:22–37.

Deal, B., J.H. Kim, G. Hewings. 2012. Complex urban systems integration: The LEAM experiences in coupling economic, land use, and transportation models in Chicago, IL. Pages 107–131 in F. Pagliara, M. de Bok, D. Simmons, A.G. Wilson, editors. *Employment Location in Cities and Regions: Models and Applications*. Springer, Heidelberg, Germany.

Dobbs, C., F.J. Escobedo, N. Clerici, F. de la Barrera, A.A. Eleuterio, I. MacGregor-Fors, S. Reyes-Paecke, A. Vásquez, J.D. Zea Camaño, J.H. Hernández. 2019. Urban ecosystem services in Latin America: Mismatch between global concepts and regional realities? *Urban Ecosystems* 22:173–187.

Dobbs, C., A. Hernández, F. de la Barrera, M.D. Miranda, S. Reyes-Paecke. 2018. Integrating urban biodiversity mapping, citizen science and technology. Pages 236–247 in A. Ossola, J. Niemelä, editors. *Urban Biodiversity: From Research to Practice*. Routledge Taylor & Francis, London.

EC. 1999. *Towards Balanced and Sustainable Development of the Territory of the European Union*. European Spatial Development Perspective, European Commission, Brussels, Belgium.

EC. 2011. *Our Life Insurance, Our Natural Capital: An EU Biodiversity Strategy to 2020*. European Commission, Brussels, Belgium.

EC. 2013. *Green Infrastructure (GI) – Enhancing Europe's Natural Capital*. European Commission, Brussels, Belgium.

EC. 2015. *Towards an EU Research and Innovation Policy Agenda for Nature-Based Solutions & Re-Naturing Cities: Final Report of the Horizon 2020 Expert Group on 'Nature-Based Solutions and Re-Naturing Cities'*. European Commission, Brussels, Belgium.

EC. 2019. *Guidance on a Strategic Framework for Further Supporting the Deployment of EU-Level Green and Blue Infrastructure*. European Commission, Brussels, Belgium.

EC. 2020. *EU Biodiversity Strategy for 2030: Bringing Nature Back into Our Lives*. European Commission, Brussels, Belgium.

EEA. 2009. *Ensuring Quality of Life in Europe's Cities and Towns – Tackling the Environmental Challenges Driven by European and Global Change*. EEA Report No5. European Environment Agence, Copenhagen, Denmark.

Elands, E., K. Vierikko, E. Andersson, L.K. Fischer, P. Goncalves, D. Haase, I. Kowarik, A.C. Luz, J. Niemela, M. Santos-Reis, K.F. Wiersum. 2018. Biocultural diversity: A novel concept to assess human-nature interrelations, nature conservation and stewardship in cities. *Urban Forestry & Urban Greening* 40:29–34.

Elmqvist, T., X. Bai, N. Frantzeskaki, C. Griffith, D. Maddox, T. McPhearson, S. Parnell, D. Roberts, L.P. Romero, D. Simon, editors. 2018. *The Urban Planet: Patterns and Pathways to the Cities We Want*. Cambridge University Press, Cambridge.

Elmqvist, T., M. Fragkias, J. Goodness, B. Guneralp, R. McDonald, P. Marcotullio, S. Parnell, M. Schewenius, M. Sendstad, K.C. Seto, C. Wilkinson, M. Alberti, C. Folke, D. Haase, M. Katti, J. Niemelä, K. Tidball, H. Nagendra, D. Pickett, C. Redman. 2013. Stewardship of the Biosphere in the Urban Europe. Pages 719–746 in T. Elmqvist, M. Fragkias, J. Goodness, B. Güneralp, P.J. Marcotullio, R.I. McDonald, S. Parnell, M. Schewenius, M. Sendstad, K.C. Seto, C. Wilkinson, editors. *Global Urbanisation, Biodiversity and Ecosystem Services: Challenges and Opportunities*. Springer Netherlands, Dordrecht, Netherlands.

Energy Sector Management Assistance Program. 2014. *Planning Energy Efficient and Livable Cities. Mayoral Guidance Note, No. 6. Energy Efficient Cities*. World Bank, Washington, DC.

Ewing, R.H. 2008. Characteristics, causes and effects of sprawl: A literature review. Pages 519–536 in J.M. Marzluff, E. Schulenberger, W. Endlicher, M. Alberti, G. Bradley, C. Ryan, U. Simon, C. ZumBrunnen, editors. *Urban Ecology*. Springer, New York.

Faludi, A. 2005. Polycentric territorial cohesion policy. *Town Planning Review* 76:107–118.

Fernández, I.C., N.S. Morales. 2019. One-class land-cover classification using MaxEnt: The effect of modelling parametrization on classification accuracy. *Peer J* 7:e7016.

Foster, K. 2020. Geodesign. Pages 7–17 in A. Kobayashi, editor. *International Encyclopedia of Human Geography*, 2nd edition. Elsevier, Oxford.

Furberg, D., Y. Ban, U. Mörtberg. 2020. Monitoring urban green infrastructure changes and impact on habitat connectivity using high-resolution satellite data. *Remote Sensing* 12:3072.

Gavrilidis, A.A., M.R. Niţă, D.A. Onose, D.L. Badiu, I.I. Năstase. 2019. Methodological framework for urban sprawl control through sustainable planning of urban green infrastructure. *Ecological Indicators* 96:67–78.

Grimm, N., S.H. Faeth, N.E. Golubiewski, C.L. Redman, J. Wu, X. Bai, J.M. Briggs. 2008. Global change and the ecology of cities. *Science* 319:756–760.

Güneralp, B., I. Güneralp, Y. Liu. 2015. Changing global patterns of urban exposure to flood and drought hazards. *Global Environmental Change* 31:217–225.

Güneralp, B., S. Lwasa, H. Masundire, S. Parnell, K.C. Seto. 2018. Urbanization in Africa: Challenges and opportunities for conservation. *Environment Research Letters* 13:015002.

Güneralp, B., M. Reba, B.U. Hales, E.A. Wentz, K.C. Seto. 2020. Trends in urban land expansion, density and land transitions from 1970 to 2010: A global synthesis. *Environmental Research Letters* 15:044015.

Haase, D. 2013. Shrinking cities, biodiversity and ecosystem services. Pages 253–274 in T. Elmqvist, M. Fragkias, J. Goodness, B. Güneralp, P.J. Marcotullio, R.I. McDonald, S. Parnell, M. Schewenius, M. Sendstad, K.C. Seto, C. Wilkinson, editors. *Global Urbanisation, Biodiversity and Ecosystem Services: Challenges and Opportunities*. Springer Netherlands, Dordrecht, Netherlands.

Haase, D. 2014. How is urban land use unique? Pages 299–312 in K.C. Seto, A. Reenberg, editors. *Rethinking Global Land Use in an Urban Era*. Strüngmann Forum Reports, Vol. 14. MIT Press, Cambridge, MA.

Haase, D. 2019. Urban telecouplings. Pages 261–280 in C. Friis, J.Ø. Nielsen, editors. *Telecoupling: Exploring Land-Use Change in a Globalised World*. Palgrave Macmillan, London.

Haase, D., J. Gläser. 2009. Determinants of floodplain forest development illustrated by the example of the floodplain forest in the District of Leipzig. *Forest Ecology and Management* 258:887–894.

Haase, D., B. Guneralp, X. Bai, T. Elmqvist, B. Dahiya, M. Fragkias, K. Gurney. 2018. Different pathways of global urbanization. Pages 19–44 in T. Elmqvist, X. Bai, N. Frantzeskaki, C. Griffith, D. Maddox, T. McPhearson, S. Parnell, D. Roberts, P. Romero Lankao, D. Simon, editors. *The Urban Planet: Patterns and Pathways to the Cities We Want*. Cambridge University Press, Cambridge.

Haase, D., C. Jänicke, T. Wellmann. 2019. Delineating private greenspaces in cities based on subpixel vegetation fractions from earth observation data using spectral unmixing. *Landscape and Urban Planning* 182:44–54.

Haase, D., H. Nuissl. 2010. The urban-to-rural gradient of land use change and impervious cover: A long-term trajectory for the city of Leipzig. *Land Use Science* 5(2):123–142.

Haase, D., N. Schwarz. 2016. Urban land use in the context of global land use. Pages 50–63 in K.C. Seto, W.D. Solecki, C. Griffith, editors. *The Routledge Handbook of Urbanization and Global Environmental Change*. Routledge Taylor & Francis, New York.

Halleux, J.M., S. Marcinczak, E. van der Krabben. 2012. The adaptive efficiency of land use planning measured by the control of urban sprawl. The cases of the Netherlands, Belgium and Poland. *Land Use Policy* 29(4):887–898.

Haraway, D. 2016. Tentacular thinking: Anthropocene, capitalocene, chthulucene. *E-Flux Journal* 75.

Hedblom, M., E. Andersson, S. Borgström. 2017. Flexible land-use and undefined governance: From threats to potentials in peri-urban landscape planning. *Land Use Policy* 63:523–527.

Hedblom, M., U. Mörtberg. 2011. Characterizing biodiversity in urban areas using remote sensing. Pages 287–304 in X. Yang, editor. *Urban Remote Sensing: Monitoring, Synthesis and Modelling in the Urban Environment*, 1st edition. John Wiley & Sons Ltd, Oxford.

Hedblom, M., B. Söderström. 2008. Woodlands across Swedish urban gradients: Status, structure and management implications. *Landscape and Urban Planning* 84(1):62–73.

Hepcan, S. 2013. Analyzing the pattern and connectivity of urban green spaces: A case study of Izmir, Turkey. *Urban Ecosystems* 16:279–293.

Hostetler, M., W. Allen, C. Meurk. 2011. Conserving urban biodiversity? Creating green infrastructure is only the first step. *Landscape and Urban Planning* 100:369–371.

Huang, C., R.I. McDonald, K.C. Seto. 2018. The importance of land governance for biodiversity conservation in an era of global urban expansion. *Landscape and Urban Planning* 173:44–50.

Huang, K., X. Li, X. Liu, K.C. Seto. 2019. Projecting global urban land expansion and heat island intensification through 2050. *Environmental Research Letters* 14:114037.

Ignatieva, M., D. Haase, D. Dushkova, A. Haase. 2020. Lawn as a unique global urban green space phenomenon: A novel way of searching for nature-based solutions in cities. *Land* 9:73.

Inostroza, L., R. Baur, E. Csaplovics. 2013. Urban sprawl and fragmentation in Latin America: A dynamic quantification and characterization of spatial patterns. *Journal of Environmental Management* 115:87–97.

Isendahl, C., M.E. Smith. 2013. Sustainable agrarian urbanism: The low-density cities of the Mayas and Aztecs. *Cities* 31:132–143.

Ives, C.D., P.-E. Lentini, C.G. Threlfall, K. Ikin, D.F. Shanahan, G.E. Garrard, S.A. Bekessy, R.A. Fuller, L. Mumaw, L. Rayner, R. Rowe, L.E. Valentine, D. Kendal. 2016. Cities are hotspots for threatened species. *Global Ecology and Biogeography* 25:117–126.

Jarvis, P.J., C.H. Young. 2005. *The Mapping of Urban Habitat and Its Evaluation*. University of Wolverhampton, Wolverhampton.

Jenerette, G.D., D. Potere. 2010. Global analysis and simulation of land-use change associated with urbanization. *Landscape Ecology* 25(5):657–670.

Kabisch, N., H. Korn, J. Stadler, A. Bonn. 2017. *Nature-based Solutions to Climate Change in Urban Areas*. Springer, Cham, Switzerland.

Karimi, A., M. Hockings. 2018. A social-ecological approach to land-use conflict to inform regional and conservation planning and management. *Landscape Ecology* 33:691–710.

Keys, P., V. Galaz, M. Dyer, N. Matthews, C. Folke, M. Nyström, S. Cornell. 2019. Anthropocene risk. *Nature Sustainability* 2:667–673.

Khoshkar, S., M. Hammer, S. Borgström, P. Dinnétz, B. Balfors. 2020. Moving from vision to action-integrating ecosystem services in the Swedish local planning context. *Land Use Policy* 97:104791.

Lausch, A., L. Bannehr, M. Beckmann, C. Boehm, H. Feilhauer, J.M. Hacker, M. Heurich, A. Jung, R. Klenke, C. Neumann, M. Pause, D. Rocchini, M.E. Schaepman, S. Schmidtlein, K. Schulz, P. Selsam, J. Settele, A.K. Skidmore, A.F. Cord. 2016a. Linking earth observation and taxonomic, structural and functional biodiversity: Local to ecosystem perspectives. *Ecological Indicators* 70:317–339.

Lausch, A., S. Erasmi, D. King, P. Magdon, M. Heurich. 2016b. Understanding forest health with remote sensing-part I – A review of spectral traits, processes and remote-sensing characteristics. *Remote Sensing* 8:1029.

Lausch, A., B. Olaf, K. Stefan, P. Leitao, A. Jung, D. Rocchini, M.E. Schaepman, A.K. Skidmore, L. Tischendorf, S. Knapp. 2018. Understanding and assessing vegetation health by in-situ species and remote sensing approaches. *Methods in Ecology and Evolution* 9:1799–1809.

Leichenko, R.M., W.D. Solecki. 2005. Exporting the American dream: The globalization of suburban consumption landscapes. *Regional Studies* 39(2):241–253.

Lenzen, M., A. Lane, A. Widmer-Cooper, M. Williams. 2009. Effects of land use on threatened species. *Conservation Biology* 23:294–306.

Li, F., G. Song, Z. Liujun, L. Di. 2017. Urban vegetation phenology analysis using high spatio-temporal NDVI time series. *Urban Forestry & Urban Greening* 25:43–57.

Liu, M., Y. Xu, Y. Hu, C. Li, F. Sun, T. Chen. 2014. A century of the evolution of the urban area in Shenyang, China. *PLoS ONE* 9:e98847.

Lwasa, S. 2014. Managing African urbanization in the context of environmental change. *INTERdisciplina* 2:263–280.

Martin, A.R., M.W. Cadotte, M.E. Isaac, R. Milla, D. Vile, C. Violle. 2019. Regional and global shifts in crop diversity through the Anthropocene. *PLoS ONE* 14:e0209788.

Martinez-Harms, M.J., S. Quijas, A.M. Merenlender, P. Balvanera. 2016. Enhancing ecosystem services maps combining field and environmental data. *Ecosystem Services* 22:32–40.

McDonald, R., A.V. Mansur, F. Ascensão, M. Colbert, K. Crossman, T. Elmqvist, A. Gonzalez, B. Güneralp, D. Haase, M. Hamann, O. Hillel, K. Huang, B. Kant, D. Maddox, A. Pacheco, H. Pereira, K. Seto, R. Simkin, B. Walsh, C. Ziter. 2019. Research gaps in knowledge of the impact of urban growth on biodiversity. *Nature Sustainability* 3:16–24.

McGranahan, G., D. Balk, B. Anderson. 2007. The rising tide: Assessing the risks of climate change and human settlements in low elevation coastal zones. *Environment and Urbanization* 19:17.

McGranahan, G., D. Mitlin, D. Satterwaite, C. Tacoli, I. Turok. 2009. *Africa's Urban Transition and the Role of Regional Collaboration*. Human Settlements Group, International Institute for Environment and Development, London.

Mell, I.C. 2014. Aligning fragmented planning structures through a green infrastructure approach to urban development in the UK and USA. *Urban Forestry & Urban Greening* 13:612–620.

Morales, N.S., I. Fernández. 2020. Land cover classification using MaxEnt: Can we trust in model quality metrics for estimating classification accuracy? *Entropy* 22:342.

Mörtberg, U., R. Goldenberg, Z. Kalantari, O. Kordas, B. Deal, B. Balfors, V. Cvekovic. 2017. Integrating ecosystem services in the assessment of urban energy trajectories – A study of the Stockholm region. *Energy Policy* 100:338–349.

Mörtberg, U., J. Haas, A. Zetterberg, J. Franklin, D. Jonsson, B. Deal. 2013. Urban ecosystems and sustainable urban development: Analysing and assessing interacting systems in the Stockholm region. *Urban Ecosystem* 16:763–782.
Mörtberg, U., H.-G.Wallentinus. 2000. Red-listed forest bird species in an urban environment – Assessment of green space corridors. *Landscape and Urban Planning* 50:215–226.
Mulligan, J., V. Bukachi, J.C. Clause, R. Jewell, F. Kirimi, C. Odbert. 2020. Hybrid infrastructures, hybrid governance: New evidence from Nairobi (Kenya) on green-blue-grey infrastructure in informal settlements. *Anthropocene* 29:100227.
Nicholls, R.J., S. Hanson, C. Herweijer, N. Patmore, S. Hallegatte, J. Corfee-Morlot, J. Chateau, R. Muir-Wood. 2008. *Ranking Port Cities with High Exposure and Vulnerability to Climate Extremes: Exposure Estimates*. OECD Environment Working Papers, No 1. OECD Publishing, Paris, France.
Nilsson, K., T.S. Nielsen, C. Aalbers, S. Bell, B. Boitier, J.P. Chery, C. Fertner, M. Groschowski, D. Haase, W. Loibl, S. Pauleit, M. Pintar, A. Piorr, J. Ravetz, M. Ristimäki, M. Rounsevell, I. Tosics, J. Westerink, I. Zasada. 2014. Strategies for sustainable urban development and urban-rural linkages. *European Journal of Spatial Development* 12(3). http://doi.org/10.5281/zenodo.5079562
Nor, A.N.M., R. Corstanje, J.A. Harris, D.R. Grafius, G.M. Siriwardena. 2017. Ecological connectivity networks in rapidly expanding cities. *Heliyon* 3(6):e00325.
OECD. 2012. The compact city concept in today's urban contexts. Page 284 in *Compact City Policies: A Comparative Assessment*. OECD Publishing, Paris, France.
Opdam, P., E. Steingröver, S. van Rooij. 2006. Ecological networks: A spatial concept for multi-actor planning of sustainable landscapes. *Landscape and Urban Planning* 75(3–4):322–332.
Page, J., U. Mörtberg, G. Destouni, C. Ferreira, H. Nässtrröm, Z. Kalantari. 2020. Open-source planning support system use for sustainable regional planning: A case study of Stockholm County, Sweden. *Environment and Planning B: Urban Analytics and City Science* 47:1508–1523.
Pauleit, S., B. Ambrose-Oji, E. Andersson, B. Anton, A. Buijs, D. Haase, B. Elands, R. Hansen, I. Kowarik, J. Kronenberg, T. Mattijssen, A. Stahl Olafsson, E. Rall, A.P.N. van der Jagt, C. Konijnendijk van den Bosch. 2019. Advancing urban green infrastructure in Europe: Outcomes and reflections from the GREEN SURGE project. *Urban Forestry & Urban Greening* 40:4–16.
Perkl, R.M. 2016. Geodesigning landscape linkages: Coupling GIS with wildlife corridor design in conservation planning. *Landscape and Urban Planning* 156:44–58.
Pieterse, E. 2011. Recasting urban sustainability in the South. *Development* 54:309–316.
Ramirez-Rubio, O., C. Daher, G. Fanjul, M. Gascon, N. Mueller, L. Pajín, A. Plasencia, D. Rojas-Rueda, M. Thondoo, M.J. Nieuwenhuijsen. 2019. Urban health: An example of a 'health in all policies' approach in the context of SDGs implementation. *Global Health* 15(1):87.
Raymond, C.M., N. Frantzeskaki, N. Kabisch, P. Berry, M. Breil, M.R. Nita, D. Geneletti, C. Calfapietra. 2017. A framework for assessing and implementing the co-benefits of nature-based solutions in urban areas. *Environmental Science & Policy* 77:15–24.
Real Estate Research Corporation. 1974. *The Costs of Sprawl: Detailed Cost Analysis*. U.S. Government Printing Office, Washington, DC.
Ricketts, T.H. 2001. The matrix matters: Effective isolation in fragmented landscapes. *The American Naturalist* 158:87–99.
Saura, S., J. Torné. 2009. Conefor Sensinode 2.2: A software package for quantifying the importance of habitat patches for landscape connectivity. *Environmental Modelling & Software* 24:135–139.
Schmidt, K.J., H.H. Poppendieck, K. Jensen. 2014. Effects of urban structure on plant species richness in a large European city. *Urban Ecosystems* 17:427–444.
Seto, K.C., M. Fragkias, B. Güneralp, M.K. Reilly. 2011. A meta-analysis of global urban land expansion. *PLoS ONE* 6(8):e23777.
Seto, K.C., A. Reenberg. 2014. *Rethinking Global Land Use in an Urban Era*. Strüngmann Forum Reports, Vol. 14. MIT Press, Cambridge, MA.
Seto, K.C., A. Reenberg, C.G. Boone, M. Fragkias, D. Haase, T. Langanke, P. Marcotullio, D.K. Munroe, B. Olah, D. Simon. 2012. Teleconnections and sustainability: New conceptualizations of global urbanization and land change. *PNAS* 109(20):7687–7692.
Steffen, W., W. Broadgate, L. Deutsch, O. Gaffney, C. Ludwig. 2015. The trajectory of the Anthropocene: The great acceleration. *The Anthropocene Review* 2(1):81–98.
Travisi, C.M., R. Camagni, P. Nijkamp. 2010. Impacts of urban sprawl and commuting: A modelling study for Italy. *Journal of Transport Geography* 18:382–392.

UN. 2011. *Urban Patterns for Sustainable Development: Towards a Green Economy*. UN Habitat, Nairobi, Kenya.

UN. 2015. *Transforming Our World: The 2030 Agenda for Sustainable Development*. United Nations, New York. https://sustainabledevelopment.un.org/post2015/transformingourworld/publication

UN. 2019. *World Urbanization Prospects: The 2019 Revision*. United Nations, Department of Economic and Social Affairs/Population Division, New York.

UN Habitat. 2011. *Cities and Climate Change: Policy Directions. United Nations Human Settlements Programme*. Earthscan, London.

UN Habitat. 2016. *Goal 11: Make Cities Inclusive, Safe, Resilient and Sustainable*. www.un.org/sustainabledevelopment/cities/

Van der Jagt, A., M. Smith, B. Ambrose-Oji, C.C. Konijnendijk, V. Giannico, D. Haase, R. Lafortezza, M. Nastran, M. Pintar, Š. Železnikar, R. Cvejić. 2018. Co-creating urban green infrastructure connecting people and nature: A guiding framework and approach. *Journal of Environmental Management* 233:757–767.

Wang, M., J. Li, S. Kuang, Y. He, G. Chen, Y. Huang, C. Song, P. Anderson, D. Lowicki. 2020. Plant diversity along the urban-rural gradient and its relationship with urbanization degree in Shanghai, China. *Forests* 11:171.

Wellmann, T., D. Haase, S. Knapp, C. Salbach, P. Selsam, A. Lausch. 2018. Urban land use intensity assessment: The potential of spatio-temporal spectral traits with remote sensing. *Ecological Indicators* 85:190–203.

Wilson, B., A. Chakraborty. 2013. The environmental impacts of sprawl: Emergent themes from the past decade of planning research. *Sustainability* 5:3302–3327.

Wu, J. 2010. Urban sustainability: An inevitable goal of landscape research. *Landscape Ecology* 25:1–4.

Zetterberg, A., U. Mörtberg, B. Balfors. 2010. Making graph theory operational for landscape ecological assessments, planning, and design. *Landscape and Urban Planning* 95:181–191.

Zhang, Z., S. Meerow, J. Newell, M. Lindquist. 2019. Enhancing landscape connectivity through multifunctional green infrastructure corridor modelling and design. *Urban Forestry & Urban Greening* 38:305–317.

Zhao, J., T. Tang, X. Wang. 2020. Effects of landscape composition on mosquito population in urban green spaces. *Urban Forestry & Urban Greening* 49:126626.

PART 2

Foundational concepts and theory in urban biodiversity research

6

THE APPLICATION OF LANDSCAPE ECOLOGY AND BIOGEOGRAPHY IN THE STUDY OF URBAN BIODIVERSITY

Christopher A. Lepczyk and Frank A. La Sorte

Introduction

Landscape ecology and biogeography both use a geographic approach to understanding ecological phenomena. Landscape ecology is traditionally defined as the effect of pattern on process (Turner 1989; Turner 2005a). That is, landscape ecology is primarily focused on spatial locations and spatial relationships between ecological objects. On the other hand, biogeography focuses on understanding the spatial distribution of species and ecosystems within geographic space (Cox et al. 2019). Both fields have a strong focus on understanding how species, patterns, and processes change over time and across space.

The spatiotemporal view of ecology at the heart of both landscape ecology and biogeography are of particular relevance to urban ecology in general and urban biodiversity specifically. For instance, the spatial arrangement of habitat patches within a city can be critical for understanding the biodiversity present. Likewise, latitude is important for understanding how biodiversity varies across different cities (Aronson et al. 2014). While not a firm rule, we tend to see a landscape ecology perspective used for understanding biodiversity within cities while we see a biogeographical perspective used for understanding biodiversity across cities. In other words, landscape ecology treats cities as replicates when assessing aspects of alpha diversity, and biogeography examines spatial gradients among cities in species turnover or beta diversity. Notably, as we will see, the two fields are inherently intertwined, and both provide valuable tools for understanding urban biodiversity. Thus, while we provide an overview of the two fields, it is important to keep in mind that for many urban biodiversity questions, we use tools and theory from both hand-in-hand.

Landscape ecology

What constitutes a landscape has long been an issue of note in landscape ecology (e.g. Wiens 1992) as different branches of the field and different regions of the world have used different definitions, particularly between Europe and North America (Wu and Hobbs 2007). For instance, landscape architects, cultural geography, and many European nations consider a landscape to be a spatially large area that is essentially defined from a human perspective. On the other hand,

 DOI: 10.4324/9781003016120-8

ecology considers landscapes to be scale independent (Turner 2005b; Lepczyk et al. 2008), which is the way we will consider them with respect to urban biodiversity. Notably, much of urban ecology considers the city area to be the bounds of the urban landscape, and as such, urban ecosystems and urban landscapes often are considered within the same physical space, even though cities may house numerous ecosystems and landscapes.

While landscapes vary in size, they all constitute an area in which pattern is affecting process. However, how that pattern and process is described and evaluated depends upon the framework used to describe the landscape. Specifically, we can consider landscapes from the *patch-matrix* framework, which describes landscapes as composed of discrete objects that have spatial relationships to one other (Forman and Godron 1986), or the *gradient* framework which describes landscapes as continuous gradients (McGarigal and Cushman 2005; Lepczyk et al. 2021). These two frameworks are similar to the geographic ideas of object versus field views of the world. Notably, there are tradeoffs between these two models, with each offering beneficial reasons for use (Lausch et al. 2015). However, while there are many valid reasons for using a gradient framework in urban ecology, what we see in reality is that most urban biodiversity research uses the patch-matrix framework (Figure 6.1), which is the focus of this chapter. That said, it is important to keep in mind that LIDAR and other three-dimensional sensors and data fusion approaches are increasingly being employed in cities to measure building and vegetation heights and volumes (Lepczyk et al. 2021), some of which often use a gradient framework.

The patch-matrix framework considers a landscape to be composed of a series of three landscape elements: *patch*, *corridor*, and *matrix* (Forman and Godron 1986). Specifically, a patch is a relatively discrete area of relatively homogeneous environmental conditions, with boundaries

Figure 6.1 Patch, corridor, and matrix elements of urban landscapes. Hamilton, NZ, Google Earth (2016).

distinguishable or relevant to the object of interest. Likewise, a corridor is a narrow, linear element of a particular type that differs from the areas on the adjacent sides. Finally, the matrix can be considered as either: (1) the most common or connected landscape element type that generally plays the dominant role in landscape function, or (2) the background cover type in a landscape that is very extensive and connected. Notably, a landscape need not contain all of these elements, but it must contain pattern, process, and area (i.e. it must be bounded; Lepczyk et al. 2008).

One of the values of describing a landscape based on the three elements is that it allows for easy quantification of a wide variety of *landscape metrics*. These metrics are often considered to be either patch-level or landscape-level, referring to whether they are describing aspects of patches specifically or the landscape overall. The metrics range from simple, such as the perimeter of a patch, to complex, such as fractal dimension. Because most metrics are just geometric measures, they are often correlated with one another (Riitters et al. 1995). In fact, there are relatively few unique metrics, and it is critical to consider specific metrics in the experimental design of the research (e.g. Flather and Sauer 1996) rather than simply running all metrics after the fact, as was the approach used in the early history of landscape analysis. While there is no single set of metrics used across the board for describing urban biodiversity, it is quite common to see a handful used regularly. This handful includes patch area, core area, perimeter (i.e. edge), shape, and isolation.

Describing the landscape in terms of both elements and a series of metrics allows for ecological data, including biodiversity metrics, to be analyzed in relation to the landscape pattern. For instance, biodiversity data can be analyzed in a spatially explicit manner that allows for specific measures, such as isolation, distance, etc., to be determined. Likewise, species data are commonly integrated with landscape data to evaluate population dynamics, including meta-populations, as well as community dynamics. Such analyses are particularly valuable where landscapes themselves can be measured over time to describe landscape change.

How urban landscapes affect biodiversity

While cities are often considered novel ecosystems, they vary markedly in the type and amounts of different land cover types and land uses present in them. Foremost amongst the land cover types that influence biodiversity are those that contain critical habitat elements, which can range from remnants habitat patches through completely artificial patches, such as green roofs (Aronson et al. 2017), but which in general are those places containing vegetation or water. Hence, urban biodiversity often correlates with the amount of area containing vegetation (or green space) and water (or blue space) in the city. The same situation holds true in the context of patches, with larger green space patches housing a greater richness of species (e.g. La Sorte et al. 2020). Other types of land cover within cities do contain species, such as streetscapes, buildings, and roads, but barring the presence of habitat attributes that support their persistence, they tend not to harbor much biodiversity.

The size and number of patches providing habitat to species are often amongst the main factors influencing the richness and abundance of species. However, patch shape, perimeter (i.e. edge), and core area, as well as interpatch distance and landscape context, also can influence both the richness and abundance of species. For instance, if two patches are too far apart from one another and have a matrix comprised of unsuitable habitat, dispersal between them may be unsuccessful. Ultimately, being able to assess the amount, location, and arrangement of different types of land cover and habitats in a spatially explicit manner is critical for understanding a city's biodiversity.

Importance of urban landscapes for maintaining biodiversity

Cities can house a wealth of species, including those that are rare, threatened, and endangered (Ives et al. 2016). As a result, cities can be viewed as sanctuaries for biodiversity (Soanes and Lentini 2019; Lepczyk et al. 2023). Hence, maintaining green and blue spaces and understanding how they occur across the urban landscape are critical for biodiversity. In particular, understanding how green and blue spaces (i.e. patches) are spatially distributed within cities, what type of land use surrounds them (i.e. matrix), and if corridors exist that connect them are all aspects that can influence biodiversity (Lepczyk et al. 2017). For instance, corridors are important for maintaining species richness (Beninde et al. 2015) and providing pathways for movement (Hale et al. 2012) and gene flow (Munshi-South 2012). Many cities have the potential to provide more habitat for species, and it is necessary to use the tools of landscape ecology in conjunction with conservation planning to identify gaps in connectivity, habitat type, etc., for developing restoration efforts and management plans. Such efforts could further improve the biodiversity found in cities.

Aside from understanding where habitat is located and what may be available for restoration, an aspect that dovetails these issues is how people intentionally and unintentionally (see Andrade et al. and Kendall et al., this volume) interact with the various land covers present in the city. People's perception of the landscape and how they engage with it will have direct repercussions for which species are present and in what numbers. Specifically, within cities, many homeowners and renters have the ability to directly interact with the land around them in ways that can alter both the biotic and the abiotic components that influence biodiversity (Lepczyk et al. 2004). For instance, recent work has demonstrated that when and how often people mow their lawns has a direct effect on species composition (Lerman et al. 2018) and that the percent of a yard planted with ornamentals can affect avian population dynamics (Narango et al. 2018). Furthermore, when multiple homeowners that manage adjacent land parcels engage in similar activities or hold similar views, their actions can affect large portions of the urban landscape (Goddard et al. 2010). Whether or not such activities occur at different spatial scales within the city (e.g. parcel, neighborhood) is also governed by local policies, including homeowner associations (HOAs) and zoning. As a result of these factors, using systems approaches, such as socioecological systems, that can integrate property owners and policy into a landscape ecology framework are important for determining how urban landscapes maintain biodiversity.

Biogeography

Biogeography applies a geographic perspective to the study of species and ecosystems, often with consideration on how these patterns have developed over geological time (Lomolino et al. 2016). The field initially focused on how ecological patterns and evolutionary processes varied as a function of geographic gradients. Classic examples include gradients of latitude, elevation, isolation, or area with variation in species richness along these gradients often used as the ecological measure. A major subfield to emerge from biogeography considers how natural populations are defined within and among insular systems. The theories underlying insular biogeography were originally developed for oceanic islands (MacArthur and Wilson 1963, 1967). These initial theories were subsequently expanded to consider any geographically isolated population or ecosystem, including anthropogenically isolated systems (Warren et al. 2015).

Until recently, the effects of human activities on species and ecosystems at geographical spatial scales were thought to be negligible (Gaston 2004). The prevailing notion was that any effects related to human activities would be overwhelmed by geographical patterns that had been shaped and molded over millions of years of geological and evolutionary history. As the impact of human activities on natural systems increased in intensity and extent across the globe (Ellis 2015), this notion was no longer viable, and biogeographers began to address the potential that the species assemblages they were studying were no longer pristine but were being increasingly shaped by human activities (Young 2014). With this growing realization, biogeographical research developed a stronger anthropogenic perspective, first applied at regional scales (e.g. La Sorte 2006; Ficetola and Padoa-Schioppa 2009) and then increasingly at global scales (e.g. Capinha et al. 2020; Triantis et al. 2022). A central goal of these efforts is to determine how different forms of land use change, including urbanization, have affected species distributions and numbers across broad geographic regions.

Urban biogeography

Urbanization dramatically alters the physical and natural environment, and these processes are intensifying worldwide. Currently, urban areas cover less than 2% of the Earth's surface (Zhou et al. 2018). This quantity is projected to increase 200% by 2100 (Li et al. 2019), resulting in the loss of 11–33 million hectares of natural habitat (Li et al. 2022). We are also witnessing the formation of multiple megacities, which are defined as cities containing greater than 10 million people (United Nations 2019). These global urbanization trends require the application of a broader spatial perspective when studying urbanization and its environmental and ecological implications.

Urban areas are unique in that they are insular systems that contain very similar anthropogenic environments and act as major focal points for species extirpations and human mediated species introductions. As cities grow in extent and density, trade and commerce also expand, further enhancing the number and diversity of non-native species introductions, which can lead to biological invasions. The field of urban ecology initially focused on ecological patterns within individual cities, often based on surveys conducted along urban-rural gradients (McDonnell and Pickett 1990). These efforts have grown over time to consider data compiled across multiple cities, first for cities located within specific countries or continents, and then for cities across continents, and then for cities globally (Rega-Brodsky et al. 2022). This expanding geographic perspective has created novel opportunities for urban ecologists to document the ecological implications of urbanization across a diverse array of environmental and anthropogenic gradients.

Studying the ecology of urban environments typically considers two patterns that can be examined individually or in combination. The first considers species richness within individual cities (alpha diversity), and the second considers turnover in species composition among cities (beta diversity). These patterns can be examined at a taxonomic, phylogenic, or functional levels. Birds and plants tend to have the greatest volume of data within urban environments, and this has driven much of the early biogeographical research. Examples based on data collected from urban-rural gradients or within and outside of urban areas includes investigations of the effects of urbanization on avian functional diversity (Sol et al. 2020), phylogenetic diversity (Sol et al. 2017), evolutionary uniqueness (Ibáñez-Álamo et al. 2017), invasion success and tolerance (González-Lagos et al. 2021), and seasonal species composition (Leveau et al. 2021). Examples of applications using citywide surveys include investigations of alpha and beta diversity of native and non-native plants and vertebrates among cities within and between continents (La Sorte et al. 2007; Winter et al. 2010). Regional estimates of species richness and species compositive have also been used as a baseline to assess how citywide estimates of taxonomic, phylogenetic, and functional diversity of birds and plants have been affected by urbanization (Aronson et al. 2014; La Sorte et al. 2018). Data for other taxa are increasingly becoming available within urban environments. Examples include data for mammals (Magura et al. 2021), terrestrial insects (Fenoglio et al. 2021), terrestrial arthropods (Fenoglio et al. 2020), reptiles (Brum et al. 2023), and amphibians (Cordier et al. 2021; Liu et al. 2021).

Other broad-scale data sources are emerging that can further support research in urban biogeography. These data sources include large-scale ecological dataset acquired through citizen science initiatives (Dickinson et al. 2010) and the development of globally comprehensive datasets on species' traits, characteristics, and behaviors, such as those that have been developed for birds (Tobias 2022) and plants (Fraser 2020). Other examples include the formation of ecological data depositories, such as the Global Biodiversity Information Facility (www.gbif.org/), the Urban Biodiversity Research Coordination Network (https://sites.rutgers.edu/urbionet/), the Projecting Responses of Ecological Diversity In Changing Terrestrial Systems (PREDICTS) project (www.nhm.ac.uk/our-science/our-work/biodiversity/predicts.html), and the Global Rural-Urban Mapping Project (GRUMP; https://sedac.ciesin.columbia.edu/data/collection/grump-v1). The proliferation of high-quality environmental data from remote sensing platforms has also helped propel research in urban biogeography. Recent examples that have leveraged these emerging resources include the use of bird occurrence data from the eBird citizen-science platform (https://ebird.org/home) to document the effects of light pollution on the seasonal species richness of nocturnally migrating birds in urban areas (La Sorte and Horton 2021), and the use of data from the PREDICTS project to assess the effects of land-use change on vertebrate functional diversity (Etard et al. 2022). In total, these resources are allowing urban ecologists to address questions at increasingly broader spatial extents with greater taxonomic and spatiotemporal detail and breadth. These efforts will inevitably expand our understanding of the current and future implications of urbanization for the world's natural systems, creating opportunities for more informed urban planning and more robust mitigation efforts designed to address the negative consequences of urbanization for species and ecosystems.

Future research needs

Both landscape ecology and biogeography have greatly aided our understanding of urban biodiversity. However, there are several areas where additional effort needs to be placed or where gaps in our knowledge exist. In looking towards the future, the following questions and topics are of particular importance to both fields and urban biodiversity:

1 *Data shortfalls.* With the increasing availability of spatially detailed data on both species and the Earth's surface, the ability to address and test questions pertaining to urban biodiversity is growing markedly. However, there remain gaps in species data for many cities outside of temperate regions. Collecting species data from poorly sampled regions of the world will be critical to test landscape and biogeographic theory and whether patterns found in cities at present are ubiquitous to cities everywhere or not. For instance, do landscape patterns in cities operate the same to filter the regional species pool? Likewise, within cities, many types of taxonomic data are poorly represented (Rega-Brodsky et al. 2022), particularly species that comprise belowground biodiversity.

2 *Three-dimensional landscape ecology.* The growing use of LIDAR, next-generation remote sensors, and the ability to fuse these datasets has allowed for the rapid expansion of three-dimensional (3D) landscape metrics and analyses to be developed that are being used in both terrestrial and aquatic systems (Lepczyk et al. 2021). While 3D building structure is already being included in many urban landscape ecology studies, the use of 3D vegetation data is still underutilized. Given the importance of vegetation volume in providing habitat (MacArthur and

MacArthur 1961), our understanding of this aspect remains underutilized and underdeveloped in cities.

3 *Migratory species*. The human alteration of the planet has had notable impacts on species migratory behavior. While artificial lights at night (ALAN) are notable for disrupting nighttime migrations of birds, we have only just started to evaluate how cities influence migration and, in particular, how this may vary in different regions of the world and at different times of the year.

4 *Synergistic effects of pollution*. While ALAN affects nighttime migration, it is also notable in disrupting behavior of resident species in cities (Pothukuchi 2021). Likewise, recent work evaluating air pollution (La Sorte et al. 2022) and noise pollution (Perillo et al. 2017) have demonstrated effects on species richness. However, how different sources of environmental pollution may act synergistically on species and ecosystems that occur within and near urban environments has only started to be evaluated, and there are no analyses that seek to disentangle the various effects of pollution on urban biodiversity.

5 *Recommendations for urban planning*. Urban planning has been integrating ecological principles into guidelines for decades. However, there is a need to develop comprehensive and reliable recommendations for urban planning that will support species survival and fitness and ecosystem function while reducing the chances of promoting population sinks (Lepczyk et al. 2017).

Conclusions

Landscape ecology and biogeography provide critical theories and tools for considering ecological phenomena from different spatiotemporal perspectives. In the case of urban biodiversity, these perspectives are critical for understanding patterns and processes both within a city and across cities. The marked increase in both species and remotely sensed surface data at fine spatial grains over time is allowing for new and exciting opportunities to understand how urban biodiversity varies both within and across cities and the factors shaping it. Furthermore, as we increasingly use systems approaches that bring together ecological and social data, we are gaining insights into the spatial variability of factors influencing urban biodiversity. However, there remain many gaps in the data needed to answer key questions if we are to use the principles of landscape ecology and biogeography for advancing urban planning in ways that further the maintenance of biodiversity.

Key companion papers

Ibáñez-Álamo, J.D., E. Rubio, Y. Benedetti, F. Morelli. 2017. Global loss of avian evolutionary uniqueness in urban areas. *Global Change Biology* 23:2990–2998.

La Sorte, F.A., M.F.J. Aronson, C.A. Lepczyk, K.G. Horton. 2020. Area is the primary correlate of annual and seasonal patterns of avian species richness in urban green spaces. *Landscape and Urban Planning* 203:103892.

Lepczyk, C.A., et al. 2017. Biodiversity in the city: Fundamental questions for understanding the ecology of urban green spaces for biodiversity conservation. *BioScience* 67(9):799–807.

Lepczyk, C.A., et al. 2021. Advancing landscape and seascape ecology from a 2D to 3D science. *BioScience* 71(6):596–608.

Leveau, L.M., J. Jokimäki, M.-L. Kaisanlahti-Jokimäki. 2021. Urbanization buffers seasonal change in composition of bird communities: A multi-continental meta-analysis. *Journal of Biogeography* 48:2391–2401.

Sol, D., et al. 2020. The worldwide impact of urbanisation on avian functional diversity. *Ecology Letters* 23:962–972.

Turner, M.G. 1989. Landscape ecology: The effect of pattern on process. *Annual Review of Ecology and Systematics* 20:171–197.

References

Aronson, M.F.J., F.A. La Sorte, C.H. Nilon, M. Katti, M.A. Goddard, C.A. Lepczyk, P.S. Warren, N.S.G. Williams, S. Cilliers, B. Clarkson, C. Dobbs, R. Dolan, M. Hedblom, S. Klotz, J.L. Kooijmans, I. Kühn, I. MacGregor-Fors, M. McDonnell, U. Mörtberg, P. Pyšek, S. Siebert, J. Sushinsky, P. Werner, M. Winter. 2014. A global analysis of the impacts of urbanization on bird and plant diversity reveals key anthropogenic drivers. *Proceedings of the Royal Society B: Biological Sciences* 281(1780):20133330.

Aronson, M.F.J., C.A. Lepczyk, K.L. Evans, M.A. Goddard, S.B. Lerman, J.S. MacIvor, C.H. Nilon, T. Vargo. 2017. Biodiversity in the city: Key challenges for urban green space management. *Frontiers in Ecology and the Environment* 15:189–196.

Beninde, J., M. Veith, A. Hochkirch. 2015. Biodiversity in cities needs space: A meta-analysis of factors determining intra-urban biodiversity variation. *Ecology Letters* 18(6):581–592.

Brum, P.H.R., S.R.A. Gonçalves, C. Strüssmann, A.L. Teixido. 2023. A global assessment of research on urban ecology of reptiles: Patterns, gaps and future directions. *Animal Conservation* 26:1–13.

Capinha, C., F. Marcolin, L. Reino. 2020. Human-induced globalization of insular herpetofaunas. *Global Ecology and Biogeography* 29(8):1328–1349.

Cordier, J.M., R. Aguilar, J.N. Lescano, G.C. Leynaud, A. Bonino, D. Miloch, R. Loyola, J. Nori. 2021. A global assessment of amphibian and reptile responses to land-use changes. *Biological Conservation* 253:108863.

Cox, C.B., R.J. Ladle, P.D. Moore, editors. 2019. *Biogeography: An Ecological and Evolutionary Approach*, 10th edition. John Wiley & Sons, Inc., Oxford, Oxfordshire.

Dickinson, J.L., B. Zuckerberg, D.N. Bonter. 2010. Citizen science as an ecological research tool: Challenges and benefits. *Annual Review of Ecology, Evolution, and Systematics* 41:149–172.

Ellis, E.C. 2015. Ecology in an anthropogenic biosphere. *Ecological Monographs* 85(3):287–331.

Etard, A., A.L. Pigot, T. Newbold. 2022. Intensive human land uses negatively affect vertebrate functional diversity. *Ecology Letters* 25(2):330–343.

Fenoglio, M.S., A. Calviño, E. González, A. Salvo, M. Videla. 2021. Urbanisation drivers and underlying mechanisms of terrestrial insect diversity loss in cities. *Ecological Entomology* 46(4):757–771.

Fenoglio, M.S., M.R. Rossetti, M. Videla. 2020. Negative effects of urbanization on terrestrial arthropod communities: A meta-analysis. *Global Ecology and Biogeography* 29(8):1412–1429.

Ficetola, G.F., E. Padoa-Schioppa. 2009. Human activities alter biogeographical patterns of reptiles on Mediterranean islands. *Global Ecology and Biogeography* 18(2):214–222.

Flather, C.H., J. R. Sauer. 1996. Using landscape ecology to test hypotheses about large scale abundance patterns in migratory birds. *Ecology* 77:28–35.

Forman, R.T.T., M. Godron, editors. 1986. *Landscape Ecology*. Cambridge University Press, Cambridge.

Fraser, L.H. 2020. TRY – A plant trait database of databases. *Global Change Biology* 26(1):189–190.

Gaston, K.J. 2004. Macroecology and people. *Basic and Applied Ecology* 5:303–307.

Goddard, M.A., A.J. Dougill, T.G. Benton. 2010. Scaling up from gardens: Biodiversity conservation in urban environments. *Trends in Ecology & Evolution* 25(2):90–98.

González-Lagos, C., L. Cardador, D. Sol. 2021. Invasion success and tolerance to urbanization in birds. *Ecography* 44(11):1642–1652.

Hale, J.D., A.J. Fairbrass, T.J. Matthews, J.P. Sadler. 2012. Habitat composition and connectivity predicts bat presence and activity at foraging sites in a large UK conurbation. *PLoS ONE* 7(3):e33300.

Ibáñez-Álamo, J.D., E. Rubio, Y. Benedetti, F. Morelli. 2017. Global loss of avian evolutionary uniqueness in urban areas. *Global Change Biology* 23(8):2990–2998.

Ives, C.D., P.E. Lentini, C.G. Threlfall, K. Ikin, D.F. Shanahan, G.E. Garrard, S.A. Bekessy, R.A. Fuller, L. Mumaw, L. Rayner, R. Rowe, L.E. Valentine, D. Kendal. 2016. The importance of cities for threatened species. *Global Ecology and Biogeography* 25(1):117–126.

La Sorte, F.A. 2006. Geographical expansion and increased prevalence of common species in avian assemblages: Implications for large-scale patterns of species richness. *Journal of Biogeography* 33:1183–1191.

La Sorte, F.A., M.F.J. Aronson, C.A. Lepczyk, K.G. Horton. 2020. Area is the primary correlate of annual and seasonal patterns of avian species richness in urban green spaces. *Landscape and Urban Planning* 203:103892.

La Sorte, F.A., K.G. Horton. 2021. Seasonal variation in the effects of artificial light at night on the occurrence of nocturnally migrating birds in urban areas. *Environmental Pollution* 270:116085.

La Sorte, F.A., C.A. Lepczyk, M.F.J. Aronson. 2022. Light pollution enhances exposure to airborne toxic chemicals for nocturnally migrating passerines. *Global Change Biology*. https://doi.org/10.1111/gcb.16443

La Sorte, F.A., C.A. Lepczyk, M.F.J. Aronson, M.A. Goddard, M. Hedblom, M. Katti, I. MacGregor-Fors, U. Mörtberg, C.H. Nilon, P.S. Warren, N.S.G. Williams, J. Yang. 2018. The phylogenetic and functional diversity of regional breeding bird assemblages is reduced and constricted through urbanization. *Diversity and Distributions* 24(7):928–938.

La Sorte, F.A., M.L. McKinney, P. Pyšek. 2007. Compositional similarity among urban floras within and across continents: Biogeographical consequences of human-mediated biotic interchange. *Global Change Biology* 13(4):913–921.

Lausch, A., T. Blaschke, D. Haase, F. Herzog, R.-U. Syrbe, L. Tischendorf, U. Walz. 2015. Understanding and quantifying landscape structure – A review on relevant process characteristics, data models and landscape metrics. *Ecological Modelling* 295(1):31–41.

Lepczyk, C.A., M.F.J. Aronson, K.L. Evans, M.A. Goddard, S.B. Lerman, J.S. MacIvor. 2017. Biodiversity in the city: Fundamental questions for understanding the ecology of urban green spaces for biodiversity conservation. *BioScience* 67(9):799–807.

Lepczyk, C.A., M.F.J. Aronson, F.A. La Sorte. 2023. Cities as sanctuaries. *Frontiers in Ecology and the Environment* 21:251–259.

Lepczyk, C.A., C.J. Lortie, L. Anderson. 2008. An ontology of landscapes. *Ecological Complexity* 5(3):272–279.

Lepczyk, C.A., A.G. Mertig, J. Liu. 2004. Assessing landowner activities that influence birds across rural-to-urban landscapes. *Environmental Management* 33(1):110–125.

Lepczyk, C.A., L.M. Wedding, G.P. Asner, S.J. Pittman, T. Goulden, M. Linderman, J. Gang, R. Wright. 2021. Advancing landscape and seascape ecology from a 2D to 3D science. *BioScience* 71(6):596–608.

Lerman, S.B., A.R. Contosta, J. Milam, C. Bang. 2018. To mow or to mow less: Lawn mowing frequency affects bee abundance and diversity in suburban yards. *Biological Conservation* 221:160–174.

Leveau, L.M., J. Jokimäki, M.-L. Kaisanlahti-Jokimäki. 2021. Urbanization buffers seasonal change in composition of bird communities: A multi-continental meta-analysis. *Journal of Biogeography* 48:2391–2401.

Li, G., C. Fang, Y. Li, Z. Wang, S. Sun, S. He, W. Qi, C. Bao, H. Ma, Y. Fan, Y. Feng, X. Liu. 2022. Global impacts of future urban expansion on terrestrial vertebrate diversity. *Nature Communications* 13:1628.

Li, X., Y. Zhou, J. Eom, S. Yu, G.R. Asrar. 2019. Projecting global urban area growth through 2100 based on historical time series data and future shared socioeconomic pathways. *Earth's Future* 7(4):351–362.

Liu, G., J.J.L. Rowley, R.T. Kingsford, C.T. Callaghan. 2021. Species' traits drive amphibian tolerance to anthropogenic habitat modification. *Global Change Biology* 27(13):3120–3132.

Lomolino, M.V., B.R. Riddle, R.J. Whittaker, editors. 2016. *Biogeography*, 5th edition. Oxford University Press, Oxford.

MacArthur, R.H., J.W. MacArthur. 1961. On bird species diversity. *Ecology* 42:594–598.

MacArthur, R.H., E.O. Wilson. 1963. An equilibrium theory of insular zoogeography. *Evolution* 17(4):373–387.

MacArthur, R.H., E.O. Wilson, editors. 1967. *The Theory of Island Biogeography*. Princeton University Press, Princeton, NJ.

Magura, T., E. Kiss, G.L. Lövei. 2021. No consistent diversity patterns in terrestrial mammal assemblages along rural-urban forest gradients. *Basic and Applied Ecology* 52:38–45.

McDonnell, M.J., S.T.A. Pickett. 1990. Ecosystem structure and function along urban rural gradients – An unexploited opportunity for ecology. *Ecology* 71(4):1232–1237.

McGarigal, K., S.A. Cushman. 2005. The gradient concept of landscape structure. Pages 112–119 in J. Wiens, M. Moss, editors. *Issues and Perspectives in Landscape Ecology*. Cambridge University Press, Cambridge.

Munshi-South, J. 2012. Urban landscape genetics: Canopy cover predicts gene flow between white-footed mouse (*Peromyscus leucopus*) populations in New York City. *Molecular Ecology* 21(6):1360–1378.

Narango, D.L., D.W. Tallamy, P.P. Marra. 2018. Nonnative plants reduce population growth of an insectivorous bird. *Proceedings of the National Academy of Sciences* 115:11549–11554.

Perillo, A., L.G. Mazzoni, L.F. Passos, V.D.L.R. Goulart, C. Duca, R.J. Young. 2017. Anthropogenic noise reduces bird species richness and diversity in urban parks. *Ibis* 159(3):638–646.

Pothukuchi, K. 2021. City light or star bright: A review of urban light pollution, impacts, and planning implications. *Journal of Planning Literature* 36(2):155–169.

Rega-Brodsky, C.C., M.F.J. Aronson, M.R. Piana, E.-S. Carpenter, A.K. Hahs, A. Herrera-Montes, S. Knapp, D.J. Kotze, C.A. Lepczyk, M. Moretti, A.B. Salisbury, N.S.G. Williams, K. Jung, M. Katti, I. MacGregor-Fors, J.S. MacIvor, F.A. La Sorte, V. Sheel, C.G. Threfall, C.H. Nilon. 2022. Urban biodiversity: State of the science and future directions. *Urban Ecosystems* 25:1083–1096.

Riitters, K.H., R.V. O'Neill, C.T. Hunsaker, J.D. Wickham, D.H. Yankee, S.P. Timmins, K.B. Jones, B.L. Jackson. 1995. A factor analysis of landscape pattern and structure metrics. *Landscape Ecology* 10:23–39.

Soanes, K., P.E. Lentini. 2019. When cities are the last chance for saving species. *Frontiers in Ecology and the Environment* 17(4):225–231.

Sol, D., I. Bartomeus, C. González-Lagos, S. Pavoine. 2017. Urbanisation and the loss of phylogenetic diversity in birds. *Ecology Letters* 20(6):721–729.

Sol, D., C. Trisos, C. Múrria, A. Jeliazkov, C. González-Lagos, A.L. Pigot, C. Ricotta, C.M. Swan, J.A. Tobias, S. Pavoine. 2020. The worldwide impact of urbanisation on avian functional diversity. *Ecology Letters* 23(6):962–972.

Tobias, J.A. 2022. A bird in the hand: Global-scale morphological trait datasets open new frontiers of ecology, evolution and ecosystem science. *Ecology Letters* 25(3):573–580.

Triantis, K.A., F. Rigal, R.J. Whittaker, J.P. Hume, C. Sheard, D. Poursanidis, J. Rolland, S. Sfenthourakis, T.J. Matthews, C. Thébaud, J.A. Tobias. 2022. Deterministic assembly and anthropogenic extinctions drive convergence of island bird communities. *Global Ecology and Biogeography* 31:1741–1755.

Turner, M.G. 1989. Landscape ecology: The effect of pattern on process. *Annual Review of Ecology and Systematics* 20:171–197.

Turner, M.G. 2005a. Landscape ecology in North America: Past, present and future. *Ecology* 86(8):1967–1974.

Turner, M.G. 2005b. Landscape ecology: What is the state of the science? *Annual Review of Ecology Evolution and Systematics* 36:319–344.

United Nations. 2019. *World Urbanization Prospects: The 2018 Revision (ST/ESA/SER.A/420)*. Department of Economic and Social Affairs, Population Division. United Nations, New York.

Warren, B.H., D. Simberloff, R.E. Ricklefs, R. Aguilée, F.L. Condamine, D. Gravel, H. Morlon, N. Mouquet, J. Rosindell, J. Casquet, E. Conti, J. Cornuault, J.M. Fernández-Palacios, T. Hengl, S.J. Norder, K.F. Rijsdijk, I. Sanmartín, D. Strasberg, K.A. Triantis, L.M. Valente, R.J. Whittaker, R.G. Gillespie, B.C. Emerson, C. Thébaud. 2015. Islands as model systems in ecology and evolution: Prospects fifty years after MacArthur-Wilson. *Ecology Letters* 18:200–217.

Wiens, J.A. 1992. What is landscape ecology, really? *Landscape Ecology* 7:149–150.

Winter, M., I. Kühn, F.A. La Sorte, O. Schweiger, W. Nentwig, S. Klotz. 2010. The role of non-native plants and vertebrates in defining patterns of compositional dissimilarity within and across continents. *Global Ecology and Biogeography* 19(3):332–342.

Wu, J., R. Hobbs. 2007. Landscape ecology: The state-of-the-science. Pages 271–287 in J. Wu, R. Hobbs, editors. *Key Topics in Landscape Ecology*. Cambridge University Press, Cambridge.

Young, K.R. 2014. Biogeography of the Anthropocene: Novel species assemblages. *Progress in Physical Geography: Earth and Environment* 38:664–673.

Zhou, Y., X. Li, G.R. Asrar, S.J. Smith, M. Imhoff. 2018. A global record of annual urban dynamics (1992–2013) from nighttime lights. *Remote Sensing of Environment* 219:206–220.

7

COMMUNITY ASSEMBLY AND FILTERING OF URBAN ECOSYSTEMS

Myla F.J. Aronson, Christopher A. Lepczyk, and Karl L. Evans

Introduction

A central question in ecology is what governs community structure, namely, the number of species, their identities, and the abundance of each of those species in the same place and at the same time (Cody and Diamond 1975; Morin 2011). In other words, why are ecological communities assembled the way they are? Attempts to answer this question typically have a strong focus on ecological filters, that is, how environmental conditions interact with species' traits to determine the suitability of that location for each of a number of potentially occurring species (Keddy 1992). These filters include abiotic and biotic factors that determine the dispersal of a species into a location and demographic traits (survival and reproduction) that control population growth rates and ultimately if a species can persist in that location and the population densities that result (Keddy 1992). Ecological filtering has a long and rich history in ecology, with a focus on the factors that shape local species assemblages relative to the global or regional pools of potential species (Morin 2011). A large body of the filter literature has focused on hypothesised theoretical frameworks (e.g. Webb et al. 2010), specific taxa (e.g. animals: Brashares 2010; plants: Funk 2021), or specific systems (e.g. the series of filters through which introduced species must pass before becoming established; Lockwood et al. 2013).

There has been an increasing focus on filters by researchers studying urban ecosystems (e.g. Williams et al. 2009; Evans et al. 2010; Aronson et al. 2016). Studying urban ecosystem filters is both complex and important due to the role of socioeconomic and cultural factors in determining the nature of biotic and abiotic filters (Aronson et al. 2016; Andrade et al. 2021; Avolio et al. 2021), coupled with the increasing numbers of people who live and work in cities and thus rely on urban biotas for delivery of ecosystem services. Given the advances in community assembly of urban ecosystems, particularly in relation to filters, our objectives in this chapter are to introduce urban community assembly, briefly discuss the principal filters in urban systems, how species traits interact with urban environments to determine communities, and important gaps in the literature.

 DOI: 10.4324/9781003016120-9

Urban community assembly

All community assembly frameworks pertaining to urban ecosystems assume that species are colonising urban areas from regional or global species pools (e.g. Williams et al. 2009; Evans et al. 2010; Aronson et al. 2016; Cadotte et al. 2017). These pools comprise a group of species that could arrive in urban areas via natural or human-assisted dispersal (La Sorte et al. 2007), either intentionally, such as horticultural plantings (van Kleunen et al. 2018) and naturalisation societies (e.g. *Hui Manu* bird societies in Hawaii; Berger 1974; Foster et al. 2009), or unintentionally, such as seed transport into urban areas from vehicles (Von der Lippe and Kowarik 2007). From these pools, species are predicted to pass through a series of filters, or sieves, if you will, in order to colonise a local site, and then survive and achieve a viable population size. Often these filters are viewed as existing within a hierarchical structure whereby a species first has to pass through filter 1 before being able to pass through filter 2, and so on (Aronson et al. 2016). Such an order is particularly pertinent in considering the set of filters that determine if a species arrives in urban areas, versus filters determining their ability to persist or even thrive there (Evans et al. 2010). Of course, some species are already present in habitat patches that become urbanised as development surrounds them or remain in seed banks after a patch was disturbed from development activities. Whilst such species will not need to overcome dispersal barriers to colonise from the regional pool into towns and cities, they do experience dispersal barriers within the city that influence the assemblages of local patches.

Filters are often categorised into two broad groups. In cities, abiotic filters, such as temperature, soil condition (moisture, nutrient content, etc.), and pollution (air, soil, water, noise) are particularly important. Biotic filters, including habitat features such as vegetation structural complexity, as well as the nature and intensity of competition and predation, also play a key role. Moreover, some filters such as urban form (i.e. the landscape pattern of the built environment, natural, and semi-natural habitats) constitute aspects of abiotic and biotic filters. In reality, these filters interact with each other; for example, urban areas with more vegetation are typically cooler (Vailshery et al. 2013). Heterogeneity in the intensity at which filters operate also arises across space as cities are highly heterogeneous at fine spatial scales (Norton et al. 2016). Of course, species traits (e.g. seed dispersal vector, body mass, wing span) and adaptive capacity (e.g. plasticity or evolution) influence which species, and how many individuals of these species, pass or fail to pass through these filters. While urban environments across the world share many characteristics, they are not identical, and thus the strength of a particular filter will vary from city to city. This variability can be imagined as variation in the size of a filter's mesh, with smaller gaps in the mesh representing stronger filters that block more species.

Notably, the main difference from traditional ecological community assembly theory and urban ecology is the predominant role that humans play in the filtering process. Human factors ultimately determine the type and intensity of filters within urban areas. These factors include current and historical land cover patterns, land use legacies, land and species management, and the socioeconomic and cultural characteristics that shape these factors.

What are the principal filters influencing community assembly in urban systems?

While there are multiple interacting filters that influence community assembly, some stand out in the literature as particularly important in determining urban communities. Here we briefly discuss those filters with the caveat that there are additional filters that may act on individual species and at different spatial and temporal scales.

Whether human assisted or natural, the ability of a species to arrive in a given location is paramount and is the result of landscape features (e.g. levels and types of habitat fragmentation, connectivity) and species dispersal traits. Dispersal is affected by natural processes, such as wind carrying a seed, and anthropogenic processes, such as the intentional introduction of an ornamental shrub into a yard or garden (Pearse et al. 2018), or the unintentional introduction of insect pests in wood-packing material (Meurisse et al. 2019). Notably, we can consider anthropogenic processes as having a strong grounding in the invasion ecology literature given its focus on dispersal and establishment (Lockwood et al. 2013).

One of the dominant filters on dispersal into and within cities and towns is urban form, that is, the way that built and non-built components are spatially arranged (see Dobbs et al. and Lepczyk and La Sorte, this volume). Specifically, urban form includes both two-dimensional considerations – such as building and urban green space density and connectedness, percent impervious surface, the area encompassed by specific land cover types – as well as three-dimensional factors, such as building height and volume (Lepczyk et al. 2021). Recent work indicates that colonisation of urban green spaces by native tropical birds is inhibited by the building height surrounding the green spaces, with higher buildings associated with less native bird species (Arévalo et al. 2022). Likewise, the way in which species move around the urban landscape is directly impacted by urban form and morphology. For example, long-distance seed dispersal of the tree-of-heaven (*Ailanthus altissma*) along road corridors in Berlin, Germany, was enabled by air flow generated from moving vehicles (Kowarik and von der Lippe 2011) and gene flow of the white-footed mouse (*Peromyscus leucopus*) in New York City was enhanced by linear corridors of forest cover (Munshi-South 2012). Most empirical tests of urban form on plant and animal dispersal have focused on general urban characteristics (e.g. predefined urban land use/cover, % impervious surface), roads and transportation networks, or connectivity of green spaces (Fusco et al. 2021). Notably, few studies focus on the 2D and 3D spatial configuration of these features in relation to each other. More studies are needed to understand how species navigate the urban landscape.

Once a species arrives at a potential habitat patch, there are then local scale filters that affect a species establishment and demographics. In urban habitats, temperature (in particular, urban heat island effects), pollution, and soil conditions appear to be the strongest abiotic filters across cities globally. For example, air pollution has long been known to influence lichen communities (Skye 1968), but research on how air pollution may influence the communities of other taxa is limited, even though it is a significant threat to bird communities across the world (La Sorte et al. 2022). Likewise, noise pollution has been shown to influence bird (Klingbeil et al. 2020) and amphibian communities (Zaffaroni-Caorsi et al. 2022). Soil conditions are particularly important for vegetation and soil communities – and can be significant filters on species persistence (Caruso et al. 2017; Tedersoo et al. 2020). As urban filters change the species composition and abundance of species, this often results in urban areas having different biotic interactions than rural ones (including changes in plant-herbivore, plant-pollinator, host-parasite, and prey-predator interactions) that in turn generates additional filters that determine urban community assembly.

At all spatial and temporal scales, human preferences, decisions, and behaviours act as strong filters on local community assemblages. Humans alter both stochastic processes, such as introductions and removals of plants, and deterministic processes, such as inputs of fertiliser, herbicides and pesticides, removal of dead organic matter, and altered soil structure (Lepczyk et al. 2004). Such actions have been shown to influence differences of both above and belowground arthropod composition and trait diversity (Birkhofer et al. 2017) and the abundance of caterpillars which affects bird population dynamics (Narango et al. 2018). People have strong preferences for how they view yards, green spaces, and the habitats that exist in urban ecosystems (see Kendal et al.

and Andrade et al., this volume). These preferences shape what species people both intentionally and unintentionally allow into urban green spaces. Such preferences can occur at a range of spatial scales, from individual parcels through neighborhoods or entire cities. Notably, preferences also act as barriers to colonisation and success with people making distinct choices on the species they remove and manage. People are the ultimate decision-makers and act as a final key filter in whether a species persists at a given site or not.

The importance of species traits in urban community assembly

Species' ecological, behavioural, and life history traits ultimately determine a species' ability to pass through the filters that form local communities. As humans, we typically observe the final results of this filtering process, rather than individual stages of the hierarchical filtering system. Consequently, much of the empirical work addressing how species' traits interact with filters takes a somewhat simplistic approach of assessing relationships between these traits and species presence/absence or abundance patterns in urban areas. Such analyses take into account the phylogenetic non-independence arising from species' shared evolutionary history. Numerous examples of such studies abound across a wide range of taxonomic groups, including plants (e.g. Knapp et al. 2008; Lopez et al. 2018); various invertebrate groups, particularly carabid beetles (e.g. Martinson and Raupp 2013); birds (e.g. La Sorte et al. 2018); and mammals (e.g. Jung and Threlfall 2018). Synthesising results across such a disparate array of studies is hindered by substantial variation in methodology. This variation includes how urban development is defined, differential focus on presence/absence, abundance, or relative density in urban versus non-urban areas (with Evans et al. 2011 advocating that relative density is the most robust approach), which types of traits are assessed, and how individual traits are measured (even within the same taxonomic group).

Despite this complexity, some rules of thumb are emerging from the literature. These generalities provide a useful guide for understanding which species from regional species pools are likely to occur, or even thrive, in urban assemblages. The most prominent rule of thumb is a species' position along the specialist-generalist continuum. Urban filtering studies have measured this trait using a variety of approaches ranging from the simple assumption (deduced from macroecological theory) that the size of a species' geographic distribution correlates positively with how generalist the species is (e.g. Bonier et al. 2007), to complex multiple dimensional definitions of niche breadth or position (e.g. how rare in the environment the resources used by a species are; Evans et al. 2011). Given the profound ecological changes imposed by urbanisation, it is perhaps no surprise that generalist species are often found to cope better with urban development than specialists (e.g. carabid beetles: Martinson and Raupp 2013; bees: Fournier et al. 2020; birds: Bonier et al. 2007; Evans et al. 2011; Callaghan et al. 2019; bats: Jung and Threlfall 2018, Threlfall and Jung, this volume). Notably, there is a lack of consistency regarding plant traits (Williams et al. 2015), but species that can colonise and persist in a variety of habitat types (i.e. annual, ruderal life histories) or that have high competitive ability (e.g. trees) and large niche breaths are often successful in cities (Williams et al. 2015).

Dispersal is another key trait that has frequently been associated with species persistence in urban environments. This association may be due to the likelihood that high dispersal capacity will facilitate a species initial arrival in urban area, re-colonisation, or immigration (facilitating persistence following a population crash), and movement among fragmented habitat patches. Robust empirical data on typical dispersal distances is rarely available for most species, not least because of a strong skew in distributions of recorded dispersal distances (with rare long distance dispersal events being difficult to document), and as dispersal varies with environmental conditions (e.g.

Hui et al. 2012). Dispersal capacity is thus often measured indirectly using information on the size and shape of morphological structures that enable dispersal. Studies taking these approaches across multiple taxa have typically found evidence that species with greater dispersal capacity (mobility) are more likely to be tolerant of urbanisation (e.g. carabid beetles: Martinson and Raupp 2013; bats: Jung and Threlfall 2018). Studies focusing on plants typically assume that dispersal capacity can be inferred from information on seed dispersal mode and often find that urban tolerant species are more likely to have greater long-range dispersal capacity (Aronson et al. 2007; Concepción et al. 2015). Furthermore, ornamental traits that promote species' selection for deliberate introductions of non-native species, such as large and colourful flowers (thus overcoming natural dispersal barriers) as well as nursery availability (Avolio et al. 2018), are also proposed to promote occurrence in urban assemblages (Williams et al. 2009; Kendal et al. 2012).

Theory suggests that numerous other traits could influence species' responses to urbanisation and have at least some empirical support. Such traits include tolerance of the warmer environments that arise from urban heat islands (e.g. *Anolis* lizards: Winchell et al. 2020; moths: Franzén et al. 2020; plants: Williams et al. 2015), behavioural flexibility which will facilitate adjustment to novel environments (e.g. birds: Callaghan et al. 2019; Sayol et al. 2020; mammals: Santini et al. 2019), higher pitched vocalisations that reduce the risk of masking from anthropogenic noise (e.g. birds: Francis et al. 2011), and reproductive investment (e.g. birds: Callaghan et al. 2019; Sayol et al. 2020; mammals: Santini et al. 2019; seed size: Duncan et al. 2011; Williams et al. 2015).

Consensuses regarding the relative importance of particular traits in enabling species to pass through urbanisation filters and the ability to generalise within and across taxa has yet to emerge. A paucity of sufficiently robust studies and contrasting methodological approaches are certainly contributing factors, but complexities arising from interactions among species traits (e.g. Aronson et al. 2007; Martinson and Raup 2013; Sayol et al. 2020) is also likely to play a role. Another contributing factor may be that studies focusing on single but contrasting taxonomic groups will sample a different section of trait space. As a simple example, avian species will typically have greater dispersal capacity than a similarly sized non-volant mammal. If traits have non-linear effects on species responses to urbanisation, sampling different sections of trait space could uncover apparently divergent trait-urbanisation relationships, leading to spurious generalisations. Furthermore, urban landscapes exhibit high habitat heterogeneity and the traits of some taxa, such as plants, are tied closely to habitat type and human influence, including historic land use, anthropogenic disturbances, and land management (Roman et al. 2018). For example, filters on forest remnant communities will be very different to those structuring yard communities, making these two not comparable for generalisations on urban-wide successful traits.

Consequences of trait filtering

The filtering of species by urban selection pressures according to their traits has consequences for the functional diversity (i.e. the diversity of traits describing the ecological function of species) of urbanised assemblages. Specifically, it is hypothesised that filtering will reduce the variety of traits represented in urban assemblages and thus functional diversity, relative to more natural areas lacking urban development. There is strong evidence that avian functional diversity in cities is lower than that of the regional pool (La Sorte et al. 2018) and that functional diversity declines along gradients of increasing urbanisation intensity (Sol et al. 2020). This loss may be primarily driven by a reduction in species richness, particularly evolutionarily distinct species, and increasing dominance of a small number of well-adapted species in the urban assemblages (La Sorte et al.

2018; Sol et al. 2020). In situations where, at city scales, urban functional diversity is not typically lower than assemblages in nearby non-urbanised areas, it is probably because functional diversity is elevated by higher habitat diversity in urban areas (Oliveira Hagen et al. 2017).

Reduced functional diversity at local scales as urbanisation intensity increases may contribute to declining provision of the ecosystem services generated by functioning ecosystems. Extreme trait filtering in the most urbanised locations (i.e. greatest building density or impervious surface cover) also generates biotic homogenisation with most city centres across the globe containing a relatively small number of similar species (McKinney 2006). Furthermore, reduced functional diversity results in the substantial proportion of people that primarily directly engage with nature in urban settings being reliant on common, widespread, and generalist species, such as the rock doves (*Columba livia*) and globally homogenised ecosystems, such as lawns (Ignatieva and Hedblom 2018). However, such interactions can play a pivotal role in generating an interest in nature and a desire to conserve it – leading to the pigeon paradox (Dunn et al. 2006), in which the ability to conserve biodiversity outside cities may be dependent on the common and unthreatened species that typically dominate cities.

Gaps in urban filtering frameworks

Considering the numerous urban filter frameworks and the overall rise in urban ecological research, there remains a surprising number of aspects about urban filtering that we have little to no knowledge of at present. As a result, there are a host of topics and subsequent questions that need answering to advance the field in any marked way. The following sections summarise the important gaps that we have identified.

Missing ecosystems

The assembly literature has been dominated by terrestrial ecologists focusing primarily on above-ground processes. However, two ecosystems stand out as particularly neglected, both in their impact on above-ground community assembly and the lack of knowledge on community assembly within these systems: soil and aquatic systems.

Given that soil is one of the primary factors influencing the base of terrestrial food webs, its lack of inclusion in filtering frameworks is particularly surprising. Considering that many cities have greatly altered physiochemical properties and high heterogeneity (Pickett and Cadenasso 2009), soils (and their associated biota) represent a critical filter in determining plant communities and ecosystem functions. Soil biota, including invertebrates, fungi (particularly mycorrhizal fungi), and microbial communities are critical to plant population and community dynamics (Tedersoo et al. 2020). However, very little work has been done testing filtering processes on these urban soil communities (but see Byrne and Szlavecz, this volume). Dispersal of soil organisms in urban ecosystems is also relatively unknown, but anthropogenic processes, such as moving soil from one place to another and introduction of horticultural plantings across urban areas from the same nursery stock, may be important drivers of soil taxa dispersal. Environmental filters, including resource availability, physiochemical conditions, land use legacies, and current management practices, are particularly strong filters on soil biota (Pavao-Zuckerman and Byrne 2009), but there is very little work testing these in urban settings.

While a large focus of urban ecosystem research has identified the impacts of pollution and excess nutrients on aquatic and riparian systems, both the role these systems may play in community assembly of adjacent ecosystems and community assembly within aquatic systems is not well understood. How aquatic systems facilitate or hinder assembly in the city is relatively unknown, particularly when we consider land-water interface (e.g. riparian areas, coastal area) or how lakes, ponds, and impounded water bodies may differ from stream and riverine systems. Riverine corridors act to facilitate both immigration into the city and emigration away from the city, as well as provide connectivity within urban areas (Aronson et al. 2017). Hence, understanding how riparian network structure and changes in disturbances, including annual flow regimes and flooding events, affect colonisation of both the aquatic and terrestrial habitats in cities is needed. Furthermore, while there is a rich body of literature focused on aquatic community assembly, how urban aquatic communities assemble is understudied (see Engman and Roy, this volume). We have only begun to evaluate the role of artificial or constructed bodies of water in cities, such as stormwater detention ponds (Sinclair et al. 2021) and wastewater treatment lakes, which can/may operate akin to brand-new islands (*sensu* Thornton and New 2007).

Multicity comparisons

Although most urban community assembly frameworks are generalised to fit the processes occurring in most cities, empirical, multicity comparisons are not common (e.g. Oliveira Hagen et al. 2017). As noted previously, cities are not homogenous, with variation in the magnitude of their heat islands, urban structure and form, governance, and cultural norms (see Hahs, this volume, for further discussion), altering the relative importance and strength of filters across cities. Multicity comparisons that take advantage of this variation to explore how these factors alter assembly processes, and their consequences, will not only capture generalisations but also how filters operate at different scales and across different environmental, anthropogenic, and social contexts, ultimately leading to a deeper understanding of community assembly processes as a whole. Crucially, such studies will also inform debates on how to grow and expand urban areas more sustainably.

Temporal dynamics

Filters are not static, and the pressures imposed by some filters can be expected to change in importance and intensity over time. The intensity of some filters typically increases as urbanisation intensity increases. For example, as cities grow and develop, urban heat island effects typically intensify, with the largest cities experiencing temperatures that are 4°C to 12°C warmer than surrounding rural areas (Phelan et al. 2015). Changes in the intensity of filters could lead to extinction debts, where some species are still present in cities but maladapted to current conditions (e.g. plants: Hahs et al. 2009), and/or functional extinctions, where populations are too low to play a significant role in ecosystem function. Alterations in human behaviour and policies over time will also lead to changes in filters. For instance, following pollution reduction and other restoration activities, otters (*Lutra lutra*) and Atlantic salmon (*Salmo salar*) have returned to cleaner urban rivers in the UK (Rotherham 2021).

One of the largest gaps is understanding and predicting how filters will change over time concerns the impacts of climate change. Climate change is resulting in the movement of species and ecosystems (Loarie et al. 2009), often resulting in no-analog communities (Stralberg et al. 2009). When we consider ranges shifts, it is important to note that these shifts will affect the regional pool of species, such that the species available to colonise a city will change. Shifting ranges will result in both additions and deletions to the regional species pool, generating dynamism at the regional scale that is rarely considered in frameworks for the assembly of urban biotas.

In addition to these regional scale shifts, climate change will impact a wide range of weather variables at the city and finer spatial scales – including precipitation regimes, seasonality, and extreme events. Extreme heat waves and drought, for example, alter tree community composition in cities by differentially causing mortality of some species and not others (Tabassum et al. 2021). The magnitude and consequences of these changes will vary with urban form. For example, extreme rainfall events cause an increase in surface water run-off from roads and other impervious surfaces, resulting in greater pollutant loads in water bodies and soil (Miller and Hutchins 2017), especially downstream of areas of high impervious surface cover. Climate change will further alter filters by leading to shifts in human behaviours, particularly those relating to urban planning and management. Many cities are planning for climate change adaptation and resilience, including nature-based solutions that may also increase habitats and connectivity to support biodiversity (Kabisch et al. 2016; Xie and Bulkeley 2020). Climate change may also bring about changes in land manager behaviour as they respond to changed climate conditions, such as droughts, by altering irrigation and choosing plants more tolerant of the novel climate conditions (Webster et al. 2017), thus changing soil conditions and vegetation composition and structure.

Lack of integration of evolutionary processes

Another major issue with static frameworks for the assembly of urban communities is their inability to consider evolutionary processes. Natural selection is an inevitable response to the strong selection pressures generated by urban filters resulting in substantial evolutionary change in urban biotas (Johnson and Munshi-South 2017). For example, assessment of population genetic structure of multiple paired urban and rural blackbird (*Turdus merula*) populations across Europe demonstrated genetic divergence and multiple independent colonisation events (Evans et al. 2009). Follow-up work using a candidate gene approach provides evidence for selection pressure on urban blackbirds that has resulted in genetic adaptation (Mueller et al. 2013) – indicating that the blackbird's spread into cities across Europe probably resulted from multiple independent genetic adaptation events. Increasing numbers of studies are assessing gene flow and phenotypic adaptation, but linkages with community assembly processes are inadequately explored. For example, the frequency of heavy, gravity-dispersed seeds increased over time in urban, fragmented populations of the forb *Crepis sancta*, indicating selection against long-distance dispersal into inhospitable locations, such as sidewalks and roads (Cheptou et al. 2008). However, how subsequent changes in *C. sancta* distribution may impact other taxa such as herbivore or pollinator communities is unknown. Similarly, urban populations of *Centaurium erythraea* have evolved to reduce their dependency on insect pollinators for seed production (Brys and Jacquemyn 2012), with potential consequences for community assembly of pollinators.

The potential for evolutionary processes to alter urban community assembly will be greatest when genetic adaptation enables a species to colonise urban areas or radically increase its population size and thus alters biotic interactions. As an example, wild pigs (*Sus scrofa*) have recently colonised many European cities (Castillo-Contreras et al. 2021). There are likely to be multiple drivers of these colonisation events, but they are associated with changes to candidate genes for controlling stress responses, implying that genetic adaptation has facilitated the arrival of this species in urban areas (Zsolnai et al. 2022). This genetic change is likely to have consequences for many other species given the role of wild pigs as ecosystem engineers that can alter plant communities and their associated insect assemblages (de Schaetzen et al. 2018). Clearly, there is a need to incorporate a temporal dimension into urban filtering that allows for species plasticity and adaptation, which can lead to colonisation and persistence in cities.

Conclusions

As is increasingly the case across the globe, human factors determine the type and intensity of filters within urban areas – but anthropogenic influence is particularly prominent in towns and cities. The planning and design of urban areas can have a large impact on the magnitude of selection/filtering pressure an assemblage is exposed to and thus determine the diversity and composition of urban biotas. Many of these urban selection pressures have adverse impacts on people, with established links, for example, between the amount of green space and risk of psychiatric disease (Engemann et al. 2019), and urban heat stress and excess mortality (Heaviside et al. 2017). Moreover, there is increasing evidence that exposure to biodiversity can enhance urban residents' mental health and wellbeing (Marselle et al. 2021; Fisher et al., this volume). A key question, which is receiving increasing attention (e.g. Lin et al. 2021), is thus how to design cities in a manner that alters filtering processes to enhance urban biodiversity – and the resultant benefits for people.

Key companion papers

Aronson, M.F.J., et al. 2016. Hierarchical filters determine community assembly of urban species pools. *Ecology* 97:2952–2963.

Evans, K.L., B.J. Hatchwell, M. Parnell, K.J. Gaston. 2010. A conceptual framework for the colonisation of urban areas: The blackbird *Turdus merula* as a case study. *Biological Reviews* 85:643–667.

Williams, N.S.G., et al. 2009. A conceptual framework for predicting the effects of urban environments on floras. *Journal of Ecology* 97:4–9.

References

Andrade, R., J. Franklin, K.L. Larson, C.M. Swan, S.B. Lerman, H.L. Bateman, P.S. Warren, A. York. 2021. Predicting the assembly of novel communities in urban ecosystems. *Landscape Ecology* 36:1–15.

Arévalo, C., J.D. Amaya-Espinel, C. Henríquez, J.T. Ibarra, C. Bonacic. 2022. Urban noise and surrounding city morphology influence green space occupancy by native birds in a Mediterranean-type South American metropolis. *Scientific Reports* 12:4471.

Aronson, M.F.J., S.N. Handel, S.E. Clemants. 2007. Fruit type, life form and origin determine the success of woody plant invaders in an urban landscape. *Biological Invasions* 9:465–475.

Aronson, M.F.J., C.H. Nilon, C.A. Lepczyk, T.S. Parker, P.S. Warren, S.S. Cilliers, M.A. Goddard, A.K. Hahs, C. Herzog, M. Katti, F.A.L. Sorte, N.S.G. Williams, W. Zipperer. 2016. Hierarchical filters determine community assembly of urban species pools. *Ecology* 97:2952–2963.

Aronson, M.F.J., M.V. Patel, K.M. O'Neill, J.G. Ehrenfeld. 2017. Urban riparian systems function as corridors for both native and invasive plant species. *Biological Invasions* 19:3645–3657.

Avolio, M.L., D.E. Pataki, T.L.E. Trammell, J. Endter-Wada. 2018. Biodiverse cities: The nursery industry, homeowners, and neighborhood differences drive urban tree composition. *Ecological Monographs* 88:259–276.

Avolio, M.L., C. Swan, D.E. Pataki, G.D. Jenerette. 2021. Incorporating human behaviors into theories of urban community assembly and species coexistence. *Oikos* 130:1849–1864.

Berger, A.J. 1974. History of exotic birds in Hawaii. *The 'Elepaio* 35:60–65.

Birkhofer, K., M.M. Gossner, T. Diekötter, C. Drees, O. Ferlian, M. Maraun, S. Scheu, W.W. Weisser, V. Wolters, S. Wurst, A.S. Zaitsev, H.G. Smith. 2017. Land-use type and intensity differentially filter traits in above- and below-ground arthropod communities. *Journal of Animal Ecology* 86:511–520.

Bonier, F., P.R. Martin, J.C. Wingfield. 2007. Urban birds have broader environmental tolerance. *Biology Letters* 3:670–673.

Brashares, J.S. 2010. Filtering wildlife. *Science* 329:402–403.

Brys, R., H. Jacquemyn. 2012. Effects of human-mediated pollinator impoverishment on floral traits and mating patterns in a short-lived herb: An experimental approach. *Functional Ecology* 26:189–197.

Cadotte, M.W., S.L.E. Yasui, S. Livingstone, J.S. MacIvor. 2017. Are urban systems beneficial, detrimental, or indifferent for biological invasion? *Biological Invasions* 19:3489–3503.

Callaghan, C.T., R.E. Major, J.H. Wilshire, J.M. Martin, R.T. Kingsford, W.K. Cornwell. 2019. Generalists are the most urban-tolerant of birds: A phylogenetically controlled analysis of ecological and life history traits using a novel continuous measure of bird responses to urbanization. *Oikos* 128:845–858.

Caruso, T., M. Migliorini, E. Rota, R. Bargagli. 2017. Highly diverse urban soil communities: Does stochasticity play a major role? *Applied Soil Ecology* 110:73–78.

Castillo-Contreras, R., G. Mentaberre, X. Fernandez Aguilar, C. Conejero, A. Colom-Cadena, A. Ráez-Bravo, C. González-Crespo, J. Espunyes, S. Lavín, J.R. López-Olvera. 2021. Wild boar in the city: Phenotypic responses to urbanisation. *Science of the Total Environment* 773:145593.

Cheptou, P.-O., O. Carrue, S. Rouifed, A. Cantarel. 2008. Rapid evolution of seed dispersal in an urban environment in the weed *Crepis sancta*. *Proceedings of the National Academy of Sciences* 105:3796–3799.

Cody, M.L., J.M. Diamond, editors. 1975. *Ecology and Evolution of Communities*. Harvard University Press, Cambridge, MA.

Concepción, E.D., M. Moretti, F. Altermatt, M.P. Nobis, M.K. Obrist. 2015. Impacts of urbanisation on biodiversity: The role of species mobility, degree of specialisation and spatial scale. *Oikos* 124:1571–1582.

de Schaetzen, F., F. van Langevelde, M.F. WallisDeVries. 2018. The influence of wild boar (Sus scrofa) on microhabitat quality for the endangered butterfly *Pyrgus malvae* in the Netherlands. *Journal of Insect Conservation* 22:51–59.

Duncan, R.P., S.E. Clemants, R.T. Corlett, A.K. Hahs, M.A. McCarthy, M.J. McDonnell, M.W. Schwartz, K. Thompson, P.A. Vesk, N.S.G. Williams. 2011. Plant traits and extinction in urban areas: A meta-analysis of 11 cities. *Global Ecology and Biogeography* 20:509–519.

Dunn, R.R., M.C. Gavin, M.C. Sanchez, J.N. Solomon. 2006. The pigeon paradox: Dependence of global conservation on urban nature. *Conservation Biology* 20:1814–1816.

Engemann, K., C.B. Pedersen, L. Arge, C. Tsirogiannis, P.B. Mortensen, J.-C. Svenning. 2019. Residential green space in childhood is associated with lower risk of psychiatric disorders from adolescence into adulthood. *Proceedings of the National Academy of Sciences* 116:5188–5193.

Evans, K.L., D.E. Chamberlain, B.J. Hatchwell, R.D. Gregory, K.J. Gaston. 2011. What makes an urban bird? *Global Change Biology* 17:32–44.

Evans, K.L., K.J. Gaston, A.C. Frantz, M. Simeoni, S.P. Sharp, A. McGowan, D.A. Dawson, K. Walasz, J. Partecke, T. Burke, B.J. Hatchwell. 2009. Independent colonization of multiple urban centres by a formerly forest specialist bird species. *Proceedings of the Royal Society B* 276:2403–2410.

Evans, K.L., B.J. Hatchwell, M. Parnell, K.J. Gaston. 2010. A conceptual framework for the colonisation of urban areas: The blackbird *Turdus merula* as a case study. *Biological Reviews* 85:643–667.

Foster, J.T. 2009. The history and impact of introduced birds. Pages 312–330 in T.K. Pratt, C.T. Atkinson, P.C. Banko, J.D. Jacobi, B.L. Woodworth, editors. *Conservation Biology of Hawaiian Forest Birds*. Yale University Press, New Haven, CT.

Fournier, B., D. Frey, M. Moretti. 2020. The origin of urban communities: From the regional species pool to community assemblages in city. *Journal of Biogeography* 47:615–629.

Francis, C.D., C.P. Ortega, A. Cruz. 2011. Noise pollution filters bird communities based on vocal frequency. *PLoS ONE* 6:e27052.

Franzén, M., P.-E. Betzholtz, L.B. Pettersson, A. Forsman. 2020. Urban moth communities suggest that life in the city favours thermophilic multi-dimensional generalists. *Proceedings of the Royal Society B: Biological Sciences* 287:20193014.

Funk, J.L. 2021. Revising the trait-based filtering framework to include interacting filters: Lessons from grassland restoration. *Journal of Ecology* 109:3466–3472.

Fusco, N.A., E.J. Carlen, J. Munshi-South. 2021. Urban landscape genetics: Are biologists keeping up with the pace of urbanization? *Current Landscape Ecology Reports* 6:35–45.

Hahs, A.K., M.J. McDonnell, M.A. McCarthy, P.A. Vesk, R.T. Corlett, B.A. Norton, S.E. Clemants, R.P. Duncan, K. Thompson, M.W. Schwartz, N.S.G. Williams. 2009. A global synthesis of plant extinction rates in urban areas. *Ecology Letters* 12:1165–1173.

Heaviside, C., H. Macintyre, S. Vardoulakis. 2017. The urban heat island: Implications for health in a changing environment. *Current Environmental Health Reports* 4:296–305.

Hui, C., N. Roura-Pascual, L. Brotons, R.A. Robinson, K.L. Evans. 2012. Flexible dispersal strategies in native and non-native ranges: Environmental quality and the 'good – stay, bad – disperse' rule. *Ecography* 35:1024–1032.

Ignatieva, M., M. Hedblom. 2018. An alternative urban green carpet. *Science* 362:148–149.

Johnson, M.T.J., J. Munshi-South. 2017. Evolution of life in urban environments. *Science* 358:eaam8327.

Jung, K., C.G. Threlfall. 2018. Trait-dependent tolerance of bats to urbanization: A global meta-analysis. *Proceedings of the Royal Society B: Biological Sciences* 285:20181222.

Kabisch, N., N. Frantzeskaki, S. Pauleit, S. Naumann, M. Davis, M. Artmann, D. Haase, S. Knapp, H. Korn, J. Stadler, K. Zaunberger, A. Bonn. 2016. Nature-based solutions to climate change mitigation and adaptation in urban areas: Perspectives on indicators, knowledge gaps, barriers, and opportunities for action. *Ecology and Society* 21.

Keddy, P.A. 1992. Assembly and response rules: Two goals for predictive community ecology. *Journal of Vegetation Science* 3:157–164.

Kendal, D., K.J.H. Williams, N.S.G. Williams. 2012. Plant traits link people's plant preferences to the composition of their gardens. *Landscape and Urban Planning* 105:34–42.

Klingbeil, B., F. La Sorte, C. Lepczyk, D. Fink, C. Flather. 2020. Geographic associations with anthropogenic noise pollution for North American breeding birds. *Global Ecology and Biogeography* 29:148–158.

Knapp, S., I. Kühn, O. Schweiger, S. Klotz. 2008. Challenging urban species diversity: Contrasting phylogenetic patterns across plant functional groups in Germany. *Ecology Letters* 11:1054–1064.

Kowarik, I., M. von der Lippe. 2011. Secondary wind dispersal enhances long-distance dispersal of an invasive species in urban road corridors. *NeoBiota* 9:49–70.

La Sorte, F.A., M.F.J. Aronson, C.A. Lepczyk, K.G. Horton. 2022. Assessing the combined threats of artificial light at night and air pollution for the world's nocturnally migrating birds. *Global Ecology and Biogeography* 31:912–924.

La Sorte, F.A., C.A. Lepczyk, M.F.J. Aronson, M.A. Goddard, M. Hedblom, M. Katti, I. MacGregor-Fors, U. Mörtberg, C.H. Nilon, P.S. Warren, N.S.G. Williams, J. Yang. 2018. The phylogenetic and functional diversity of regional breeding bird assemblages is reduced and constricted through urbanization. *Diversity and Distributions* 24:928–938.

La Sorte, F.A., M.L. McKinney, P. Pyšek. 2007. Compositional similarity among urban floras within and across continents: Biogeographical consequences of human-mediated biotic interchange. *Global Change Biology* 13:913–921.

Lepczyk, C.A., A.G. Mertig, J. Liu. 2004. Assessing landowner activities that influence birds across rural-to-urban landscapes. *Environmental Management* 33:110–125.

Lepczyk, C.A., L.M. Wedding, G.P. Asner, S.J. Pittman, T. Goulden, M.A. Linderman, J. Gang, R. Wright. 2021. Advancing landscape and seascape ecology from a 2D to a 3D science. *BioScience* 71:596–608.

Lin, B.B., A. Ossola, M. Alberti, E. Andersson, X. Bai, C. Dobbs, T. Elmqvist, K.L. Evans, N. Frantzeskaki, R.A. Fuller, K.J. Gaston, D. Haase, C.Y. Jim, C. Konijnendijk, H. Nagendra, J. Niemelä, T. McPhearson, W.R. Moomaw, S. Parnell, D. Pataki, W.J. Ripple, P.Y. Tan. 2021. Integrating solutions to adapt cities for climate change. *The Lancet Planetary Health* 5:e479–e486.

Loarie, S.R., P.B. Duffy, H. Hamilton, G.P. Asner, C.B. Field, D.D. Ackerly. 2009. The velocity of climate change. *Nature* 462:1052–1055.

Lockwood, J.L., M.F. Hoopes, M. Marchetti. 2013. *Invasion Ecology*, 2nd edition. Wiley-Blackwell, Hoboken, NJ.

Lopez, B.E., D. Urban, P.S. White. 2018. Testing the effects of four urbanization filters on forest plant taxonomic, functional, and phylogenetic diversity. *Ecological Applications* 28:2197–2205.

Marselle, M.R., S.J. Lindley, P.A. Cook, A. Bonn. 2021. Biodiversity and health in the urban environment. *Current Environmental Health Reports* 8:146–156.

Martinson, H.M., M.J. Raupp. 2013. A meta-analysis of the effects of urbanization on ground beetle communities. *Ecosphere* 4:art60.

McKinney, M.L. 2006. Urbanization as a major cause of biotic homogenization. *Biological Conservation* 127:247–260.

Meurisse, N., D. Rassati, B.P. Hurley, E.G. Brockerhoff, R.A. Haack. 2019. Common pathways by which non-native forest insects move internationally and domestically. *Journal of Pest Science* 92:13–27.

Miller, J.D., M. Hutchins. 2017. The impacts of urbanisation and climate change on urban flooding and urban water quality: A review of the evidence concerning the United Kingdom. *Journal of Hydrology: Regional Studies* 12:345–362.

Morin, P.J. 2011. *Community Ecology*, 2nd edition. Wiley-Blackwell, Hoboken, NJ.

Mueller, J.C., J. Partecke, B.J. Hatchwell, K.J. Gaston, K.L. Evans. 2013. Candidate gene polymorphisms for behavioural adaptations during urbanization in blackbirds. *Molecular Ecology* 22:3629–3637.

Munshi-South, J. 2012. Urban landscape genetics: Canopy cover predicts gene flow between white-footed mouse (*Peromyscus leucopus*) populations in New York City. *Molecular Ecology* 21:1360–1378.

Narango, D.L., D.W. Tallamy, P.P. Marra. 2018. Nonnative plants reduce population growth of an insectivorous bird. *Proceedings of the National Academy of Sciences* 115:11549–11554.

Norton, B.A., K.L. Evans, P.H. Warren. 2016. Urban biodiversity and landscape ecology: Patterns, processes and planning. *Current Landscape Ecology Reports* 1:178–192.

Oliveira Hagen, E., O. Hagen, J.D. Ibáñez-Álamo, O.L. Petchey, K.L. Evans. 2017. Impacts of urban areas and their characteristics on avian functional diversity. *Frontiers in Ecology and Evolution* 5.

Pavao-Zuckerman, M.A., L.B. Byrne. 2009. Scratching the surface and digging deeper: Exploring ecological theories in urban soils. *Urban Ecosystems* 12:9–20.

Pearse, W.D., J. Cavender-Bares, S.E. Hobbie, M.L. Avolio, N. Bettez, R. Roy Chowdhury, L.E. Darling, P.M. Groffman, J.M. Grove, S.J. Hall, J.B. Heffernan, J. Learned, C. Neill, K.C. Nelson, D.E. Pataki, B.L. Ruddell, M.K. Steele, T.L.E. Trammell. 2018. Homogenization of plant diversity, composition, and structure in North American urban yards. *Ecosphere* 9:e02105.

Phelan, P.E., K. Kaloush, M. Miner, J. Golden, B. Phelan, H. Silva, R.A. Taylor. 2015. Urban heat island: Mechanisms, implications, and possible remedies. *Annual Review of Environment and Resources* 40:285–307.

Pickett, S.T.A., M.L. Cadenasso. 2009. Altered resources, disturbance, and heterogeneity: A framework for comparing urban and non-urban soils. *Urban Ecosystems* 12:23–44.

Roman, L.A., H. Pearsall, T.S. Eisenman, T.M. Conway, R.T. Fahey, S. Landry, J. Vogt, N.S. van Doorn, J.M. Grove, D.H. Locke, A.C. Bardekjian, J.J. Battles, M.L. Cadenasso, C.C.K. van den Bosch, M. Avolio, A. Berland, G.D. Jenerette, S.K. Mincey, D.E. Pataki, C. Staudhammer. 2018. Human and biophysical legacies shape contemporary urban forests: A literature synthesis. *Urban Forestry & Urban Greening* 31:157–168.

Rotherham, I.D. 2021. The impacts of recolonisation of an urbanised river by native and non-native species. *Frontiers in Ecology and Evolution* 9.

Santini, L., M. González-Suárez, D. Russo, A. Gonzalez-Voyer, A. von Hardenberg, L. Ancillotto. 2019. One strategy does not fit all: Determinants of urban adaptation in mammals. *Ecology Letters* 22:365–376.

Sayol, F., D. Sol, A.L. Pigot. 2020. Brain size and life history interact to predict urban tolerance in birds. *Frontiers in Ecology and Evolution* 8.

Sinclair, J.S., L.S. Reisinger, C.R. Adams, E. Bean, A.J. Reisinger, B.V. Iannone. 2021. Vegetation management and benthic macroinvertebrate communities in urban stormwater ponds: Implications for regional biodiversity. *Urban Ecosystems* 24:725–735.

Skye, E. 1968. Lichens and air pollution: A study of cryptogamic epiphytes and environment in the Stockholm region. *Acta Phytogeographica Suecica* 52.

Sol, D., C. Trisos, C. Múrria, A. Jeliazkov, C. González-Lagos, A.L. Pigot, C. Ricotta, C.M. Swan, J.A. Tobias, S. Pavoine. 2020. The worldwide impact of urbanisation on avian functional diversity. *Ecology Letters* 23:962–972.

Stralberg, D., D. Jongsomjit, C.A. Howell, M.A. Snyder, J.D. Alexander, J.A. Wiens, T.L. Root. 2009. Re-shuffling of species with climate disruption: A no-analog future for California birds? *PLoS ONE* 4(9):e6825.

Tabassum, S., A. Manea, A. Ossola, B. Thomy, D. Blackham, M.R. Leishman. 2021. The angriest summer on record: Assessing canopy damage and economic costs of an extreme climatic event. *Urban Forestry & Urban Greening* 63:127221.

Tedersoo, L., M. Bahram, M. Zobel. 2020. How mycorrhizal associations drive plant population and community biology. *Science* 367:eaba1223.

Thornton, I., T. New, editors. 2007. *Island Colonization: The Origin and Development of Island Communities*. Cambridge University Press, New York.

Vailshery, L.S., M. Jaganmohan, H. Nagendra. 2013. Effect of street trees on microclimate and air pollution in a tropical city. *Urban Forestry & Urban Greening* 12:408–415.

Van Kleunen, M., F. Essl, J. Pergl, G. Brundu, M. Carboni, S. Dullinger, R. Early, P. González-Moreno, Q.J. Groom, P.E. Hulme, C. Kueffer, I. Kühn, C. Máguas, N. Maurel, A. Novoa, M. Parepa, P. Pyšek, H. Seebens, R. Tanner, J. Touza, L. Verbrugge, E. Weber, W. Dawson, H. Kreft, P. Weigelt, M. Winter, G. Klonner, M.V. Talluto, K. Dehnen-Schmutz. 2018. The changing role of ornamental horticulture in alien plant invasions. *Biological Reviews* 93:1421–1437.

Von Der Lippe, M., I. Kowarik. 2007. Long-distance dispersal of plants by vehicles as a driver of plant invasions. *Conservation Biology* 21:986–996.

Webb, C.T., J.A. Hoeting, G.M. Ames, M.I. Pyne, N. LeRoy Poff. 2010. A structured and dynamic framework to advance traits-based theory and prediction in ecology. *Ecology Letters* 13:267–283.

Webster, E., R.W.F. Cameron, A. Culham. 2017. *Gardening in a Changing Climate*. Royal Horticultural Society, London.

Williams, N.S.G., A.K. Hahs, P.A. Vesk. 2015. Urbanisation, plant traits and the composition of urban floras. *Perspectives in Plant Ecology, Evolution and Systematics* 17:78–86.

Williams, N.S.G., M.W. Schwartz, P.A. Vesk, M.A. McCarthy, A.K. Hahs, S.E. Clemants, R.T. Corlett, R.P. Duncan, B.A. Norton, K. Thompson, M.J. McDonnell. 2009. A conceptual framework for predicting the effects of urban environments on floras. *Journal of Ecology* 97:4–9.

Winchell, K.M., K.P. Schliep, D.L. Mahler, L.J. Revell. 2020. Phylogenetic signal and evolutionary correlates of urban tolerance in a widespread neotropical lizard clade*. *Evolution* 74:1274–1288.

Xie, L., H. Bulkeley. 2020. Nature-based solutions for urban biodiversity governance. *Environmental Science & Policy* 110:77–87.

Zaffaroni-Caorsi, V., C. Both, R. Márquez, D. Llusia, P. Narins, M. Debon, M. Borges-Martins. 2022. Effects of anthropogenic noise on anuran amphibians. *Bioacoustics*. http://doi.org/10.1080/09524622.2022.2070543

Zsolnai, A., A. Csókás, L. Szabó, L. Patkó, S. Csányi, M. Márton, E.A. Lakatos, I. Anton, F. Deutsch, M. Heltai. 2022. Genetic adaptation to urban living: Molecular DNA analyses of wild boar populations in Budapest and surrounding area. *Mammalian Biology* 102:221–234.

8

FROM FILTERS TO FEEDBACKS

How do people influence biodiversity across scales?

Riley Andrade, Paige S. Warren, and Sarel S. Cilliers

Introduction

Prominent scholars in the field of urban ecology have called for unifying frameworks that explicitly consider the effects of human activity on biodiversity and thus also the need to consider people as integrated components of ecosystems (Groffman et al. 2016). One of the key approaches to do so is through the consideration of biodiversity filters and feedbacks across scales – known as the **multi-scalar framework**. The following chapter uses the multi-scalar framework to synthesize sociocultural, political, and economic filters – here termed "social filters" – on biodiversity across the patchy landscape mosaics that characterize cities (Cook et al. 2011; Aronson et al. 2016; Andrade et al. 2020; Anderson et al. 2021). Broadly, the multi-scalar framework can be used to explain how biodiversity is spatially structured within urban landscapes, with varying effects of social, biophysical, and spatiotemporal filters at each scale. The framework places social-ecological dynamics into a spatial context and describes them as a nested hierarchy of local to regional scales. The concept was adapted from the hierarchical system concept (Urban et al. 1987) by Grove et al. (2005) to describe the linkages between human and forest communities and has since been used to describe social-ecological dynamics in residential landscapes (Chowdhury et al. 2011; Cook et al. 2011). Aronson et al. (2016) and Andrade et al. (2020) further advance the concept by considering the urban drivers of biodiversity across multiple spatial scales.

The use of spatial scales to study urban landscapes is grounded in the assumption of some form of hierarchical structure across levels in the system. Thinking about scale in urban landscapes is useful because the effects of human society, organizations, communities, and individual people on biodiversity are spatially structured within nested levels. Hierarchy theory grounds the concept of multi-scalar filters by defining the complexity of landscapes as having vertical and horizontal structure (Wu 1999). Vertical structure defines systems as having multiple levels composed of subsystems as finer-scale components, each nested within a broader level. Horizontal structure describes the connections and/or interactions between subsystem-level components. Thus, research approaches that apply hierarchy theory to urban ecosystems consider the patterns and processes that shape biodiversity across multiple levels through the use of a multi-scalar framework (Figure 8.1).

DOI: 10.4324/9781003016120-10

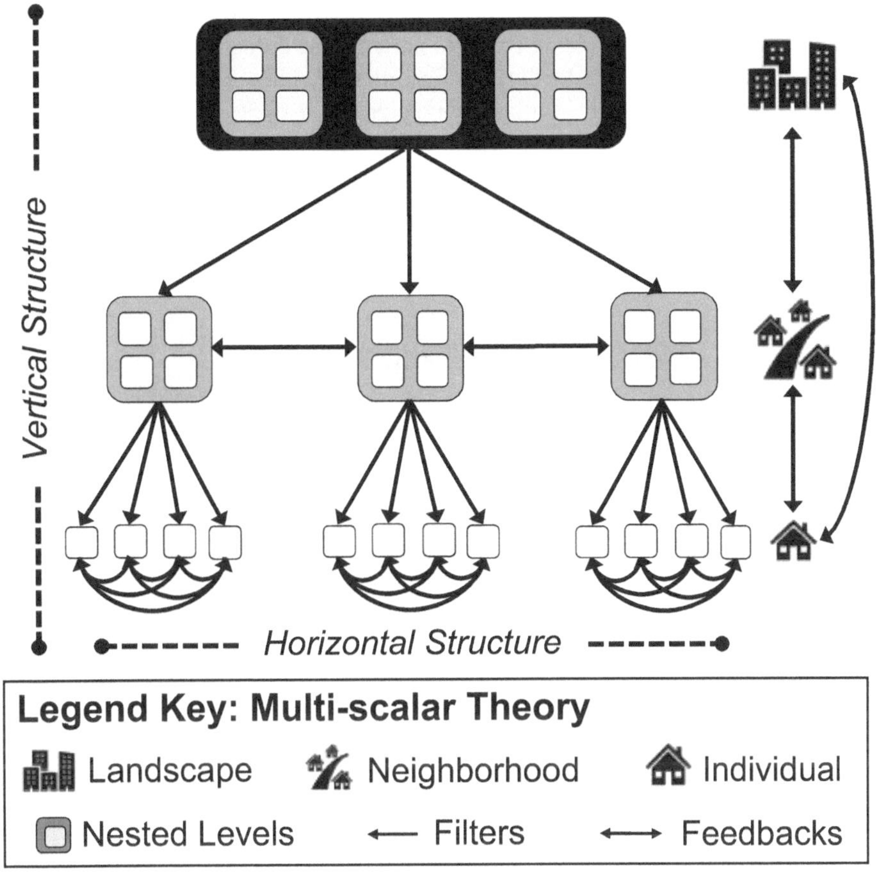

Figure 8.1 Multi-scalar framework of the sociocultural, political, and economic filters of biodiversity across various levels of human organization in urban landscapes. Biodiversity filters at local levels are nested into broader levels, known as vertical structure. The interactions that occur within a level, or horizontal structure, create feedbacks between people and biodiversity. Similarly, social-ecological dynamics that filter biodiversity also scale up or down to influence one another to create spatiotemporal feedbacks across levels.

State of knowledge

In the following sections, we outline important social filters of biodiversity from regional to local scales, organized into three main levels as indicated in the conceptual figure: (1) regional filters influencing biodiversity at broad, citywide scales (urban landscape); (2) neighborhood-level filters influencing biodiversity across collectives of land parcels that are spatially organized into socially defined clusters (neighborhood); and (3) local filters imposed by individual landowners and managers within a single parcel (individual). Regional filters of biodiversity include climate and biogeography, broad socioeconomic gradients, temporal legacies, as well as the sociocultural and political backdrop of a particular metropolitan area. The social expectations imposed by informal institutions and the codified regulations of formal institutions dominate as social filters at the neighborhood level. At local scales, people's internal motivations interact with structural constraints to influence how the actions of individual landowners and managers filter biodiversity.

As we review social filters of biodiversity across nested levels, it is worth specifically noting that the surrounding environment, including the biota present, also influences management decisions and human well-being. As a result, biodiversity, in part, shapes how social filters are formed and enforced across scales. Instead of detailing the relationship between any given social filter and biodiversity across a variety of taxa or geographies, we largely focus on reviewing the social filters themselves, as well as the underlying mechanisms that have been demonstrated to influence biodiversity across diverse global study areas.

Global, regional, and urban filters

Biogeographical and morphological filters

Urbanization is a primary social filter on global biodiversity patterns by creating landscapes characterized by human-built structures and activities (Piano et al. 2020). However, the particular effects of urbanization on biodiversity are highly variable across scales and ecological taxon (Knapp et al. 2017; Fournier et al. 2020). In general, urbanization acts as a broad filter on the regional species pool, where some specialist species that would be found within a geographical extent are unable to colonize or persist within metropolitan areas. Human activities also introduce some new species over time to the urban species pool. As a result, urbanization is often considered to be a homogenizing force on biodiversity because urban species pools are often similar to one another, despite geographical separation (Pearse et al. 2018). The turfgrass species found in lawns, Rock Pigeon (*Columba livia*), and House Sparrow (*Passer domesticus*) are just some examples representing notorious species associated with urbanization nearly everywhere on the globe. However, urban species pools can also harbor many species unique to the ecoregion in which the city is located (Aronson et al. 2014).

Climate and interrelated geographical filters – such as latitudinal gradients and resource availability – are regional filters of biodiversity that dictate the spatial distributions of species. For example, tree cover doubles along the evapotranspirative gradient of the continental United States, from North Dakota to Florida, with forest structure varying along the same gradient (Ossola and Hopton 2018a). Globally, tropical cities typically boast a higher native diversity than temperate cities, but as a general trend, cities typically have lower native biodiversity than the surrounding rural areas for a number of wildlife species (Aronson et al. 2014; Norton et al. 2016; Leveau et al. 2017). However, for other biota such as plants, urban landscapes can actually support greater levels of biodiversity (McKinney 2008). Studies on bird diversity have indicated that cities contain a lower proportion of the regional species pool as the regional diversity increases, probably due to the presence of a higher number of specialist species in highly speciose regions, which may be less able to adapt to urban conditions (Evans et al. 2011; Ferenc et al. 2014). Still other studies have shown that certain bird species associated with aquatic environments have greater diversity in urban areas in arid regions, highlighting the importance of biogeographical context when considering the effects of urbanization as a broad filter (Andrade et al. 2018).

Municipal filters: legacies and socioeconomics

Patterns of biodiversity are shaped by both current socioeconomics and demographic trends, but also the historical social-ecological legacies defined as "the impacts that previous conditions have on current processes" (Monger et al. 2015). Although urbanization is often viewed as an endpoint of ecological disturbance, cities are spatially and temporally dynamic (Pickett et al. 2016). In particular, biophysical conditions, cultural trends, and municipal policies that determine urban form and function shift over time, with legacy effects from past conditions (Roman et al. 2018). As a result, sociocultural and ecological legacies are an important filter of biodiversity over time and space. Historical land uses, ecological disturbances, biological invasions, colonialism, and racial segregation are all examples of how history plays a prominent role in structuring current biodiversity patterns at regional and urban/municipal scales (Foster et al. 2003; du Toit et al. 2016; Roman et al. 2018). Here, we focus on the legacies of colonialism and racism as broad, but understudied, social filters that interconnectedly influence biodiversity and human well-being across scales (Schell et al. 2020).

The "legacy of cultural and political symbolism" is a good example of how the history of colonialism influences the current biodiversity of tree species in the Global North (Roman et al. 2018). For example, palm species introduced by Spanish missionaries in California, USA, still dominate urban landscapes today, and in Canada, maple tree species have a long association with national identity. Likewise, histories of colonialism in cities of the Global South are reflected in non-native species richness, introduced from the colonizing countries in Europe. For example, 65% of the urban tree flora in Bangalore, India, are non-native (Nagendra and Gopal 2010), planted to improve living conditions during the colonial era (Anderson et al. 2021). European settlers have also introduced many plant species into South Africa that became highly invasive (Potgieter et al. 2020). However, urban residents often oppose the removal of these species due to their perceived cultural and heritage value (van Wilgen 2012).

Racial segregation and the inequitable consolidation of resources into specific areas within the city is another cultural legacy that influences both current biodiversity and human well-being (Schell et al. 2020). Overall, discriminatory policies in cities worldwide have resulted in the spatial allocation and consolidation of resources – as well as inequitable access to these resources – such as trees and parks (Hoover and Lim 2021; Locke et al. 2021). As a result, racially discriminatory housing policies in the United States like redlining, restrictive covenants, and discriminatory policies across multiple government levels have acted as prominent biodiversity filters, which also affect ecosystem service provisioning (Pulido 2000; Watkins and Gerrish 2018; Yi et al. 2019). For example, Hoffman et al. (2020) found that 94% of neighborhoods that were redlined are now hotter by as much as 7°C than non-redlined neighborhoods. Legacies of racism are also a dominant social filter on biodiversity in the Global South, where racial segregation has created steep socioeconomic gradients that influence local patterns of biodiversity (Cilliers et al. 2014; du Toit 2015) and is referred to as "green apartheid" in South Africa (Venter et al. 2020). As a result, the legacies of sociocultural and political histories are often part of the mechanisms underlying the often-studied relationship between socioeconomics and biodiversity (Leong et al. 2018; Kuras et al. 2020).

Studies have posited that higher socioeconomic status is associated with greater biodiversity (Hope et al. 2003), largely through municipal policies and resulting spatial structure of social variables (Perkins et al. 2004). In general, socioeconomic status is increasingly considered as a proxy for other social or biophysical filters that are actually driving biodiversity patterns, such as urban density and legacies of land use (Ramalho and Hobbs 2012). Shifting climatic factors across latitudinal gradients also confound the relationship between socioeconomic status and biodiversity. Research has found that socioeconomic status can be a pronounced social filter in arid regions, where people provide limiting resources such as water (Chamberlain et al. 2020). Conversely, water appears to be less of a limiting factor in tropical regions (Leong et al. 2018). Due to these complex factors, many recent studies have found either no relationship or reverse relationships between socioeconomic status and biodiversity. For example, Chamberlain et al. (2019) found that income was negatively associated with bird biodiversity in highly urbanized landscapes in South Africa due to the associated biophysical features of the urban landscape. Likewise, Ossola and Hopton (2018b) found that income did not influence tree cover loss in cities across multiple climatic zones. Overall, these conflicting results demonstrate that the complex filter of socioeconomic status at regional scales is shaped by a number of underlying factors, which also influences the local scale relationships between wealth (resources) and biodiversity (Kuras et al. 2020).

Neighborhoods link regional and local filters

Cities are spatially constructed along socially defined boundaries, such as land parcels (Pincetl 2012). As a result, the spatial stratification of resources – such as wealth, property ownership, and open space – is a product of broader social and political forces that determine finer scale social filters (Martin 2003). In this regard, neighborhoods act as a spatial link between broader social filters and the individual interactions between people and biota at local scales. We use the term "neighborhood" with caution to encapsulate the geographical construct of social-ecologically distinct clusters that arise from regional sociocultural and political forces to form human communities (Galster 2001). Social filters of biodiversity at the neighborhood scale are relevant over a range of private and public green infrastructure in cities, including community parks, nature preserves/reserves, cemeteries, golf courses, commercial development, and residential yards/gardens. Additionally, landowners and managers act both individually and collectively from both within and outside of the neighborhood to influence biodiversity (Goddard et al. 2010).

Initial development sets the stage for both local and regional social filters of biodiversity in cities because it shapes what people expect from the area and because people are less likely to dramatically alter the landscaping conditions they inherit (Larson et al. 2017). Linking to legacy effects reviewed in the previous section, the historical period in which development takes place (e.g. the age of development) shapes the subsequent social and environmental conditions that filter biodiversity (Aronson et al. 2016). For example, the number of tree species offered by nurseries in California increased over a 110-year period, shifting local species pools over time (Pincetl et al. 2013). As a result of historical trends of nursery availability and people's preferences, vegetation communities often differ between neighborhoods of varying ages (Avolio et al. 2018), with the potential for cascading effects on other species (Burghardt et al. 2009; Narango et al. 2017).

Considering formal and informal institutions as social filters of biodiversity

After initial development, formal and informal institutions act as two general categories for social filters of biodiversity at the neighborhood scale through deterministic (niche-based) processes (Andrade et al. 2020). Municipal policies and ordinances, homeowner's associations (Lerman et al. 2012), and legalized property rights are all codified rules and regulations to govern decision-making, known as formal institutions (Larson et al. 2020). In contrast, informal institutions influence management decisions made through the adherence to belief systems, values, and social norms (Ostrom 2000). We situate both formal and informal institutions at the neighborhood scale because they filter biodiversity through collective processes that occur across multiple land parcels through social interactions (Sisser et al. 2016).

Adherence to the guidelines set forth by formal and informal institutions filter biodiversity through land management practices that result in habitat preservation, creation, or destruction (Andrade et al. 2020). As a result, neighborhood-scale social filters are those that structure individual landscaping decisions, as well as create habitat patches through scaled up environmental properties (Goddard et al. 2010). The guidelines and regulations put forth by formal institutions can be enforced through methods such as citations and fines; however, they are also often upheld and created by the presence of informal expectations known as social norms (Larson and Brumand 2014; Marshall et al. 2020). In the Global South, customs define obligations and collective practices not legally enforced but upheld through the actions of individuals adhering to established social norms (Ostrom 2000). Informality in the Global South should not be seen as a lack of structure but rather a "dilution" of formal authorities and should be regarded as an "inseparable" part of urbanization (du Toit et al. 2021). Additionally, we see social norms in the Global North through mimicry and neighbor expectations in residential landscaping types creating similar decisions across multiple land parcels (Minor et al. 2016).

The institutions that act as social filters of biodiversity depend on the type of green infrastructure (e.g. private versus public) and its intended use, biological taxa of interest, and broader geographical context of a city (Aronson et al. 2017). In residential landscapes, social expectations to maintain well-kept yards are often codified through entities such as homeowner associations (HOAs) in the United States, which dictate specific landscaping regulations (Sisser et al. 2016). Stipulations put in place by HOAs, such as requirements for native landscaping, can either support or decrease plant, insect, and bird biodiversity in neighborhoods (Lerman et al. 2012). Similar to residential yards, plant communities in commercial developments are filtered by landscaping regulations, such as the required amount of vegetation cover, as well as the perceived expectations of the building tenants (Wolf 2009). As a result, the biodiversity on commercial property is largely driven by development patterns and social norms, versus the surrounding socioeconomics (Dyson 2019). In public parks and open space, land managers typically balance the values of multiple users that drive their expectations while providing a safe environment that is not costly to maintain (Carreiro and Zipperer 2011).

Local filters of biodiversity

Individual land managers and owners largely act as social filters of biodiversity at local scales of parcels and households. For individuals, internal motivators (e.g. psychological factors) interact with external constraints to create social filters of biodiversity (Andrade et al. 2020). External constraints here include individual capacity, such as having the time and money to engage in conservation or stewardship activities. However, broader scale filters (e.g. municipal or local ordinances reviewed previously in regional and neighborhood sections) can also constrain individual's actions at the local scale (Larson et al. 2017). Therefore, fine-scale social filters of biodiversity are typically nested within the broader social and biophysical context of the urban system (Aronson et al. 2016). Likewise, regional factors that influence biodiversity at broader scales, such as legacy effects, also emerge locally (Avolio et al. 2018). However, these multi-scalar social filters are spatially and temporally heterogeneous. As a result, the assemblage of species found at any particular location within a city are dependent on a unique suite of social factors that filter biodiversity from the regional into the local species pool (Pearse et al. 2018).

The interplay between external constraints and internal motivations

Structural constraints, such as legacy effects, often create mismatches between preferences and biodiversity at local scales (Wheeler et al. 2020). In particular, the legacy effect also acts as a strong structural constraint at local scales because people are largely beholden to maintain pre-existing standards (Larson and Brumand 2014; Larson et al. 2017). In general, legacy effects may stabilize or shift biodiversity over time through landscape change. Additionally, time and money often both act as constraining resources that filter local biodiversity, with socioeconomic status being tied to the ability to install and maintain greener, more extravagant landscaping to match personal preferences (Avolio et al. 2018). This was confirmed by a South African study that indicates a change in garden design with wealth, leading to a shift from a culturally driven garden design including plant species to a Westernized garden design with a dominance of non-native species (Davoren et al. 2016). There has been some support that preferences influence biodiversity, such as park managers and householders selecting native plants to match ecological functionality or showy plants for aesthetic expectations (Goodness 2018). For example, people's preferences for specific plant traits, such as flower size and color, predict the actual plants occurring in Australian gardens (Kendal et al. 2012). However, other studies have found that land managers' preferences are only partially related to the actual vegetation structure or plant species present (Wheeler et al. 2020; Padullés Cubino et al. 2020).

In addition to structural constraints, social filters of biodiversity at local scales are often related to the form and function of the urban landscape at broader levels. For example, urban form can act as a strong filter on local level biodiversity patterns in residential yards (Ossola et al. 2019). Likewise, social-ecological dynamics at local scales can also scale up to create "bottom-up" effects, often linked to informality as a distinctive feature in the Global South (du Toit et al. 2021). People who value open space, and have social capital, have been shown to work together to preserve land from further development through collective governance (York et al. 2017). This is an example that highlights that the hierarchical structure of multi-scalar filters does not only function in a top-down manner but that local levels can also scale from the bottom-up to influence broader biodiversity filters (Pearse et al. 2018). For example, intervention by a citizen group in a Cape Town township to prevent the development of a shopping mall in a wetland drawing the attention of professional landscape planners, eventually leading to planning for the restoration of the native vegetation, is another good example of "bottom-up" approach for biological conservation (Aalto and Ernstson 2017).

Local feedbacks between people and biodiversity

The social filters we review in this chapter are not unidirectional but instead act as feedbacks between people and biodiversity (Hedblom et al. 2017). People actively respond to and interact with the biota in their yards, neighborhoods, and cities (Head and Muir 2006). For example, Belaire et al. (2016) found that habitat stewardship in Chicago yards is partly based on people's perceptions about the local bird community. In turn, the actual species richness in the same neighborhoods is supported by yard stewardship decisions that are driven by people's perceptions about local bird communities (Belaire et al. 2014). Therefore, we can envision social filters of biodiversity and the interactions between biota and people as a dynamic, iterative loop across local, neighborhood, and regional scales (Figure 8.1).

Biodiversity promotes human well-being by providing ecosystem services, but urban biota can also spark ecosystem disservices and human-wildlife conflict (Soulsbury and White 2016). Hearing bird songs can elicit an emotional response in people to provide mental and spiritual rejuvenation (Ratcliffe et al. 2016), and the diversity of birdsong heard at urban parks increases their social value for youths (Hedblom et al. 2017). However, other times the sounds birds make are perceived as "annoying" (Cox and Gaston 2015). Some people dislike encountering urban wildlife that they view as damaging personal property or otherwise present a risk to their livelihood, which is often a reality as the human-baboon conflict in Cape Town, South Africa (Hoffman and O'Riain 2012), and human-deer conflicts in the northeastern United States illustrate (Curtis 2020). Urbanites who feel disconnected from nature often hold negative environmental attitudes that may influence land stewardship (Soga et al. 2016). Negative environmental attitudes caused by an indifference toward nature experiences can mediate the degree to which people engage in both private behaviors, such as engaging in yard stewardship activities, as well as their broader support for biodiversity conservation initiatives through civic engagement (Soga and Gaston 2016). However, exposure to fascinating and unfamiliar wildlife, such as insects, can provoke ecological interest in people about the natural world to support biodiversity conservation (Head and Muir 2006; Shipley and Bixler 2017). Understanding how people interact and perceive wildlife, especially the everyday encounters in cities between people and biota close to home, offers an exciting but relatively unexplored area of investigation (Head and Muir 2006).

Key questions and future research needs

Social filters influence urban biodiversity in every metropolitan area. Yet the multi-scalar effects of people and institutions, as well as the interrelationships between social filters, are dependent on specific and unique sets of external factors across various cities. As a result, we can expect to see mechanistic differences that are dependent on both taxa and geographical context. In general, global comparisons of the social filters of biodiversity are a largely understudied area and ripe for future research. The literature to date is largely biased in the Global North and differences from the Global South highlight the importance of considering geographical and socioeconomic context (Anderson et al. 2021; du Toit et al. 2021). For example, the idea that economic filters act as a strong control on biodiversity through income has been recently brought into question due to the extreme variation in the direction and strength of the relationship with factors such as geographical and climatic gradients, as well as the scale of observation (Kuras et al. 2020). Another key research gap is how social filters interact with one another or how their relative importance influences biodiversity differently across scales. There is often a mismatch between the processes that structure social filters and the resulting ecological patterns and processes (Borgström et al. 2006). Therefore, the question remains how to understand and study social filters at ecologically meaningful scales, or vice versa. Overall, a better understanding of multi-scalar filters of biodiversity may provide insight on how to more effectively conserve biodiversity across the multiple interacting social levels that define urban landscapes. Key questions to expand our understanding include the following:

- How do the effects of sociocultural, political, and economic filters vary across scales (e.g. local to regional), taxa (e.g. plants, arthropods, mammals, birds, herpetofauna), green space type (e.g. private yards, cemeteries, public parks, natural and semi-natural areas), and biogeographical context (e.g. Global North versus South)?
- How can we best resolve social-ecological scale mismatches to understand the iterative relationships between multi-scalar social filters and urban biodiversity?
- How should we measure, balance, and prioritize the sometimes-synergistic but sometimes-competing requirements to support human well-being and biodiversity conservation?
- What is the best way to include biodiversity and ecosystem services/disservices in urban planning, design, and governance in ways that are relevant, inclusive, equitable, and context-specific?

Key companion papers

Several key papers exist that conceptually theorize the concept of multi-scalar filters of biodiversity (Grove et al. 2005; Chowdhury et al. 2011). Other papers synthesize current research on multi-scalar filters to explain why we see specific biodiversity patterns in cities (Cook et al. 2011; Aronson et al. 2016; Andrade et al. 2020). However, the applications of this spatial, systems-level understanding have yet to be fully explored (but see Goddard et al. 2010).

Andrade, R., et al. 2020. Predicting the assembly of novel communities in urban ecosystems. *Landscape Ecology* 36:1–15.
Aronson, M.F.J., et al. 2016. Hierarchical filters determine community assembly of urban species pools. *Ecology* 97(11):2952–2963.
Chowdhury, R.R., et al. 2011. A multi-scalar approach to theorizing socio-ecological dynamics of urban residential landscapes. *Cities and the Environment* 4(1):6.
Cook, E.M., S.J. Hall, K.L. Larson. 2012. Residential landscapes as social-ecological systems: A synthesis of multi-scalar interactions between people and their home environment. *Urban Ecosystems* 15(1):19–52.
Goddard, M.A., A.J. Dougill, T.G. Benton. 2010. Scaling up from gardens: Biodiversity conservation in urban environments. *Trends in Ecology & Evolution* 25(2):90–98.
Grove, J.M., W.R. Burch, S.T.A. Pickett. 2005. Social mosaics and urban forestry in Baltimore, Maryland. Pages 250–274 in R.G. Lee, D.R. Field, editors. *Communities and Forests: Where People Meet the Land*. Oregon State University Press, Corvallis, OR.

References

Aalto, H.E., H. Ernstson. 2017. Of plants, high lines and horses: Civic groups and designers in the relational articulation of values of urban natures. *Landscape and Urban Planning* 157:309–321.
Anderson, P.M.L., L.J. Potgieter, L. Chan, S.S. Cilliers, H. Nagendra. 2021. Urban plant diversity: Understanding informing processes and emerging trends. Pages 145–168 in C.M. Shackleton, S.S. Cilliers, E. Davoren, M.J. du Toit, editors. *Urban Ecology in the Global South*. Springer, Cham, Switzerland.
Andrade, R., H.L. Bateman, J. Franklin, D. Allen. 2018. Waterbird community composition, abundance, and diversity along an urban gradient. *Landscape and Urban Planning* 170:103–111.
Andrade, R., J. Franklin, K.L. Larson, C.M. Swan, S.B. Lerman, H.L. Bateman, P.S. Warren, A. York. 2020. Predicting the assembly of novel communities in urban ecosystems. *Landscape Ecology* 36:1–15.
Aronson, M.F.J., F.A. La Sorte, C.H. Nilon, M. Katti, M.A. Goddard, C.A. Lepczyk, P.S. Warren, N.S.G. Williams, S. Cilliers, B. Clarkson, C. Dobbs, R. Dolan, M. Hedblom, S. Klotz, J.L. Kooijmans, I. Kühn, I. MacGregor-Fors, M. McDonnell, U. Mörtberg, P. Pyšek, S. Siebert, J. Sushinsky, P. Werner, M. Winter. 2014. A global analysis of the impacts of urbanization on bird and plant diversity reveals key anthropogenic drivers. *Proceedings of the Royal Society B: Biological Sciences* 281.
Aronson, M.F.J., C.A. Lepczyk, K.L. Evans, M.A. Goddard, S.B. Lerman, J.S. MacIvor, C.H. Nilon, T. Vargo. 2017. Biodiversity in the city: Key challenges for urban green space management. *Frontiers in Ecology and the Environment* 15:189–196.
Aronson, M.F.J., C.H. Nilon, C.A. Lepczyk, T.S. Parker, P.S. Warren, S.S. Cilliers, M.A. Goddard, A.K. Hahs, C. Herzog, M. Katti, F.A. La Sorte, N.S.G. Williams, W. Zipperer. 2016. Hierarchical filters determine community assembly of urban species pools. *Ecology* 97:2952–2963.
Avolio, M.L., D.E. Pataki, T.L.E. Trammell, J. Endter-Wada. 2018. Biodiverse cities: The nursery industry, homeowners, and neighborhood differences drive urban tree composition. *Ecological Monographs* 88:259–276.
Belaire, J.A., L.M. Westphal, E.S. Minor. 2016. Different social drivers, including perceptions of urban wildlife, explain the ecological resources in residential landscapes. *Landscape Ecology* 31:401–413.
Belaire, J.A., C.J. Whelan, E.S. Minor. 2014. Having our yards and sharing them too: The collective effects of yards on native bird species in an urban landscape. *Ecological Applications* 24:2132–2143.
Borgström, S.T., T. Elmqvist, P. Angelstam, C. Alfsen-Norodom. 2006. Scale mismatches in management of urban landscapes. *Ecology and Society* 11(2):16.
Burghardt, K.T., D.W. Tallamy, W.G. Shriver. 2009. Impact of native plants on bird and butterfly biodiversity in suburban landscapes. *Conservation Biology* 23:219–224.
Carreiro, M.M., W.C. Zipperer. 2011. Co-adapting societal and ecological interactions following large disturbances in urban park woodlands. *Austral Ecology* 36:904–915.
Chamberlain, D.E., D.A.W. Henry, C. Reynolds, E. Caprio, A. Amar. 2019. The relationship between wealth and biodiversity: A test of the luxury effect on bird species richness in the developing world. *Global Change Biology* 25:3045–3055.

Chamberlain, D.E., C. Reynolds, A. Amar, D. Henry, E. Caprio, P. Batáry. 2020. Wealth, water and wildlife: Landscape aridity intensifies the urban luxury effect. *Global Ecology and Biogeography* 29:1595–1605.

Chowdhury, R.R., K. Larson, M. Grove, C. Polsky, E. Cook, J. Onsted, L. Ogden. 2011. A multi-scalar approach to theorizing socio-ecological dynamics of urban residential landscapes. *Cities and the Environment (CATE)* 4(1):Article 6.

Cilliers, S., M. du Toit, J. Cilliers, E. Drewes, F. Retief. 2014. Sustainable urban landscapes: South African perspectives on transdisciplinary possibilities. *Landscape and Urban Planning* 125:260–270.

Cook, E.M., S.J. Hall, K.L. Larson. 2011. Residential landscapes as social-ecological systems: A synthesis of multi-scalar interactions between people and their home environment. *Urban Ecosystems* 15:19–52.

Cox, D.T.C., K.J. Gaston. 2015. Likeability of garden birds: Importance of species knowledge & richness in connecting people to nature. *PLoS ONE* 10:e0141505.

Curtis, P.D. 2020. After decades of suburban deer research and management in the eastern United States: Where do we go from here? *Human-Wildlife Interactions* 14:111–128.

Davoren, E., S. Siebert, S. Cilliers, M.J. du Toit. 2016. Influence of socioeconomic status on design of Batswana home gardens and associated plant diversity patterns in northern South Africa. *Landscape and Ecological Engineering* 12:129–139.

du Toit, M.J. 2015. *An Urban Ecological Synthesis of Socio-Ecological Systems Dynamics in Potchefstroom, South Africa*. Ph.D. Dissertation, North-West University, Potchefstroom, South Africa.

du Toit, M.J., D.J. Kotze, S.S. Cilliers. 2016. Landscape history, time lags and drivers of change: Urban natural grassland remnants in Potchefstroom, South Africa. *Landscape Ecology* 31:2133–2150.

du Toit, M.J., C.M. Shackleton, S.S. Cilliers, E. Davoren. 2021. Advancing urban ecology in the Global South: Emerging themes and proposed future research directions. Pages 433–461 in C.M. Shackleton, S.S. Cilliers, E. Davoren, M.J. du Toit, editors. *Urban Ecology in the Global South*. Springer, Cham, Switzerland.

Dyson, K. 2019. Vegetation communities on commercial developments are heterogenous and determined by development and landscaping decisions, not socioeconomics. *PLoS ONE* 14:e0222069.

Evans, K.L., D.E. Chamberlain, B.J. Hatchwell, R.D. Gregory, K.J. Gaston. 2011. What makes an urban bird? *Global Change Biology* 17:32–44.

Ferenc, M., O. Sedláček, R. Fuchs, M. Dinetti, M. Fraissinet, D. Storch. 2014. Are cities different? Patterns of species richness and beta diversity of urban bird communities and regional species assemblages in Europe. *Global Ecology and Biogeography* 23:479–489.

Foster, D., F. Swanson, J. Aber, I. Burke, N. Brokaw, D. Tilman, A. Knapp. 2003. The importance of land-use legacies to ecology and conservation. *BioScience* 53:77–88.

Fournier, B., D. Frey, M. Moretti. 2020. The origin of urban communities: From the regional species pool to community assemblages in city. *Journal of Biogeography* 47:615–629.

Galster, G. 2001. On the nature of neighbourhood. *Urban Studies* 38:2111–2124.

Goddard, M.A., A.J. Dougill, T.G. Benton. 2010. Scaling up from gardens: Biodiversity conservation in urban environments. *Trends in Ecology & Evolution* 25:90–98.

Goodness, J. 2018. Urban landscaping choices and people's selection of plant traits in Cape Town, South Africa. *Environmental Science & Policy* 85:182–192.

Groffman, P.M., M.L. Cadenasso, J. Cavender-Bares, D.L. Childers, N.B. Grimm, J.M. Grove, S.E. Hobbie, L.R. Hutyra, G.D. Jenerette, T. McPhearson, D.E. Pataki, S.T.A. Pickett, R.V. Pouyat, E. Rosi-Marshall, B.L. Ruddell. 2016. Moving towards a new urban systems science. *Ecosystems* 20:38–43.

Grove, J.M., W.R. Burch Jr., S.T.A. Pickett. 2005. Social mosaics and urban forestry in Baltimore, Maryland. Pages 250–274 in R.G. Lee, D.R. Field, editors. *Communities and Forests: Where People Meet the Land*. Oregon State University Press, Corvallis, OR.

Head, L., P. Muir. 2006. Suburban life and the boundaries of nature: Resilience and rupture in Australian backyard gardens. *Transactions of the Institute of British Geographers* 31:505–524.

Hedblom, M., I. Knez, B. Gunnarsson. 2017. Bird diversity improves the well-being of city residents. Pages 287–306 in E. Murgui, M. Hedblom, editors. *Ecology and Conservation of Birds in Urban Environments*. Springer, Cham, Switzerland.

Hoffman, J.S., V. Shandas, N. Pendleton. 2020. The effects of historical housing policies on resident exposure to intra-urban heat: A study of 108 US urban areas. *Climate* 8:12.

Hoffman, T.S., M.J. O'Riain. 2012. Monkey management: Using spatial ecology to understand the extent and severity of human-baboon conflict in the Cape Peninsula, South Africa. *Ecology and Society* 17.

Hoover, F.-A., T.C. Lim. 2021. Examining privilege and power in US urban parks and open space during the double crises of antiblack racism and COVID-19. *Socio-Ecological Practice Research* 3:55–70.

Hope, D., C. Gries, W. Zhu, W.F. Fagan, C.L. Redman, N.B. Grimm, A.L. Nelson, C. Martin, A. Kinzig. 2003. Socioeconomics drive urban plant diversity. *Proceedings of the National Academy of Sciences* 100:8788–8792.

Kendal, D., K.J.H. Williams, N.S.G. Williams. 2012. Plant traits link people's plant preferences to the composition of their gardens. *Landscape and Urban Planning* 105:34–42.

Knapp, S., M. Winter, S. Klotz. 2017. Increasing species richness but decreasing phylogenetic richness and divergence over a 320-year period of urbanization. *Journal of Applied Ecology* 54:1152–1160.

Kuras, E.R., P.S. Warren, J.A. Zinda, M.F.J. Aronson, S. Cilliers, M.A. Goddard, C.H. Nilon, R. Winkler. 2020. Urban socioeconomic inequality and biodiversity often converge, but not always: A global meta-analysis. *Landscape and Urban Planning* 198:103799.

Larson, K.L., R. Andrade, K.C. Nelson, M.M. Wheeler, J.M. Engebreston, S.J. Hall, M.L. Avolio, P.M. Groffman, M. Grove, J.B. Heffernan, S.E. Hobbie, S.B. Lerman, D.H. Locke, C. Neill, R.R. Chowdhury, T.L.E. Trammell. 2020. Municipal regulation of residential landscapes across US cities: Patterns and implications for landscape sustainability. *Journal of Environmental Management* 275:111132.

Larson, K.L., J. Brumand. 2014. Paradoxes in landscape management and water conservation: Examining neighborhood norms and institutional forces. *Cities and the Environment (CATE)* 7.

Larson, K.L., J. Hoffman, J. Ripplinger. 2017. Legacy effects and landscape choices in a desert city. *Landscape and Urban Planning* 165:22–29.

Leong, M., R.R. Dunn, M.D. Trautwein. 2018. Biodiversity and socioeconomics in the city: A review of the luxury effect. *Biology Letters* 14:20180082.

Lerman, S.B., V.K. Turner, C. Bang. 2012. Homeowner associations as a vehicle for promoting native urban biodiversity. *Ecology and Society* 17(4):45.

Leveau, L.M., C.M. Leveau, M. Villegas, J.A. Cursach, C.G. Suazo. 2017. Bird communities along urbanization gradients: A comparative analysis among three neotropical cities. *Ornitología Neotropical* 28:77–87.

Locke, D.H., B. Hall, J.M. Grove, S.T.A. Pickett, L.A. Ogden, C. Aoki, C.G. Boone, J.P.M. O'Neil-Dunne. 2021. Residential housing segregation and urban tree canopy in 37 US cities. *Urban Sustainability* 1:15.

Marshall, A.J., M.J. Grose, N.S.G. Williams. 2020. Of mowers and growers: Perceived social norms strongly influence verge gardening, a distinctive civic greening practice. *Landscape and Urban Planning* 198:103795.

Martin, D.G. 2003. Enacting Neighborhood1. *Urban Geography* 24:361–385.

McKinney, M.L. 2008. Effects of urbanization on species richness: A review of plants and animals. *Urban Ecosystems* 11:161–176.

Minor, E., A.J. Belaire, A. Davis, M. Franco. 2016. Socioeconomics and neighbor mimicry drive yard and neighborhood vegetation patterns. Pages 74–92 in R.A. Francis, J.D.A. Millington, M.A. Chadwick, editors. *Urban Landscape Ecology*. Routledge, London.

Monger, C., O.E. Sala, M.C. Duniway, H. Goldfus, I.A. Meir, R.M. Poch, H.L. Throop, E.R. Vivoni. 2015. Legacy effects in linked ecological – soil – geomorphic systems of drylands. *Frontiers in Ecology and the Environment* 13:13–19.

Nagendra, H., D. Gopal. 2010. Tree diversity, distribution, history and change in urban parks: Studies in Bangalore, India. *Urban Ecosystems* 14:211–223.

Narango, D.L., D.W. Tallamy, P.P. Marra. 2017. Native plants improve breeding and foraging habitat for an insectivorous bird. *Biological Conservation* 213:42–50.

Norton, B.A., K.L. Evans, P.H. Warren. 2016. Urban biodiversity and landscape ecology: Patterns, processes and planning. *Current Landscape Ecology Reports* 1:178–192.

Ossola, A., M.E. Hopton. 2018a. Measuring urban tree loss dynamics across residential landscapes. *Science of The Total Environment* 612:940–949.

Ossola, A., M.E. Hopton. 2018b. Climate differentiates forest structure across a residential macrosystem. *Science of the Total Environment* 639:1164–1174.

Ossola, A., D. Locke, B. Lin, E. Minor. 2019. Greening in style: Urban form, architecture and the structure of front and backyard vegetation. *Landscape and Urban Planning* 185:141–157.

Ostrom, E. 2000. Collective action and the evolution of social norms. *Journal of Economic Perspectives* 14:137–158.

Padullés Cubino, J., M.L. Avolio, M.M. Wheeler, K.L. Larson, S.E. Hobbie, J. Cavender-Bares, S.J. Hall, K.C. Nelson, T.L.E. Trammell, C. Neill, D.E. Pataki, J.M. Grove, P.M. Groffman. 2020. Linking yard plant diversity to homeowners' landscaping priorities across the U.S. *Landscape and Urban Planning* 196:103730.

Pearse, W.D., J. Cavender-Bares, S.E. Hobbie, M.L. Avolio, N. Bettez, R.R. Chowdhury, L.E. Darling, P.M. Groffman, J.M. Grove, S.J. Hall, J.B. Heffernan, J. Learned, C. Neill, K.C. Nelson, D.E. Pataki, B.L. Ruddell, M.K. Steele, T.L.E. Trammell. 2018. Homogenization of plant diversity, composition, and structure in North American urban yards. *Ecosphere* 9(2):ecs2.2105.

Perkins, H.A., N. Heynen, J. Wilson. 2004. Inequitable access to urban reforestation: The impact of urban political economy on housing tenure and urban forests. *Cities* 21:291–299.

Piano, E., C. Souffreau, T. Merckx, L.F. Baardsen, T. Backeljau, D. Bonte, K.I. Brans, M. Cours, M. Dahirel, N. Debortoli, E. Decaestecker, K. de Wolf, J.M.T. Engelen, D. Fontaneto, A.T. Gianuca, L. Govaert, F.T.T. Hanashiro, J. Higuti, L. Lens, K. Martens, H. Matheve, E. Matthysen, E. Pinseel, R. Sablon, I. Schön, R. Stoks, K. Van Doninck, H. Van Dyck, P. Vanormelingen, J. Van Wichelen, W. Vyverman, L. De Meester, F. Hendrickx. 2020. Urbanization drives cross-taxon declines in abundance and diversity at multiple spatial scales. *Global Change Biology* 26:1196–1211.

Pickett, S.T.A., M.L. Cadenasso, E.J. Rosi-Marshall, K.T. Belt, P.M. Groffman, J.M. Grove, E.G. Irwin, S.S. Kaushal, S.L. LaDeau, C.H. Nilon, C.M. Swan, P.S. Warren. 2016. Dynamic heterogeneity: A framework to promote ecological integration and hypothesis generation in urban systems. *Urban Ecosystems* 20:1–14.

Pincetl, S. 2012. Nature, urban development and sustainability – What new elements are needed for a more comprehensive understanding? *Cities* 29:32–37.

Pincetl, S., S.S. Prabhu, T.W. Gillespie, G.D. Jenerette, D.E. Pataki. 2013. The evolution of tree nursery offerings in Los Angeles county over the last 110 years. *Landscape and Urban Planning* 118:10–17.

Potgieter, L.J., E. Douwes, M. Gaertner, J. Measey, T. Paap, D.M. Richardson. 2020. Biological invasions in South Africa's urban ecosystems: Patterns, processes, impacts, and management. Pages 275–309 in B.W. van Wilgen, J. Measey, D.M. Richardson, J.R. Wilson, T.A. Zengeya, editors. *Biological Invasions in South Africa*. Springer International Publishing, Berlin, Germany.

Pulido, L. 2000. Rethinking environmental racism: White privilege and urban development in southern California. *Annals of the Association of American Geographers* 90(1):12–40.

Ramalho, C.E., R.J. Hobbs. 2012. Time for a change: Dynamic urban ecology. *Trends in Ecology & Evolution* 27(3):179–188.

Ratcliffe, E., B. Gatersleben, P.T. Sowden. 2016. Associations with bird sounds: How do they relate to perceived restorative potential? *Journal of Environmental Psychology* 47:136–144.

Roman, L.A., H. Pearsall, T.S. Eisenman, T.M. Conway, R.T. Fahey, S. Landry, J. Vogt, N.S. van Doorn, J.M. Grove, D.H. Locke, A.C. Bardekjian, J.J. Battles, M.L. Cadenasso, C.C.K. van den Bosch, M. Avolio, A. Berland, G.D. Jenerette, S.K. Mincey, D.E. Pataki, C. Staudhammer. 2018. Human and biophysical legacies shape contemporary urban forests: A literature synthesis. *Urban Forestry & Urban Greening* 31:157–168.

Schell, C.J., K. Dyson, T.L. Fuentes, S. des Roches, N.C. Harris, D.S. Miller, C.A. Woelfle-Erskine, M.R. Lambert. 2020. The ecological and evolutionary consequences of systemic racism in urban environments. *Science* 2020:369.

Shipley, N.J., R.D. Bixler. 2017. Beautiful bugs, bothersome bugs, and FUN bugs: Examining human interactions with insects and other arthropods. *Anthrozoös* 30:357–372.

Sisser, J.M., K.C. Nelson, K.L. Larson, L.A. Ogden, C. Polsky, R.R. Chowdhury. 2016. Lawn enforcement: How municipal policies and neighborhood norms influence homeowner residential landscape management. *Landscape and Urban Planning* 150:16–25.

Soga, M., K.J. Gaston. 2016. Extinction of experience: The loss of human – nature interactions. *Frontiers in Ecology and the Environment* 14:94–101.

Soga, M., K.J. Gaston, Y. Yamaura, K. Kurisu, K. Hanaki. 2016. Both direct and vicarious experiences of nature affect children's willingness to conserve biodiversity. *International Journal of Environmental Research and Public Health* 13(6):529.

Soulsbury, C.D., P.C.L. White. 2016. Human – wildlife interactions in urban areas: A review of conflicts, benefits and opportunities. *Wildlife Research* 42:541–553.

Urban, D.L., R.V. O'Neill, H.H. Shugart Jr. 1987. A hierarchical perspective can help scientists understand spatial patterns. *BioScience* 37:119–127.

van Wilgen, B. 2012. Evidence, perceptions, and trade-offs associated with invasive alien plant control in the Table Mountain National Park, South Africa. *Ecology and Society* 17(2).

Venter, Z.S., C.M. Shackleton, F. Van Staden, O. Selomane, V.A. Masterson. 2020. Green Apartheid: Urban green infrastructure remains unequally distributed across income and race geographies in South Africa. *Landscape and Urban Planning* 203:103889.

Watkins, S.L., E. Gerrish. 2018. The relationship between urban forests and race: A meta-analysis. *Journal of Environmental Management* 209:152–168.

Wheeler, M.M., K.L. Larson, R. Andrade. 2020. Attitudinal and structural drivers of preferred versus actual residential landscapes in a desert city. *Urban Ecosystems* 23:659–673.

Wolf, K.L. 2009. Strip malls, city trees, and community values. *Arboriculture and Urban Forestry* 35:33–40.

Wu, J. 1999. Hierarchy and scaling: Extrapolating information along a scaling ladder. *Canadian Journal of Remote Sensing* 25:367–380.

Yi, H., U.P. Kreuter, D. Han, B. Güneralp. 2019. Social segregation of ecosystem services delivery in the San Antonio region, Texas, through 2050. *Science of the Total Environment* 667:234–247.

York, A.M., K. Kane, C.M. Clark, L.E. Gentile, A. Wutich, S.L. Harlan. 2017. What determines public support for graduated development impact fees? *State and Local Government Review* 49:15–26.

9

PEOPLE'S DECISIONS SHAPE URBAN HABITATS

Dave Kendal, Emily Minor, and Monika Egerer

Introduction

There is increasing awareness that the patterns and mechanisms explaining urban ecosystems and the biodiversity therein can only be understood by including people in urban ecological explorations. People's decisions shape the planning, design, and management of urban habitats and cityscapes. Selecting trees to grow along roadsides, leaving space for parks in new residential development, cultivating soils in garden beds, lawn mowing, and placing bird seed or water baths in private yards are all examples of decisions made by urban planners, managers, and residents at various spatial scales that have direct and indirect effects on urban biodiversity and the ecology of urban habitats (Goddard et al. 2010; Aronson et al. 2017a). The integration of social theory around human decision-making is needed in contemporary research that aims to understand the ecology 'in, of and for' cities (McPhearson et al. 2016).

In this chapter, we introduce typologies of decision-making mechanisms, attributes, and actors to help make sense of decision-making related to urban habitats. In addition, we review some of the main concepts and theories in the social sciences that have been used to explain human decision-making in this context. We describe how humans can shape urban habitats as decision-makers, how their values and preferences and political processes interact with structural constraints and biophysical factors to shape these decisions. In doing so, we use some case studies and examples from across the world to provide a broad picture of diverse management decisions, actions, and outcomes for urban habitats. We have three main objectives: to (1) identify the major mechanisms by which human decisions shape urban habitats; (2) explain the attributes of human decisions, including spatial and temporal dimensions and intentional versus unintentional effects; and (3) describe the main drivers of human decision-making.

Key decision-making mechanisms that influence urban habitats

Millions of decisions that influence urban habitats are made every day by people in different capacities, all over the world. Mechanisms by which people influence urban habitats include governance, policy and regulation, planning and design, and management.

 DOI: 10.4324/9781003016120-11

Governance

Governance can be defined as 'the collection of institutions, rules, and processes of collective decision-making that allows stakeholders to influence and coordinate their needs' (Ordóñez et al. 2019). The concept of governance includes government, private enterprise, community actors, tools such as policy and legislation, community engagement, risk and project management, accountability, and the reproduction or reinforcement of social norms. While some environmental governance directly engages with urban habitats (e.g. threatened species legislation, community stewardship of local parks), governance strongly shapes the way cities and towns are designed, planned and managed, in turn indirectly shaping habitats and biodiversity conservation outcomes (Puppim de Oliveira et al. 2011). Governance arrangements can vary greatly across countries with different political and economic systems. They can be formally codified in processes, policies, rules, and regulations, or more informal with a greater dependence on social relationships. They can change over time in ways that affect biodiversity outcomes, such as the recent shifts in the postindustrial West to greater community participation in decisions traditionally made by technical experts on the design and management of urban landscapes (Gulsrud et al. 2018). Understanding governance can help to explain how power is accumulated and wielded in cities and how this translates to habitat management.

Policy and regulation

City policies and regulations help to formalise governance arrangements. Policies are statements of intent by particular institutions describing how they would like the world to be, for example, urban biodiversity strategies by local governments (Nilon et al. 2017). Regulations are enforceable rules such as legislation, ordinances, and bylaws that constrain decisions and actions to achieve desired outcomes and implement policy. Policy and regulation can direct planning and design of green spaces (e.g. mandating certain plant species; mandating amount of minimum green space areas), constrain adverse behaviours (e.g. restricting cat roaming or the cultivation of invasive species), identify desirable outcomes (e.g. canopy cover targets), identify which habitats need to be preserved and which are permitted to be cleared in urban development, protect threatened species, and specify rules for biodiversity offsets.

Planning and design

In urban areas, planning heavily influences the arrangement of space and therefore habitat structure, patch size, and connectivity. Planning decisions apply regulation and policy to shape the size and location of parks and the amount of space left for vegetation along roadsides and in residential gardens (Larson et al. 2020). In a review of 135 plans from 40 cities around the world, more than 80% of plans incorporated at least one goal for enhancing ecosystem services (Nilon et al. 2017). Most plans also included goals for conserving biodiversity and habitats, but few plans included quantitative targets (e.g. increasing amount of area under protection), reducing the likelihood of plans being achieved (Nilon et al. 2017).

Design specifies the initial configuration of built and natural elements within a site. Important aspects of design include tree species selection, which can have long-term effects on biodiversity (White et al. 2005), spatial arrangement of natural elements that define connectivity and structural complexity (Peng et al. 2020), and edge treatments that connect biodiverse areas to their surrounding urban landscapes (Nassauer 1995). Design can be formalised in landscape plans and architectural drawings, such as in urban parks, or can be informally implemented based on spontaneous decisions by the relevant landholder/manager, such as in many residential gardens.

Planning and design decisions work together to influence biodiversity. For example, trees can exist in cities where planning allows enough space for trees to grow and designs choose trees over alternative treatments such as turf. Planning and design for multifunctionality is increasingly occurring in urban green spaces, where the same space can provide ecosystem service benefits such as cooling and flood reduction, biodiversity outcomes, amenity, and recreation (Lovell and Taylor 2013). There are often trade-offs and synergies across these multiple objectives that must be negotiated in the planning and design processes (Dobbs et al. 2014).

Management practices

Management practices include the planned and unplanned activities that implement and maintain designs within urban habitats (Aronson et al. 2017a). This includes regular maintenance such as mowing, replanting, and removing undesirable species and populations, and interventions such as installing artificial habitat including 'bat boxes' and bird baths. Management activities can be conducted by different actors, from municipal arborists to home gardeners. Management can also incorporate public involvement in urban habitats (Andersson et al. 2014) and ecological restoration (Clarkson and Kirby 2016). Management regimes, such as frequency of mowing, can have significant ecological effects on the composition and abundance of species at different trophic levels (Threlfall et al. 2016b; Lerman et al. 2018). Other management decisions can have unexpected impacts on ecological communities. For example, fencing the edge of vacant lots in Chicago (USA) resulted in the creation of distinct plant communities along the fence line (Anderson and Minor 2020).

Monitoring the ecological effects of management decisions is an often-overlooked area of management. It has been adopted in some areas of management, such as street trees, primarily to manage risk and costs. Increasingly, citizen science programs are being used to monitor urban biodiversity to collect data and for community engagement and education.

Attributes of human decisions

Intentional and unintentional, positive and negative effects

Human decisions have intentional and unintentional effects on urban habitats and urban biodiversity. Some decisions are intended to directly affect particular habitats or species. For example, when urban residents place bird feeders in their yards, this decision has the intentional effect of increasing resources for a small group of bird species. Decisions to manipulate urban habitat, such as planting or removing urban trees, and starting wildlife-friendly gardening for birds and bees (Apfelbeck et al. 2020; Goddard et al. 2013), are decisions with intentional effects. Decisions that add or remove wildlife from the habitat, such as using pesticide to eliminate arthropod or rodent pests, also have intentional effects.

Many decisions may have large unintended consequences on urban habitats and biodiversity. For example, urban planning and design decisions that maximise human population density and profit for developers may leave little room for large trees in the landscape or may heavily fragment existing vegetation. The aforementioned bird feeder could reduce natural insect or seed predation if bird seed is favoured over 'natural' prey, and the pesticides could eliminate non-target species. The timing of mowing can allow grassland species to set seed (or not), affecting the composition of species in neighbouring landscapes. Simplifying urban yard vegetation due to preferences for 'tidy' landscapes (Nassauer 1995) can reduce wildlife species richness and abundance. Allowing house cats to roam outside may decrease parental feeding of chicks and increase nest predation for urban songbirds (Bonnington et al. 2013).

As these examples show, decisions can have either positive or negative outcomes on urban biodiversity, often without explicitly considering impacts on biodiversity. Sometimes the same decision can have both positive and negative effects on different species or taxonomic groups. When considering the effects of decisions on urban biodiversity, it must be recognised that decisions can affect multiple species, taxonomic groups, and ecological communities, sometimes in different ways. While it is possible to plan for multi-species outcomes, this is difficult, particularly in complex social-ecological systems such as urban areas.

Spatial and temporal attributes of the effects of decisions

The effects of all human decisions have temporal and spatial extents (Figure 9.1). Some decisions, such as planting a tree, have a small spatial extent but a long temporal extent (possibly hundreds of years). Other decisions, such as municipal mowing, may have a direct effect that only lasts a few days or weeks but can cover many hectares. Decisions such as city planning and state or national laws may have particularly large spatial and long temporal effects. Decisions may also vary in frequency. Lawn irrigation may be very frequent, even daily, which can cumulatively lead to long temporal effects. Decisions with small spatial and short temporal effects, such as residential lawn mowing, can be repeated frequently by many different decision-makers, leading to large cumulative effects across space and time. Decisions have legacies that may lead to large temporal

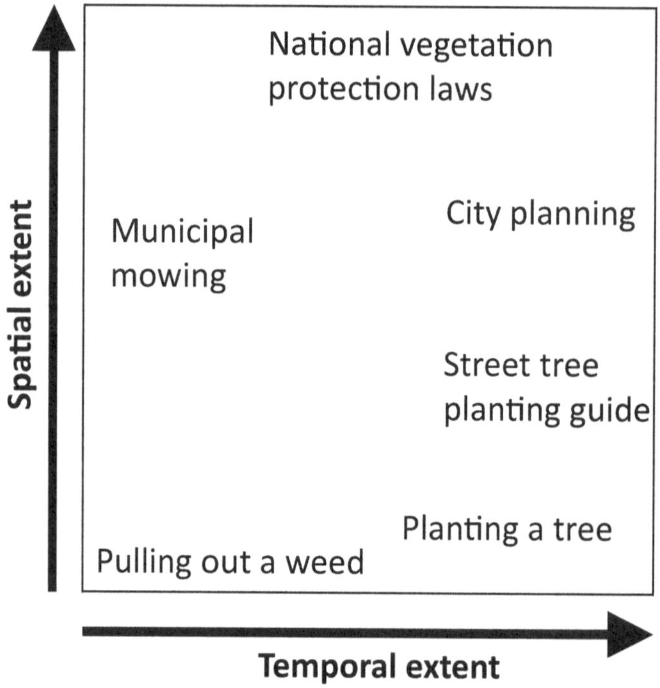

Figure 9.1 The temporal and spatial extent of different decisions.

extents. Urban tree diversity and structural complexity are good examples of decision-making at different temporal scales: long-term landscape-scale planning around forest diversity and structure is needed, while removal and pruning decisions may be very short term and sporadic (Stagoll et al. 2010). All these management decisions are important for influencing habitat availability.

Land use types and their decision-makers

In many cities, land use planning spatially segregates the effects of decision-making through regulation. In developing countries, there is often less strict segregation of land uses and a greater emphasis on informal planning, including green spaces (Chishaleshale et al. 2015). Decisions in different land uses are the responsibility of different actors encompassing individuals or organisations responsible for different spatial extents (Table 9.1). For example, municipal authorities may be responsible for public land, including parks and streetscapes, while residents, businesses, and homeowners' associations may have greater responsibility in the private realm of residential or commercial landscaping. Some actors, such as landscape architects, may make decisions that influence both public and private landscapes. However, decisions are fluid, and decisions made about a particular land use can affect others, and many different actors may be involved in decisions within a particular land use. In the next section, we outline some important land use types shaping urban habitats, the decision-making mechanisms that influence them, and how these land uses are typically associated with different actors.

Understanding drivers of human decision-making

Much exploration of decision-making related to urban biodiversity assumes that decision-making is a rational process, based on predictable choices people make in response to evidence and evaluation of alternatives (Laurans et al. 2020). Within this rational framework, key influences on decision-making include policies and planning, individual cognitive factors such as values and preferences, sociocultural drivers such as demographic variables, economic factors, and knowledge of the biophysical environment (Sutherland and Freckleton 2012). To support better decision-making within this framework, more information is collected, and a range of decision-support systems are implemented. Information can include spatial information on biodiversity assets (e.g. occurrence, species distribution models), as well as information on individual factors (e.g. preferences for particular species) and sociocultural data (e.g. demographics and cultural background of residents).

However, decision-making is not always rational. Individuals can make decisions that are inconsistent with the information available and their own values and preferences. Decision-making can also be seen as a political process. For example, theories in political ecology and environmental justice show how decision-making can favour powerful interests over those less powerful, leading to the unequal distribution of environmental goods (Heynen et al. 2006). A rich body of literature has demonstrated patterns of biodiversity, and tree canopy are associated with patterns of socioeconomic advantage. This has been dubbed the 'luxury effect' and has been explained using rational approaches, such as 'economic wherewithal' of residents to plant vegetation or move to vegetated areas (Hope et al. 2008). Yet individual actions are not the only explanation for this association. Political processes also play an important role in many places. Public decision-making can lead to private benefit from public goods (such as street trees or biodiversity) for advantaged sections of the community (Landry and Chakraborty 2009). Policies, ordinances, and rules that restrict resident behaviours in private landscapes (e.g. when renters are unable to plant trees, or

Table 9.1 Some of the actors involved in decisions across different land use types

Land use type	*Description*	*Householder*	*Arborist/ Horticulturist*	*Municipal planner/ policymaker*	*Landscaper*	*Landscape architect*	*Property developer*	*Nursery manager*	*Ecological restorationist*	*Conservation planner*
Public parks	Planning influences size, connectivity, and human use of parks, and design influences species composition and habitat structure. Parks are infrequently developed but are likely to have fairly stable management after establishment. They often comprise larger patches, and have larger trees, than other urban habitats.		X	X	X	X	X	X		
Streetscapes	Trees are often the main element in this land use (although turf or understorey vegetation can be important), and tree management for safety and amenity has historically been a key driver of decision-making (Ordóñez et al. 2019). Initial planning and design are important, but decision-making is frequent due to the stressful environment, public risk, and high mortality of street trees (Roman et al. 2014). These areas are individually small but cumulatively can cover a large area (Marshall et al. 2019).		X	X	X	X	X	X		
Conservation areas	Urban expansion has resulted in patches of ecologically-important communities being preserved in cities (Kendal et al. 2017b). Planning affects patch size, quality, connectivity, and surrounding land uses. Some ecosystems require disturbance regimes (e.g. fire) that have been difficult to implement in urban areas due to public opinions, although there is increasing evidence that the public can support these management			X			X		X	X

	approaches (Farrar et al. 2020). Design, particularly around edges and entrances, can shape public opinion towards these patches (Nassauer 1995).								
Riparian zones	Areas subject to flooding are spared from development in many cities. These can be important areas for habitat and landscape connectivity, although they can also contribute to connectivity for non-native species (Aronson et al. 2017b). The planning of riparian corridors is increasingly regulated to allow for multifunctional use e.g. linear paths, recreational areas.		X	X	X	X	X	X	X
Cemeteries	These are an important refuge for many species with different patterns of management (and therefore disturbance) than many other urban green spaces (Kowarik et al. 2016; Smith and Minor 2019)		X					X	X
Institutional/ Botanical gardens	Many cities have botanic gardens, natural history museums, zoos, and other institutional landscapes, which may be relatively small but contain very high levels of plant and animal species diversity. Botanic gardens can play an important role in conservation and education for the public to learn about plants and the role they play in ecosystems (Ballantyne et al. 2008).	X	X		X				

(Continued)

Table 9.1 (Continued)

Land use type	*Description*	*Householder*	*Arborist/ Horticulturist*	*Municipal planner/ policymaker*	*Landscaper*	*Landscape architect*	*Property developer*	*Nursery manager*	*Ecological restorationist*	*Conservation planner*
Residential gardens	These green spaces are typically small but cumulatively can cover a large proportion (35–50%) of Western cities (Kendal et al. 2012b; Loram et al. 2007), although often much less in large, high density cities. Planning and regulation shape the size and arrangement of spaces. Frequent, uncoordinated, small-scale decisions by homeowners can lead to very high levels of plant diversity (Kendal et al. 2010). While these decisions can be lightly regulated, in some places a range of policies and regulations by local municipal authorities or subdivision developers can constrain decision-making (Sisser et al. 2016)	X		X	X	X	X	X		
Constructed green spaces	Green roofs, facades, and green interiors can be important for biodiversity conservation (Oberndorfer et al. 2007). Habitat design and management decisions, e.g. plant species diversity and ground cover, may influence arthropod diversity and community composition (e.g. Braaker et al. 2014; Kratschmer et al. 2018). The decision to install green roofs and facades may be driven by climate mitigation policies (Norton et al. 2015).			X	X	X	X	X		
Commercial landscapes	Shopping areas and industrial precincts are often managed for low maintenance and safety rather than for biodiversity. Nevertheless, they can still be habitats for biodiversity, though these sites often support generalist taxa and not habitat specialists (Cook and Faeth 2006).				X	X		X		

Golf courses	These spaces can be relatively large in area but homogenous and simplified in their ecological structure: for example, turf grass is often the dominant plant species, and there is relatively little canopy cover (Threlfall et al. 2016a). However, in some regions, golf courses support endangered habitats (e.g. wetlands in England) and can be essential habitats for supporting urban biodiversity if carefully designed (Colding and Folke 2009). The design of golf courses requires some areas (e.g. greens, fairways) to be intensively managed turf grass, yet other areas can support greater vegetation complexity and provide habitat for biodiversity (Threlfall et al. 2016b).	w	X		X
Community and allotment gardens	These spaces are often managed by multiple individuals who have been allocated a space within the garden, although sometimes the whole garden is communally managed (Egerer et al. 2018). Management is often intense and diverse due to the many gardeners involved, leading to high levels of habitat heterogeneity and species diversity. Policies can restrict decisions in some gardens.	X		X	
Informal green spaces	These areas are not subject to traditional land use planning and management, and include vacant or abandoned lots, street and railway verges (Rupprecht et al. 2015), and "wastelands" (Kowarik 2018). They may be publicly or privately owned but are subject to little organised planning, design, or management. Spontaneous ecosystems can emerge from these systems.				

particular species are labelled as weeds) can also reinforce the political drivers of urban landscape configuration (Rappaport and Horn 1998). In this model, additional information does not necessarily lead to 'better' decisions but instead tends to reinforce inequality in outcomes as information is also unequally distributed.

Individual preferences, beliefs, and values

In psychology, people's values, beliefs, and preferences are seen as key determinants of decision-making (Ives and Kendal 2014). **Transcendental values** are the broad, guiding principles that shape the way people interact with the world (Schwartz 1994). While there has been much research on people's environmental values and how they shape pro-environmental behaviours (Stern et al. 1995), research has also identified a 'value-action' gap that highlights that many people act inconsistently with their values (Shove 2010). However, in rational decision-making processes, values are seen as important in shaping objectives and priorities and are increasingly being elicited from stakeholders through activities such as structured decision-making that explicitly seek to understand stakeholder values and objectives (Martin et al. 2018). **Beliefs** are the things that people hold to be true (although they may not be supported by evidence). A variety of beliefs are important in decision-making, including beliefs around the consequences of decisions and normative beliefs around the way people think the world should be, and are influenced by environmental concerns and awareness. **Preferences** are a type of attitude that describes people's judgments about how much they like a thing, often from a set of alternatives. In urban ecology, preferences have been identified as a key filter operating in urban ecosystems to determine species composition (Williams et al. 2009). A range of theories in environmental psychology have been used to explain people's preferences. Evolutionary theories argue that humans have an evolved preference for landscapes that provide good habitat, while cultural approaches show that preferences can vary between different cultural groups. Values, beliefs, and attitudes are often placed in a hierarchy, where values are seen to influence beliefs, which in turn are seen to influence preferences/attitudes (Ives and Kendal 2014).

Psychological theories based on individual values, beliefs, and preferences have been used to explain decision-making and behaviours around urban habitat. The theory of planned behaviour (Ajzen 1991) proposes that people's beliefs (e.g. about their ability to take an action), personal and social norms, and attitudes influence behavioural intentions. Value-belief-norm theory (Stern et al. 1995) argues that values and beliefs about the consequences of actions are important predictors of decision-making. Theories from sociology and human geography tend to focus less on the individual and more on the political and social structures individuals are acting within. An important theory bridging social structures and individuals is practice theory – rather than focussing on individual agency. as psychological approaches do, practice theory focuses on behaviours that are reproduced and negotiated by individuals as part of their sense-making in the world (Hargreaves 2011).

Social/cultural and evolutionary drivers

As well as individual factors, cultural and social factors can be strong drivers of decision-making around urban habitats. Social norms are the informal rules that shape everyday decisions and behaviours within social or cultural groups. A range of studies have shown that people's gardening decisions and behaviours around plant selection are shaped by their cultural background. For example, people from Mediterranean cultures are more likely to plant food plants, while people from British cultures are more likely to plant shade trees (Fraser and Kenney 2000). Food preferences and cultural associations with specific plant species and varieties can shape the flora of a garden habitat (Glowa et al. 2019). Social norms around lawn care can lead to the homogenisation of urban residential habitats (Goddard et al. 2010). Another group of theories suggest that people's response to the environment around them is genetically determined – humans have evolved to prefer particular landscapes and live in the landscape in particular ways that maximise genetic fitness (Orians and Heerwagen 1992). For example, people liking plants with large green leaves can be seen as an evolutionary response to an environment – humans are programmed to prefer signs of high nutrient and water availability (Kendal et al. 2012a).

Environmental drivers

While social drivers are an important influence on decision-making, contextual environmental considerations are also important constraints on decision-making. For example, the climate of an urban region can shape urban forestry and greening plans or home gardeners' plant selection. Irrigation or shading may allow urban residents to grow plant species beyond their climate niche thresholds, though temperature variation can still be a challenge to manage (Egerer et al. 2019). Climate change and increasing urban heat presents many new challenges to decision-makers, from increasing temperature to precipitation extremes. Forecasted climatic conditions are often unfamiliar; city climates are forecast to shift the equivalent of hundreds or even thousands of kilometres towards the equator. It remains to be seen what plant species will be selected, how changing species selection will influence genotypic and phenotypic diversity, and how these unfamiliar species will be managed in urban habitats by human actors worldwide. A range of other important environmental drivers include soil geochemistry, wind flow, and topography.

Emerging trends in decision-making affecting urban habitats

Decision-support tools, data, and techniques

Many realms of professional decision-making are increasingly being supported by tools and technologies, interdisciplinary research methods, and citizen-science interfaces. There is increasing use of geographic information systems (GIS) to capture and disseminate information on urban habitats, such as tree cover, tree species inventories, species observation records, and the location of different land uses, including protected areas. Spatially explicit data is being collected by community members through tools such as Public Participatory GIS (PPGIS: Gulsrud et al. 2018). Some governments at municipal, state, and federal levels are starting to make data more available through open data platforms online. This allows novel data analyses to be conducted to inform decision-making (Kendal et al. 2017a). Structured decision-making, a process for capturing objectives and values and modelled predictions (e.g. ecosystem services) of the outcomes of decisions, including uncertainty, is beginning to be used more widely in ecosystem management (Martin et al. 2018).

Citizen science

Citizen science methods can affordably and effectively expand the capacity and breadth of urban ecological monitoring in relation to management decisions by involving the decision-makers in the research process, which can also help build community participants (Bonney et al. 2009). This is particularly effective for broad biodiversity assessments or hard-to-access sites in urban areas. Urban ecology research using citizen science studied residential plant management in backyard gardens and parks (Cooper et al. 2007) and garden water management (Lin et al. 2018). Participation in citizen science programs can improve environmental education in society and community engagement in science to affect management decisions (van Heezik et al. 2012).

Future research

There are many key questions that still must be answered in future research to improve human decision-making around urban habitats. This is particularly important for urban conservation and biodiversity management, the focus of this edited book. A key challenge remains to better understand the effects of decision-making scale on ecosystem properties, including diversity of different taxonomic groups, structural heterogeneity, functional redundancy, and response diversity. Many different decision-makers working in small patches from allotment plots to yards may lead to increases in plant species richness and structural complexity at neighbourhood and city scales (Kendal et al. 2010). Yet coordinating decision-making across larger scales could lead to better conservation planning for particular species (e.g. birds; Belaire et al. 2014; Goddard et al. 2010). Another key challenge is understanding how first nations perspectives can be incorporated into better decision making in urban ecosystems, an issue that is of growing importance in international policy frameworks. Research that explores the broad range of actors that influence urban habitats and their different motivations, preferences, and behaviours is needed to make sense of decision-making in complex urban social-ecological systems.

Conclusions

In this chapter, we have shown that there are a myriad of ways that human decision-making influences urban habitats. There are a range of ways of making sense of this complexity. Key mechanisms by which people influence urban habitats include governance, policy and regulation, planning and design, and management. Decisions can have both intended and unintended effects, and these effects can be positive or negative (and sometimes both for different taxonomic groups). Decisions can have very different spatial and temporal effects, and some decisions accumulate over space and/or time. There are many different actors that influence different land uses. A diverse range of social theories can be used to explore and explain human decision-making. These theories suggest that individual (e.g. values, beliefs, and preferences), cultural, and even genetic factors can influence human decision-making and behaviours. Urban habitats that support complex species assemblages, and high levels of ecosystem function and ecosystem service provision, require improved decision-making. Decision-making continues to evolve through the explosion of increasingly available data, improved tools and technologies, and processes that incorporate community views, citizen science, and scientific evidence. Such progress will improve decision-making and community support for decisions of urban habitat management.

Key companion papers

Aronson, M.F.J., et al. 2017. Biodiversity in the city: Key challenges for urban green space management. *Frontiers in Ecology and the Environment* 15(4):189–196.

Belaire, J.A., C.J. Whelan, E.S. Minor. 2014. Having our yards and sharing them too: The collective effects of yards on native bird species in an urban landscape. *Ecological Applications* 24:2132–2143.

Egerer, M., A. Ossola, B.B. Lin. 2018. Creating socioecological novelty in urban agroecosystems from the ground up. *BioScience* 68(1):25–34.

Goddard, M.A., A.J. Dougill, T.G. Benton. 2010. Scaling up from gardens: Biodiversity conservation in urban environments. *Trends in Ecology and Evolution* 25(2):90–98.
Ives, C.D., D. Kendal. 2014. The role of social values in the management of ecological systems. *Journal of Environmental Management* 144:67–72.
McPhearson, T., et al. 2016. Advancing urban ecology toward a science of cities. *BioScience* 66:198–212.
Nilon, C.H., et al. 2017. Planning for the future of urban biodiversity: A global review of city-scale initiatives. *BioScience* 67:332–342.

References

Ajzen, I. 1991. The theory of planned behavior. *Orgnizational Behavior and Human Decision Processes* 50(2):179–211.
Anderson, E.C., E.S. Minor. 2020. Management effects on plant community and functional assemblages in Chicago's vacant lots. *Applied Vegetation Science* 23(2):266–276.
Andersson, E., S. Barthel, S. Borgström, J. Colding, T. Elmqvist, C. Folke, Å. Gren. 2014. Reconnecting cities to the biosphere: Stewardship of green infrastructure and urban ecosystem services. *Ambio* 43:445–453.
Apfelbeck, B., E.P.H. Snep, T.E. Hauck, J. Ferguson, M. Holy, C. Jakoby, J.S. MacIvor, L. Schär, M. Taylor, W.W. Weisser. 2020. Designing wildlife-inclusive cities that support human-animal co-existence. *Landscape and Urban Planning* 200:103817.
Aronson, M.F.J., C.A. Lepczyk, K.L. Evans, M.A. Goddard, S.B. Lerman, J.S. MacIvor, C.H. Nilon, T. Vargo. 2017a. Biodiversity in the city: Key challenges for urban green space management. *Frontiers in Ecology and the Environment* 15:189–196.
Aronson, M.F.J., M.V. Patel, K.M. O'Neill, J.G. Ehrenfeld. 2017b. Urban riparian systems function as corridors for both native and invasive plant species. *Biological Invasions* 19:3645–3657.
Ballantyne, R., J. Packer, K. Hughes. 2008. Environmental awareness, interests and motives of botanic gardens visitors: Implications for interpretive practice. *Tourism Management* 29:439–444.
Belaire, J.A., C.J. Whelan, E.S. Minor. 2014. Having our yards and sharing them too: The collective effects of yards on native bird species in an urban landscape. *Ecological Applications* 24(8):2132–2143.
Bonney, R., C.B. Cooper, J. Dickinson, S. Kelling, T. Phillips, K.V. Rosenberg, J. Shirk. 2009. Citizen science: A developing tool for expanding science knowledge and scientific literacy. *BioScience* 59(11):977–984.
Bonnington, C., K.J. Gaston, K.L. Evans. 2013. Fearing the feline: Domestic cats reduce avian fecundity through trait-mediated indirect effects that increase nest predation by other species. *Journal of Applied Ecology* 50(1):15–24.
Braaker, S., J. Ghazoul, M.K. Obrist, M. Moretti. 2014. Habitat connectivity shapes urban arthropod communities: The key role of green roofs. *Ecology* 95(4):1010–1021.
Chishaleshale, M., C.M. Shackleton, J. Gambiza, D. Gumbo. 2015. The prevalence of planning and management frameworks for trees and green spaces in urban areas of South Africa. *Urban Forestry & Urban Greening* 14(4):817–825.
Clarkson, B.D., C.L. Kirby. 2016. Ecological restoration in urban environments in New Zealand. *Ecological Management and Restoration* 17(3):180–190.
Colding, J., C. Folke. 2009. The role of golf courses in biodiversity conservation and ecosystem management. *Ecosystems* 12:191–206.
Cook, W.M., S.H. Faeth. 2006. Irrigation and land use drive ground arthropod community patterns in an urban desert. *Environmental Entomology* 35:1532–1540.
Cooper, C.B., J. Dickinson, T. Phillips, R. Bonney. 2007. Citizen science as a tool for conservation in residential ecosystems. *Ecology and Society* 12(2):11.
Dobbs, C., C.R. Nitschke, D. Kendal. 2014. Global drivers and tradeoffs of three urban vegetation ecosystem services. *PLoS ONE* 9:e113000.
Egerer, M.H., B.B. Lin, C.G. Threlfall, D. Kendal. 2019. Temperature variability influences urban garden plant richness and gardener water use behavior, but not planting decisions. *Science of the Total Environment* 646:111–120.
Egerer, M.H., A. Ossola, B.B. Lin. 2018. Creating socioecological novelty in urban agroecosystems from the ground up. *BioScience* 68:25–34.

Farrar, A., D. Kendal, K.J.H. Williams, B.J. Zeeman. 2020. Social and ecological dimensions of urban conservation grasslands and their management through prescribed burning and woody vegetation removal. *Sustainability* 12(8):3461.

Fraser, E.D.G., W.A. Kenney. 2000. Cultural background and landscape history as factors affecting perceptions of the urban forest. *Journal of Arboriculture* 26:106–113.

Glowa, K.M., M. Egerer, V. Jones. 2019. Agroecologies of displacement: A study of land access, dislocation, and migration in relation to sustainable food production in the Beach Flats Community Garden. *Agroecology and Sustainable Food Systems* 43(1):92–115.

Goddard, M.A., A.J. Dougill, T.G. Benton. 2010. Scaling up from gardens: Biodiversity conservation in urban environments. *Trends in Ecology and Evolution* 25(2):90–98.

Goddard, M.A., A.J. Dougill, T.G. Benton. 2013. Why garden for wildlife? Social and ecological drivers, motivations and barriers for biodiversity management in residential landscapes. *Ecological Economics* 86:258–273.

Gulsrud, N.M., K. Hertzog, I. Shears. 2018. Innovative urban forestry governance in Melbourne?: Investigating 'green placemaking' as a nature-based solution. *Environmental Research* 161:158–167.

Hargreaves, T. 2011. Practice-ing behaviour change: Applying social practice theory to pro-environmental behaviour change. *Journal of Consumer Culture* 11(1):79–99.

Heynen, N., H.A. Perkins, P. Roy. 2006. The political ecology of uneven urban green space: The impact of political economy on race and ethnicity in producing environmental inequality in Milwaukee. *Urban Affairs Review* 42(1):3–25.

Hope, D., C. Gries, W. Zhu, W.F. Fagan, C.L. Redman, N.B. Grimm, A.L. Nelson, C. Martin, A. Kinzig. 2008. Socioeconomics drive urban plant diversity. Pages 339–347 in J.M. Marzluff, E. Shulenberger, W. Endlicher, M. Alberti, G. Bradley, C. Ryan, U. Simon, C. ZumBrunnen, editors. *Urban Ecology: An International Perspective on the Interaction Between Humans and Nature*. Springer, New York.

Ives, C.D., D. Kendal. 2014. The role of social values in the management of ecological systems. *Journal of Environmental Management* 144:67–72.

Kendal, D., A. Farrar, L. Plant, C.G. Threlfall, J. Bush, J. Baumann. 2017a. *Risks to Australia's Urban Forest from Climate Change and Urban Heat*. University of Melbourne, Melbourne, Australia.

Kendal, D., K.J.H. Williams, N.S.G. Williams. 2012a. Plant traits link people's plant preferences to the composition of their gardens. *Landscape and Urban Planning* 105(1–2):34–42.

Kendal, D., N.S.G. Williams, K.J.H. Williams. 2010. Harnessing diversity in gardens through individual decision makers. *Trends in Ecology and Evolution* 25(4):201–202.

Kendal, D., N.S.G. Williams, K.J.H. Williams. 2012b. Drivers of diversity and tree cover in gardens, parks and streetscapes in an Australian city. *Urban Forestry & Urban Greening* 11(3):257–265.

Kendal, D., B.J. Zeeman, K. Ikin, I.D. Lunt, M.J. McDonnell, A. Farrar, L.M. Pearce, J.W. Morgan. 2017b. The importance of small urban reserves for plant conservation. *Biological Conservation* 213:146–153.

Kowarik, I. 2018. Urban wilderness: Supply, demand, and access. *Urban Forestry & Urban Greening* 29:336–347.

Kowarik, I., S. Buchholz, M. von der Lippe, B. Seitz. 2016. Biodiversity functions of urban cemeteries: Evidence from one of the largest Jewish cemeteries in Europe. *Urban Forestry & Urban Greening* 19:68–78.

Kratschmer, S., M. Kriechbaum, B. Pachinger. 2018. Buzzing on top: Linking wild bee diversity, abundance and traits with green roof qualities. *Urban Ecosystems* 21:429–446.

Landry, S.M., J. Chakraborty. 2009. Street trees and equity: Evaluating the spatial distribution of an urban amenity. *Environment and Planning A: Economy and Space* 41(11):2651–2670.

Larson, K.L., R. Andrade, K.C. Nelson, M.M. Wheeler, J.M. Engebreston, S.J. Hall, M.L. Avolio, P.M. Groffman, M. Grove, J.B. Heffernan, S.E. Hobbie, S.B. Lerman, D.H. Locke, C. Neill, R.R. Chowdhury, T.L.E. Trammell. 2020. Municipal regulation of residential landscapes across US cities: Patterns and implications for landscape sustainability. *Journal of Environmental Management* 275:111132.

Laurans, Y., X. Leflaive, A. Rankovic. 2020. Decision-making, now in 3D: Exploring three dimensions of decision-making processes and their consequences for biodiversity research. *Environmental Science & Policy* 113:31–38.

Lerman, S.B., A.R. Contosta, J. Milam, C. Bang. 2018. To mow or to mow less: Lawn mowing frequency affects bee abundance and diversity in suburban yards. *Biological Conservation* 221:160–174.

Lin, B.B., M.H. Egerer, H. Liere, S. Jha, P. Bichier, S.M. Philpott. 2018. Local- and landscape-scale land cover affects microclimate and water use in urban gardens. *Science of the Total Environment* 610–611:570–575.

Loram, A., J. Tratalos, P.H. Warren, K.J. Gaston. 2007. Urban domestic gardens (X): The extent & structure of the resource in five major cities. *Landscape Ecology* 22:601–615.
Lovell, S.T., J.R. Taylor. 2013. Supplying urban ecosystem services through multifunctional green infrastructure in the United States. *Landscape Ecology* 28:1447–1463.
Marshall, A.J., M.J. Grose, N.S.G. Williams. 2019. From little things: More than a third of public green space is road verge. *Urban Forestry & Urban Greening* 44:126423.
Martin, D.M., M. Mazzotta, J. Bousquin. 2018. Combining ecosystem services assessment with structured decision making to support ecological restoration planning. *Environmental Management* 62:608–618.
McPhearson, T., S.T.A. Pickett, N.B. Grimm, J. Niemelä, M. Alberti, T. Elmqvist, C. Weber, D. Haase, J. Breuste, S. Qureshi. 2016. Advancing urban ecology toward a science of cities. *BioScience* 66(3):198–212.
Nassauer, J.I. 1995. Messy ecosystems, orderly frames. *Landscape Journal* 14(2):161–170.
Nilon, C.H., M.F.J. Aronson, S.S. Cilliers, C. Dobbs, L.J. Frazee, M.A. Goddard, K.M. O'Neill, D. Roberts, E.K. Stander, P. Werner, M. Winter, K.P. Yocom. 2017. Planning for the future of urban biodiversity: A global review of city-scale initiatives. *BioScience* 67(4):332–342.
Norton, B.A., A.M. Coutts, S.J. Livesley, R.J. Harris, A.M. Hunter, N.S.G. Williams. 2015. Planning for cooler cities: A framework to prioritise green infrastructure to mitigate high temperatures in urban landscapes. *Landscape and Urban Planning* 134:127–138.
Oberndorfer, E., J. Lundholm, B. Bass, R.R. Coffman, H. Doshi, N. Dunnett, S. Gaffin, M. Köhler, K.K.Y. Liu, B. Rowe. 2007. Green roofs as urban ecosystems: Ecological structures, functions, and services. *BioScience* 57(10):823–833.
Ordóñez, C., C.G. Threlfall, D. Kendal, D.F. Hochuli, M. Davern, R.A. Fuller, R. van der Ree, S.J. Livesley. 2019. Urban forest governance and decision-making: A systematic review and synthesis of the perspectives of municipal managers. *Landscape and Urban Planning* 189:166–180.
Orians, G.H., J.H. Heerwagen. 1992. Evolved responses to landscapes. Pages 555–579 in J.H. Barkow, L. Cosmides, J. Tooby, editors. *The Adapted Mind: Evolutionary Psychology and the Generation of Culture*. Oxford University Press, Oxford.
Peng, M.-H., Y.-C.Hung, K.-L. Liu, K.-B. Neoh. 2020. Landscape configuration and habitat complexity shape arthropod assemblage in urban parks. *Scientific Reports* 10:16043.
Puppim de Oliveira, J.A., O. Balaban, C.N.H. Doll, R. Moreno-Peñaranda, A. Gasparatos, D. Iossifova, A. Suwa. 2011. Cities and biodiversity: Perspectives and governance challenges for implementing the convention on biological diversity (CBD) at the city level. *Biological Conservation* 144(5):1302–1313.
Rappaport, B., B. Horn. 1998. Weeding out bad vegetation-control ordinances. *Restoration & Management Notes* 16(1):51–58.
Roman, L.A., J.J. Battles, J.R. McBride. 2014. The balance of planting and mortality in a street tree population. *Urban Ecosystems* 17:387–404.
Rupprecht, C.D.D., J.A. Byrne, J.G. Garden, J.M. Hero. 2015. Informal urban green space: A trilingual systematic review of its role for biodiversity and trends in the literature. *Urban Forestry & Urban Greening* 14(4):883–908.
Schwartz, S.H. 1994. Are there universal aspects in the structure and contents of human values? *Journal of Social Issues* 50(4):19–45.
Shove, E. 2010. Beyond the ABC: Climate change policy and theories of social change. *Environment and Planning A: Economy and Space*. 42(6):1273–1285.
Sisser, J.M., K.C. Nelson, K.L. Larson, L.A. Ogden, C. Polsky, R.R. Chowdhury. 2016. Lawn enforcement: How municipal policies and neighborhood norms influence homeowner residential landscape management. *Landscape and Urban Planning* 150:16–25.
Smith, A.D., E. Minor. 2019. Chicago's urban cemeteries as habitat for cavity-nesting birds. *Sustainability* 11:1–16.
Stagoll, K., A.D. Manning, E. Knight, J. Fischer, D.B. Lindenmayer. 2010. Using bird-habitat relationships to inform urban planning. *Landscape and Urban Planning* 98(1):13–25.
Stern, P.C., L. Kalof, T. Dietz, G.A. Guagnano. 1995. Values, beliefs, and proenvironmental action: Attitude formation toward emergent attitude objects. *Journal of Applied Social Psychology* 25(18):1611–1636.
Sutherland, W.J., R.P. Freckleton. 2012. Making predictive ecology more relevant to policy makers and practitioners. *Philosophical Transactions of the Royal Society B: Biological Sciences* 36(1586):322–330.
Threlfall, C.G., A. Ossola, A.K. Hahs, N.S.G. Williams, L. Wilson, S.J. Livesley. 2016a. Variation in vegetation structure and composition across urban green space types. *Frontiers in Ecology and Evolution* 4:1–12.

Threlfall, C.G., N.S.G. Williams, A.K. Hahs, S.J. Livesley. 2016b. Approaches to urban vegetation management and the impacts on urban bird and bat assemblages. *Landscape and Urban Planning* 153:28–39.
van Heezik, Y.M., K.J.M. Dickinson, C. Freeman. 2012. Closing the gap: Communicating to change gardening practices in support of native biodiversity in urban private gardens. *Ecology and Society* 17(1):34.
White, J.G., M.J. Antos, J.A. Fitzsimons, G.C. Palmer. 2005. Non-uniform bird assemblages in urban environments: The influence of streetscape vegetation. *Landscape and Urban Planning* 71(2–4):123–135.
Williams, N.S.G., M.W. Schwartz, P.A. Vesk, M.A. McCarthy, A.K. Hahs, S.E. Clemants, R.T. Corlett, R.P. Duncan, B.A. Norton, K. Thompson, M.J. McDonnell. 2009. A conceptual framework for predicting the effects of urban environments on floras. *Journal of Ecology* 97(1):4–9.

10

SCALING-UP THE MANAGEMENT AND CONSERVATION OF URBAN BIODIVERSITY

Susannah B. Lerman, Josep Padullés Cubino, and Mark A. Goddard

Introduction

Urban landscapes often constitute a complex mosaic of vegetation or green spaces. These green spaces act as habitat patches and include residential yards, gardens, recreational parks, cemeteries, vacant lots, golf courses, and protected parks. The vegetation within these habitat patches includes swaths of lawns, isolated trees, and natural areas of various sizes. These relatively small, fragmented, and often isolated habitat patches are interspersed with roads, buildings, and other impervious surfaces. Although altered and disconnected from the wildlands, habitat patches within the urban matrix support the persistence of native biodiversity (Aronson et al. 2014; Lepczyk et al. 2017). However, the quality, composition, and size of habitat patches varies tremendously (Lepczyk et al. 2017).

Urban biodiversity patterns are driven by ecological and social factors that operate at multiple spatial scales (Goddard et al. 2017; Swan et al. 2021), ranging from different features within a habitat patch (e.g. flower beds or hedgerows), to residential yards, neighborhoods, cities, or regions (Figure 10.1). The relative importance of these factors and the most important scale at which they operate varies with taxa and across an organism's life cycle (Beninde et al. 2015; Lepczyk et al. 2017). For example, foraging distances for small-bodied bees likely cover a single (or several, depending on parcel size) suburban area (i.e. ~50 m) whereas foraging distances for large-bodied bees could span an entire neighborhood (i.e. ~1–2 km; Greenleaf et al. 2007; Lerman and Milam 2016). In general, larger habitat patches will support greater biodiversity (Beninde et al. 2015), and the species-area relationship is especially pertinent in urban ecosystems for birds (Evans et al. 2009), amphibians (Parris 2006), mammals (Magle et al. 2009), and carabid beetles (Sadler et al. 2006).

Habitat connectivity also shapes biodiversity patterns (Ossola et al. 2019). This connectivity can be greatly enhanced by management of private yards, especially when coordination among multiple landowners can increase the availability and quality of habitat. In this way, residential yards play a key role in maintaining the ecological integrity of urban landscapes (i.e. the ability of a system to support diverse communities and ecosystem processes; Pimentel et al. 2000). Although this chapter is applicable to the many different urban land-uses, we focus on residential yards and gardens (hereafter yards) because of their dominance throughout the urban and suburban landscape (Nowak et al. 2001; Loram et al. 2007), and their relevance to the majority of urban and

 DOI: 10.4324/9781003016120-12

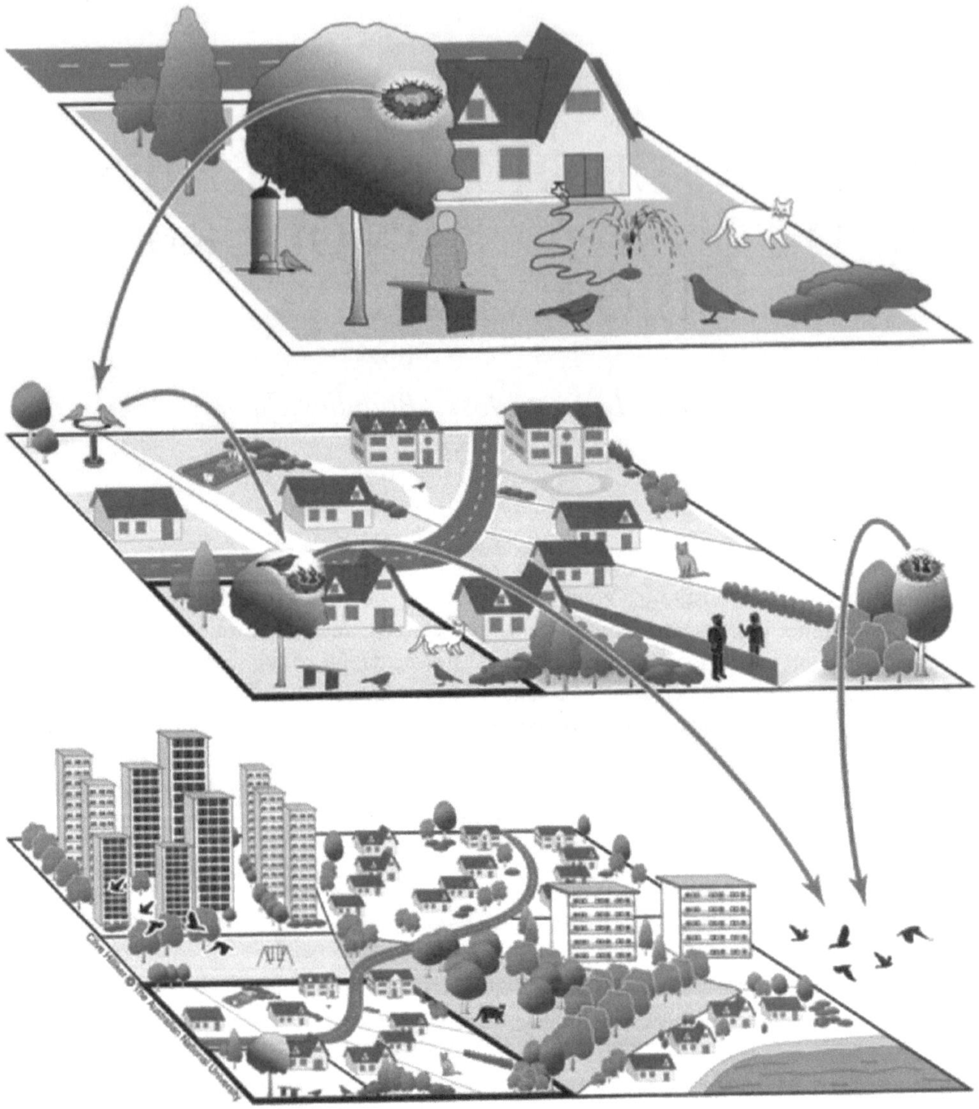

Figure 10.1 Ecological and social factors drive urban biodiversity patterns at nested spatial scales. At the individual yard (top panel), specific habitat features (e.g. trees and shrubs) and householder decisions (e.g. to provide supplementary food, to irrigate the yard, and have an outdoor cat) affect the ability of wildlife, for example, birds, to find suitable forage and nest sites. At the neighborhood scale (middle panel), variation in socioeconomic status and social norms in yard management practices determine the composition and configuration of habitat in groups of adjacent yards. At the city scale (lower panel), the clustering of residential habitat with other urban green spaces reflects urban planning and green infrastructure strategies. Arrows represent the different life stages of backyard birds at different scales in the residential yard hierarchy, from eggs in an individual nest at the yard scale (top panel) to nestlings in multiple yards (middle panel) and fledglings dispersing within and beyond the city (lower panel). From Goddard et al. (2017).

suburban households. In this chapter, we focus on how spatial and temporal scales influence urban biodiversity patterns, the composition and configuration of yards, their connection throughout the urban matrix, how scaling informs urban planning, and future directions for urban biodiversity research.

The scaling of biodiversity in the urban landscape

Scaling up from an individual yard to the city and broader landscape has relevance for understanding broad patterns of urban biodiversity and helps inform management. However, urban landscapes are both homogenous and heterogeneous, depending on the spatial scale of investigation, complicating broad generalizations (Lerman et al. 2021b). At the city/landscape scale, urbanization usually has a homogenizing effect: ecological form and function among cities are more similar to each other than native ecosystems that they replaced (Groffman et al. 2014; Hall et al. 2016; Wheeler et al. 2017). This homogenizing force can lead to more depauperate native animal communities, less taxonomically diverse (Aronson et al. 2014; Batáry et al. 2018) and less evolutionarily distinct (Morelli et al. 2016). Scaling down to the individual yard, socioeconomic processes, existing regulations, and legacies of different land uses and systemic racism influence the management of and preferences for different green features (Larson et al. 2020; Schell et al. 2020; Cook et al. 2012), which in turn structure the ecological communities (Padullés Cubino et al. 2020a). The "tyranny of small decision", whereby millions of individual households make independent management decisions and actions (Odum 1982), generates fragmented and discontinuous landscapes that constrain the ability of many wildlife species to maintain populations and thus has local, regional, and global implications for biodiversity (Blair 1996; Butchart et al. 2010; Shochat et al. 2010).

The importance and influence of temporal scales are less studied (though see Ramalho and Hobbs 2012; Warren et al. 2019; Niemiec et al. 2020; Ossola et al. 2021), and temporal effects can vary with spatial scale (Allen et al. 2019; Swan et al. 2021). For example, long-term studies might place higher emphasis on the influence of local-scale environmental changes. This could include urban intensification or infill at the parcel or neighborhood scale or species turnover (both losses and gains) without considering regional-level population trends (Warren et al. 2019). The contribution of local factors to urban biodiversity can vary seasonally (Van Helden et al. 2021), underlying the importance of multi-seasonal sampling to enhance understanding of the mechanisms driving biodiversity patterns. Temporal scales also have implications for management (Wilson et al. 2016): ecological processes operate on different temporal scales (shorter or longer) than decision-making (Luck et al. 2009; Borgstrom et al. 2006). These differences might be exacerbated in cities, where urban development and policy decisions have even shorter timeframes or cycle faster. This presents a challenge for aligning traditional management timeframes, commonly used in non-urban (often a single landowner) systems to urban ecosystems (Piana et al. 2021).

A focus on temporal scales in urban biodiversity research can contextualize the legacies of national policies that have current ecological consequences. For example, the US policy of redlining (1933–1966), which strategically and systematically denied minoritized people access to mortgages and other financial investments. The policy perpetuated segregated neighborhoods based on race (Schell et al. 2020) and have shaped the vegetation in today's residential landscape. Specific legacies today include significantly reduced tree canopy cover in the redlined neighborhoods (average of 23% compared with 43%; Locke et al. 2021) and perpetual environmental inequities (Wolch et al. 2014). For example, households in these neighborhoods likely have lower biodiversity (Leong et al. 2018) due to the strong relationship with canopy cover (Lerman et al. 2021b). However, access to urban green spaces amidst racial and socioeconomic status can be complex.

For example, in Bradford, UK, predominantly Asian/Asian British neighborhoods had higher street tree density, yet households in predominantly white and higher income neighborhoods had better access to public green spaces. The latter might reflect a perception of certain ethno-racial groups being excluded from public green spaces as well as the legacies of immigration patterns whereby immigrants settle in high-density and low-income neighborhoods with little open space (Ferguson et al. 2018). Clearly, the interactions amongst spatial and temporal scales, along with social and ecological variables contribute to urban biodiversity (Andrade et al. 2021a).

Residential yards scale-up to support biodiversity

Residential yards are social-ecological systems that represent a significant component of the urban green infrastructure (Gaston et al. 2005; Mathieu et al. 2007). Their contribution to biodiversity, ecosystem services, and sustainability spans a wide range of spatial scales, from individual parcels and neighborhoods to watersheds and larger-scale regions (Cameron et al. 2012). While evidence of the value of residential yards for supporting biodiversity of multiple taxa at the local scale continues to emerge from studies worldwide (e.g. Thompson et al. 2004; Goddard et al. 2010; Threlfall et al. 2016; Lerman et al. 2018), few studies have explicitly examined the contribution of residential yards at a broader scale (e.g. neighborhood to city).

Residential yards account for a large part of urban green spaces. Estimated areas in cities vary from 16% in Stockholm, Sweden (Colding et al. 2006), to 22–27% in the UK (Loram et al. 2007), and 36% in Dunedin, New Zealand (Mathieu et al. 2007). Because of their collective extent, yards also host a large number of plant species (Smith et al. 2006b), including a mix of native and ornamental exotics (Bigirimana et al. 2012; Avolio et al. 2019; Padullés Cubino et al. 2019), that can sustain many functions for wildlife, such as habitat or food provision (Davies et al. 2009; Cameron et al. 2012), and tree biomass. In the UK, yards host just under a quarter of all trees occurring outside woodlands in more natural areas (Davies et al. 2009). When compared to reference natural areas and interstitial patches of native vegetation of equivalent size, yards generally host a higher number of plant species at the individual yard (Avolio et al. 2019; Padullés Cubino et al. 2020b), neighborhood (Threlfall et al. 2016), and city scale (Pearse et al. 2018). However, plants in yards are more functionally similar across cities than those in reference natural areas leading to continental-scale homogenization of yard plant communities (Pearse et al. 2018; Padullés Cubino et al. 2020b), and decreases in regional plant diversity (McKinney 2006). Despite this general pattern of homogenization in the US, plant communities in yards managed to support wildlife appear more functionally heterogeneous than those managed for other goals (e.g. lawns; Padullés Cubino et al. 2020b). However, this does not necessarily hold for higher trophic levels; reference natural areas and interstitial patches of native vegetation often support higher bird richness than yards (e.g. Lerman et al. 2021b).

An individual yard is intricately related to a network of yards and scales up to the neighborhood and city (Goddard et al. 2010). The functional connectivity of these individual yards depends upon the ecological features and management practices across multiple yards that facilitates species movement throughout the urban matrix. These features influence community structure (Lerman and Warren 2011; Pearse et al. 2018) and wildlife population dynamics (Narango et al. 2018). Population persistence depends upon the available resources and scale at which wildlife species respond to habitat composition and structure. For example, habitat modification on the scale of an individual yard may be inadequate to satisfy the life history requirements for some species, particularly those with large territories that span beyond yard boundaries (e.g. birds and large mammals). Identifying the relative importance of local versus landscape variables, in addition to the specific habitat features present, might help with targeting management recommendations but dismisses the complex ways local and landscape variables interact. At yard scales, native avian diversity increases with yard size (Daniels and Kirkpatrick 2006; van Heezik et al. 2013), the presence of native vegetation (Lerman and Warren 2011; McCaffrey and Mannan 2012), and the inclusion of wildlife-friendly features such as bird feeders (Fuller et al. 2008; Belaire et al. 2014). In New Zealand, cultivation of popular landscaping exotic plant species also promotes bird diversity as they provide important food sources across all seasons (Gray and van Heezik 2016). However, non-native tree species can also reduce the availability of food resources for habitat specialists (Narango et al. 2018). At the neighborhood scale, bird richness can increase with the amount of open space (McCaffrey and Mannan 2012), the number of trees (Gonçalves et al. 2021), the amount of fruit and berries, or the proportion of evergreen to deciduous trees (Belaire et al. 2014).

For smaller and less vagile species, such as invertebrates, local and landscape scales have complex correlations with diversity and abundance (Smith et al. 2006a; Baldock et al. 2019). For example, studies have shown that the number of woody plants and the structural complexity of the vegetation in yards predicts invertebrate abundance (Sperling and Lortie 2010; van Heezik et al. 2016). Soil fauna and soil multifunctionality (i.e. the provision of multiple ecological functions) also increases with high plant diversity in yards (Tresch et al. 2019), and there is evidence that human-mediated dispersal is a key factor shaping belowground urban fauna, both at local and regional scales (Tóth et al. 2020). In addition to actively increasing vegetation, specific vegetation management practices also influence biodiversity. For example, mowing lawns less frequently results in an increase of floral resources and, subsequently, higher bee abundance (Lerman et al. 2018). These management practices suggest that simple changes in yards can enhance their capacity to provide suitable habitat. Nonetheless, some groups of invertebrates, like moths, only increase in abundance with complex yard features at a scale that extends beyond the yard boundary (Ellis and Wilkinson 2021). Communities of snails in yards also respond to soil factors at a scale larger than individual yards (Bergey and Figueroa 2016), and plant-pollinator networks do not always scale-up with the number of plant taxa in yards (Lowenstein et al. 2019).

The widespread distribution of yards also provides habitat connectivity within the urban matrix and thus a key link that connects local and landscape scales. For example, having diverse vegetation communities within yards improves habitat availability for bats (Mimet et al. 2020), largely driven by linking tree canopy throughout backyards across the city (Ossola et al. 2019). These habitat corridors also increase the connectivity to source habitats for arthropods. In Paris, France, arthropod species richness and abundance was higher in yards connected to woody corridors (i.e. within 10 m) compared with yards disconnected to these corridors (i.e. >300 m; Vergnes et al. 2012). Wildlife visitation rates also increase in larger yards, those with greater vegetation structure, and when connected to natural habitats (Baker and Harris 2007). Therefore, yards have the potential to mitigate the negative effects of habitat fragmentation in the urban landscape (Baker et al. 2003). Increasing habitat quality in yards and other urban green spaces scales up and can increase the connectivity and patch size throughout the urban matrix (Lerman et al. 2021b), even though the net contributions of yards to ecological connectivity are expected to be species dependent (Lepczyk et al. 2017). Increased habitat connectivity between yards can also occur through social contagion that is a key mechanism for spreading sustainable gardening practices across neighborhoods (Hunter and Brown 2012).

Overall, yards, in conjunction with local open green spaces, contribute to form a large urban matrix of habitat patches, corridors, and resources that can provide beneficial services not only to the urban biota but also to people (Goddard et al. 2010). Yards within a city are connected through both biological and social mechanisms (Egerer et al. 2020a), and the latter can have positive effects on social connections such as community cohesion, social capital, and networking (Gray et al. 2014). Across cities, different frameworks have been proposed to better understand how ecological and sociocultural factors interact to affect biodiversity in yards (e.g. Cook et al. 2012; Beumer and Martens 2016). The results from the studies summarized in this section highlight the interconnection between local and landscape scales, and the potential for cross-scale initiatives to support urban biodiversity through private landscaping. Because yards are usually designed and managed to favor particular groups of species, this gives householders an important role, as well as the nursery industry they rely upon (Cavender-Bares et al. 2020), in the conservation of urban biodiversity. The effective conservation of urban biodiversity requires coordinated management decisions across different stakeholders and spatial scales.

Implications for urban planning and policy

Urban planning for biodiversity often considers habitat quantity or quality (Nilon et al. 2017) and is implemented at various spatial scales with contrasting management strategies (Figure 10.2). At the landscape-scale, this is often pitched as "land sparing" for example, preserving and protecting existing native habitat patches, promoting native vegetation, and implementing pro-greening policies (Aronson et al. 2017). Research supports this strategy, particularly for species sensitive to fragmentation, and it can reduce biotic homogenization (Sushinsky et al. 2013; Ibáñez-Álamo et al. 2020). At the parcel scale, "land sharing", for example managing yards as habitat also has implications for biodiversity conservation (Goddard et al. 2010), particularly when yards include native plants and limit the use of pesticides and herbicides (Lerman and Warren 2011; Belaire et al. 2014; Muratet and Fontaine 2015; Narango et al. 2018). Rather than an "either or" strategy, considering the value of both scales can have broader impacts on biodiversity. Neighborhoods that belong to homeowner associations (HOAs) or are part of conservation developments can also support higher biodiversity, partially explained by more consistent (Lerman et al. 2012) and intentional (Farr et al. 2018) management. Further, it has been proposed that neighbor mimicry and wildlife-oriented neighborhood standards can result in the diffusion of wildlife-friendly practices across individual residential yards (Goddard et al. 2013). And finally, enhancing habitat in yards provides opportunities for the public to interact with and appreciate biodiversity at home (Luck et al. 2011; Sushinsky et al. 2013), with the potential to increase support for additional conservation action (Hughes et al. 2018).

Although biodiversity patterns reflect the spatial, temporal, and functional scales in which they are measured, management action seldom considers the cross-scale interactions, perhaps due to lack of guidance, and hence, they are not prioritized in urban planning (Borgstrom et al. 2006). Broad-scale recommendations to increase vegetation cover across an entire city generally have net positive impacts on biodiversity (Aronson et al. 2017). However, these sweeping gestures rarely consider social factors, budgetary constraints, legacy effects of land use, and structural racism that has shaped the urban environment (Schell et al. 2020). Therefore, scale mismatches (whereby the scale of management does not match the scale of ecological processes; Borgstrom et al. 2006) might lead to a reduction in ecological function and ecosystem services. Counteracting these scale mismatches requires a holistic multi-scale approach to urban biodiversity management, policy, and planning that coordinates these three arms of decision-making during goal setting, implementation, and monitoring (Nilon et al. 2017).

Further directions

The importance of the concept of scale for enhancing our understanding of urban biodiversity patterns and management is increasingly recognized (Uchida et al. 2021), but nevertheless, key

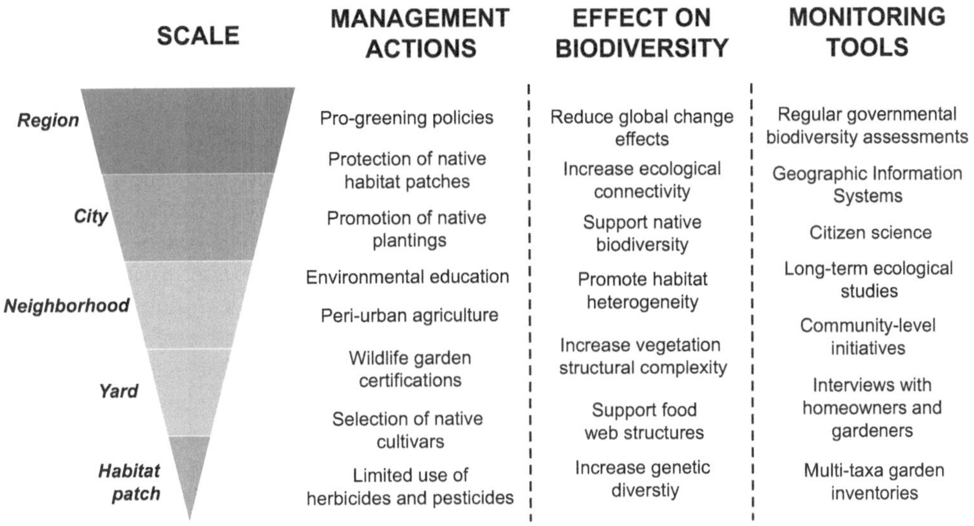

Figure 10.2 Schematic representation showing how management of residential landscapes can support biodiversity at different spatial scales: from individual habitat patches within yards (e.g. lawns, flower beds, hedges) to residential macro-systems across cities and regions. A list of potential monitoring tools across a spatial scale is also provided.

knowledge gaps remain. Yards and other urban green spaces are important places for people to connect with urban nature (e.g. Cox and Gaston 2016; McEwan et al. 2020), and realizing the health and well-being benefits associated with access to nature is now high on the research and political agendas (Shanahan et al. 2015; Public Health England 2020; Marselle et al. 2021). Indeed, the value of nearby urban green space, both private and public, for health and well-being has been greatly intensified during the COVID-19 pandemic (Lehberger et al. 2021; Poortinga et al. 2021). Given that an increasingly important motivation of urban biodiversity management is to enhance human well-being, research is needed to examine how biodiversity management decisions influence human well-being at multiple scales, from individual yards, through neighborhoods, to entire cities (Andrade et al. 2021b; Uchida et al. 2021). Interventions that seek to increase the availability, accessibility, and quality of biodiversity in urban green spaces must also address environmental justice problems to ensure that well-being benefits are distributed equitably across socioeconomic status levels and ethno-racial groups (Ferguson et al. 2018; Kuras et al. 2020).

In conjunction with an increased emphasis on health and well-being in urban biodiversity research, there is also a growing awareness that yards and residential landscapes could be designed and managed to provide a range of other ecosystem services (Beumer and Martens 2016; Larson et al. 2016). For instance, tree planting efforts in yards have the capacity for increasing the resilience and ability of these landscapes to endure disturbances such as storms and climate change (e.g. floods, droughts, and extreme heat episodes; Cameron et al. 2012). However, some efforts might increase housing prices, resulting in displacement of the community originally intended to serve (Wolch et al. 2014). A better understanding of how to balance the complex drivers and inherent trade-offs that inform yard management decisions can greatly enhance landscape planning. Further, identifying opportunities for creating multifunctional residential landscapes that advance biodiversity conservation, provide ecosystem services, and support health and well-being at multiple scales can improve urban sustainability (McDonald et al. 2014). Finally, managing for urban biodiversity in yards provides immediate and local opportunities to connect with nature. These connections have the potential to inspire additional stewardship action to further increase urban biodiversity (Soga and Gaston 2020; Lerman et al. 2021b)

In the past decade, urban biodiversity research has progressed from predominantly single city studies focusing on one or a few species or habitats, to more comprehensive investigations at regional (e.g. Padullés Cubino et al. 2019; Baldock et al. 2019) and global (e.g. Aronson et al. 2014; Tóth et al. 2020) scales. However, multicity and multi-taxa collaborations that involve the systematic and standardized collection of comparable long-term datasets remain nonexistent. In an attempt to address this research gap, Magle et al. (2019) introduced the Urban Wildlife Information Network (https://urbanwildlifeinfo.org/) as a tool for facilitating the collection of a dataset that will ultimately allow a more nuanced understanding of how local social factors interact with city size, urban form, and geographic location to affect spatiotemporal patterns in biodiversity. Such networks can operate alongside citizen science programs that have a global reach, including eBird (www.eBird.org) and iNaturalist (www.inaturalist.org). Moreover, technological advancements, such as the deployment of robotics and autonomous systems (RAS), enable rapid, real-time monitoring of urban biodiversity over large spatial and temporal scales (e.g. Unmanned Aerial Vehicles (UAVs) surveying yard vegetation; Egerer et al. 2020b), although the impact of RAS themselves on species and habitats remains uncertain (Goddard et al. 2021). Multicity, standardized monitoring could also be combined with experimental approaches to help elucidate the mechanisms underpinning these patterns (Shochat et al. 2006). To date, urban biodiversity experiments at the global scale are still in their infancy, for example, the Global Urban Soil Ecology and Education Network (GLUSEEN; Pouyat et al. 2017; www.gluseen.org). We suggest they expand to consider multiple taxa and their interactions and how urban biodiversity relates to ecosystem function and service. Indeed, the concept of multi-scalar species interactions is central to recent frameworks that propose the integration of a metacommunity perspective into urban biodiversity research (Andrade et al. 2021a). Further interdisciplinary studies can increase our understanding of how human decisions interact with environmental and spatial heterogeneity at multiple scales and how these decisions scale up and shape urban biodiversity (Andrade et al. 2021a).

Key companion papers

Belaire, J.A., C.J. Whelan, E.S. Minor. 2014. Having our yards and sharing them too: The collective effects of yards on native bird species in an urban landscape. *Ecological Applications* 24(8):2132–2143.

Goddard, M.A., A.J. Dougill, T.G. Benton. 2010. Scaling up from gardens: Biodiversity conservation in urban environments. *Trends in Ecology and Evolution* 25(2):90–98.

Lerman, S.B., P.S. Warren. 2011. The conservation value of residential yards: Linking birds and people. *Ecological Applications* 21(4):1327–1339.

Swan, C., et al. 2021. A framework for understanding how biodiversity patterns unfold across multiple spatial scales in urban ecosystems. *Ecosphere* 12(7):e03650.

Uchida, K., et al. 2021. Urban biodiversity and the importance of scale. *Trends in Ecology and Evolution* 36(2):123–131.

References

Allen, D.C., H.L. Bateman, P.S. Warren, F.S. de Albuquerque, S. Arnett-Romero, B. Harding. 2019. Long-term effects of land-use change on bird communities depend on spatial scale and land-use type. *Ecosphere* 10(11):e02952.

Andrade, R., J. Franklin, K.L. Larson, C.M. Swan, S.B. Lerman, H.L. Bateman, P.S. Warren, A. York. 2021a. Predicting the assembly of novel communities in urban ecosystems. *Landscape Ecology* 36:1–15.

Andrade, R., D.M. Hondula, K.L. Larson, S.B. Lerman. 2021b. Landscaping preferences influence neighborhood satisfaction and yard management decisions. *Urban Forestry & Urban Greening* 59:126983.

Aronson, M.F.J., F.A. La Sorte, C.H. Nilon, M. Katti, M.A. Goddard, C.A. Lepczyk, P.S. Warren, N.S.G. Williams, S. Cilliers, B. Clarkson, C. Dobbs, R. Dolan, M. Hedblom, S. Klotz, J.L. Kooijmans, I. Kühn, I. MacGregor-Fors, M. McDonnell, U. Mörtberg, P. Pyšek, S. Siebert, J. Sushinsky, P. Werner, M. Winter. 2014. A global analysis of the impacts of urbanization on bird and plant diversity reveals key anthropogenic drivers. *Proceedings of the Royal Society B: Biological Sciences* 281(281):20133330.

Aronson, M.F.J., C.A. Lepczyk, K.L. Evans, M.A. Goddard, S.B. Lerman, J.S. MacIvor, C.H. Nilon, T. Vargo. 2017. Biodiversity in the city: Key challenges for urban green space management. *Frontiers in Ecology and the Environment* 15(4):189–196.

Avolio, M., D.E. Pataki, G.D. Jenerette, S. Pincetl., L.W. Clarke, J. Cavender-Bares, T.W. Gillespie, S.E. Hobbie, K.L. Larson, H.R. McCarthy, T.L E. Trammell. 2019. Urban plant diversity in Los Angeles, California: Species and functional type turnover in cultivated landscapes. *Plants, People, Plants* 2(2):144–156.

Baker, P.J., R.J. Ansell, P.A.A. Dodds, C.E. Webber, S. Harris. 2003. Factors affecting the distribution of small mammals in an urban area. *Mammal Review* 33:95–100.

Baker, P.J., S. Harris. 2007. Urban mammals: What does the future hold? An analysis of the factors affecting patterns of use of residential gardens in Great Britain. *Mammal Review* 37(4):297–315.

Baldock, K.C.R., M.A. Goddard, D.M. Hicks, W.E. Kunin, N. Mitschunas, H. Morse, L.M. Osgathorpe, S.G. Potts, K.M. Robertson, A.V. Scott, P.P.A. Staniczenko, G.N. Stone, I.P. Vaughan, J. Memmott. 2019. A systems approach reveals urban pollinator hotspots and conservation opportunities. *Nature Ecology and Evolution* 3:363–373.

Batáry, P., K. Kurucz, M. Suarez-Rubio, D.E. Chamberlain. 2018. Non-linearities in bird responses across urbanization gradients: A meta-analysis. *Global Change Biology* 24(3):1046–1054.

Belaire, J.A., C.J. Whelan, E.S. Minor. 2014. Having our yards and sharing them too: The collective effects of yards on native bird species in an urban landscape. *Ecological Applications* 24(8):2132–2143.

Beninde, J., M. Veith, A. Hochkirch. 2015. Biodiversity in cities needs space: A meta-analysis of factors determining intra-urban biodiversity variation. *Ecology Letters* 18(6):581–592.

Bergey, E., L.L. Figueroa. 2016. Residential yards as designer ecosystems: Effects of yard management on land snail species composition. *Ecological Applications* 26(8):1538–2547.

Beumer, C., P. Martens. 2016. BIMBY's first steps: A pilot study on the contribution of residential front-yards in Phoenix and Maastricht to biodiversity, ecosystem services and urban sustainability. *Urban Ecosystems* 10(1):87–100.

Bigirimana, J., J. Bogaert, C. de Cannière, M.-J. Bigendako, I. Parmentier. 2012. Domestic garden plant diversity in Bujumbura, Burundi: Role of the socio-economical status of the neighborhood and alien species invasion risk. *Landscape and Urban Planning* 107(2):118–126.

Blair, R.B. 1996. Land use and avian species diversity along an urban gradient. *Ecological Applications* 6(2):506–519.

Borgstrom, S.T., T. Elmqvist, P. Angelstam, C. Alfsen-Norodom. 2006. Scale mismatches in management of urban landscapes. *Ecology and Society* 11(2):16.

Butchart, S.H.M., M. Walpole, B. Collen, A. van Strien, J.P.W. Scharlemann, R.E.A. Almond, J.E.M. Baillie, B. Bomhard, C. Brown, J. Bruno, K.E. Carpenter, G.M. Carr, J. Chanson, A.M. Chenery, J. Csirke, N.C. Davidson, F. Dentener, M. Foster, A. Galli, J.N. Galloway, P. Genovesi, R.D. Gregory, M. Hockings, V. Kapos, J.-F. Lamarque, F. Leverington, J. Loh, M.A. McGeoch, L. McRae, A. Minasyan, M. Hernández Morcillo, T.E.E. Oldfield, D. Pauly, S. Quader, C. Revenga, J.R. Sauer, B. Skolnik, D. Spear, D. Stanwell-Smith, S.N. Stuart, A. Symes, M. Tierne, T.D. Tyrrell, J.-C. Vié, R. Watson. 2010. Global biodiversity: Indicators of recent declines. *Science* 328(5982):1164–1168.

Cameron, R.W.F., T. Blanusa, J.E. Taylor, A. Salisbury, A.J. Halstead, B. Henricot, K. Thompson. 2012. The domestic garden – its contribution to urban green infrastructure. *Urban Forestry & Urban Greening* 11(2):129–137.

Cavender-Bares, J., J. Padullés Cubino, W.D. Pearse, S.E. Hobbie, A.J. Lange, S. Knapp, K.C. Nelson. 2020. Horticultural availability and homeowner preferences drive diversity and composition in urban yards. *Ecological Applications* 30:e02082.

Colding, J., J. Lundberg, C. Folke. 2006. Incorporating green-area user groups in urban ecosystem management. *AMBIO* 35(5):237–244.
Cook, E., S. Hall, K. Larson. 2012. Residential landscapes as social-ecological systems: A synthesis of multi-scalar interactions between people and their home environment. *Urban Ecosystems* 15:19–52.
Cox, D.T.C., K.J. Gaston. 2016. Urban bird feeding: Connecting people with nature. *PLoS ONE* 11:e0158717.
Daniels, G.D., J.B. Kirkpatrick. 2006. Does variation in garden characteristics influence the conservation of birds in suburbia? *Biodiversity Conservation* 133:326–335.
Davies, Z.G., R.A. Fuller, A. Loram, K.N. Irvine, V. Sims, K.J. Gaston. 2009. A national scale inventory of resource provision for biodiversity within domestic gardens. *Biological Conservation* 142:761–771.
Egerer, M.H., N. Fouch, E.C. Anderson, M. Clarke. 2020a. Socio-ecological connectivity differs in magnitude and direction across urban landscapes. *Scientific Reports* 10:4252.
Egerer, M.H., B. Wagner, B.B. Lin, D. Kendal, K. Zhu. 2020b. New methods of spatial analysis in urban gardens inform future vegetation surveying. *Landscape Ecology* 35:761–778.
Ellis, E.E., T.L. Wilkinson. 2021. Moth assemblages within urban domestic gardens respond positively to habitat complexity, but only at a scale that extends beyond the garden boundary. *Urban Ecosystems* 24(3):469–479.
Evans, K.L., S.E. Newson, K.J. Gaston. 2009. Habitat influences on urban avian assemblages. *Ibis* 151(1):19–39.
Farr, C.M., S.E. Reed, L. Pejchar. 2018. How often are conservation developments managed for biodiversity protection? A case study in Colorado, USA. *Landscape and Urban Planning* 169:105–114.
Ferguson, M., H.E. Roberts, R.R.C. McEachan, M. Dallimer. 2018. Contrasting distributions of urban green infrastructure across social and ethno-racial groups. *Landscape and Urban Planning* 175:136–148.
Fuller, R.A., P.H. Warren, P.R. Armsworth, O. Barbosa, K.J. Gaston. 2008. Garden bird feeding predicts the structure of urban avian assemblages. *Diversity and Distributions* 14:131–137.
Gaston, K.J., R.M. Smith, K. Thompson, P.H. Warren. 2005. Urban Domestic gardens (II): Experimental tests of methods for increasing biodiversity. *Biodiversity and Conservation* 14:395–413.
Goddard, M.A., Z.G. Davies, S. Guenat, M.J. Ferguson, J.C. Fisher, A. Akanni, T. Ahjokoski, P.M.L. Anderson, F. Angeoletto, C. Antoniou, A.J. Bates, A. Barkwith, A. Berland, C.J. Bouch, C.C. Rega-Brodsky, L.B. Byrne, D. Cameron, R. Canavan, T. Chapman, S. Connop, S. Crossland, M.C. Dade, D.A. Dawson, C. Dobbs, C.T. Downs, E.C. Ellis, F.J. Escobedo, P. Gobster, N.M. Gulsrud, B. Guneralp, A.K. Hahs, J.D. Hale, C. Hassall, M. Hedblom, D.F. Hochuli, T. Inkinen, I.-C. Ioja, D. Kendal, T. Knowland, I. Kowarik, S.J. Langdale, S.B. Lerman, I. MacGregor-Fors, P. Manning, P. Massini, S. McLean, D.D. Mkwambisi, A. Ossola, G.P. Luque, L. Pérez-Urrestarazu, K. Perini, G. Perry, T.J. Pett, K.E. Plummer, R.A. Radji, U. Roll, S.G. Potts, H. Rumble, J.P. Sadler, S. de Saille, S. Sautter, C.E. Scott, A. Shwartz, T. Smith, R.P.H. Snep, C.D. Soulsbury, M.C. Stanley, T. Van de Voorde, S.J. Venn, P.H. Warren, C.-L. Washbourne, M. Whitling, N.S.G. Williams, J. Yang, K. Yeshitela, K.P. Yocom, M. Dallimer. 2021. A global horizon scan of the future impacts of robotics and autonomous systems on urban ecosystems. *Nature Ecology & Evolution* 5:219–230.
Goddard, M.A., A.J. Dougill, T.G. Benton. 2010. Scaling up from gardens: Biodiversity conservation in urban environments. *Trends in Ecology and Evolution* 25(2):90–98.
Goddard, M.A., A.J. Dougill, T.G. Benton. 2013. Why garden for wildlife? Social and ecological drivers, motivations and barriers for biodiversity management in residential landscapes. *Ecological Economics* 86:258–273.
Goddard, M.A., K. Ikin, S.B. Lerman. 2017. Ecological and social factors determining the diversity of birds in residential yards and gardens. Pages 371–397 in E. Murgui, M. Hedblom, editors. *Ecology and Conservation of Birds in Urban Environments*. Springer International Publishing, Cham, Switzerland.
Gonçalves, S.F., A.C. de Paula Lourenço, J.S. de Sousa Bueno Filho, M.C.B. de Toledo. 2021. Characteristics of residential backyards that contribute to conservation and diversity of urban birds: A case study in a Southeastern Brazilian city. *Urban Forestry & Urban Greening* 61:127095.
Gray, E.R., Y. van Heezik. 2016. Exotic trees can sustain native birds in urban woodlands. *Urban Ecosystems* 19(1):315–329.
Gray, L., P. Guzman, K.M. Glowa, A.G. Drevno. 2014. Can home gardens scale up into movements for social change? The role of home gardens in providing food security and community change in San Jose, California. *Local Environment* 19(2):187–203.
Greenleaf, S.S., N.M. Williams, R. Winfree, C. Kremen. 2007. Bee foraging ranges and their relationship to body size. *Oecologia* 153:589–596.

Groffman, P.M., J. Cavender-Bares, N.D. Bettez, J.M. Grove, S.J. Hall, J.B. Heffernan, S.E. Hobbie, K.L. Larson, J.L. Morse, C. Neill, K. Nelson, J. O'Neil-Dunne, L. Ogden, D.E. Pataki, C. Polsky, R.R. Chowdhury, M.K. Steele. 2014. Ecological homogenization of urban USA. *Frontiers in Ecology and the Environment* 12(1):74–81.

Hall, S.J., J. Learned, B. Ruddell, K.L. Larson, J. Cavender-Bares, N. Bettez, P.M. Groffman, J.M. Grove, J.B. Heffernan, S.E. Hobbie, J.L. Morse, C. Neill, K.C. Nelson, J.P.M. O'Neil-Dunne, L. Ogden, D.E. Pataki, W.D. Pearse, C. Polsky, R. Roy Chowdhury, M.K. Steele, T.L.E. Trammell. 2016. Convergence of microclimate in residential landscapes across diverse cities in the United States. *Landscape Ecology* 31(1):101–117.

Hughes, J., M. Richardson, R. Lumber. 2018. Evaluating connection to nature and the relationship with conservation behaviour in children. *Journal for Nature Conservation* 45:11–19.

Hunter, M.C.R., D.G. Brown. 2012. Spatial contagion: Gardening along the street in residential neighborhoods. *Landscape and Urban Planning* 105:407–416.

Ibáñez-Álamo, J.D., F. Morelli, Y. Benedetti, E. Rubio, J. Jokimäki, T. Pérez-Contreras, P. Sprau, J. Suhonen, P. Tryjanowski, M.-L. Kaisanlahti-Jokimäki, A. Pape Møller, M. Díaz. 2020. Biodiversity within the city: Effects of land sharing and land sparing urban development on avian diversity. *Science of the Total Environment* 707:135477.

Kuras, E.R., P.S. Warren, J.A. Zinda, M.F.J. Aronson, S. Cilliers, M.A. Goddard, C.H. Nilon, R. Winkler. 2020. Urban socioeconomic inequality and biodiversity often converge, but not always: A global meta-analysis. *Landscape and Urban Planning* 198:103799.

Larson, K.L., R. Andrade, K.C. Nelson, M.M. Wheeler, J.M. Engebreston, S.J. Hall, M.L. Avolio, P.M. Groffman, M. Grove, J.B. Heffernan, S.E. Hobbie, S.B. Lerman, D.H. Locke, C. Neill, R.R. Chowdhury, T.L.E. Trammell. 2020. Municipal regulation of residential landscapes across US cities: Patterns and implications for landscape sustainability. *Journal of Environmental Management* 275:111132.

Larson, K.L., K.C. Nelson, S.R. Samples, S.J. Hall, N. Bettez, J. Cavender-Bares, P.M. Groffman, M. Grove, J.B. Heffernan, S.E. Hobbie, J. Learned, J.L. Morse, C. Neill, L.A. Ogden, J. O'Neil-Dunne, D.E. Pataki, C. Polsky, R.R. Chowdhury, M. Steele, T.L.E. Trammell. 2016. Ecosystem services in managing residential landscapes: Priorities, value dimensions, and cross-regional patterns. *Urban Ecosystems* 19:95–113.

Lehberger, M., A.-K. Kleih, K. Sparke. 2021. Self-reported well-being and the importance of green spaces – A comparison of garden owners and non-garden owners in times of COVID-19. *Landscape and Urban Planning* 212(4):104108.

Leong, M., R.R. Dunn, M.D. Trautwein. 2018. Biodiversity and socioeconomics in the city: A review of the luxury effect. *Biology Letters* 14(5):20180082.

Lepczyk, C.A., M.F.J. Aronson, K.L. Evans, M.A. Goddard, S.B. Lerman, J.S. MacIvor. 2017. Biodiversity in the city: Fundamental questions for understanding the ecology of urban green spaces for biodiversity conservation. *BioScience* 67(9):799–807.

Lerman, S.B., A.R. Contosta, J. Milam, C. Bang. 2018. To mow or to mow less: Lawn mowing frequency affects bee abundance and diversity in suburban yards. *Biological Conservation* 221:160–174.

Lerman, S.B., J. Milam. 2016. Bee fauna and floral abundance within lawn-dominated suburban yards in Springfield, MA. *Annals of the Entomological Society of America* 109(5):713–723.

Lerman, S.B., D.L. Narango, R. Andrade, P.S. Warren, A.M. Grade, K. Straley. 2021a. Wildlife in the city: Human drivers and human consequences. Pages 37–66 in P. Barbosa, editor. *Urban Ecology: Its Nature and Challenges*. CABI Publishing, Boston, MA.

Lerman, S.B., D.L. Narango, M.L. Avolio, A.R. Bratt, J.M. Engebreston, P.M. Groffman, S.J. Hall, J.B. Heffernan, S.E. Hobbie, K.L. Larson, D.H. Locke, C. Neill, K.C. Nelson, J. Padullés Cubino, T.L.E. Trammell. 2021b. Residential yard management and landscape cover affect urban bird community diversity across the continental USA. *Ecological Applications* 3(8):e02455.

Lerman, S.B., V.K. Turner, C. Bang. 2012. Homeowner associations as a vehicle for promoting native urban biodiversity. *Ecology and Society* 17(4):45.

Lerman, S.B., P.S. Warren. 2011. The conservation value of residential yards: Linking birds and people. *Ecological Applications* 21(4):1327–1339.

Locke, D.H., B. Hall, J.M. Grove, S.T.A. Pickett, L.A. Ogden, C. Aoki, C.G. Boone, J.P.M. O'Neil-Dunne. 2021. Residential housing segregation and urban tree canopy in 37 US cities. *NPJ Urban Sustainability* 1:15.

Loram, A., J. Tratalos, P.H. Warren, K.J. Gaston. 2007. Urban domestic gardens (X): The extent & structure of the resource in five major cities. *Landscape Ecology* 22:601–615.

Lowenstein, D.M., K.C. Matteson, E.S. Minor. 2019. Evaluating the dependence of urban pollinators on ornamental, non-native, and 'weedy' floral resources. *Urban Ecosystems* 22(2):293–302.

Luck, G.W., P. Davidson, D. Boxall, L. Smallbone. 2011. Relations between urban bird and plant communities and human well-being and connection to nature. *Conservation Biology* 25(4):816–826.

Luck, G.W., L.T. Smallbone, R. O'Brien. 2009. Socio-economics and vegetation change in urban ecosystems: Patterns in space and time. *Ecosystems* 12(4):604–620.

Magle, S.B., M. Fidino, E.W. Lehrer, T. Gallo, M.P. Mulligan, M.J. Ríos, A.A. Ahlers, J.L. Angstmann, A. Belaire, B. Dugelby, A. Gramza, L. Hartley, B. MacDougall, T.J. Ryan, C.M. Salsbury, H. Sander, C. Schell, K. Simon, S. St. Onge, D. Drake. 2019. Advancing urban wildlife research through a multi-city collaboration. *Frontiers in Ecology and the Environment* 17(4):232–239.

Magle, S.B., V.M. Hunt, M. Vernon, K.R. Crooks. 2009. Urban wildlife research: Past, present, and future. *Biological Conservation* 155:23–32.

Marselle, M.R., S.J. Lindley, P.A. Cook, A. Bonn. 2021. Biodiversity and health in the urban environment. *Current Environmental Health Reports* 8:146–156.

Mathieu, R., C. Freeman, J. Aryal. 2007. Mapping private gardens in urban areas using object-oriented techniques and very high-resolution satellite imagery. *Landscape and Urban Planning* 81:179–192.

McCaffrey, R.E., R.W. Mannan. 2012. How scale influences birds' responses to habitat features in urban residential areas. *Landscape and Urban Planning* 105(3):274–280.

McEwan, K., F.J. Ferguson, M. Richardson, R. Cameron. 2020. The good things in urban nature: A thematic framework for optimising urban planning for nature connectedness. *Landscape and Urban Planning* 194:103687.

McKinney, M.L. 2006. Urbanization as a major cause of biotic homogenization. *Biological Conservation* 127(3):247–260.

Mimet, A., C. Kerbiriou, L. Simon, J.-F. Julien, R. Raymond. 2020. Contribution of private gardens to habitat availability, connectivity and conservation of the common pipistrelle in Paris. *Landscape and Urban Planning* 193:103671.

Morelli, F., Y. Benedetti, J.D. Ibáñez-Álamo, J. Jokimäki, R. Mänd, P. Tryjanowski, A.P. Møller. 2016. Evidence of evolutionary homogenization of bird communities in urban environments across Europe. *Global Ecology and Biogeography* 25(11):1284–1293.

Muratet, A., B. Fontaine. 2015. Contrasting impacts of pesticides on butterflies and bumblebees in private gardens in France. *Biological Conservation* 182:148–154.

Narango, D.L., D.W. Tallamy, P.P. Marra. 2018. Nonnative plants reduce population growth of an insectivorous bird. *Proceedings of the National Academy of Sciences of USA* 115(45):11549–11554.

Niemiec, R.M., G.P. Asner, J.A. Gaertner, P.G. Brodrick, N. Vaughn, J. Heckler, F. Hughes, L. Keith, T. Matsumoto. 2020. Using spatially explicit, time-dependent analysis to understand how social factors influence conservation outcomes. *Conservation Biology* 34(2):505–514.

Nilon, C.H., M.F.J. Aronson, S.S. Cilliers, C. Dobbs, L.J. Frazee, M.A. Goddard, K.M. O'Neill, D. Roberts, E.K. Stander, P. Werner, M. Winter, K.P. Yocom. 2017. Planning for the future of urban biodiversity: A global review of city-scale initiatives. *BioScience* 67(4):332–342.

Nowak, D.J., M.H. Noble, S.M. Sisinni, J.F. Dwyer. 2001. People and trees: Assessing the US urban forest resource. *Journal of Forestry* 99(3):37–42.

Odum, W.E. 1982. Environmental degradation and the tyranny of small decisions. *BioScience* 32(9):728–729.

Ossola, A., M.L. Cadenasso, E.K. Meineke. 2021. Valuing the role of time in urban ecology. *Frontiers in Ecology and Evolution* 9:620620.

Ossola, A., D. Locke, B. Lin, E. Minor. 2019. Yards increase forest connectivity in urban landscapes. *Landscape Ecology* 34:2935–2948.

Padullés Cubino, J., M.L. Avolio, M.M. Wheeler, K.L. Larson, S.E. Hobbie, J. Cavender-Bares, S.J. Hall, K.C. Nelson, T.L.E. Trammell, C. Neill, D.E. Pataki, J. Morgan Grove, P.M Groffman. 2020a. Linking yard plant diversity to homeowners' priorities across the U.S. *Landscape and Urban Planning* 196:103730.

Padullés Cubino, J., J. Cavender-Bares, P.M. Groffman, M.L. Avolio, A.R. Bratt, S.J. Hall, K.L. Larson, S.B. Lerman, D.L. Narango, C. Neill, T.L.E. Trammell, M.M. Wheeler, S.E. Hobbie. 2020b. Taxonomic, phylogenetic, and functional composition and homogenization of residential yard vegetation with contrasting management. *Landscape & Urban Planning* 202:103877.

Padullés Cubino, J., J. Cavender-Bares, S.E. Hobbie, D.E. Pataki, M.L. Avolio, L.E. Darling, K.L. Larson, S.J. Hall, P.M. Groffman, T.L.E. Trammell, M.K. Steele, J.M. Grove, C. Neill. 2019. Drivers of plant

species richness and phylogenetic composition in urban yards at the continental scale. *Landscape Ecology* 34:63–77.
Parris, K.M. 2006. Urban amphibian assemblages as metacommunities. *Journal of Animal Ecology* 75(3):757–764.
Pearse, W.D., J. Cavender-Bares, S.E. Hobbie, M.L. Avolio, N. Bettez, R. Roy Chowdhury, L.E. Darling, P.M. Groffman, J.M. Gove, S.J. Hall, J.B. Heffernan, J. Learned, C. Neill, K.C. Nelson, D.E. Pataki, B.L. Ruddell, M.K. Steele, T.L.E. Trammell. 2018. Homogenization of plant diversity, composition, and structure in North American urban yards. *Ecosphere* 9:e02105.
Piana, M.R., C.C. Pregitzer, R.A. Hallett. 2021. Advancing management of urban forested natural areas: Toward an urban silviculture? *Frontiers in Ecology and the Environment* 19(9):526–535.
Pimentel, D., L. Lach, R. Zuniga, D. Morrison. 2000. Environmental and economic costs of non-indigenous species in the United States. *BioScience* 50(1):53–65.
Poortinga, W., N. Bird, B. Hallingberg, R. Phillips, D. Williams. 2021. The role of perceived public and private green space in subjective health and wellbeing during and after the first peak of the COVID-19 outbreak. *Landscape and Urban Planning* 211:104092.
Pouyat, R.V., H. Setälä, K. Szlavecz, I.D. Yesilonis, S. Cilliers, E. Hornung, S. Yarwood, D.J. Kotze, M. Dombos, M.P. McGuire, T.H. Whitlow. 2017. Introducing GLUSEEN: A new open access and experimental network in urban soil ecology. *Journal of Urban Ecology* 3(1):jux002.
Public Health England. 2020. *Improving Access to Greenspace: A New Review for 2020*. PHE Publications, London.
Ramalho, C.E., R.J. Hobbs. 2012. Time for a change: Dynamic urban ecology. *Trends in Ecology and Evolution* 27(3):179–188.
Sadler, J.P., E.C. Small, H. Fiszpan, M.G. Telfer, J. Niemelä. 2006. Investigating environmental variation and landscape characteristics of an urban-rural gradient using woodland carabid assemblages. *Journal of Biogeography* 33(6):1126–1138.
Schell, C.J., K. Dyson, T.L. Fuentes, S. Des Roches, N.C. Harris, D.S. Miller, C.A. Woelfle-Erskine, M.R. Lambert. 2020. The ecological and evolutionary consequences of systemic racism in urban environments. *Science* 369:eaay4497.
Shanahan, D.F., R.A. Fuller, R. Bush, B.B. Lin, K.J. Gaston. 2015. The health benefits of urban nature: How much do we need? *BioScience* 65(5):476–485.
Shochat, E., S.B. Lerman, J.M. Anderies, P.S. Warren, S.H. Faeth, C.H. Nilon. 2010. Invasion, competition, and biodiversity loss in urban ecosystems. *BioScience* 60(3):199–208.
Shochat, E., P.S. Warren, S.H. Faeth, N.E. McIntyre, D. Hope. 2006. From patterns to emerging processes in mechanistic urban ecology. *Trends in Ecology and Evolution* 21(4):186–191.
Smith, R.M., K.J. Gaston, P.H. Warren, K. Thompson. 2006a. Urban domestic gardens (VIII): Environmental correlates of invertebrate abundance. *Biodiversity and Conservation* 15:2515–2545.
Smith, R.M., K. Thompson, J.G. Hodgson, P.H. Warren, K.J. Gaston. 2006b. Urban domestic gardens (IX): Composition and richness of the vascular plant flora, and implications for native biodiversity. *Biological Conservation* 129:312–322.
Soga, M., K.J. Gaston. 2020. The ecology of human – nature interactions. *Proceedings of the Royal Society B: Biological Sciences* 287(19198):20191882.
Sperling, C.D., C.J. Lortie. 2010. The importance of urban backgardens on plant and invertebrate recruitment: A field microcosm experiment. *Urban Ecosystems* 13(2):223–235.
Sushinsky, J.R., J.R. Rhodes, H.P. Possingham, T.K. Gill, R.A. Fuller. 2013. How should we grow cities to minimize their biodiversity impacts? *Global Change Biology* 19:401–410.
Swan, C.M., B. Brown, D. Borowy, J. Cavender-Bares, A. Jeliazkov, S. Knapp, Z. Lososová, J. Padullés Cubino, S. Pavoine, C. Ricotta, D. Sol. 2021. A framework for understanding how biodiversity patterns unfold across multiple spatial scales in urban ecosystems. *Ecosphere* 12(7):e03650.
Thompson, K., J.G. Hodgson, R.M. Smith, P.H. Warren, K.J. Gaston. 2004. Urban domestic gardens (III). Composition and diversity of lawn floras. *Journal of Vegetation Science* 15(3):373–378.
Threlfall, C.G., A. Ossola, A.K. Hahs, N.S.G. Williams, L. Wilson, S.J. Livesley. 2016. Variation in vegetation structure and composition across urban green space types. *Frontiers in Ecology and Evolution* 4:66.
Tóth, Z., K. Szlavecz, D.J. Epp Schmidt, E. Hornung, H. Setälä, I.D. Yesilonis, D.J. Kotze, M. Dombos, R. Pouyat, S. Mishra, S. Cilliers, S. Yarwood, C. Csuzdi. 2020. Earthworm assemblages in urban habitats across biogeographical regions. *Applied Soil Ecology* 151:103530.

Tresch, S., D. Frey, R.-C. Le Bayon, P. Mäuder, B. Stehle, A. Fliessbach, M. Moretti. 2019. Direct and indirect effects of urban gardening on aboveground and belowground diversity influencing soil multifunctionality. *Scientific Reports* 9(1):9769.

Uchida, K., R.V. Blakey, J.R. Burger, D.S. Cooper, C.A. Niesner, D.T. Blumstein. 2021. Urban biodiversity and the importance of scale. *Trends in Ecology and Evolution* 36(2):123–131.

van Heezik, Y., K.J.M. Dickinson, C. Freeman, S. Porter, J. Wing, B.I.P. Barratt. 2016. To what extent does vegetation composition and structure influence beetle communities and species richness in private gardens in New Zealand? *Landscape and Urban Planning* 151:79–88.

van Heezik, Y., C. Freeman, S. Porter, K.J.M. Dickinson. 2013. Garden size, householder knowledge, and socio-economic status influence plant and bird diversity at the scale of individual gardens. *Ecosystems* 16(8):1442–1454.

Van Helden, B.E., P.G. Close, B.A. Stewart, P.C. Speldewinde. 2021. Managing gardens for wildlife: Features that predict mammal presence and abundance in gardens vary seasonally. *Ecosphere* 12(3):e03453.

Vergnes, A., I. Le Viol, P. Clergeau. 2012. Green corridors in urban landscapes affect the arthropod communities of domestic gardens. *Biological Conservation* 145(1):171–178.

Warren, P.S., S.B. Lerman, R. Andrade, K.L. Larson, H.L. Bateman. 2019. The more things change: Species losses detected in Phoenix despite stability in bird – socioeconomic relationships. *Ecosphere* 10(3):e02624.

Wheeler, M.M., C. Neill, P.M. Groffman, M. Avolio, N. Bettez, J. Cavender-Bares, R. Roy Chowdhury, L. Darling, J.M. Grove, S.J. Hall, J.B. Heffernan, S.E. Hobbie, K.L. Larson, J.L. Morse, K.C. Nelson, L.A. Ogden, J. O'Neil-Dunne, D.E. Pataki, T.L.E. Trammell. 2017. Continental-scale homogenization of residential lawn plant communities. *Landscape and Urban Planning* 165:54–63.

Wilson, R.S., D.J. Hardisty, R.S. Epanchin-Niell, M.C. Runge, K.L. Cottingham, D.L. Urban, L.A. Maguire, A. Hastings, P.J. Mumby, D.P.C. Peters. 2016. A typology of time-scale mismatches and behavioral interventions to diagnose and solve conservation problems. *Conservation Biology* 30:42–49.

Wolch, J., J. Byrne, J. Newell. 2014. Urban green space, public health, and environmental justice: The challenge of making cities 'just green enough'. *Landscape and Urban Planning* 125:234–244.

11

CONNECTING BIODIVERSITY AND HUMAN WELLBEING

Jessica C. Fisher, Maximilian Nawrath, Martin Dallimer, Katherine N. Irvine, and Zoe G. Davies

Introduction

Approximately 55% of the world's human population live in urban areas (UN 2018). By 2050, that figure will be nearly 70% (UN 2018). People migrate to cities for better housing, health care, and education opportunities (UN 2018). Yet poor urban planning and ineffectively managed expansion can lead to overcrowding, pollution, and inadequate sanitation (UN-Habitat 2016). For urban populations, the multiple and synergistic impacts of these stressors can lead to physical (e.g. respiratory illnesses, cardiovascular disease) and mental ill health (e.g. anxiety, depression) (Peen et al. 2010; Abbot 2012; WHO 2016). Indeed, the prevalence and costs allied with treating mental ill health and associated non-communicable diseases (e.g. diabetes, cardiovascular disease, depression) are rising worldwide (WHO 2014). Given the projected expansion of urban populations, maintaining good health and wellbeing for people living in cities is a global priority, reflected in international policy agreements such as the Sustainable Development Goals (SDGs; Griggs et al. 2013).

Urban green (e.g. parks, gardens, roadside verges) and blue (e.g. rivers, waterways, lakes) space is a critical centrepiece of urban design due to the contribution it makes to improving human health and wellbeing (WHO 2017). Highlighting how urban biodiversity can benefit people living in cities is important, particularly where low budget and competing land-uses are prioritised over high-quality green and blue space provision in planning and management decision-making (Abbot 2012; van den Bosch and Nieuwenhuijsen 2016). Compared to high-income countries, very little is known about the health-promoting ('salutogenic') qualities of urban green and blue space in low- and middle-income countries (Nawrath et al. 2021), yet almost all of the projected growth in urban populations will take place in these nations (Angel et al. 2011). This is particularly concerning, given that they are also in the world's most biodiverse regions (Barlow et al. 2018; Simkin et al. 2022).

Ecosystem services and disservices

Biodiversity underpins the functioning of Earth's natural systems and the support they provide to humanity, conceptualised as 'ecosystem services' (Reid et al. 2005). Urban green and blue space deliver a wide range of ecosystem services in cities (TEEB 2011), such as agriculture (provisioning services), reducing pollution and improving air quality (regulating services), and promoting tourism (cultural services). Vegetation in these spaces can play roles such as acting as a flood

 DOI: 10.4324/9781003016120-13

defence, or absorbing rainfall that is deflected off impervious surfaces (e.g. buildings, roads; Silvennoinen et al. 2017). Likewise, shade-giving trees can contribute to temperature cooling at local scales, helping mitigate the effects of heatwaves and local heat-island effects (Ward et al. 2016).

Urban biodiversity can also lead to ecosystem disservices, negatively impacting human health and wellbeing. For instance, in Europe, the likelihood of allergic and respiratory disorders can be greatly increased by living near green space (Parmes et al. 2020). Green space can provide habitat for vectors of disease (e.g. mosquitoes; Zhao et al. 2020), especially in equatorial regions (Allen et al. 2017), which can lead to people worrying about disease risks (e.g. Kathmandu, Nepal; Nawrath et al. 2022). Similarly, dense vegetation can be perceived as a hotspot for criminal activity in both low/middle-income (Fisher et al. 2021c) and high-income nations (Kondo et al. 2017), as well as potentially harbouring species thought to be dangerous or unappealing (Bixler and Floyd 1997). In Mumbai, India, leopards enter residential areas and pose a potential health risk (Nulkar 2017). From a wider perspective, the spatial distribution of urban green and blue space can enhance social inequities through gentrification and the displacement of low-income residents (Wolch et al. 2014). As such, understanding both the benefits and disservices underpinned by urban biodiversity is fundamental to the sustainable design and management of cities.

Ecosystem service assessments are now one of the prominent lenses through which nature is valued and incorporated into decision-making. Nonetheless, our knowledge of how biodiversity underpins cultural ecosystem services and disservices is still relatively limited for human health and wellbeing (Bratman et al. 2019).

Theories and mechanisms underpinning biodiversity-wellbeing relationships

Defining human wellbeing

Wellbeing is a multidimensional concept, encompassing different contributions to human quality of life (Stiglitz et al. 2010). The World Health Organisation (WHO) conceptualises wellbeing as 'a state of complete, physical, mental and social well-being and not merely the absence of disease or infirmity' (WHO and CBD 2015). Multiple domains of wellbeing are encompassed in this definition – bio, which is physical; psycho, which is mental, consisting of both cognitive and emotional (Andrews and McKennell 1980); and social – comprise the 'biopsychosocial' model of wellbeing that originates from medicine (Engel 1977). An expanded version of this model, called the biopsychosocial-spiritual model of human wellbeing (Linton et al. 2016; Irvine et al. 2019), also includes a spiritual domain, conceived as including a connection to something greater than oneself.

Many traditional measures of quality of life (e.g. life expectancy, social support) correlate with self-reported evaluations of life satisfaction (Helliwell et al. 2016). However, evidence shows that life satisfaction does not improve with increases in GDP (gross domestic product), beyond a threshold of meeting a person's basic needs (Diener et al. 2018). As such, self-reported, subjective measures of human wellbeing such as life satisfaction are now being used by some policymakers as important indicators of economic and social progress (Costanza et al. 2014). They have been shown to be comparable across continents (Dolan and Metcalfe 2012), a determinant of mortality (Steptoe et al. 2015) and interlinked with life expectancy (Diener and Chan 2011).

Theoretical underpinnings

Several prominent theories underpin the widely accepted fact that human wellbeing improves in natural environments. Attention restoration theory (ART) postulates that nature restores people's ability to concentrate, focus their attention, process information, and problem-solve (Kaplan and Kaplan 1989; Kaplan 1995). The 'perceived restorativeness' of nature is thought to be based on four qualities (Kaplan and Kaplan 1989; Kaplan 1995): (1) 'fascination' (interesting stimuli that effortlessly attract attention); (2) 'coherence' (arrangement of stimuli); (3) 'compatibility' (the ability to carry out purposes freely); and (4) 'being away' (distance from everyday tasks or those that demand directed attention). Additionally, natural environments facilitate recovery from stress. The psychoevolutionary theory (Ulrich et al. 1991) posits that initial emotive reactions to natural environments (e.g. positive emotion – also known as 'affect'), can stimulate reduce physiological activity (e.g. cardiovascular, skeletomuscular; Ulrich 1983).

Kellert and Wilson (1993) popularised the term 'biophilia' (Fromm 1964), whereby people are presumed to have evolved to be inherently emotionally affiliated and drawn to natural environments. Nonetheless, genetic adaptations predispose people to exhibit negative responses to certain stimuli ('biophobia'), such as snakes. There remains contention as to whether biophilia is too broad a concept to be considered a valid explanation, and whether biophobia is in fact a contradiction of the concept itself (Joye and de Block 2011; Clayton et al. 2017). Biophilia theory extends to 'topophilia', which posits that people's emotional affiliation with and attachment to their local environment occurs via both biological selection and cultural learning (Tuan 1974; Beery et al. 2015).

Human perceptions of the environment can also be influenced by memories, ideas, and conceptions (sense of place; i.e. place attachment, place identity, meanings; Proshansky et al. 1983). However, the contribution that biodiversity makes to a sense of place remains not well understood (Hausmann et al. 2016).

People's preferences for specific attributes in nature determine the benefits they obtain. Theory proposes that moderately complex environments are most preferred (Ulrich 1983; Kaplan and Kaplan 1989). Ecologists have demonstrated that geometric fractals (a shape comprised of similar copies of the whole, determined by a D score) are indicative of biodiversity in an environment (e.g. habitat complexity or species richness; Dibble and Thomaz 2009; Stevens 2018). A mid-range fractal D score is most prevalent in nature and species-rich habitats (Hägerhäll et al. 2015; Stevens 2018). Concurrently, studies have shown that mid-range D scores are aesthetically preferred by people (Spehar et al. 2003; Bies et al. 2016) as they simplify the ease with which the human brain can process its surroundings (Joye et al. 2016).

Frameworks, mechanisms, and pathways

Several frameworks have sought to structure our understanding of how nature/biodiversity relates to different aspects of human wellbeing (Hartig et al. 2014; Markevych et al. 2017; Bratman et al. 2019; Marselle et al. 2021). Within the pathways and mechanisms, these frameworks incorporate the biopsychosocial (but not biopsychosocial-spiritual) model of health, as well as many of the prominent theories (e.g. ART, psychoevolutionary theory). Recent work suggests that biodiversity (actual or perceived, measured as the organism or trait, abundance, or richness) influences health and wellbeing (e.g. improved self-reported health, greater risk of allergic rhinitis) via four pathways (Marselle et al. 2021): (1) reducing harm (e.g. reducing heat exposure); (2) restoring capacities (e.g. stress recovery); (3) building capacities (e.g. encouraging physical activity); and (4) causing harm (e.g. harmful microbiota). Much of the empirical research substantiating these frameworks originates from high-income countries and may therefore mis- or under-represent the diverse ways in which people across the world experience or relate to urban biodiversity. As our knowledge of biodiversity-wellbeing relationships develop, these frameworks will likely undergo revision.

Biodiversity is a multisensory experience (Franco et al. 2017; Marselle et al. 2021). Each species may comprise a variety of morphologies (e.g. body size, eye position, flower structure), colours (e.g. red plumage, pink flowers, yellow leaves), sounds (e.g. wind blowing through trees, insects buzzing, bird song), smells (e.g. pine sap, mammal faeces, wild garlic), textures (e.g. dead/living leaves, stinging plants, furry animals), behaviours (e.g. parental care of young, elusive, predatory), and cultural meanings (e.g. emblematic, historical, spiritual). Research into how the multisensory aspects of biodiversity affect human wellbeing is only just emerging (e.g. Ferraro et al. 2020; Austen et al. 2021).

Empirical research examining biodiversity-wellbeing relationships

Much of the available evidence linking urban natural environments to human wellbeing focusses on the concepts of nature, green/blue space, or remotely sensed greenness from satellite imagery (e.g. normalised difference vegetation index [NDVI]). For instance, exposure to green space has been linked to improved quality of life (Stigsdotter et al. 2010), life satisfaction (White et al. 2013), and happiness (MacKerron and Mourato 2013). Higher residential greenness has been associated with reduced use of depression and anxiety-related medication in Spain (Gascón et al. 2018) and Scotland (Roberts et al. 2021), as well as higher subjective wellbeing in Australia (Mavoa et al. 2019). However, these approaches assume that urban green space is homogeneous (Eigenbrod et al. 2010) and do not elucidate which specific features are needed to promote health/wellbeing, despite this information being crucial to improving the quality of such environments for people. Certainly, specific characteristics like cleanliness, maintenance, and adequate facilities are important in urban green spaces (Giles-Corti et al. 2005; McCormack et al. 2010; Madureira et al. 2018). In low- to middle-income countries, this list includes crime, pollution, and sanitation more prominently (Amano et al. 2018). In urban blue space, features like the linearity of canals and motion and fluidity of the water can also influence people's wellbeing (Fisher et al. 2021a; Völker and Kistemann 2011).

Dissecting nature, green/blue space, and greenness into their constituent parts (i.e. biodiversity) provides a better assessment of quality from an ecological perspective and enables researchers to ask whether more biodiverse natural environments lead to enhanced benefits to human wellbeing (Wheeler et al. 2015; Wood et al. 2018). However, comparisons between existing biodiversity-wellbeing studies are complicated by the use of a diverse range of human wellbeing measures (e.g. anxiety or psychological wellbeing), biodiversity metrics (e.g. species richness or abundance), focal taxonomic groups (e.g. birds or plants), geographic locations (e.g. high-, middle-, or low-income countries), and sampling methodologies (e.g. proxy-measures or point counts for biodiversity; cross-sectional or momentary assessments of people). Disparities can also exist between people's perceptions of biodiversity and objective reality (Pett et al. 2016). In a seminal study in England by Dallimer et al. (2012), perceived butterfly, bird, and plant species richness all had a positive effect on wellbeing, whereas actual species richness had a variable impact, differing for each of the three taxa (Dallimer et al. 2012). In an experimental manipulation of green space in France, Shwartz et al. (2014) showed that although people preferred sites that were species-rich, they were unable to detect changes in richness of plants, birds, and pollinators after manipulation, and species richness was largely underestimated.

Nonetheless, the evidence-base exploring links between biodiversity and wellbeing is diverse and growing. In three English towns, vegetation cover and bird abundances were positively associated with lower depression prevalence (Cox et al. 2017). In perennial urban meadows in southern England, high perceived plant, bird, and butterfly species richness was linked to higher psychological wellbeing (Southon et al. 2017). Elsewhere in England, the introduction of potted plants to front gardens was linked to reduced stress (Chalmin-Pui et al. 2021). In Italy, Carrus et al. (2015) showed that a more biodiverse urban woodland stimulates greater perceived restorativeness, while Marselle et al. (2020) found that street tree density in residential urban areas in Leipzig in Germany was inversely associated with antidepressant prescription rates. At a countrywide scale across Germany, Methorst et al. (2021) demonstrated that bird and plant species richness of plants and birds was positively associated with mental health, and across 26 countries in Europe, bird species richness was associated with higher life satisfaction (Methorst et al. 2020).

Although, much of the empirical evidence investigating biodiversity and wellbeing in urban green space originates from Europe, a nascent body of biodiversity-wellbeing literature is emerging from Asia and South America. In Hong Kong, people were more likely to visit urban blue space and have good mental wellbeing if there were wildlife to see (Garrett et al. 2019b). Across urban trails in Singapore, higher perceived animal diversity (mammals, amphibians, birds, insects) was positively linked to perceived restorativeness and, subsequently, greater positive and lower negative emotion (Nghiem et al. 2021). Contrary to this, Chang et al. (2016) showed that urban green space with higher insect species richness, abundance, and Simpson diversity in Taiwan was not correlated with any physiological responses. Similarly, in Guyana, there was no relationship between actual measures of bird diversity and anxiety, positive or negative emotion (Fisher et al. 2021b). However, perceived biodiversity was mediated by perceived restorativeness, which then improved measures of anxiety and positive emotion (Fisher et al. 2021c). Across low-, middle-, and high-income countries, consistent patterns are yet to emerge.

People perceive biodiversity in rich and complex ways, responding to specific attributes (e.g. colours, textures) and relating them to past experiences (e.g. culture, media; Austen et al. 2021; Bell et al. 2019). In a mobile phone app-based study in the UK, Bakolis et al. (2018) showed people who saw trees and heard birds reported higher momentary wellbeing. Elsewhere in the UK, birdsong improved attention restoration and reduced stress (Ratcliffe et al. 2013). These findings are corroborated by similar studies in Guyana (Fisher et al. 2021c) and Ecuador (Moscoso et al. 2018). Other work in England also shows more colourful planting regimes provide greater aesthetic enjoyment (Hoyle et al. 2018). People also tend to recognise species that they are familiar with in other situations, such as domestic gardens (Austen et al. 2021), or misidentify species due to what they see represented in the media (Celis-Diez et al. 2017). Perceptions can be both positive and negative (Austen et al. 2021). For instance, people have positive associations with culturally important songbirds (Clucas et al. 2015; Brock et al. 2017) and negative associations with local wildlife thought to be dangerous (Schuttler et al. 2019).

Implications for policy and practice

Practitioners and policy-makers tasked with managing urban landscapes must deliver, and trade-off among, multiple biodiversity, individual, and societal benefits and disservices. Interventions that produce mutually reinforcing positive outcomes for both humans and conservation are highly desirable. Nonetheless, we cannot optimise these crucial trade-offs until we understand how biodiversity underpins health and wellbeing, which requires transcending terminology, methods, and paradigms championed by different disciplines (Sandifer et al. 2015). This evidence can then be incorporated into urban design, landscape planning, and architecture to create new, and retrofit old, spaces to maximise the occurrence of positive human-biodiversity interactions (Davies et al. 2019).

Biodiversity conservation

Understanding biodiversity-wellbeing linkages could lead to positive outcomes for conservation through two main pathways (Davies et al. 2019). The first is through win-win scenarios, whereby green/blue space that contain high levels of biodiversity (e.g. including threatened or specialist species) are also beneficial for people (e.g. gardens designed for recreation or aesthetic value). These include space primarily managed for people, such as botanical gardens, which are popular recreational spaces and simultaneously support high levels of species richness (Westwood et al. 2021), and allotments, which are known to be particularly beneficial for pollinating insects (Baldock 2020), as well as for human health and food sustainability (Soga et al. 2016). Win-wins are also seen in green space managed for biodiversity, such as urban protected areas, which protect habitat for species while also attracting tourists, bringing employment opportunities and other benefits (ten Brink et al. 2016).

The second pathway through which biodiversity-wellbeing research could benefit conservation is through altering people's perceptions of biodiversity so that their interest and concern for it may stimulate conservation gains. Urban biodiversity has been associated with '(re)connecting people with nature' at a time when urban populations are increasingly devoid of nature experiences (Soga and Gaston 2016). The literature is expanding rapidly on the associations between nature connection, pro-environmental behaviours, and human wellbeing (Rogerson et al. 2017; Alcock et al. 2020; Martin et al. 2020). Yet these concepts are still muddled by terminological plurality (i.e. what exactly constitutes 'nature', an 'experience', or a 'connection'), thereby limiting the scientific rigour and policy-impact of the research (Clayton et al. 2017; Ives et al. 2017). Moreover, the imagining and visualisation of nature ('*thereness*') is important for people's wellbeing but is independent of any actual interaction with it (Kaplan and Kaplan 1989). Fundamentally, acknowledging that people's experiences of biodiversity differ between individuals is essential to delivering win-wins for both humans and non-human sssspecies.

Public health

Through carefully targeted interventions, such as strategically optimising the quality of green/blue space (e.g. urban parks, woodlands, wetlands, rivers) within heavily populated landscapes, relatively small health and wellbeing gains at an individual level could scale-up to substantial benefits across entire populations, generating substantial healthcare cost-savings (DEFRA 2011). Indeed, government policies searching for low-cost, non-medical health interventions have fuelled the growth of social-prescribing (pathways linking individuals to activities that improve health and wellbeing) in countries such as the UK (e.g. Garside et al. 2020) and Canada (e.g. www.parkprescriptions.ca/). 'Green' or 'nature-based' social prescribing encompasses interventions that harness the natural environment. These may constitute changes to spaces where people visit (e.g. provision of hospital gardens) or changes to people's behaviour (e.g. enrolling people in programmes; Shanahan et al. 2019). Evidence is still needed about what qualities of these spaces might be the most beneficial, and for whom, during interventions (Shanahan et al. 2019).

Likewise, the benefits of biodiversity are not equal across different sectors of society. For instance, urban populations from deprived households have been shown to benefit most from interactions with biodiversity (Garrett et al. 2019a; Roberts et al. 2021). We also know relatively little about the physical and mental barriers that facilitate the (in)accessibility of urban biodiversity (Cronin-de-chavez et al. 2019; Ibes et al. 2021). Indeed, our growing knowledge about the rich and varied ways in which different people relate to biodiversity (Austen et al. 2021) could be seen to complicate the emerging literature about what 'doses' are needed to optimise health and wellbeing gains (e.g. 120 minutes, White et al. 2019; 25 minutes, Fisher et al. 2021d; 10 minutes, Meredith et al. 2020) Nonetheless, it is clear that biodiversity could become a powerful tool through which significant public health benefits could become available for all.

Aligning global challenges

International, national, and local government support is required for the effective implementation of policies that support both biodiversity conservation and human wellbeing. The linkage between biodiversity and human wellbeing has been endorsed by the WHO and the Convention on Biological Diversity (CBD), stating that human health and wellbeing are basic human rights and essential to securing longer-term insurance and resilience for future generations (WHO and CBD 2015). The International Union for the Conservation of Nature (IUCN) have promoted urban-protected areas, arguing that alongside contributing toward urban ecosystem service delivery, climate change resilience, and protection of vulnerable species, they also serve to change attitudes towards nature for vast numbers of visitors (Trzyna et al. 2014).

There will be instances where biodiversity is detrimental to human wellbeing. As such, a mismatch can exist between the biodiversity that conservationists seek to support and those that actually have beneficial influence (Pett et al. 2016). While some trade-offs will be necessary to avoid disservices, these will be lessened by uncovering instances where win-wins can be achieved (e.g. botanical gardens; Gobster et al. 2007; Adams 2014). Ultimately, decisions about which biodiversity attributes of the urban environment are to be conserved for what purpose will be influenced by people's individual values and motivations, but a balance must be struck between prioritising people or biodiversity (Dearborn and Kark 2010).

Conclusions

Recognising the benefits and disservices that biodiversity can bring to human wellbeing is crucial in a world of fast-changing urban landscapes. As the building block that underpins ecosystem service delivery and the promotion of human wellbeing, biodiversity must be an essential part of the toolkit in the design and management of cities (Giles-Corti et al. 2016; Hartig and Kahn 2016). However, while current evidence exploring biodiversity-wellbeing linkages is promising, more is needed. This is particularly true because of the complexity of people-biodiversity relationships (Austen et al. 2021) and geographical skew in the current evidence-base (Nawrath et al. 2021). Given that improving the quality of urban green/blue space is likely to be synergistically advantageous to both people and biodiversity, human experiences of nature need to be explored empirically, systematically, and with more nuance in relation to biodiversity specifically.

Key companion papers

Austen, G.E., et al. 2021. Exploring shared public perspectives on biodiversity attributes. *People and Nature* 3(4):901–913.

Dallimer, M., et al. 2012. Biodiversity and the feel-good factor: Understanding associations between self-reported human well-being and species richness. *BioScience* 62:47–55.

Marselle, M.R., et al. 2021. Pathways linking biodiversity to human health: A conceptual framework. *Environment International* 150:106420.

Nawrath, M., S. Guenat, H. Elsey, M. Dallimer. 2021. Exploring uncharted territory: Do urban greenspaces support mental health in low- and middle-income countries? *Environmental Research* 194:110625.

Pett, T.J., et al. 2016. Unpacking the people – Biodiversity paradox: A conceptual framework. *BioScience* 66(7):576–583.

Romanelli, C., et al. 2015. *Connecting Global Priorities: Biodiversity and Human Health: A State of Knowledge Report.* World Health Organization/Convention on Biological Diversity, Geneva, Switzerland.

References

Abbot, A. 2012. Urban decay. *Nature* 490(7419):162–164.

Adams, W.M. 2014. The value of valuing nature. *Science* 346(6209):549–551.

Alcock, I., M.P. White, S. Pahl, R. Duarte-Davidson, L.E. Fleming. 2020. Associations between pro-environmental behaviour and neighbourhood nature, nature visit frequency and nature appreciation: Evidence from a nationally representative survey in England. *Environment International* 136:105441.

Allen, T., K.A. Murray, C. Zambrana-Torrelio, S.S. Morse, C. Rondinini, M. Di Marco, N. Breit, K.J. Olival, P. Daszak. 2017. Global hotspots and correlates of emerging zoonotic diseases. *Nature Communications* 8:1124.

Amano, T., I. Butt, K.S.-H. Peh. 2018. The importance of green spaces to public health: A multi-continental analysis. *Ecological Applications* 28(6):1473–1480.

Andrews, F.M., A.C. McKennell. 1980. Measures of self-reported well-being: Their affective, cognitive, and other components. *Social Indicators Research* 8:127–155.

Angel, S., J. Parent, D.L. Civco, A. Blei, D. Potere. 2011. The dimensions of global urban expansion: Estimates and projections for all countries, 2000–2050. *Progress in Planning* 75(2):53–107.

Austen, G.E., M. Dallimer, K.N. Irvine, P.R. Maund, R.D. Fish, Z.G. Davies. 2021. Exploring shared public perspectives on biodiversity attributes. *People and Nature* 3:901–913.

Bakolis, I., R. Hammoud, M. Smythe, J. Gibbons, N. Davidson, S. Tognin, A. Mechelli. 2018. Urban mind: Using smartphone technologies to investigate the impact of nature on mental well-being in real time. *BioScience* 68(2):134–145.

Baldock, K.C.R. 2020. Opportunities and threats for pollinator conservation in global towns and cities. *Current Opinion in Insect Science* 38:63–71.

Barlow, J., F. França, T.A. Gardner, C.C. Hicks, G.D. Lennox, E. Berenguer, L. Castello, E.P. Economo, J. Ferreira, B. Guénard, C. Gontijo Leal, V. Isaac, A.C. Lees, C.L. Parr, S.K. Wilson, P.J. Young, N.A.J. Graham. 2018. The future of hyperdiverse tropical ecosystems. *Nature* 559:517–526.

Beery, T.H., K.I. Jönsson, J. Elmberg. 2015. From environmental connectedness to sustainable futures: Topophilia and human affiliation with nature. *Sustainability* 7(7):8837–8854.

Bell, S.L., C. Leyshon, R. Foley, R.A. Kearns. 2019. The 'healthy dose' of nature: A cautionary tale. *Geography Compass* 13(1):1–14.

Bies, A.J., D.R. Blanc-Goldhammer, C.R. Boydston, R.P. Taylor, M.E. Sereno. 2016. Aesthetic responses to exact fractals driven by physical complexity. *Frontiers in Human Neuroscience* 10:210.

Bixler, R.D., M.F. Floyd. 1997. Nature is scary, disgusting, and uncomfortable. *Environment and Behavior* 29:443–467.

Bratman, G.N., C.B. Anderson, M.G. Berman, B. Cochran, S. de Vries, J. Flanders, C. Folke, H. Frumkin, J.J. Gross, T. Hartig, P.H. Kahn, M. Kuo, J.J. Lawler, P.S. Levin, T. Lindahl, A. Meyer-Lindenberg, R. Mitchell, Z. Ouyang, J. Roe, L. Scarlett, J.R. Smith, M. van den Bosch, B.W. Wheeler, M.P. White, H. Zheng, G.C. Daily. 2019. Nature and mental health: An ecosystem service perspective. *Science Advances* 5(7):eaax0903.

Brock, M., G. Perino, R. Sugden. 2017. The warden attitude: An investigation of the value of interaction with everyday wildlife. *Environmental and Resource Economics* 67(1):127–155.

Carrus, G., M. Scopelliti, R. Lafortezza, G. Colangelo, F. Ferrini, F. Salbitano, M. Agrimi, L. Portoghesi, P. Semenzato, G. Sanesi. 2015. Go greener, feel better? The positive effects of biodiversity on the well-being of individuals visiting urban and peri-urban green areas. *Landscape and Urban Planning* 134:221–228.

Celis-Diez, J.L., C.E. Muñoz, S. Abades, P.A. Marquet, J.J. Armesto. 2017. Biocultural homogenization in urban settings: Public knowledge of birds in City Parks of Santiago, Chile. *Sustainability* 9:485.

Chalmin-Pui, L.S., J. Roe, A. Griffiths, N. Smyth, T. Heaton, A. Clayden, R. Cameron. 2021. 'It made me feel brighter in myself' – The health and well-being impacts of a residential front garden horticultural intervention. *Landscape and Urban Planning* 205:103958.

Chang, K.G., W.C. Sullivan, Y.-H. Lin, W. Su, C.-Y. Chang. 2016. The effect of biodiversity on green space users' wellbeing – An empirical investigation using physiological evidence. *Sustainability* 8(10):1049.

Clayton, S., A. Colléony, P. Conversy, E. Maclouf, L. Martin, A.-C. Torres, M.-X. Truong, A.-C. Prévot. 2017. Transformation of experience: Toward a new relationship with nature. *Conservation Letters* 10(5):645–651.

Clucas, B., S. Rabotyagov, J.M. Marzluff. 2015. How much is that birdie in my backyard? A cross-continental economic valuation of native urban songbirds. *Urban Ecosystems* 18:251–266.

Costanza, R., I. Kubiszewski, E. Giovannini, H. Lovins, J. McGlade, K.E. Pickett, K.V. Ragnarsdóttir, D. Roberts, R. De Vogli, R. Wilkinson. 2014. Time to leave GDP behind. *Nature* 505(7483):283–285.

Cox, D.T.C., D.F. Shanahan, H.L. Hudson, K.E. Plummer, G.M. Siriwardena, R.A. Fuller, K. Anderson, S. Hancock, K.J. Gaston. 2017. Doses of neighborhood nature: The benefits for mental health of living with nature. *BioScience* 67(2):147–155.

Cronin-de-chavez, A., S. Islam, R.R.C. McEachan. 2019. Not a level playing field: A qualitative study exploring structural, community and individual determinants of greenspace use amongst low-income multi-ethnic families. *Health & Place* 56:118–126.

Dallimer, M., K.N. Irvine, A.M.J. Skinner, Z.G. Davies, J.R. Rouquette, L.M. Maltby, P.H. Warren, P.R. Armsworth, K.J. Gaston. 2012. Biodiversity and the feel-good factor: Understanding associations between self-reported human well-being and species richness. *BioScience* 62:47–55.

Davies, Z.G., M. Dallimer, J.C. Fisher, R.A. Fuller. 2019. Biodiversity and health: Implications for conservation. Pages 283–294 in M. Marselle, J. Stadler, H. Korn, K. Irvine, A. Bonn, editors. *Biodiversity and Health in the Face of Climate Change*. Springer, Cham, Switzerland.

Dearborn, D.C., S. Kark. 2010. Motivations for conserving urban biodiversity. *Conservation Biology* 24(2):432–440.

DEFRA (Department for Environment Food and Rural Affairs). 2011. The natural choice: Securing the value of nature. The Stationery Office, London, United Kingdom.

Dibble, E.D., S.M. Thomaz. 2009. Use of fractal dimension to assess habitat complexity and its influence on dominant invertebrates inhabiting tropical and temperate macrophytes. *Journal of Freshwater Ecology* 24(1):93–102.

Diener, E., M.Y. Chan. 2011. Happy people live longer: Subjective well-being contributes to health and longevity. *Applied Psychology: Health and Well-Being* 3(1):1–43.

Diener, E., S. Oishi, L. Tay. 2018. Advances in subjective well-being research. *Nature Human Behaviour* 2:253–260.

Dolan, P., R. Metcalfe. 2012. Measuring subjective wellbeing: Recommendations on measures for use by national governments. *Journal of Social Policy* 41(2):409–427.

Eigenbrod, F., P.R. Armsworth, B.J. Anderson, A. Heinemeyer, S. Gillings, D.B. Roy, C.D. Thomas, K.J. Gaston. 2010. The impact of proxy-based methods on mapping the distribution of ecosystem services. *Journal of Applied Ecology* 47(2):377–385.

Engel, G.L. 1977. The need for a new medical model: A challenge for biomecine. *Science* 196(4286):129–136.

Ferraro, D.M., Z.D. Miller, L.A. Ferguson, B.D. Taff, J.R. Barber, P. Newman, C.D. Francis. 2020. The phantom chorus: Birdsong boosts human well-being in protected areas. *Proceedings of the Royal Society B: Biological Sciences* 287(1941):20201811.

Fisher, J.C., J.E. Bicknell, K.N. Irvine, D. Fernandes, J. Mistry, Z.G. Davies. 2021a. Exploring how urban nature is associated with human wellbeing in a neotropical city. *Landscape and Urban Planning* 212:104119.

Fisher, J.C., J.E. Bicknell, K.N. Irvine, W.M. Hayes, D. Fernandes, J. Mistry, Z.G. Davies. 2021b. Bird diversity and psychological wellbeing: A comparison of green and coastal blue space in a neotropical city. *Science of the Total Environment* 793:148653.

Fisher, J.C., K.N. Irvine, J.E. Bicknell, W.M. Hayes, D. Fernandes, J. Mistry, Z.G. Davies. 2021c. Perceived biodiversity, sound, naturalness and safety enhance the restorative quality and wellbeing benefits of green and blue space in a neotropical city. *Science of the Total Environment* 755(2):143095.

Fisher, J.C., J. Mistry, M. Pierre, Y. Huichang, A. Harris, N. Hunte, D. Fernandes, J.E. Bicknell, Z.G. Davies. 2021d. Using participatory video to share people's experiences of neotropical urban green and blue spaces with decision-makers. *Geographical Journal* 187:346–360.

Franco, L.S., D.F. Shanahan, R.A. Fuller. 2017. A review of the benefits of nature experiences: More than meets the eye. *International Journal of Environmental Research and Public Health* 14(8):864.

Fromm, E. 1964. *The Heart of Man: Its Genius for Good and Evil*. Harper & Row Publishers, London.

Garrett, J.K., T.J. Clitherow, M.P. White, B.W. Wheeler, L.E. Fleming. 2019a. Coastal proximity and mental health among urban adults in England: The moderating effect of household income. *Health & Place* 59:102200.

Garrett, J.K., M.P. White, J. Huang, S. Ng, Z. Hui, C. Leung, L.A. Tse, F. Fung, L.R. Elliott, M.H. Depledge, M.C.S. Wong. 2019b. Urban blue space and health and wellbeing in Hong Kong: Results from a survey of older adults. *Health & Place* 55:100–110.

Garside, R., N. Orr, R. Short, B. Lovell, K. Husk, R. McEachan, R. Rashid, I. Dickie. 2020. *Therapeutic Nature: Nature-Based Social Prescribing for Diagnosed Mental Health Conditions in the UK.* Final Report for DEFRA. European Centre for Environment and Human Health, University of Exeter medical School, Truro.

Gascón, M., G. Sánchez-Benavides, P. Dadvand, D. Martínez, N. Gramunt, X. Gotsens, M. Cirach, C. Vert, J.L. Molinuevo, M. Crous-Bou, M.J. Nieuwenhuijsen. 2018. Long-term exposure to residential green and blue spaces and anxiety and depression in adults: A cross-sectional study. *Environmental Research* 162:231–239.

Giles-Corti, B., M.H. Broomhall, M. Knuiman, C. Collins, K. Douglas, K. Ng, A. Lange, R.J. Donovan. 2005. Increasing walking: How important is distance to, attractiveness, and size of public open space? *American Journal of Preventive Medicine* 28(2 Suppl 2):169–176.

Giles-Corti, B., A. Vernez-Moudon, R. Reis, G. Turrell, A.L. Dannenberg, H. Badland, S. Foster, M. Lowe, J.F. Sallis, M. Stevenson, N. Owen. 2016. City planning and population health: A global challenge. *The Lancet* 388(10062):2912–2924.

Gobster, P.H., J.I. Nassauer, T.C. Daniel, G. Fry. 2007. The shared landscape: What does aesthetics have to do with ecology? *Landscape Ecology* 22:959–972.

Griggs, D., M. Stafford-Smith, O. Gaffney, J. Rockström, M.C. Öhman, P. Shyamsundar, W. Steffen, G. Glaser, N. Kanie, I. Noble. 2013. Sustainable development goals for people and planet. *Nature* 495:305–307.

Hägerhäll, C.M., T. Laike, M. Küller, E. Marcheschi, C. Boydston, R.P. Taylor. 2015. Human physiological benefits of viewing nature: EEG responses to exact and statistical fractal patterns. *Nonlinear Dynamics, Psychology, and Life Sciences* 19:1–12.

Hartig, T., P.H.J. Kahn. 2016. Living in cities, naturally. *Science* 352(6288):938–940.

Hartig, T., R. Mitchell, S. de Vries, H. Frumkin. 2014. Nature and health. *Annual Review of Public Health* 35:207–228.

Hausmann, A., R. Slotow, J.K. Burns, E. Di Minin. 2016. The ecosystem service of sense of place: Benefits for human well-being and biodiversity conservation. *Environmental Conservation* 43(2):117–127.

Helliwell, J., R. Layard, J. Sachs. 2016. *World Happiness Report 2016, Update (Vol. I)*. Sustainable Development Solutions Network, New York.

Hoyle, H., B. Norton, N. Dunnett, J.P. Richards, J.M. Russell, P. Warren. 2018. Plant species or flower colour diversity? Identifying the drivers of public and invertebrate response to designed annual meadows. *Landscape and Urban Planning* 180:103–113.

Ibes, D.C., D.A. Rakow, C.H. Kim. 2021. Barriers to nature engagement for youth of colour. *Children, Youth, and Environments* 31:49–73.

Irvine, K.N., D. Hoesly, R. Bell-Williams, S.L. Warber. 2019. Biodiversity and spiritual well-being. Pages 213–247 in M. Marselle, J. Stadler, H. Korn, K. Irvine, A. Bonn, editors. *Biodiversity and Health in the Face of Climate Change*. Springer, Cham, Switzerland.

Ives, C.D., M. Giusti, J. Fischer, D.J. Abson, K. Klaniecki, C. Dorninger, J. Laudan, S. Barthel, P. Abernethy, B. Martín-López, C.M. Raymond, D. Kendal, H. von Wehrden. 2017. Human – nature connection: A multidisciplinary review. *Current Opinion in Environmental Sustainability* 26–27:106–113.

Joye, Y., A. de Block. 2011. 'Nature and I are two': A critical examination of the biophilia hypothesis. *Environmental Values* 20(2):189–215.

Joye, Y., L. Steg, A.B. Ünal, R. Pals. 2016. When complex is easy on the mind: Internal repetition of visual information in complex objects is a source of perceptual fluency. *Journal of Experimental Psychology: Human Perception and Performance* 42(1):103–114.

Kaplan, R., S. Kaplan. 1989. *The Experience of Nature: A Psychological Perspective*. Cambridge University Press, Cambridge.

Kaplan, S. 1995. The restorative benefits of nature: Toward an integrative framework. *Journal of Environmental Psychology* 15(3):169–182.

Kellert, S.R., E.O. Wilson. 1993. *The Biophilia Hypothesis*. Island Press, Washington, DC.

Kondo, M.C., S.H. Han, G.H. Donovan, J.M. MacDonald. 2017. The association between urban trees and crime: Evidence from the spread of the emerald ash borer in Cincinnati. *Landscape and Urban Planning* 157:193–199.
Linton, M.-J., P. Dieppe, A. Medina-Lara. 2016. Review of 99 self-report measures for assessing well-being in adults: Exploring dimensions of well-being and developments over time. *BMJ Open* 6:e010641.
MacKerron, G., S. Mourato. 2013. Happiness is greater in natural environments. *Global Environmental Change* 23(5):992–1000.
Madureira, H., F. Nunes, J.V. Oliveira, T. Madureira. 2018. Preferences for urban green space characteristics: A comparative study in three Portuguese cities. *Environments* 5(2):23.
Markevych, I., J. Schoierer, T. Hartig, A. Chudnovsky, P. Hystad, A.M. Dzhambov, S. de Vries, M. Triguero-Mas, M. Brauer, M.J. Nieuwenhuijsen, G. Lupp, E.A. Richardson, T. Astell-Burt, D. Dimitrova, X. Feng, M. Sadeh, M. Standl, J. Heinrich, E. Fuertes. 2017. Exploring pathways linking greenspace to health: Theoretical and methodological guidance. *Environmental Research* 158:301–317.
Marselle, M.R., D.E. Bowler, J. Watzema, D. Eichenberg, T. Kirsten, A. Bonn. 2020. Urban street tree biodiversity and antidepressant prescriptions. *Scientific Reports* 10:22445.
Marselle, M.R., T. Hartig, D.T.C. Cox, S. de Bell, S. Knapp, S. Lindley, M. Triguero-Mas, K. Böhning-Gaese, M. Braubach, P.A. Cook, S. de Vries, A. Heintz-Buschart, M. Hofmann, K.N. Irvine, N. Kabisch, F. Kolek, R. Kraemer, I. Markevych, D. Martens, R. Müller, M. Nieuwenhuijsen, J.M. Potts, J. Stadler, S. Walton, S.L. Warber, A. Bonn. 2021. Pathways linking biodiversity to human health: A conceptual framework. *Environment International* 150:106420.
Martin, L., M.P. White, A. Hunt, M. Richardson, S. Pahl, J. Burt. 2020. Nature contact, nature connectedness and associations with health, wellbeing and pro-environmental behaviours. *Journal of Environmental Psychology* 68:101389.
Mavoa, S., M. Davern, M. Breed, A. Hahs. 2019. Higher levels of greenness and biodiversity associate with greater subjective wellbeing in adults living in Melbourne, Australia. *Health & Place* 57:321–329.
McCormack, G.R., M. Rock, A.M. Toohey, D. Hignell. 2010. Characteristics of urban parks associated with park use and physical activity: A review of qualitative research. *Health & Place* 16:712–726.
Meredith, G.R., D.A. Rakow, E.R.B. Eldermire, C.G. Madsen, S.P. Shelley, N.A. Sachs. 2020. Minimum time dose in nature to positively impact the mental health of college-aged students, and how to measure it: A scoping review. *Frontiers in Psychology* 10:2942.
Methorst, J., U. Arbieu, A. Bonn, K. Boehning-Gaese, T. Mueller. 2020. Non-material contributions of wildlife to human well-being: A systematic review. *Environmental Research Letters* 15:093005.
Methorst, J., K. Rehdanz, T. Mueller, B. Hansjürgens, A. Bonn, K. Böhning-Gaese. 2021. The importance of species diversity for human well-being in Europe. *Ecological Economics* 181:106917.
Moscoso, P., M. Peck, A. Eldridge. 2018. Emotional associations with soundscape reflect human-environment relationships. *Journal of Ecoacoustics* 2(1):1.
Nawrath, M., H. Elsey, M. Dallimer. 2022. Why cultural ecosystem services matter most: Exploring the pathways linking greenspaces and mental health in a low-income country. *Science of the Total Environment* 806(pt 3):150551.
Nawrath, M., S. Guenat, H. Elsey, M. Dallimer. 2021. Exploring uncharted territory: Do urban greenspaces support mental health in low- and middle-income countries? *Environmental Research* 194:110625.
Nghiem, T.P.L., K.L. Wong, L. Jeevanandam, C.C. Chang, L.Y.C. Tan, Y. Goh, L.R. Carrasco. 2021. Biodiverse urban forests, happy people: Experimental evidence linking perceived biodiversity, restoration, and emotional wellbeing. *Urban Forestry & Urban Greening* 59:127030.
Nulkar, G. 2017. Silent conflicts – Human-wildlife interactions in urban spaces. *Journal of Ecological Society* 29(1):34–43.
Parmes, E., G. Pesce, C.E. Sabel, S. Baldacci, R. Bono, S. Brescianini, C. D'Ippolito, W. Hanke, M. Horvat, H. Liedes, S. Maio, P. Marchetti, A. Marcon, E. Medda, M. Molinier, S. Panunzi, J. Pärkkä, K. Polańska, J. Prud'homme, P. Ricci, J. Snoj Tratnik, G. Squillacioti, M.A. Stazi, C.N. Maesano, I. Annesi-Maesano. 2020. Influence of residential land cover on childhood allergic and respiratory symptoms and diseases: Evidence from 9 European cohorts. *Environmental Research* 183:108953.
Peen, J., R.A. Schoevers, A.T. Beekman, J. Dekker. 2010. The current status of urban-rural differences in psychiatric disorders. *Acta Psychiatrica Scandinavica* 121(2):84–93.
Pett, T.J., A. Shwartz, K.N. Irvine, M. Dallimer, Z.G. Davies. 2016. Unpacking the people – biodiversity paradox: A conceptual framework. *BioScience* 66(7):576–583.

Proshansky, H.M., A.K. Fabian, R. Kaminoff. 1983. Place-identity: Physical world socialization of the self. *Journal of Environmental Psychology* 3(1):57–83.
Ratcliffe, E., B. Gatersleben, P.T. Sowden. 2013. Bird sounds and their contributions to perceived attention restoration and stress recovery. *Journal of Environmental Psychology* 36:221–228.
Reid, W.V., H.A. Mooney, A. Cropper, D. Capistrano, S.R. Carpenter, K. Chopra, P. Dasgupta, T. Dietz, A.K. Duraiappah, R. Hassan, R. Kasperson, R. Leemans, R.M. May, T. McMichael, P. Pingali, C. Samper, R. Scholes, R.T. Watson, A.H. Zakri, Z. Shidong, N.J. Ash, E. Bennett, P. Kumar, M.J. Lee, C. Raudsepp-Hearne, H. Simons, J. Thonell, M.B. Zurek. 2005. *Ecosystems and Human Well-Being Synthesis: A Report of the Millennium Ecosystem Assessment*. Island Press, Washington, DC.
Roberts, M., K.N. Irvine, A. McVittie. 2021. Associations between greenspace and mental health prescription rates in urban areas. *Urban Forestry & Urban Greening* 64:127301.
Rogerson, M., J. Barton, R. Bragg, J. Pretty. 2017. *The Health and Wellbeing Impacts of Volunteering with The Wildlife Trusts*. The Wildlife Trusts, Newark, UK.
Sandifer, P.A., A.E. Sutton-Grier, B.P. Ward. 2015. Exploring connections among nature, biodiversity, ecosystem services, and human health and well-being: Opportunities to enhance health and biodiversity conservation. *Ecosystem Services* 12:1–15.
Schuttler, S.G., K. Stevenson, R. Kays, R.R. Dunn. 2019. Children's attitudes towards animals are similar across suburban, exurban, and rural areas. *PeerJ* 7:e7328.
Shanahan, D.F., T. Astell – Burt, E.A. Barber, E. Brymer, D.T.C. Cox, J. Dean, M. Depledge, R.A. Fuller, T. Hartig, K.N. Irvine, A. Jones, H. Kikillus, R. Lovell, R. Mitchell, J. Niemelä, M. Nieuwenhuijsen, J. Pretty, M. Townsend, Y. van Heezik, S. Warber, K.J. Gaston. 2019. Nature-based interventions for improving health and wellbeing: The purpose, the people and the outcomes. *Sports* 7(6):141.
Shwartz, A., A. Turbé, L. Simon, R. Julliard. 2014. Enhancing urban biodiversity and its influence on city-dwellers: An experiment. *Biological Conservation* 171:82–90.
Silvennoinen, S., M. Taka, V. Yli-pelkonen, H. Koivusalo, M. Ollikainen, H. Setälä. 2017. Monetary value of urban green space as an ecosystem service provider: A case study of urban runoff management in Finland. *Ecosystem Services* 28(A):17–27.
Simkin, R.D., K.C. Seto, R.I. McDonald, W. Jetz. 2022. Biodiversity impacts and conservation implications of urban land expansion projected to 2050. *Proceedings of the National Academy of Sciences of the United States of America* 119:1–10.
Soga, M., K.J. Gaston. 2016. Extinction of experience: The loss of human-nature interactions. *Frontiers in Ecology and the Environment* 14(2):94–101.
Soga, M., K.J. Gaston, Y. Yamaura. 2016. Gardening is beneficial for health: A meta-analysis. *Preventive Medicine Reports* 5:92–99.
Southon, G.E., A. Jorgensen, N. Dunnett, H. Hoyle, K.L. Evans. 2017. Biodiverse perennial meadows have aesthetic value and increase residents' perceptions of site quality in urban green-space. *Landscape and Urban Planning* 158:105–118.
Spehar, B., C.W.G. Clifford, B.R. Newell, R.P. Taylor. 2003. Universal aesthetic of fractals. *Computers & Graphics* 27(5):813–820.
Steptoe, A., A. Deaton, A.A. Stone. 2015. Subjective wellbeing, health, and ageing. *The Lancet* 385(9968):640–648.
Stevens, P. 2018. Fractal dimension links responses to a visual scene to its biodiversity. *Ecopsychology* 10(2):89–96.
Stiglitz, J.E., A. Sen, J.-P. Fitoussi. 2010. *Report by the Commission on the Measurement of Economic Performance and Social Progress*. Commission on the Measurement of Economic Performance and Social Progress, Paris, France.
Stigsdotter, U.K., O. Ekholm, J. Schipperijn, M. Toftager, F. Kamper-Jørgensen, T.B. Randrup. 2010. Health promoting outdoor environments – Associations between green space, and health, health-related quality of life and stress based on a Danish national representative survey. *Scandinavian Journal of Public Health* 38(4):411–417.
TEEB (The Economics of Ecosystems and Biodiversity for National and International Policy Makers). 2011. *TEEB Manual for Cities: Ecosystem Services in Urban Management*. TEEB, Geneva, Switzerland.
ten Brink, P., K. Mutafoglu, J.-P. Schweitzer, M. Kettunen, C. Twigger-Ross, J. Baker, Y. Kuipers, M. Emonts, L. Tyrväinen, T. Hujala, A. Ojala. 2016. *Health and Social Benefits of Nature and Biodiversity Protection. A Report for the European Commission*. Institute for European Environmental Policy, London/Brussels.

Trzyna, T., J.T. Edmiston, G. Hyman, J.A. Mcneely, P. da Cunha e Menezes, B. Myrdal, A. Phillips. 2014. *Urban Protected Areas: Profiles and Best Practice Guidelines*. Best Practices Protected Area Guidelines Series No. 22. IUCN, Gland, Switzerland.

Tuan, Y.-F. 1974. *Topophilia: A Study of Environmental Perception, Attitudes, and Values*. Columbia University Press, New York.

Ulrich, R.S. 1983. Aesthetic and affective response to natural environment. Pages 85–125 in I. Altman, J. Wohlwill, editors. *Behavior and the Natural Environment*. Springer, Boston, MA.

Ulrich, R.S., R.F. Simons, B.D. Losito, E. Fiorito, M.A. Miles, M. Zelson. 1991. Stress recovery during exposure to natural and urban environments. *Journal of Environmental Psychology* 11(3):201–230.

UN-Habitat. 2016. *Urbanization and Development: Emerging Futures*. UN Habitat World Cities Report 2016, Nairobi, Kenya.

United Nations. 2018. *World Urbanization Prospects: The 2018 Revision*. Demographic Research, United Nations, New York.

van den Bosch, M.A., M.J. Nieuwenhuijsen. 2016. No time to lose – Green the cities now. *Environment International* 99:343–350.

Völker, S., T. Kistemann. 2011. The impact of blue space on human health and well-being – Salutogenetic health effects of inland surface waters: A review. *International Journal of Hygiene and Environmental Health* 214(6):449–460.

Ward, K., S. Lauf, B. Kleinschmit, W. Endlicher. 2016. Heat waves and urban heat islands in Europe: A review of relevant drivers. *Science of the Total Environment* 569–570:527–539.

Westwood, M., N. Cavender, A. Meyer, P. Smith. 2021. Botanic garden solutions to the plant extinction crisis. *Plants People Planet* 3(1):22–32.

Wheeler, B.W., R. Lovell, S.L. Higgins, M.P. White, I. Alcock, N.J. Osborne, K. Husk, C.E. Sabel, M.H. Depledge. 2015. Beyond greenspace: An ecological study of population general health and indicators of natural environment type and quality. *International Journal of Health Geographics* 14:17.

White, M.P., I. Alcock, J. Grellier, B.W. Wheeler, T. Hartig, S.L. Warber, A. Bone, M.H. Depledge, L.E. Fleming. 2019. Spending at least 120 minutes a week in nature is associated with good health and wellbeing. *Scientific Reports* 9:7730.

White, M.P., I. Alcock, B.W. Wheeler, M.H. Depledge. 2013. Would you be happier living in a greener urban area? A fixed-effects analysis of panel data. *Psychological Science* 24:920–928.

Wolch, J.R., J. Byrne, J.P. Newell. 2014. Urban green space, public health, and environmental justice: The challenge of making cities 'just green enough'. *Landscape and Urban Planning* 125:234–244.

Wood, E., A. Harsant, M. Dallimer, A.C. de Chavez, R.R.C. McEachan, C. Hassall. 2018. Not all green space is created equal: Biodiversity predicts psychological restorative benefits from urban green space. *Frontiers in Psychology* 9:2320.

World Health Organization (WHO). 2014. *Preventing Suicide: A Global Imperative*. WHO Press, Geneva, Switzerland.

World Health Organization (WHO). 2016. *Global Report on Urban Health: Equitable, Healthier Cities for Sustainable Development*. WHO Press, Geneva, Switzerland.

World Health Organization (WHO). 2017. *Urban Green Spaces: A Brief for Action*. WHO Regional Office for Europe, Copenhagen, Denmark.

World Health Organization (WHO), Secretariat of the Convention on Biological Diversity (CBD). 2015. *Connecting Global Priorities: Biodiversity and Human Health*. WHO Press, Geneva, Switzerland.

Zhao, J., T. Tang, X. Wang. 2020. Effects of landscape composition on mosquito population in urban green spaces. Urban Forestry & Urban Greening 49:126626.

12

MULTIDIMENSIONALITY OF BIODIVERSITY IN URBAN SOCIAL-ECOLOGICAL SOIL SYSTEMS

Loren B. Byrne and Katalin Szlavecz

Introduction

In urban social-ecological systems, humans exert a dominant influence on biodiversity patterns, including through land cover management decisions across many independently owned properties. Because these decisions are spatiotemporally variable, they lead to the creation of highly heterogeneous and dynamic landscapes composed of diverse gray (impervious surfaces, buildings, pipes, etc.), blue (ponds, streams, etc.), and green infrastructure (lawns, gardens, parks, street trees, etc.) that are interspersed with remnant patches of urban-impacted native habitats (e.g. forest, grassland, desert, wetlands). Although management decisions mostly target aboveground conditions, the landscape mosaic that emerges has many direct and indirect effects on spatiotemporal patterns of the belowground "brown infrastructure" – the soil – including its hyper-diverse microbial and faunal inhabitants (Table 12.1) (Szlavecz et al. 2018; Pouyat et al. 2019). Soil organisms provide many ecosystem services that support human and societal well-being, such as pest control, water regulation, carbon sequestration, and increasingly, food production through urban agriculture (Pavao-Zuckerman and Pouyat 2017; Szlavecz et al. 2018; O'Riordan et al. 2021). Therefore, soil biodiversity is an important focus for social-ecological research and environmental management decisions that aim to improve the sustainability of urban systems.

Although soils and their biodiversity are foundational to the existence and well-being of urban systems, their complexity, social-ecological value, and fragility are often overlooked, which can lead to their degradation. In an urban context, soil is often seen as a simple, inert, abiotic substrate that can be removed, transported, replaced, fertilized, and even created (e.g. on green roofs) for desired infrastructure and management goals. These and many other human activities and the resulting urban conditions (e.g. pollution, urban heat island effect, invasive species) have many direct and indirect impacts on soil abiotic and biological properties that are rarely considered (Pavao-Zuckerman and Pouyat 2017). When soil organisms are a focus of management (e.g. plant and human pests and pathogens, other perceived nuisance species), they are often viewed in isolation, apart from the complex soil habitat, communities, and landscapes in which they reside. In contrast to this urban context – and largely based on research in non-urban settings – knowledge about and societal appreciation for relationships among soil biodiversity, human health, and long-term societal security (for food, water, etc.) have grown enormously over the past several decades

DOI: 10.4324/9781003016120-14

Table 12.1 Overview of major taxa inhabiting soils and their ecological roles (not a complete list; see Orgiazzi et al. 2016 and FAO et al. 2020 for images and additional descriptions including sizes and natural histories)

Taxa				*Common names and select descriptions*	*Key ecological relationships and functions**
Bacteria and Archaea				microbial prokaryotes, many taxa	decomposers, pathogens, nitrogen fixers and mineralizers; methane cycling, aggregate formation
Fungi				mushrooms, yeasts; hyphal growth, single-celled, lichen forming (with algae and bacteria)	saprophages, decomposers, mycorhizzae, pathogens, aggregate formation
Protists (Protozoans)				amoeba, cilliates, flagellates, algae, etc.; mostly single celled, water-dwelling	decomposers, pathogens, photosynthesis
Animals					
	Nematoda			roundworms; water-dwelling	decomposers, plant and insect parasites, predators, microbivores, omnivores
	Rotifera			rotifers; water-dwelling	decomposers
	Tardigrada			water bears, moss bears; water-dwelling	decomposers
	Annelida				
			Enchytraeidae	pot worms, white worms	detritivores
			Lumbricidae	earthworms	detritivores, ecosystem engineers, pests in some contexts (e.g. golf courses)
	Mollusca			snails, slugs	mostly herbivores, some detritivores, garden pests
	Arthropoda			exoskeletons and jointed legs	
		Hexapoda		6 legs	
			Collembola	springtails	detritivores, microbivores
			Insecta	beetles, ants, bees, termites, grubs, etc.	detritivores, predators, herbivores, ecosystem engineers, garden and structural pests
		Myriapoda		many legs	
			Chilopoda	centipedes	predators
			Diplopoda	millipedes	detritivores
		Arachnida		8 legs	
			Pseudoscorpiones	pseudoscorpions	predators
			Opiliones	daddy longlegs, harvestmen	predators
			Araneae	spiders	predators

(*Continued*)

Table 12.1 (Continued)

Taxa			*Common names and select descriptions*	*Key ecological relationships and functions**
		Acari	mites (oribatids, mesostigmatids, etc.)	predators, herbivores, detritivores; garden pests
	Crustacea			
		Isopoda	woodlice, pillbugs, sowbugs	detritivores; sometimes become pests
Vertebrata			with backbones: prairie dogs, moles, gophers, burrowing owls, cecaelians, salamanders, etc.	predators, ecosystem engineering; pests in some contexts

* All organisms are consumed by others, but this is not included here. Organisms that consume detritus (dead organic matter) are referred to broadly as decomposers because they contribute to this ecosystem function. Detritivores are animal decomposers that chew organic matter, which also has the function of fenestration (creating holes in plant leaves) and comminution (breaking detritus into small pieces), which can accelerate microbially mediated decomposition. Detritivores often also consume microbes on the detritus.

(FAO et al. 2020). It is not an exaggeration to say that humanity cannot afford to continue ignoring and thus unintentionally mismanaging soils and their organisms. This is especially true in urban systems where the majority of humans around the world now reside and whose well-being depend on locally derived ecosystem services to increase and maintain environmental quality.

To enhance the sustainability of urban societies, the brown infrastructure of urbanized landscapes needs to be viewed very differently than it commonly has been in the past. In short, all urban soils – even significantly altered and degraded ones – must be seen as dynamic living systems that provide homes for valuable organisms that perform critical ecosystem functions (e.g. Singh et al. 2019). Further, the reciprocal, feedback relationships between soil systems and humans in urbanized places suggests a more holistic perspective, that of "social-ecological urban soil systems". With this perspective, human activities that degrade urban soils, reduce services, and possibly generate ecosystem disservices must be reconsidered in light of the need to conserve and restore beneficial biodiversity (Byrne 2021). Developing this social-ecological view is important because urban soil systems have many unique characteristics that cannot be understood through transfer of knowledge generated from studying other ecosystems (Pavao-Zuckerman and Byrne 2009). Thus, advancing urban-specific soil systems research will support development of holistic understanding to more effectively guide sustainable management of belowground urban ecosystems for diverse societal goals (Friedrichsen et al. 2022).

In this chapter, we use this coupled social-ecological perspective to reflect on the state and future of urban soil biodiversity science. We first summarize a history of urban soil biology research and its associated challenges. Then, we briefly discuss key relationships within soil systems, emphasizing the coupled interactions among soil biodiversity, abiotic variables, and ecosystem functions that influence urban soil formation and resulting spatiotemporal heterogeneity. A full examination of the biological and ecological knowledge relevant to understanding the complexity of urban soil systems is outside the scope of this chapter. (Table 12.1 provides an introductory overview of main taxa and their associated functional roles; for details, including photos, see Turbé et al. 2010; Orgiazzi et al. 2016; and FAO et al. 2020). However, we emphasize that basic knowledge (e.g. how to identify soil organisms, life history strategies, ecophysiology, trophic and functional relationships) is important because it has enormous direct and tangible relationships with planning and managing urban landscapes for conservation and restoration of ecosystem services. This argument is developed in two sections in which we describe an integrated, multidimensional framework to advance the study and management of urban social-ecological soil systems and discuss its relevance to varied dimensions of social systems. Overall, our goal is to convince readers about the enormous importance and value of soil biodiversity to urban soil systems, aboveground environments, and humanity. As such, society needs urban soil organisms to become a much bigger focus of future research, teaching, policies, urban planning, decision-making, and environmental management, all of which will support efforts to improve the sustainability of urban systems and their residents' well-being everywhere (Byrne 2022; Friedrichsen et al. 2022).

History of urban soil biodiversity science and ongoing challenges

Historically, urban soil biodiversity and soil science research efforts were done independently. Traditional soil fauna surveys in urban areas began two centuries ago, when studies of urban spaces, such as parks, botanical gardens, and greenhouses, identified undescribed species and ones to add to regional lists (Szlavecz et al. 2021). Urban soil science is much younger with the first soil surveys starting in the 1950s in Germany (Burghardt et al. 2015). Most early studies were descriptive with mapping and classification based on physicochemical variables. In the 1970s, with growing interest in pollution and its effects on human health, studies investigating the ecotoxicology of soil biota emerged (Guilland et al. 2018). At the same time, city-scale soil fauna surveys to describe population and community patterns were initiated, exemplified by pioneering studies in Warsaw, Poland (reviewed in Szlavecz et al. 2021). In the 1980–1990s, the scope of urban soil research expanded and became more integrated as studies examined abiotic-biotic relationships and a broader set of ecological topics including effects of disturbance, fragmentation, and varied land uses; species losses and introductions; biogeochemical processes; patterns along urban-rural gradients; and applying soil ecological knowledge to environmental (e.g. lawn) management (Byrne 2007; Guilland et al. 2018; Szlavecz et al. 2021). The number of publications about urban soil ecosystem services increased from 1 or 2 per year from 1997 to 2006 to more than 15 from 2014 to 2019 (O'Riordan et al. 2021).

More recent studies on urban soil communities have gone beyond descriptive approaches to test ecological hypotheses about community assembly, ecological convergence, and other topics across spatial scales (Pavao-Zuckerman and Byrne 2009). Such work reveals important drivers of urban soil community composition and how their relative importance varies by taxonomic group, climate, site history, and other factors. Though few, multicity comparisons support initial identification of generalized patterns and local-scale differences. For instance, Tóth et al. (2020) found that, while earthworm assemblages were similar across five cities, the mechanisms leading to this convergence varied based on land use and human-mediated species dispersal. In contrast, Braschler et al. (2020) emphasized that locally unique ground-dwelling invertebrate communities formed within different gardens due to interactions between site-specific characteristics and varied landscape contexts. An important insight emerging from this growing body of research is that urban soils can support surprisingly diverse communities, possibly with more individuals and species than some agricultural habitats (e.g. Joimel et al. 2017).

Despite substantial progress, our knowledge of urban soil biodiversity remains limited for many reasons. First, research to date is geographically constrained to mostly temperate regions and developed countries (Europe, North America, Australia; Guilland et al. 2018; O'Riordan et al. 2021), preventing a global outlook on how urban soil ecology patterns vary among environmental and social contexts. Synthesis of research and even simple cross-site comparisons are also challenged by a lack of standardized protocols; methods of studying urban soils are often tailored to specific sites, often for logistical reasons (Guilland et al. 2018). Attempts have been made to overcome this (e.g. Niemelä et al. 2000; Pouyat et al. 2017), but the proposed approaches have yet to be widely adopted. A third challenge arises from the difficulty of connecting ecological and socioeconomic variables in science and management; soil is rarely, if at all, considered in green space planning, management, and restoration (da Silva et al. 2018), which may limit demand (and funding) for relevant science. Further, societal awareness of soil organisms – including among scientists – is often low, making it difficult to convince people to care about and focus on them more. Thus, progress in urban soil biodiversity science and management depends on educational efforts to expand people's knowledge and appreciation of "life in the brown infrastructure" (see Social dimensions section).

Additional challenges relate to the soil organisms themselves (White et al. 2020). Because they live in an opaque, dark, structurally complex habitat, they cannot be directly observed *in situ*; studying them necessitates disrupting their "homes" and activities. Their enormous ecological and taxonomic diversity, often small (including microscopic) size and insufficient taxonomic resources (identification guides, local experts, time) make it very difficult to "know your species" in research and management contexts (for introductory overviews, see Table 12.1, Turbé et al. 2010; Orgiazzi et al. 2016; FAO et al. 2020). Further, their sizes, and thus effects on ecological patterns, vary across four orders of magnitude, which makes it difficult to examine many groups simultaneously.

As a result, studies are often completed using one or a few focal groups with identification to coarse taxonomic, morphospecies or functional levels, especially for hyper-diverse invertebrates like mites and nematodes. Depending on the goals of a study, this might be sufficient; however, such lumping generates limited insights about the full complexity of belowground communities and their relationships with ecosystem functions. In addition to these challenges, taxonomic and sampling biases arise from researchers' and managers' focus on larger, "attractive" organisms that are easier to collect and identify; ones that are economically important (e.g. pests); and ones that are already well-studied (e.g. ground beetles). Advances in molecular identification techniques enable some challenges to be partially overcome, especially for microbes, leading to new insights about the diversity and spatial patterns of urban soil communities, their drivers, and influences on functions (e.g. Ramirez et al. 2014; Epp Schmidt et al. 2017; Singh et al. 2019; White et al. 2020). To expand the field, future urban soil ecologists should focus on developing large, multi-taxon collaborative projects that employ a diversity of methodological approaches (including mesocosm and field experiments, and before-after comparisons in urbanizing landscapes) to yield more holistic discoveries about urban soil communities, their macroecological patterns, and relationships with multiple ecosystem services (Guilland et al. 2018; Eisenhauer et al. 2021; White et al. 2020; O'Riordan et al. 2021).

Urban soil organisms inhabit and influence spatiotemporally dynamic soil systems

Soils are complex, dynamic systems with solid, liquid, and gaseous phases interacting in a three-dimensional matrix. A well-established principle of soil ecology is that soils' abiotic conditions (including disturbance regimes) affect soil organisms' abundance and distribution and therefore patterns of ecosystem functions (Pavao-Zuckerman and Byrne 2009; Orgiazzi et al. 2016; Eisenhauer et al. 2021). This applies to urban landscapes where investigating detailed, mechanistic influences of soil physicochemical properties on biodiversity and their interactive effects on functions are important areas of ongoing basic research (Byrne 2007). Less well understood is how biota affect urban soil structure and chemistry, especially in different environmental contexts. Nonetheless, soil organisms must be considered an important urban soil forming factor (Pouyat et al. 2019). For instance, as in non-urban soils, many organisms consume detritus and mix organic matter with the mineral fraction; this affects surface detritus layers, amount of soil organic matter, and the formation of aggregates and pore space, all of which influence the movement and retention of water. In turn, organisms' influences on abiotic conditions affect other species' dispersal, distribution, and trophic interactions, which all interact to mediate patterns of soil nutrients, organic matter, plant growth and health, and thus continued soil formation processes and patterns (Bray and Wickings 2019; FAO et al. 2020).

In addition to soil organisms, humans strongly influence temporal patterns of urban soil formation through construction and soil management activities that frequently alter soil structure, chemistry, and biota (e.g. gardening activities; Egerer et al. 2018; Pouyat et al. 2019; Tresch et al. 2019). Further, humans can essentially cause soil formation to start anew after severe disturbances (e.g. removing topsoil and exposing subsoil at the surface) and creation of new substrates such as fill added to lots following demolition of buildings and "engineered soils" for street tree pits and green roofs. These "anthrosols" can be quickly colonized by organisms who influence how the soil properties change over time (Byrne 2021). Further, by intentionally and unintentionally enhancing or eliminating populations of selected species (by altering habitat conditions, using pesticides, introducing non-native species, etc.), humans create novel assemblages, including ones with species that have not evolved together. Unpredictable interactions in such communities may result in soil formation trajectories that differ from naturally occurring pedogenesis. Studies of urban soils as complex social-ecological systems with coupled abiotic, biotic, and anthropogenic dimensions and feedback relationships among them have only just begun. Understanding of how urban soil formation affects spatiotemporal biodiversity patterns, and vice versa, will be significantly advanced by long-term studies that examine a wide range of urban-impacted and human-created soils.

The specific nature of abiotic-biotic-functional relationships in urban soil systems, as in all soils, is affected by site-specific biodiversity patterns (Eisenhauer et al. 2021; White et al. 2020). In this context, biodiversity must be considered as more than species richness in a place (alpha diversity; Byrne 2022); other dimensions, such as evenness, diversity of functional groups, trophic and other interspecific relationships, and spatial turnover of communities from patch to patch (beta diversity) must also be examined. For instance, the use of functional traits has been proposed to assess responses of communities to land use changes (e.g. Vandewalle et al. 2010). Though this is a common practice in vegetation studies, applying such approaches to soil organisms will require integration of enormous, fragmented datasets and development of unified terminology (White et al. 2020). Nonetheless, future development in this area for urban soils will likely be useful because functional diversity can directly and indirectly affect soil ecosystem dynamics and services. For instance, presence and changing abundances of earthworms, which are ecosystem engineers, has been shown to decrease bulk density (Byrne et al. 2008) and increase nitrogen mineralization rates (Szlavecz et al. 2006). Some evidence, including in urban systems, indicates that different taxa within the same functional group (e.g. micro vs macro detritivores) have differing, perhaps complimentary, effects on ecosystem functions like decomposition (Bray and Wickings 2019). However, although existing research clearly shows that soil communities are dynamic – as population sizes change and species are lost and added to communities – not all biodiversity changes should be expected to cause functional changes. Due to the high number of species within many soil taxa, soil communities are often considered to be functionally redundant, leading to high resistance and resilience of functions to environmental changes (Setälä et al. 2005). Untangling how different measures of biodiversity relate to ecosystem functions and services both above and belowground, especially how and when certain taxa are important, is an important frontier for urban soil biodiversity research (Guilland et al. 2018).

The coupled, co-developing nature of abiotic-biotic feedback relationships in soil systems means that they are highly heterogeneous over space (from μm through km and global scales) (Ossola and Livesley 2016; Eisenhauer et al. 2021). Further, the perspective of social-ecological soil systems emphasizes that we cannot ignore the diverse human activities that add additional layers of spatial variation (Pouyat et al. 2019; see next section). Thus, a major challenge to advancing urban soil biodiversity science is conducting studies that integrate social and ecological dimensions across a full range of scales, across small patches with different land covers, differentially managed parcels, and cities with contrasting histories and biome contexts (Szlavecz et al. 2018; Guilland et al. 2018). Spatial considerations include examining patterns at a range of soil depth, a dimension which is often overlooked in soil ecology research (White et al. 2020). Many aspects of urbanization alter soil profiles (Herrmann et al. 2018), which suggests the need for investigations of how biodiversity varies throughout vertical space in urban soils, perhaps in unexpected but functionally important ways.

Across urbanized landscapes, investigating connections among soil abiotic properties, biota, functions, and social dimensions in one urban place (i.e. a point in space and time) can be facilitated by delineating discrete patches based on belowground patterns (e.g. based on soil maps), aboveground habitat structure (land cover, Byrne 2007), disturbance regimes (Pouyat et al. 2017), or other criteria, and characterizing their average conditions from multiple samples across many patches (Grove et al. 2015). Such habitat structure- and patch-based studies have revealed novel relationships within urban soil systems such as how application of wood mulch on the soil surface can increase earthworm abundances, which then reduces bulk density (Byrne et al. 2008). A comparative patch-based approach (using four general land cover types characterized by soil structure, disturbance, and management regimes) is the basis for the approach advocated by the Global Urban Soil Ecology and Education Network (GLUSEEN, Pouyat et al. 2017). Studies using this approach have found that microbial and earthworm communities were more similar across cities than to other locations within the same city (Epp Schmidt et al. 2017; Tóth et al. 2020). To expand such approaches, additional research is needed about soil system properties and dynamics under a greater diversity of uniquely urban land cover types (e.g. gardens, mulches, impervious surfaces) in different combinations with other variables such as soil texture, plant diversity, and biomes.

Basic patch-based data are needed to support spatially explicit analysis of soil biodiversity across landscapes (gamma diversity) and its link to emergent patterns of ecosystem functions and services over space (e.g. Ziter and Turner 2018). Further, soil conditions and biodiversity in a given patch are affected by patch characteristics (size, shape, isolation) and the social-environmental landscape context (e.g. conditions in adjacent patches; Byrne 2007). Foundational research using urban-rural gradients has shown that regional-scale conditions affect urban soil systems in a patch, including in remnant native green spaces (Guilland et al. 2018). At larger scales, landscape composition and structure interact with local, patch-scale conditions to affect the dispersal, colonization, and persistence of soil organisms and thus community structure, through a process known as filtering (Aronson et al. 2016). For example, Tóth et al. (2021) found that the degree to which urbanization reduced species richness of isopods and millipedes depended on climate-related variables. However, in general, very little is known about how such filtering mechanisms affect soil taxa. This relates, in part, to the lack of studies about dispersal (including human-facilitated) of soil organisms among patches in urban systems. How inter-patch metapopulation and metacommunity dynamics interact with social system conditions to affect soil biodiversity patterns across biogeographical boundaries is an exciting frontier for urban ecologists (Swan et al. 2011; Szlavecz et al. 2021). In addition, multiscale aboveground-belowground relationships are an important context for studying urban soils, though they remain largely unexplored. For instance, some soil organisms are important food for birds so higher beta diversity of soil communities may support higher avian diversity across urbanized landscapes.

The landscape-scale, patch-dynamics approach (Grove et al. 2015) helps connect soil systems to sociocultural variables through a focus on individually owned parcels. As important economic and legal entities, parcels – defined by property boundaries – are the scale at which many planning and management decisions are made (see next section). Across neighborhoods, cities, and urban regions, different parcels composed of diverse land cover patches in varied spatial configurations, and with different management inputs, can be expected to have different soil biodiversity patterns (Ossola and Livesley 2016). Thus, alpha, beta, and gamma urban soil biodiversity patterns are ultimately affected by multiscale (individual, household, town, regional) social drivers (e.g. aesthetic values, laws, wealth) that affect soil-influencing activities of parcel managers. More studies that use this multiscale coupled social-ecological view are needed to advance a holistic approach to sustainable planning and management of soil biodiversity within urbanized landscape mosaics (Guilland et al. 2018).

In sum, the spatiotemporal complexity and heterogeneity of urban soil systems arise from interactions among changing abiotic, biotic, and social conditions within and around patches (Ossola and Livesley 2016). Urban soils may especially be affected by historical conditions through strong legacy influences of pollutants and previous land uses (e.g. Lewis et al. 2006; Clarke et al. 2015; Ziter and Turner 2018; Singh et al. 2019). As in other environments, urban soil populations exhibit seasonal variation (e.g. Huang et al. 2020), though multi-year studies and comparisons with non-urban settings are rare. Reflecting longer time scales, a few studies have found evidence for adaptive evolution in soil fungi, isopods, snails, and ants in response to urban heat island effects (reviewed in Lambert et al. 2021). Otherwise, current knowledge about the spatiotemporal dynamics of urban soil biodiversity patterns is sparse, limiting our abilities to predict how urban soil systems will respond to future social and environmental conditions, including continued urbanization, varied management regimes, and climate change. Investigating how coupled social-ecological dimensions of soil systems affect their emergent patterns, stability, resilience, and thus abilities to provide ecosystem services across many spatiotemporal scales are important goals for future research (Pavao-Zuckerman and Byrne 2009).

A multidimensional framework for urban soil biodiversity science

Scientific advances are often facilitated by conceptual frameworks that help organize knowledge and inspire new insights and questions. This is especially true for complex, coupled human-environmental systems (e.g. Liu et al. 2007). To guide future urban soil biodiversity science and its application to sustainable planning and management, we developed a holistic framework for social-ecological soil systems (Figure 12.1). For this, we integrated previous urban soil ecology frameworks that emphasized using habitat structure (Byrne 2007), heterogeneity of resources, disturbance and abiotic conditions (Pavao-Zuckerman and Byrne 2009), and a land cover typology (Pouyat et al. 2017) to investigate biodiversity patterns. Further, we adapt the multi-spatiotemporal scale view of soil systems (as described in the previous section) advocated by Eisenhauer et al. (2021) to generate a fuller, integrated, multidimensional view of how ecological and social dimensions of urban soil systems interact across spatial and temporal scales to affect biodiversity patterns (Figure 12.2).

A logical starting point for the framework is the role of social systems in "urbanizing" soil systems through many strong direct (Figure 12.1A) and indirect (Figure 12.1B) effects on abiotic and biotic variables (Pavao-Zuckerman and Pouyat 2017; Pouyat et al. 2019). Many of these effects are caused by human decisions based on perceptions, values (e.g. cultural, aesthetic), and other sociocultural variables (e.g. research, communication, education, policies; see next section) that determine, at many scales, activities that change aboveground environmental conditions (for reviews of sociocultural drivers, see Andrade et al., Kendal et al., this volume). Direct impacts include physical disturbances to the soil matrix (e.g. from transport, digging, compaction), chemical additions (fertilizers, pesticides), irrigation, introduction of non-native species, management of surface detritus, and installing plants. Indirect impacts arise from activities such as management of habitat structure (e.g. mowing lawns, planting trees for shade), land use planning, transportation, and industrial processes, among many others, that alter (mostly aboveground) environmental conditions (e.g. temperature, surface hydrology) and inputs to soils (e.g. atmospheric deposition). Interactions between such sociocultural drivers and environmental variables vary over space and time (Figure 12.2) and act as filters to regional species pools that generate soil community patterns within and between patches (Aronson et al. 2016). As discussed in the previous section, these patterns then affect ecosystem functions and soil abiotic variables, which in turn feedback to further influence biodiversity (Figure 12.1C).

Though human activities strongly impact current urban soil patterns and future changes, the longer history of pre-urban soil formation is just as important to consider. The "native" soil template

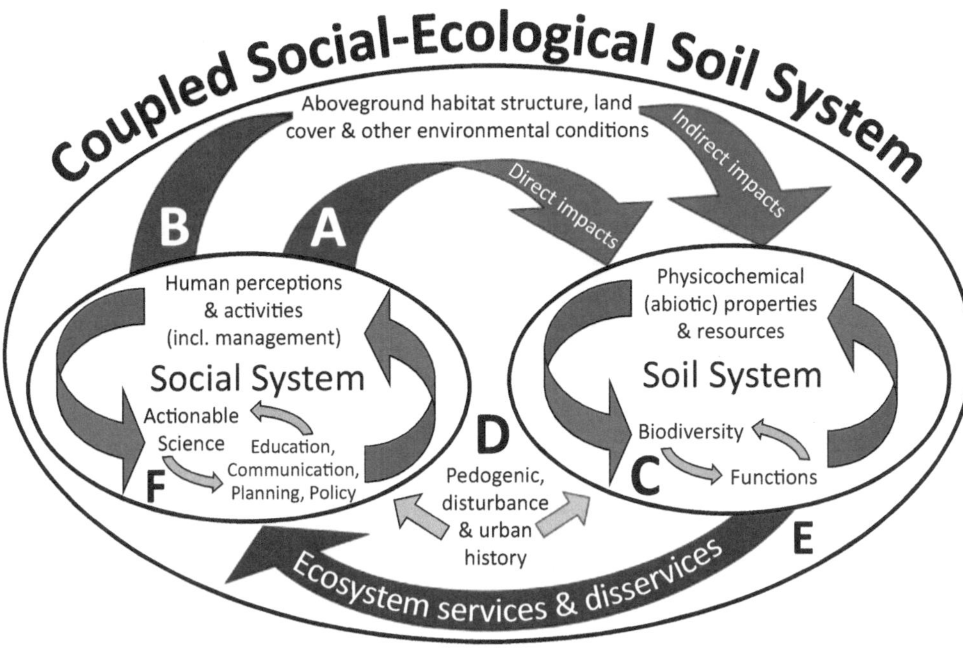

Figure 12.1 A framework for examining soil biodiversity patterns in coupled urban social-ecological soil systems. Within social systems, human perceptions and activities have (A) direct and (B) indirect impacts on soil systems, including (C) feedback relationships between biodiversity and ecosystem functions. (D) The histories of pedogenesis, disturbance, and urbanization have legacy influences on conditions of the soil and social systems. (E) Ecosystem services and disservices emerging from soil systems affect social systems resulting in a coupled relationship. (F) Within social systems, scientists can generate actionable scientific knowledge that can be used in diverse contexts to guide actions that affect soils; in turn, societal needs for knowledge about urban soils can inspire actionable research. See the main text for additional descriptions.

on which all urban development has occurred was created through past pedogenic processes that determine the baseline spatial patterns of physical and chemical characteristics (e.g. texture, mineralogy, depth, profiles), especially in remnant habitat patches, that also influence soil biodiversity (Ossola and Livesley 2016; Pouyat et al. 2019). Though outside our focus, we propose that the resultant landscape-scale soil heterogeneity has likely influenced urbanization patterns in many places (another dimension to urban social-ecological soil systems that needs additional study). As such, urban soil mosaics emerge from interactive influences of non-anthropogenic pedogenic processes, historical urbanization patterns (e.g. development of transportation networks, green space planning), and past social-ecological system dynamics (e.g. turnover of parcel ownership and management). Thus, understanding drivers of site-specific urban soil biodiversity requires placing pedogenic and urbanization history as central concerns in our framework (Figure 12.1D). Although identifying and untangling complex coupled social and pedogenic histories is a major challenge, they are essential to include, as much as possible, in descriptions of study sites, and mechanistic, predictive models and general explanations of urban soil biodiversity patterns.

The view of coupled social-ecological soil system dynamics points to the utility of the term "anthropedogenesis" to describe the temporal dimension of our framework (Figure 12.2A). This

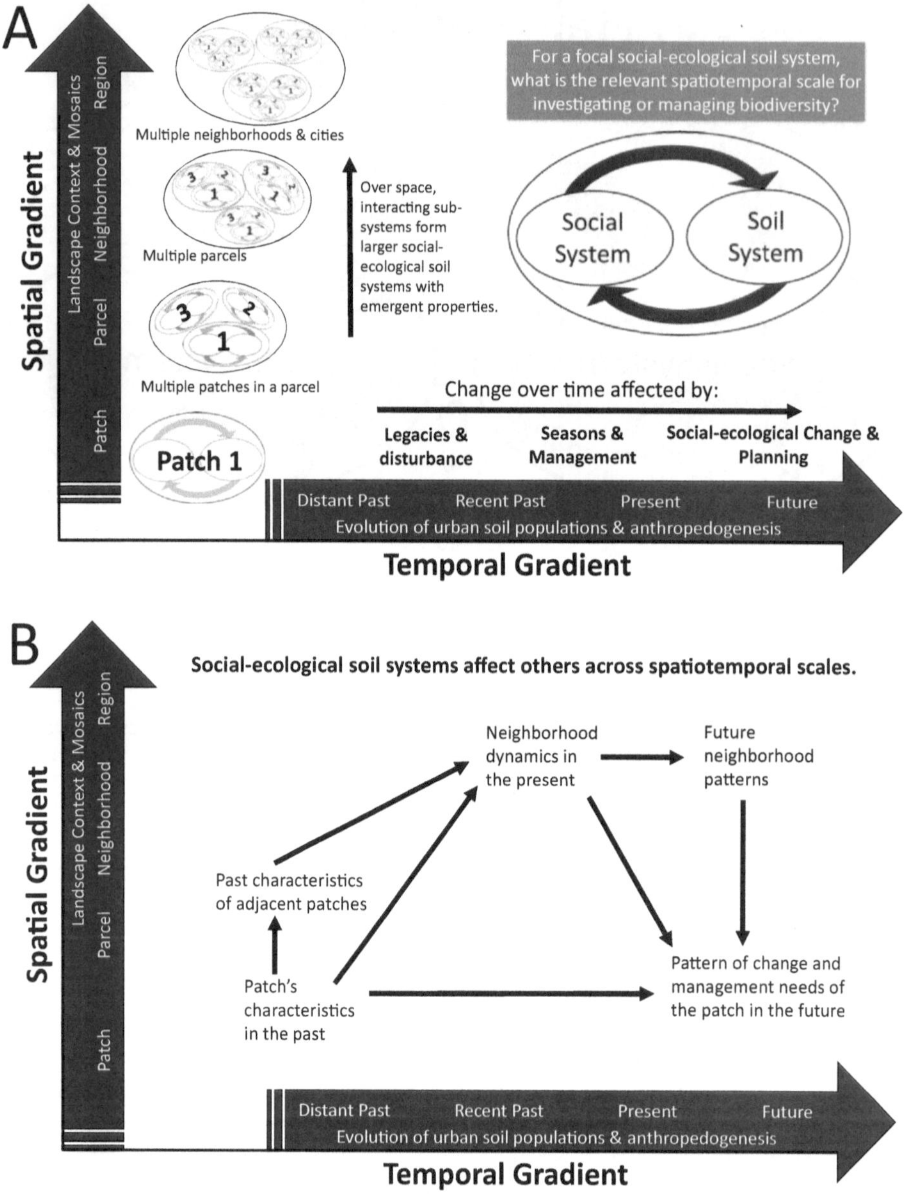

Figure 12.2 Biodiversity in social-ecological soil systems exists across many spatiotemporal scales. (A) Across the spatial gradient, larger social-ecological soil systems emerge from interacting subsystems (e.g. multiple patches form a parcel-scale system). Across the temporal gradient, past social and environmental conditions (including pedogenesis and disturbances) create legacies that interact with seasonal changes and management to determine current patterns. Planning can influence future social-ecological changes to create more sustainable soils. (B) Various direct and indirect interactions across spatial and temporal scales for a focal system need to be examined and compared for a holistic understanding of patterns and drivers. This is illustrated here in the coordinate space between the axes with example influences (as arrows) of a patch's past characteristics on present and future variables across spatial scales. See the main text for additional discussion.

term emphasizes that urban soil systems are dynamically heterogenous over time due to the coupled effects of abiotic, biotic, and social variables, while still being affected by the traditional soil formation factors of climate, parent material, and topography (Ossola and Livesley 2016; Pouyat et al. 2019). Human factors can be strong enough to erase native pedogenic legacies (e.g. through complete alteration of soil structure, chemistry, and profiles; Herrmann et al. 2018) and create "novel soil ecosystems" with combinations of taxa and pedogenic patterns that do not exist in non-urban systems (e.g. due to introduction of non-natives species and gardening and restoration activities; Egerer et al. 2018; Byrne 2021). Examining the role of biota in anthropedogenesis is especially needed for "young" soils developing after recent disturbances and within novel, human-derived, and mixed parent material (e.g. dredged materials, infrastructure-derived mineral debris), newly created by humans for green roofs and sites undergoing pollution remediation (i.e. anthrosols, technosols; Byrne 2021). In these and all urban-impacted soils, urban ecologists can help generate cutting-edge insights about anthropedogenesis by examining successional patterns of soil communities and their resistance and resilience to ongoing disturbance and management regimes, alongside urban-influenced genetic evolution of soil populations, both of which may feedback to affect pedogenic processes.

Increased understanding of urban anthropedogenesis is important because it can inform long-term management plans and actions needed to conserve both above- and belowground biodiversity and ecosystem services. This points to a second coupling relationship in our framework: how the brown infrastructure generates ecosystem services and disservices that affect humans (Figure 12.1E). Urban soil biodiversity and its ecological functions generate a wide range of benefits, which are often taken for granted and only become explicitly valued when they are degraded or lost (Pavao-Zuckerman and Pouyat 2017). Less well studied are how some soil organisms (e.g. soil borne pests and diseases) create disservices that reduce environmental or human health. Managing urban soil systems to maximize many services (i.e. a multifunctional view) while minimizing disservices is challenging because of trade-offs; for instance, interactions among pedogenic and ecological variables may simultaneously cause desired and harmful outcomes, and societal goals for specific land uses and management inputs may conflict with what is needed to support services (see Setälä et al. 2014 for detailed discussion). Whether the state and understanding of soil-derived ecosystem services and disservices can "close the feedback loop" by changing how humans perceive and manage urban soils through education, planning, and policy is an open question (Figure 12.1F). Nonetheless, it is clearly an issue that demands greater attention by scientists and other stakeholders (see next section).

As emphasized earlier, urban soil systems, including their biodiversity patterns, are spatiotemporally heterogeneous due to varied combinations of social and ecological conditions across scales. This key insight suggests the need to situate a focal social-ecological soil system in a matrix formed by axes of spatial and temporal gradients to generate a more complete framework (Figure 12.2A). Spatial scales extend from small patch-based (cm-m) through multi-patch parcels, which combine to form urban soil mosaics across neighborhoods, cities, and regions (km). Over time, distant and recent historical conditions (legacies and disturbances) affect current soil biodiversity, along with seasonal variables and present management. Past and current conditions can create legacies that, in part, affect future soil communities and their functions and services. Because of this spatiotemporal complexity, science and management projects should explicitly identify meaningful focal scales that are aligned with project questions and goals. Further, to support syntheses and theory development, research is needed that spans many combinations of space and time scales (e.g. long and short term in one patch and across large spatial scales; Figure 12.2B). This is especially important for examining the responses of soil organisms to urbanization because of the enormous range of scales their natural histories encompass (White et al. 2020). What may seem relevant to humans, especially aboveground, may have little relevance to microscopic lives belowground. Reconciling these soil biota-scales and human social-scales to advance research and sustainable soil management is a challenge that will require creative, collaborative, multidimensional science and thoughtful, effective education and partnerships with many stakeholders, including policymakers and planners. Because these social system variables are so important to the maturation of urban soil biodiversity science and its application, they deserve their own section.

Social dimensions of urban soil biodiversity: linking research and application

Relationships between soils and societies are diverse, multidimensional, and complex, with long histories and many psychological and cultural aspects (Warkenten 2006; Puig de la Bellacasa 2019; McElwee 2021; Friedrichsen et al. 2022). We are only beginning to discover how urbanization and human activities interact to shape soil systems and how soils affect urban societies. However, current knowledge indicates that soils and their biodiversity directly and indirectly influence urban ecosystem services and disservices (e.g. green spaces like parks and yards; air quality; pollution remediation; pest and disease control; stormwater flow; Figure 12.1E) that impact many aspects of people's daily lives and overall well-being, a perspective that Byrne (2022) referred to as the "essence of soil biodiversity". Identifying the specific nature and dynamics of such soil-society relationships has relevance to urban economics, public health, the aesthetic value of urban systems, and many other social concerns, from local through global scales. In this section, we highlight the essential place of social dimensions (research, education, communication, urban planning and policy) within our framework (Figure 12.1F) for advancing urban soil biodiversity science and its value for environmental management.

Future knowledge about urban soil biodiversity depends on expanding the scope of research topics and approaches, as described in previous sections. For example, connecting urban soil microbial diversity to human immune systems (Mills et al. 2020) exemplifies important linkages that need to be made between basic and applied research. Such "actionable science" can contribute to evidence-based decision-making to achieve societal goals for healthier landscape management and urban planning. In particular, long-term monitoring of urban soil communities and ecosystem functions is needed across diverse places, land cover types, and spatial scales (as in Figure 12.2A). This is crucial for improving biologically-based urban soil quality indices (including for pollutants) which are currently underdeveloped (e.g. Santorufo et al. 2012; Anne et al. 2018). Given the many challenges of studying urban soil biodiversity and wide-ranging, sometimes conflicting views of what constitutes "high quality" urban soils, we are skeptical that "one size fits all" indices can be created and used widely (Hartley et al. 2008). Nevertheless, demand for such actionable scientific knowledge is likely to grow as it relates to informing site-specific conservation, restoration, and management decisions (Byrne 2021). Urban ecologists should collaborate with end users of such information to develop applied urban-specific "soil ecological knowledge" (*sensu* Heneghan et al. 2008) and multifunctional, context-dependent approaches to guide sustainable urban soil management practices and policies (e.g. Tresch et al. 2019; O'Riordan et al. 2021; Byrne 2022; Creamer et al. 2022)..

End users of soil ecological knowledge include homeowners, gardeners, park managers, lawn care companies, and environmental consultants. Two other key stakeholder groups that should be engaged more in thinking about soils are urban planners and policymakers. To date, soils – especially their ecological properties that impact services – do not feature prominently in planning and policies (Anne et al. 2018; da Silva et al. 2018). This suggests enormous opportunities for urban ecologists and other scientists to communicate and collaborate with public officials to raise the profile of urban soil biodiversity, and advocate for including its protection and sustainable management within environmental laws, development plans, and especially, budgets. To that end, we echo previous calls for the expansion of careers and associated training programs for urban soil outreach specialists that coordinate multi-stakeholder conversations about landscape-scale urban soil management plans and policymaking (Pavao-Zuckerman and Byrne 2009; Byrne 2021; Friedrichsen et al. 2022). Such work is difficult given the many stakeholders that need to be involved and trade-offs among their competing goals for land uses, all of which will vary among sociocultural contexts around the world. Outreach coordinators could facilitate feedback to scientists about what stakeholders want to know to influence actionable science research (Figure 12.1F). Admittedly, this is a daunting vision, but the multidimensional view emphasized in our framework can help everyone (including scientists) think more broadly about the larger social-ecological systems they are part of and become more aware of the diverse interconnections that need to be considered to make these systems more sustainable. To this end, seeing soils as part of multifunctional cultural landscapes in which multiple management plans are needed for different parcels could help identify ways to "optimize" soil biodiversity patterns and associated ecosystem services across spatiotemporal scales (Setälä et al. 2014; Anne et al. 2018; Byrne 2021).

All social dimensions of urban soil biodiversity – research, management, planning, and policy – depend on a stronger foundation of expanded education and communication efforts (Byrne 2022). These are needed to both train the next generations of urban soil scientists and raise stakeholders' "soil awareness" and support for generating and using scientific evidence (Siebe et al. 2017). All people already have some relationship with and view of urban soils, even if the connections are often unrecognized, little appreciated, or taken for granted (e.g. "Sure, soil exists but why should I care?"). General (mis)perceptions of urban soils include that they are "dirty", unhealthy, devoid of life, and inhabited by harmful, fear-inducing "creepy bugs" and pests, all of which may lead to disrespect, mismanagement, and unwillingness to consider their beneficial organisms and functions. Lack of knowledge may be remedied in part with engaging, hands-on urban soil education and community (citizen) science programs (especially in K-16 settings and nature centers) that connect people of all ages to the beauty, wonder, fun, and value of soil organisms, an approach referred to as "pedagogy for the pedosphere" (Byrne et al. 2016). In this way, soil organisms provide the cultural ecosystem service of environmental education (Johnson and Catley 2009), a benefit that should be more widely used given that urban soils and organisms are easily accessible in most places, including school grounds, and can quickly and inexpensively be explored, as exemplified by community-based programs to sample earthworms in Britain (www.earthwormsoc.org.uk/earthwormwatch) and Minnesota, USA (Ziter et al. 2021). Such efforts may cultivate more "soil literate" citizens who help generate and implement actionable soil biodiversity knowledge for sustainable outcomes. As suggested by Szlavecz et al. (2018), framing people's relationships with soils through positive terms such as "stewards of brown infrastructure" may help "unhide" soil organisms and increase societal enthusiasm and care for protecting urban soils. Sustainable gardening and "ecological landscaping" also provide management contexts to foster personal connections between people and soil organisms (e.g. detritus in my yard is food, pesticides I apply are poisons) as well as larger social-environmental issues such as climate change, solid waste management, and water pollution (Byrne and Grewal 2008). If scientists and educators can help people appreciate soil organisms as valuable parts of their surrounding environments, this could "scale-up" to increased support for soil-based planning and policies and ultimately the creation of more biodiverse, healthy, and sustainable urban systems (Byrne 2022).

Conclusions

Although soil organisms don't receive the same level of appreciation as other biota (a case of "out of sight, out of mind"), they are diverse, fascinating, ubiquitous, and ecologically important (Table 12.1). They play key roles in forming and maintaining healthy soil systems and essential ecosystem services that support human well-being. This is true for urban soils, which, contrary to popular perception, are often inhabited by species- and functionally-rich communities.

Knowledge about urban soil biodiversity has increased dramatically, but much remains unknown. Throughout this chapter we have highlighted many future research opportunities that can help generate basic and applied understanding of urban soil systems. Actionable science is especially needed about species' basic biological and ecological characteristics (e.g. population dynamics, functional roles, trophic and biogeochemical linkages) to inform conservation and restoration decisions for diverse management goals. As emphasized by our framework (Figure 12.2), such knowledge is needed across a range of spatiotemporal scales because urban soils are highly heterogenous and dynamic.

Much of the variation across urban soil mosaics is created by diverse human activities that directly and indirectly affect soil patterns and processes. All patches of urban soil are connected to aboveground environments and societies. As such, we have advocated a view of urban soils as complex social-ecological systems in which organisms, abiotic conditions, and humans interact and affect each other in many ways (Figure 12.1) to create novel, emergent spatiotemporal patterns (Figure 12.2). Identifying the nature and importance of these relationships and patterns is a broader thematic frontier for advancing a holistic understanding of urban soil biodiversity – and urban systems more generally given that soil systems are the brown infrastructure that literally supports their existence. Such knowledge of urban soil systems provides a valuable basis for urban planning, policies, and management that seek to improve many dimensions of urban systems, including aesthetics of the aboveground environment, water and air quality, public health, and their economic, cultural, and environmental resilience. To this end, it is imperative that scientists communicate with stakeholders regarding their needs for actionable urban soil ecological knowledge (Figure 12.1F).

Finally, we can't stress enough how important education and communication are to facilitating a successful future for urban soil biodiversity science and management. How can urban soil organisms be integrated more widely into school curricula across grade levels and other public outreach places such as museums and zoos? What messages and communication methods work best to capture the attention of different audiences? How can planners, policymakers, managers, and even other scientists be motivated to consider soil biota seriously and thoroughly in their work? Because soil organisms live nearly everywhere within urban landscapes, they have enormous potential to generate the ecosystem service of education more widely and in diverse contexts. This underexploited dimension of urban soil biodiversity may be the most important one to focus on for ensuring a more sustainable future for the brown infrastructure of urban social-ecological systems.

Key Companion Papers

Janzen, H.H., D.W. Janzen, E.G. Gregorich. 2021. The 'soil health' metaphor: Illuminating or illusory? *Soil Biology and Biochemistry* 159:108167.

Lal, R., B.A. Stewart, editors. 2017. *Urban Soils*. CRC Press, Boca Raton, FL.

Li, G., et al. 2018. Urban soil and human health: A review. *European Journal of Soil Science* 69: 196–215.

Pouyat, R.V., et al. 2020. Urban soils. Pages 127–143 in R.V. Pouyat, D.S. Page-Dumroese, T. Patel-Weynand, L.H. Geiser, editors. *Forest and Rangeland Soils of the United States Under Changing Conditions*. Springer, New York.

Rate, A.W., editor. 2022. *Urban Soils: Principles and Practice*. Springer, Cham.

Schittko, C., et al. 2022. Biodiversity maintains soil multifunctionality and soil organic carbon in novel urban ecosystems. *Journal of Ecology* 110:916–934.

Acknowledgements

We thank Jennifer Adams Krumins, Myla Aronson, and Charles Nilon for their comments that helped improve this chapter.

References

Anne, B., S. Geoffroy, J. Cherel, G. Warot, S. Marie, C.J. Noël, M.J. Louis, S. Christophe. 2018. Towards an operational methodology to optimize ecosystem services provided by urban soils. *Landscape and Urban Planning* 176:1–9.

Aronson, M.F.J., C.H. Nilon, C.A. Lepczyk, T.S. Parker, P.S. Warren, S.S. Cilliers, M.A. Goddard, A.H. Hahs, C. Herzog, M. Katti, F.A. La Sorte, N.S.G. Williams, W. Zipperer. 2016. Hierarchical filters determine community assembly of urban species pools. *Ecology* 97(11):2952–2963.

Braschler, B., J.D. Gilgado, V. Zwahlen, H.-P. Rusterholz, S. Buchholz, B. Baur. 2020. Ground-dwelling invertebrate diversity in domestic gardens along a rural-urban gradient: Landscape characteristics are more important than garden characteristics. *PLoS ONE* 15:e0240061.

Bray, N., K. Wickings. 2019. The roles of invertebrates in the urban soil microbiome. *Frontiers in Ecology and Evolution* 7:359.

Burghardt, W., J.L. Morel, G. Zhang. 2015. Development of the soil research about urban, industrial, traffic, mining and military areas (SUITMA). *Soil Science and Plant Nutrition* 61(sup1):3–21.

Byrne, L.B. 2007. Habitat structure: A fundamental concept and framework for urban soil ecology. *Urban Ecosystems* 10:255–274.

Byrne, L.B. 2021. Socioecological soil restoration in urban cultural landscapes. Pages 373–410 in J.A. Stanturf, M.A. Callaham, editors. *Soils and Landscape Restoration*. Academic Press, London.

Byrne, L.B. 2022. The essence of soil biodiversity. *Conservation Letters*:e412900.

Byrne, L.B., M.A. Bruns, K.C. Kim. 2008. Ecosystem properties of urban land covers at the aboveground – belowground interface. *Ecosystems* 11:1065–1077.

Byrne, L.B., P. Grewal. 2008. Introduction to ecological landscaping: A holistic description and framework to guide the study and management of urban landscape parcels. *Cities and the Environment* 1(2). http://doi.org/10.15365/cate.1232008

Byrne, L.B., R.K. Thiet, V.B. Chaudhary. 2016. Pedagogy for the pedosphere. *Frontiers in Ecology and the Environment* 14:238–240.

Clarke, L.W., G.D. Jenerette, D.J. Bain. 2015. Urban legacies and soil management affect the concentration and speciation of trace metals in Los Angeles community garden soils. *Environmental Pollution* 197:1–12.

Creamer, R.E., J.M. Barel, G. Bongiorno, M.J. Zwetsloot. 2022. The life of soils: Integrating the who and how of multifunctionality. *Soil Biology and Biochemistry* 166:108561.

da Silva, R.T., L. Fleskens, H. van Delden, M. van der Ploeg. 2018. Incorporating soil ecosystem services into urban planning: Status, challenges and opportunities. *Landscape Ecology* 33:1087–1102.

Egerer, M., A. Ossola, B.B. Lin. 2018. Creating socioecological novelty in urban agroecosystems from the ground up. *BioScience* 68:25–34.

Eisenhauer, N., F. Buscot, A. Heintz-Buschart, S.D. Jurburg, K. Küsel, J. Sikorski, H.-J. Vogel, C.A. Guerra. 2021. The multidimensionality of soil macroecology. *Global Ecology and Biogeography* 30:4–10.

Epp Schmidt, D.J., R. Pouyat, K. Szlavecz, H. Setälä, D.J. Kotze, I. Yesilonis, S. Cilliers, E. Hornung, M. Dombos, S.A. Yarwood. 2017. Urbanization leads to the loss of ectomycorrhizal fungal diversity and the convergence of archaeal and fungal soil communities. *Nature Ecology and Evolution* 1(5):123.

FAO, ITPS, GSBI, SCBD, EC. 2020. *State of Knowledge of Soil Biodiversity: Status, Challenges and Potentialities*. Food and Agriculture Organization of the United Nations, Rome, Italy.

Friedrichsen, C.N., K. Mizuta, J.D. Wulfhorst. 2022. Advancing the intersection of soil and well-being systems science. *Soil Security* 6:100036.

Grove, J.M., M.L. Cadenasso, S.T.A. Pickett, G.E. Machlis, W.R. Burch. 2015. *The Baltimore School of Urban Ecology: Space, Scale, and Time for the Study of Cities*. Yale University Press, New Haven, CT.

Guilland, C., P.-A. Maron, O. Damas, L. Ranjard. 2018. Biodiversity of urban soils for sustainable cities. *Environmental Chemistry Letters* 16:1267–1282.

Hartley, W., L. Uffindell, A. Plumb, H.A. Rawlinson, P. Putwain, N.M. Dickinson. 2008. Assessing biological indicators for remediated anthropogenic urban soils. *Science of the Total Environment* 405(1–3):358–369.

Heneghan, L., S.P. Miller, S. Baer, M.A. Callaham, J. Montgomery, M. Pavao-Zuckerman, C.C. Rhoades, S. Richardson. 2008. Integrating soil ecological knowledge into restoration management. *Restoration Ecology* 16:608–617.

Herrmann, D.L., L.A. Schifman, W.D. Shuster. 2018. Widespread loss of intermediate soil horizons in urban landscapes. *Proceedings of the National Academy of Sciences* 115(26):6751–6755.

Huang, Y., I. Yesilonis, K. Szlavecz. 2020. Soil microarthropod communities of urban green spaces in Baltimore, Maryland, USA. *Urban Forestry and Urban Greening* 53:126676.

Johnson, E.A., K.M. Catley. 2009. Urban soil ecology as a focal point for environmental education. *Urban Ecosystems* 12:79–93.

Joimel, S., C. Schwartz, M. Hedde, S. Kiyota, P.H. Krogh, J. Nahmani, G. Pérès, A. Vergnes, J. Cortet. 2017. Urban and industrial land uses have a higher soil biological quality than expected from physicochemical quality. *Science of the Total Environment* 584:614–621.

Lambert, M.R., K.I. Brans, S. Des Roches, C.M. Donihue, S.E. Diamond. 2021. Adaptive evolution in cities: Progress and misconceptions. *Trends in Ecology and Evolution* 35:239–257.

Lewis, D.B., J.P. Kaye, C. Gries, A.P. Kinzig, C.L. Redman. 2006. Agrarian legacy in soil nutrient pools of urbanizing arid lands. *Global Change Biology* 12(4):703–709.

Liu, J., T. Dietz, S.R. Carpenter, M. Alberti, C. Folke, E. Moran, A.N. Pell, P. Deadman, T. Kratz, J. Lubchenco, E. Ostrom, Z. Ouyang, W. Provencher, C.L. Redman, S.H. Schneider, W.W. Taylor. 2007. Complexity of coupled human and natural systems. *Science* 317:1513–1516.

McElwee, P. 2021. The role of soils in learning and inspiration, physical and psychological experiences, and in supporting identities. *Philosophical Transactions of the Royal Society B* 376:20200184.

Mills, J.G., A. Bissett, N.J.C. Gellie, A.J. Lowe, C.A. Selway, T. Thomas, P. Weinstein, L.S. Weyrich, M.F. Breed. 2020. Revegetation of urban green space rewilds soil microbiotas with implications for human health and urban design. *Restoration Ecology* 28:S322–S334.

Niemelä, J., J. Kotze, A. Ashworth, P. Brandmayr, K. Desender, T. New, L. Penev, M. Samways, J. Spence. 2000. The search for common anthropogenic impacts on biodiversity: A global network. *Journal of Insect Conservation* 4:3–9.

Orgiazzi, A., R.D. Bardgett, E. Barrios, V. Behan-Pelletier, M.J.I. Briones, J.L. Chotte, et al. 2016. *Global Soil Biodiversity Atlas*. European Commission, Publications Office of the European Union, Luxembourg.

O'Riordan, R., J. Davies, C. Stevens, J.N. Quinton, C. Boyko. 2021. The ecosystem services of urban soils: A review. *Geoderma* 395:115076.

Ossola, A., S.J. Livesley. 2016. Drivers of soil heterogeneity in the urban landscape. Pages 37–59 in R.A. Francis, J.D.A. Millington, M.A. Chadwick, editors. *Urban Landscape Ecology*. Routledge, London.

Pavao-Zuckerman, M.A., L.B. Byrne. 2009. Scratching the surface and digging deeper: Exploring ecological theories in urban soils. *Urban Ecosystems* 12:9–20.

Pavao-Zuckerman, M.A., R.V. Pouyat. 2017. The effects of urban expansion on soil health and ecosystem services: An overview. Pages 123–145 in C. Gardi, editor. *Urban Expansion, Land Cover and Soil Ecosystem Services*. Routledge, London.

Pouyat, R.V., H. Setälä, K. Szlavecz, I.D. Yesilonis, S. Cilliers, E. Hornung, S. Yarwood, D.J. Kotze, M. Dombos, M.P. McGuire, T.H. Whitlow. 2017. Introducing GLUSEEN: A new open access and experimental network in urban soil ecology. *Journal of Urban Ecology* 3(1):jux002.

Pouyat, R.V., K. Szlavecz, I.D. Yesilonis. 2019. Human influences in urban soil development. Pages 132–154 in S.T.A. Pickett, M.L. Cadenasso, J.M. Grove, editors. *Science for the Sustainable City: Empirical Insights from the Baltimore School of Urban Ecology*. Yale University Press, New Haven, CT.

Puig de la Bellacasa, M. 2019. Re-animating soils: Transforming human – soil affections through science, culture and community. *The Sociological Review* 67:391–407.

Ramirez, K.S., J.W. Leff, A. Barberán, S.T. Bates, J. Betley, T.W. Crowther, E.F. Kelly, E.E. Oldfield, E.A. Shaw, C. Steenbock, M.A. Bradford, D.H. Wall, N. Fierer. 2014. Biogeographic patterns in below-ground diversity in New York City's Central Park are similar to those observed globally. *Proceedings of the Royal Society B* 28:20141988.

Santorufo, L., C.A. Van Gestel, A. Rocco, G. Maisto. 2012. Soil invertebrates as bioindicators of urban soil quality. *Environmental Pollution* 161:57–63.

Setälä, H., R.D. Bardgett, K. Birkhofer, M. Brady, L. Byrne, P.C. de Ruiter, F.T. de Vries, C. Gardi, K. Hedlund, L. Hemerik, S. Hotels, M. Liiri, S.R. Mortimer, M. Pavao-Zuckerman, R. Pouyat, M. Tsiafouli, W.H. van der Putten. 2014. Urban and agricultural soils: Conflicts and trade-offs in the optimization of ecosystem services. *Urban Ecosystems* 17:239–253.

Setälä, H., M.P. Berg, T.H. Jones. 2005. Trophic structure and functional redundancy in soil communities. Pages 236–249 in R.D. Bardgett, M.D. Usher, D.W. Hopkins, editors. *Biological Diversity and Function in Soils*. Cambridge University Press, Cambridge.

Siebe, C., S. Cram, L.M. Palomino. 2017. Enhancing awareness about the importance of urban soils. Pages 351–374 in R. Lal, B.A. Stewart, editors. *Urban Soils*. CRC Press, Boca Raton, FL.

Singh, J.P., B.P. Vaidya, N.M. Goodey, J.A. Krumins. 2019. Soil microbial response to metal contamination in a vegetated and urban brownfield. *Journal of Environmental Management* 244:313–319.

Swan, C.M., S.T.A. Pickett, K. Szlavecz, P. Warren, K.T. Willey. 2011. Biodiversity and community composition in urban ecosystems: Coupled human, spatial, and metacommunity processes. Pages 179–186 in J. Niemala, editor. *Handbook of Urban Ecology*. Oxford University Press, New York.

Szlavecz, K., C. Csuzdi, E. Hornung, Z. Korsós. 2021. Urban soil fauna. Pages 425–439 in I. Douglas, P. Anderson, D. Goode, M. Houc, D. Maddox, H. Nagendra, P.Y. Tan, editors. *Routledge Handbook of Urban Ecology*, 2nd edition. Taylor and Francis, London.

Szlavecz, K., S.A. Placella, R.V. Pouyat, P.M. Groffman, C. Csuzdi, I. Yesilonis. 2006. Invasive earthworm species and nitrogen cycling in remnant forest patches. *Applied Soil Ecology* 32:54–62.

Szlavecz, K., I. Yesilonis, R. Pouyat. 2018. Soil as a foundation to urban biodiversity. Pages 18–53 in A. Ossola, J. Niemelä, editors. *Urban Biodiversity*. Routledge, New York.

Tóth, Z., E. Hornung, K. Szlavecz. 2021. Urban effects on saprophagous macroarthropods are mainly driven by climate: A global meta-analysis. *Science of the Total Environment* 797:149182.

Tóth, Z., K. Szlavecz, D.J. Epp Schmidt, E. Hornung, H. Setälä, I.D. Yesilonis, D.J. Kotze, M. Dombos, R. Pouyat, S. Mishra, S. Cilliers, S. Yarhood, C. Csuzdi. 2020. Earthworm assemblages in urban habitats across biogeographical regions. *Applied Soil Ecology* 151:103530.

Tresch, S., D. Frey, R. Le Bayon, P. Mäder, B. Stehle, A. Fliessbach, M. Moretti. 2019. Direct and indirect effects of urban gardening on aboveground and belowground diversity influencing soil multifunctionality. *Scientific Reports* 9:1–13.

Turbé, A., A. De Toni, P. Benito, P. Lavelle, P. Lavelle, N. Ruiz, W.H. Van der Putten, E. Labouze, S. Mudgal. 2010. *Soil Biodiversity: Functions, Threats and Tools for Policy Makers*. Bio Intelligence Service, IRD, and NIOO, Report for European Commission (DG Environment).

Vandewalle, M., F. De Bello, M.P. Berg, T. Bolger, S. Doledec, F. Dubs, C.K. Feld, R. Harrington, S. Lavorel, P. Martins da Silva, M. Moretti, J. Niemela, P. Santos, T. Sattler, J.P. Sousa, M.T. Skyes, A.J. Vandergen, B.A. Woodcock. 2010. Functional traits as indicators of biodiversity response to land use changes across ecosystems and organisms. *Biodiversity and Conservation* 19:2921–2947.

Warkenten, B.P. 2006. *Footprints in the Soil: People and Ideas in Soil History*. Elsevier, Amsterdam, Netherlands.

White, H.J., L. León-Sánchez, V.J. Burton, E.K. Cameron, T. Caruso, L. Cunha, T. Dirilgen, S.D. Jurburg, R. Kelly, D. Kumaresan, R. Ochoa-Hueso, A. Ordonez, H.R.P. Phillips, I. Prieto, O. Schmidt, P. Caplat. 2020. Methods and approaches to advance soil macroecology. *Global Ecology and Biogeography* 29:1674–1690.

Ziter, C.D., B.M. Herrick, M.R. Johnston, M.G. Turner. 2021. Ready, set, go: Community science field campaign reveals habitat preferences of nonnative Asian earthworms in an urban landscape. *BioScience* 71:280–291.

Ziter, C.D., M.G. Turner. 2018. Current and historical land use influence soil-based ecosystem services in an urban landscape. Ecological Applications 28:643–654.

13

PLANT INVASIONS IN CITIES

Patterns, processes, consequences, and management

Mirijam Gaertner and Ingo Kowarik

Introduction

Cities around the globe harbor high numbers of non-native plant species (Aronson et al. 2014). Non-native species are defined as species not indigenous to an area the presence of which is due to direct or indirect consequences of human activities (Essl et al. 2018). Given the role of cities as hotspots of plant invasions (Pyšek 1998; Kowarik 2008), a better understanding of urban plant invasions and associated consequences is a timely challenge for research – and also needed for developing adequate responses in urban environmental policies and management (Gaertner et al. 2017a).

A wide range of habitats are created in urban environments, which are home to a great diversity of plant species, many of which are of alien origin (Sukopp and Werner 1983; Štajerová et al. 2017). The increase of settlement size is often met with a significant increase of non-native species (Pyšek 1998; Čeplová et al. 2017). This is related to a high environmental heterogeneity of cities, including the availability of natural resources, habitats, and climate (Géron et al. 2021).

Plant invasions in cities are a challenging issue for research and management because they are often related to both risks and benefits. Invasive plant species are often assumed to displace native species and contribute to biotic homogenization within cities (Burton et al. 2005; Kühn and Klotz 2006; McKinney 2006; Trentanovi et al. 2013) and may also disrupt important ecosystem services (van Wilgen and Scott 2001; Potgieter et al. 2017). On the other hand, invasive plant species have been described for their beneficial ecological, economic, and sociocultural influences in cities as they are often better adapted to harsh urban conditions and to climate change (Kowarik and Fischer 2020; Schlaepfer et al. 2020).

Whether non-native plant species have positive or negative effects on their environment depends largely on species identity and the specific environmental context, from local to biogeographic scales. The same species in the same (or another) city may induce positive, negative, or neutral ecological effects. Whether, and in which context, these effects can be addressed as benefits or adverse impacts depends on the underlying value system that usually varies within and across urban societies (Bartz et al. 2010; Estévez et al. 2015; Bartz and Kowarik 2019). Considering the various dimensions of context dependency is thus essential for understanding patterns – and related consequences – of non-native species in cities (Kowarik and Fischer 2020). The aims of our chapter are to (1) provide an overview over the specific situation of plant invasions in urban

DOI: 10.4324/9781003016120-15

areas globally and (2) discuss the context-specific pros and cons of non-native plant species management in urban settings using Cape Town, South Africa, and Berlin, Germany, as case studies.

Urban plant invasions: patterns and processes

Invasion patterns differ across cities in different biogeographical contexts. La Sorte et al. (2014) investigated numbers of native and non-native species in 110 European and non-European cities. They found that European cities had a greater proportion of archaeophyte (i.e. species introduced into Europe before 1492) and invasive species relative to non-European cities. Non-European cities had greater numbers of native species and non-native (but non-invasive) species relative to European cities. Interestingly, cities in disparate regions still show regionally distinct assemblages of native and non-native plant species. In a study on 54 European cities, both the number and relative contribution of neophytes (i.e. post 1492 introductions) to the total flora increased with city size (Pyšek 1998), which reflects the increasing role of humans in fostering plant invasions through introductions and habitat modifications. Accordingly, the gain of non-native species accelerates with increasing urbanization since the 19th century in cities around the globe (Chocholoušková and Pyšek 2003; Tait et al. 2005; Knapp et al. 2010).

Invasion processes start with the human-mediated transport and introduction of a species to an area beyond its natural range and may in some cases further proceed to the establishment and spread of this species in the new range (Blackburn et al. 2011). An array of pathways facilitates initial introductions and the subsequent spread of non-native species at different spatial scales (Hulme 2009). In cities, as human population hubs, deliberate and accidental introductions exist. This is one of the main reasons why patterns and processes of urban plant invasions differ in important ways from invasions in other contexts (Kühn et al. 2004; Cadotte et al. 2017). Transport linkages (e.g. airports, harbors, road and rail networks) foster the human-mediated movement of commodities and culminates in the introduction and dissemination of non-native species through these dispersal pathways and vectors (Gaertner et al. 2017a).

Padayachee et al. (2017) compiled a global database of cities with more than one million inhabitants and identified the most prominent pathways of introduction and vectors of the spread of non-native species within these cities. Their results show that the most prominent pathway for plants was the escape pathway. La Sorte et al. (2014) identified escaped ornamentals as the biggest proportion of 634 invasive species in 110 European cities. As early as 1912, Albert Thellung conducted a first analysis on the relative contribution of introduction pathways for the establishment success of introduced species in an urban region (Kowarik and Pyšek 2012). He showed that, in Montpellier, France, 78% out of 800 non-native species had been introduced accidentally, with wool, seed and feed grains, grain crops, ballast materials, and with traffic. Yet intentionally introduced species, such as ornamentals, were more successful in establishing self-replacing populations (57%; Thellung 1912).

Human influence is not only decisive for the initial introduction of non-native species to cities. Subsequently, deliberate or accidental secondary releases by humans often support the further spread of non-native species in urban regions (Kowarik 2003), for example, by plantings in various urban green spaces, seedings of bee plants, and dumping of garden waste (Rusterholz et al. 2012). While the introduction of ornamentals is a key pathway for initial introductions, the number of ornamental plantings, time since introduction (Pyšek et al. 2009), and the availability of species on the market (Dehnen-Schmutz et al. 2007) are important predictors of invasion success; all of these drive the propagule pressure of introduced plant species. Human-mediated dispersal by attachment of seed to vehicles can also facilitate the long-distance dispersal of introduced species within transportation corridors within cities and along urban-rural gradients (von der Lippe and Kowarik 2007; Taylor et al. 2012).

Cities are not only hubs of the human-mediated introduction and dispersal of non-native species but also support biological invasions by providing an array of resources that benefit resource acquisition both by non-native and native plant species (Stohlgren et al. 2003). Urban regions comprise often topographically heterogeneous areas (Kühn et al. 2017) and include a range of habitats that reflect different transition stages from natural remnants to novel urban ecosystems (Kowarik 2011). With urbanization, substantial changes in climatic conditions, soils, and hydrology modify the availability and distribution of resources (Grimm et al. 2008). Moreover, many sites in urban areas are greatly disturbed by human action, which enhances their invasibility (Kowarik 2008; Kueffer and Kull 2017). The manifold changes to urban environments lead to a wealth of ecological niches that support a high native and non-native species richness at the city scale but also filters for the composition of species assemblages and species' establishment success in different urban ecosystems. That is why abundance and numbers of non-native plant species usually decreases along an urban to rural gradient (Kunick 1974; Kühn et al. 2017) and why in cities, many non-native species establish more successfully in anthropogenic ecosystems than in natural remnants (Kowarik and von der Lippe 2018). In the same vein, the functional composition of the non-native flora changes with urbanization (Knapp et al. 2010; Šilc 2010; Wolf et al. 2020).

Consequences of urban plant invasions

The abundance of non-native plant species in cities has been often related to an array of environmental and socioeconomic consequences that may be both positive or negative. These consequences can be illustrated/described as provisioning of ecosystem services or disservices (see Table 13.1 for a comprehensive overview).

Environmental consequences

Plant invasions can induce changes in urban biodiversity at all levels, from the genetic to the landscape level. Many non-native plant species hybridize with other non-native and native congeners introduced by horticulturalists. This process can compromise the genetic integrity of native taxa and/or enhance the invasive ability of their hybrid offspring (Williamson and Fitter 1996). A classic example is the hybridization of two Asian knotweed species (*Fallopia japonica* and *F. sachalinsens*) in Europe which resulted in a hybrid (*F. x bohemica*) with higher regeneration potential and competitive strength (Bímová et al. 2003; Parepa et al. 2014).

Non-native plant species shape many urban plant communities. An analysis of the complete Berlin, Germany vegetation, for example, revealed a share of 30–60% of non-native species in plant communities of disturbed anthropogenic sites; the shares for forest and wetland vegetation, on the other hand, were much lower (Kowarik 2008). A high presence of non-native species is often associated with the decline of native species. It is often unclear, though, whether non-native species are the trigger for native species declines or a symptom of previous urbanization impacts that disadvantage native species while promoting non-native species by providing novel environmental conditions (Didham et al. 2005). However, there is evidence that non-native species can lead to a decline of native species, particularly when invading near-natural ecosystems in the urban fringe (Gaertner et al. 2017a).

The example of the tree *Robinia pseudoacacia* illustrates the context dependence of invasion impacts. *Robinia* often invades dry grassland of conservation concern in Europe, within and outside cities, which leads to a replacement of many species of conservation concern (Cierjacks et al. 2013; Vítková et al. 2017). In *Robinia* forests emerging on urban sites, however, native and non-native

Table 13.1 Environmental and socioeconomic consequences of plant invasions in cities

Socioeconomic consequences			
Ecosystem service or disservice	*Examples*		*Reference*
Cultural and Aesthetic	Recreation	Picnicking under tall shade-providing trees (e.g. *Pinus pinea*)	Potgieter et al. (2019c)
	Physical, intellectual, and spiritual interactions with nature, including aesthetic values, inspiration and cognitive development, and spiritual enrichment.	Well-managed urban green spaces with abundant vegetation; aesthetically attractive invasive species such as *Trachycarpus fortunei*	Bastian et al. (2012), Dobbs et al. (2011), Lindemann-Matthies (2016)
	Sense of place and aesthetic values	Sense of place and aesthetic values due to the presence of invasive alien plant species (for example, Jacarantha trees in Pretoria, South Africa)	Novoa et al. (2017), Lindemann-Matthies (2016)
Provisioning	Firewood	Trees such as *Acacia*, *Eucalyptus*, or *Pinus* spp. can be used for firewood	Dickie et al. (2014)
	Construction material	Trees such as *Eucalyptus* or *Pinus* spp. can be used for poles	Dickie et al. (2014)
	Edible plants	Non-native plants as components of urban agriculture and as forgeable wild species	Shackleton and Mograbi (2020), Fischer and Kowarik (2020)
	Biochemical resources	*Ailanthus altissima* provides biochemical components for pharmaceutical research	Kowarik and Säumel (2007), Sladonja et al. (2015)
Regulating	Shade	Shade from tall trees with wide canopy such as *Pinus pinea*	Pooley (2014), Potgieter et al. (2019c)
	Climate regulation	Cooling effects (by transpiration) of street trees such as *Platanus x acerifolia*	Jim and Chen (2009), Schlaepfer et al. (2020)
	Air quality	Atmospheric pollutant removal by trees such as *P. x acerifolia*	McPherson (2003), Schlaepfer et al. (2020)
	Carbon sequestration	Trees such as *Morus alba* sequester carbon	Potgieter et al. (2017), Riley et al. (2018)
	Increase of soil fertility	*Acacia, Robinia, Ailanthus* spp. enrich the soil	Castro-Díez et al. (2021)
	Erosion control	Erosion control by trees such *A. altissima*	Kowarik and Säumel (2007), Sladonja et al. (2015)

(*Continued*)

Table 13.1 (Continued)

Socioeconomic consequences			
Ecosystem service or disservice	*Examples*		*Reference*
Ecosystem disservices			
Cultural	Loss of sense of place and aesthetic values	Loss of sense of place and aesthetic values due to the presence of invasive alien plant species	de Wit et al. (2001), Le Maitre et al. (2011)
	Unattractive species or landscapes	Ugly landscapes dominated by *Acacia* spp. Neglected vacant lots overgrown with "weedy" vegetation	Carruthers et al. (2011)
Economic Problems	Increased maintenance costs	Increased maintenance due to the spread and vegetative propagation of *A. altissima* in urban spaces	Sladonja et al. (2015)
	Infrastructural damage	Roots of *A. altissima* damaging paved surfaces and antique remains	Celesti-Grapow and Ricotta (2020)
Health Risks	Reduced air quality	Emissions of biogenic volatile organic compounds reducing air quality	Potgieter et al. (2017), von Döhren and Haase (2019)
	Increasing attack by associated animals	Areas with dense vegetation can harbor potentially dangerous animals such as venomous snakes	Roy et al. (2012)
	Pollen allergies	Pollen allergy and/or dermatitis caused by *A. altissima*, *Acacia dealbata*, *Cortaderia selloana*, and *Schinus terebinthifolius*	Pyšek and Richardson (2010), von Döhren and Haase (2019)
	Poisoning	Cardiac problems and poisoning from *Echium plantagineum*	Pyšek and Richardson (2010)
	Physical injury	Physical injury through contact with plant spines or thorns	Pyšek and Richardson (2010), Shackleton et al. (2014)

(*Continued*)

Table 13.1 (Continued)

Socioeconomic consequences			
Ecosystem service or disservice	*Examples*		*Reference*
Safety and Security	Fears of insects and other animals	Areas with dense vegetation can invoke fear of feared animals such as poisonous snakes	Vaz et al. (2017a)
	Increased crime risk	Criminal activity in dense vegetation close to informal settlement	Potgieter et al. (2019b)
	Increased fire risk (safety risk to infrastructure, but also impact on native plants due to increased frequency and intensity of fires)	Increased fire risk due to trees invasions along the urban edge	Gaertner et al. (2014), Le Maitre et al. (2011), van Wilgen and Richardson (2014), Potgieter et al. (2019c)
Environmental consequences			
Ecosystem services			
Supporting	Habitat provision	Tall alien trees (i.e. *Eucalyptus*, *Pinus* spp.) provide nesting sites for birds with which many urban dwellers can enjoy encounters.	McPherson et al. (2011)
	Species richness	Non-native species increase overall biodiversity in cities	Knapp et al. (2010), Schlaepfer et al. (2020)
	Ecosystem functioning	Increase of soil formation and soil fertility, climate regulation	Castro-Díez et al. (2021)
Ecosystem disservices			
Ecosystem Functioning	Reduced soil quality	Decrease of soil quality and promotion of soil erosion	de Wit et al. (2001), Shackleton et al. (2014), Castro-Díez et al. (2021)
	Disruption of soil-nutrient cycling, carbon and nitrogen fixation*	Nutrient influx in naturally nutrient-poor ecosystems of conservation concern due to nitrogen fixation by Invasive alien trees such as *Acacia* spp.	Yelenik (2007), Gaertner et al. (2014), Qiu (2015)

(*Continued*)

Table 13.1 (Continued)

Socioeconomic consequences			
Ecosystem service or disservice	*Examples*		*Reference*
Diversity	Triggering invasions into adjacent landscapes/ Displacement of native plant species/Reduced species richness*	Invasive plants spreading into natural areas can disrupt native plant species, and continued spread may reduce native species richness	Carruthers et al. (2011), Dickie et al. (2014), Kull et al. (2011), Le Maitre et al. (2011), Shackleton et al. (2016), van Wilgen and Richardson (2014), Vicente et al. (2013), McLean et al. (2017)
	Reduction of functional or phylogenetic diversity	Introduction of invasive alien species reduced phylogenetic diversity in urban plant communities	Čeplová et al. (2015)
Hybridization	Compromise of genetic integrity of native taxa and/or enhance the invasive ability of their hybrid offspring	Hybridization of two Asian knotweed species (*Fallopia japonica, F. sachalinsens*) in Europe which resulted in a hybrid (*F. x bohemica*) with a higher regeneration potential and competitive strength	Bímová et al. (2003), Parepa et al. (2014)

species richness is positively correlated, and these stands harbor a similar number of endangered plant and invertebrate species as native emerging forests do (Kowarik et al. 2019). Consequently, strategies to manage *Robinia* invasions should consider the specific environmental contexts (Sádlo et al. 2017).

Cities can also act as launching sites for invasions into adjacent landscapes with natural and near-natural ecosystems. While large cities are usually the first sites of introduction, small towns or housing areas are more numerous and have a greater chance of launching invasions into natural areas as they have proportionally larger interfaces with their surroundings (McLean et al. 2017). Studies from New Zealand (Sullivan et al. 2005) and North America (Gavier-Pizarro et al. 2010), for example, showed a positive correlation between invasion of semi-natural habitats and proximity to gardens. However, this risk cannot be generalized across different environmental contexts. For example, a study from Great Britain suggests that non-native plant species do not spread out of cities into the wider countryside (Botham et al. 2009). The predominant agricultural and semi-natural ecosystems here were obviously less susceptible to invasions than, for example, the endemic fynbos vegetation close to some South African cities, which is highly susceptible to plant invasions (Gaertner et al. 2016; McLean et al. 2017).

At the city level, the influx of non-native plant species has led to an increase in species richness, as mentioned previously. This increase, in combination with a decline of native species, has been hypothesized to cause a biotic homogenization of cities or urban ecosystems (McKinney 2006). Yet urban homogenization studies often provide weak evidence (Hahs and McDonnell 2016), with studies demonstrating both existing (e.g. Zeeman et al. 2017) and missing homogenization effects (e.g. Kühn and Klotz 2006; Trentanovi et al. 2013; Aronson et al. 2015). A study on 32 European cities (Lososová et al. 2012), for example, found an increased differentiation of urban floras through non-native plant species, particularly through neophytes. In the same vein, a global analysis of urban floras revealed that despite some non-native species are shared across many cities, urban floras have not yet become taxonomically homogenized (Aronson et al. 2014).

Socioeconomic consequences

Non-native plant species can have positive or negative socioeconomic impacts. Urban green spaces benefit urban societies by providing an array of ecosystem services such as cooling, flood prevention, air filtration, noise reduction, carbon sequestration, and a wealth of cultural services, including recreation (Haase et al. 2014). However, urban nature can contribute to disservices, such as health risks induced by plant-born allergens (von Döhren and Haase 2015; Davoren and Shackleton 2021). Generally, both native and non-native plant species can contribute to urban ecosystem services and to disservices as well. The importance of non-native plants as components of urban green spaces directly reflects how people appreciate them for aesthetical and other reasons (Lindemann-Matthies 2016; Vaz et al. 2018). Urban lawns in cities outside Europe, for example, are often dominated by formerly sown European grassland species (Fischer et al. 2016) which support an array of cultural services associated with lawns (Ignatieva et al. 2015). In South African settlements, non-native species are frequently used as food resources and for other uses (Shackleton and Mograbi 2020), and non-native tree species support a range of regulating services in cities in the USA (Kim 2016; Riley et al. 2018).

On the other hand, non-native plant species can also exert a range of negative socioeconomic impacts through ecosystem disservices in cities, including health risks or damage to infrastructure. A prominent example is *Ambrosia artemisiifolia*, which releases highly allergic pollen that induce asthma attacks (D'Amato et al. 1998). The accelerating spread of this North American species in Europe results in significant costs to the health system (Essl et al. 2015). The high vegetative propagation of the Chinese tree, *Ailanthus altissima* leads to increased maintenance efforts in urban open spaces and can damage built structures and antique remnants, mostly in Southern Europe (Casella and Vurro 2013; Celesti-Grapow and Ricotta 2020). At the same time, *Ailanthus altissima* supports a range of regulating and cultural ecosystem services in cities (Kowarik and Säumel 2007; Sladonja et al. 2015), in particular on sites where native species are less successful. *Ailanthus* is well adapted to the urban heat island (Roloff et al. 2018), and its contribution to air pollution removal, stormwater management, and carbon sequestration has been quantified for some cities (Kim 2016; Riley et al. 2018; Arrington 2021). The classic challenge thus is to balance the negative and positive effects that can be associated with non-native plants (van Wilgen 2012; Potgieter et al. 2019a; Shackleton and Mograbi 2020) – which may lead to different assessments from different contexts. Accordingly, a recent study found that how urban residents disapprove or approve of *Ailanthus* trees depends on the specific urban environment (Kowarik et al. 2021).

While the provisioning of ecosystem services and disservices in cities is an important issue for urban livability, our understanding of the specific role of non-native species is still limited. A first review of studies from 58 cities in developed and developing countries (Potgieter et al. 2017), reports 307 non-native plant species as contributing to 39 different ecosystem services (most prominently to food provisioning and cultural services) and 53 non-native species associated with ecosystem disservices (mostly health risks). Most of the reviewed studies are from single cities, however, do not quantify a comprehensive set of services and disservices provided by the same species or compare the relative contribution of native and non-native species to the same service or disservice. As an exception, Schlaepfer et al. (2020) quantified a range of regulating and cultural ecosystem services and disservices for the urban region of Geneva, Switzerland – and found similar contributions of native and non-native trees.

There is no tool to adequately measure socioeconomic impacts. Many schemes try to monetize the impact (e.g. Reinhardt et al. 2003; Born et al. 2005), but they do not allow a straightforward assessment of species' impacts before or without control. and they are highly context-dependent (e.g. wages vary in different regions or parts of the world). Socioeconomic impacts of non-native taxa can be more appropriately reduced by technology or adaptive behavior in affluent countries as opposed to poor countries, where non-native taxa can, in extreme cases, lead to the collapse of socioeconomic sectors, thereby causing irreversible societal changes (Bacher et al. 2018).

Invasive plant species may hence provide benefits or nuisances in the social realm, without necessarily being considered a prior beneficial or detrimental asset in the ecological realm (Vaz et al. 2017a). For instance, people will value invasions differently depending on available management options and the capacity to use ecosystem services and mitigate ecosystem disservices provided by them, at certain geographical and temporal contexts (Kueffer 2013). Also, the perceived impact of a species can change over time (Strayer et al. 2006).

Management of non-native plant species in urban areas

Non-native species increasingly dominate "novel urban ecosystems" as defined by Kowarik (2011), which leads to the insight that introduced species have become a characteristic component of many urban ecosystems. Some authors therefore call for new, more objective, approaches for managing non-native species in cities, as urban ecosystems are already considered to be "degraded" with major human-induced changes in the abiotic environment and significant changes in the species pool (Hobbs et al. 2006; Kowarik 2011; Francis and Chadwick 2015). Also, the demand and pressure for recreational green space (not necessarily comprising native species only) is increasing in many cities. This means that enhancement of cultural ecosystem services (irrespective whether provided by native or non-native species) is being prioritized. However, the threat of invasive non-native species escaping into surrounding natural areas where their disservices might outweigh their services has to always be considered.

As a consequence, cities are often not seen as priority areas for non-native species management but also as areas where non-native species can support many ecosystem services. Interestingly, cities harbor not only high numbers of introduced species but also of endangered species (e.g. Ives et al. 2016; Kowarik and von der Lippe 2018) and sometimes more native species than adjacent rural areas (Kühn et al. 2004). Yet endangered native species and invasive non-native species do not necessarily thrive in the same urban ecosystems. Invasive species thus do not necessarily conflict with the conservation of endangered native species in cities. However, since such cases do exist, information about site-specific impacts of non-native species on species or habitats of conservation concern is a key requirement for management efforts.

To manage plant invasions effectively, we need to understand what conflicts of interest they generate. Important arguments for and against managing invasive species in urban areas increasingly hinge on their contributions to the delivery of ecosystem services and disservices (Vaz et al. 2017b). Engagement and citizen science is crucial for avoiding, reducing, or resolving such conflicts (Crowley et al. 2017; Novoa et al. 2018). Better collaborations and more bottom-up management should be sought to the uptake of policy and management practices and will help to address the fundamental challenges relating to conflicts of interest, aiding with prioritization and cooperative management implementation (Shackleton et al. 2019). Citizen science tools can be applied to manage invasive plants (e.g. for identifying and reporting the presence of alien plants), especially in urban areas where the potential of these tools is much higher than non-urban areas (Novoa et al. 2018). The need for inter- and transdisciplinary understanding is slowly being recognized, but biological research still dominates this field, and further cross-collaboration is urgently needed (Vaz et al. 2017a).

To illustrate how differing impacts and benefits of non-native species invasions can influence perceptions and management approaches in different socio-ecological systems, we describe two case studies – Cape Town, South Africa, and Berlin, Germany – in the following section.

Urban invasions in different biogeographical settings: the case of Cape Town, South Africa, and Berlin, Germany

Patterns and consequences of urban plant invasions differ markedly in different parts of the world and so do management approaches urban societies perceive as being adequate. We use the example of Cape Town and Berlin to illustrate the paramount importance of context-dependence in consequences of urban plant invasions and how to respond to them adequately.

Cape Town, South Africa

Cape Town has an outstanding conservation significance, which derives from its location in the Cape Floristic Region, a global centre of plant endemism (Cowling et al. 1996). The city (2445 km^2) includes Table Mountain National Park (221 km^2), as well as 17 smaller nature reserves and 500 biodiversity network sites that together cover 270 km^2. Cape Town has a Mediterranean climate with hot, dry summers and cool, wet winters. Cape Town has a population of 3.8 million people and is growing more rapidly than any other southern African metropolis on a per capita basis (Boraine et al. 2006), especially within its poorer suburbs ("townships") which have experienced an influx of citizens following the collapse of apartheid.

Cape Town has a long history of European colonization, and the associated introductions of non-native species present a significant challenge to people and landscapes (van Wilgen 2012). For example, invasive tree species such as pines (*Pinus* species), grown in plantations, and Australian wattles (*Acacia* species), planted mainly along the coast for dune stabilization, have spread widely into natural vegetation. Aquatic invasive species such as water hyacinth (*Eichhornia crassipes*) block waterways and affect water quality (Richardson and van Wilgen 2004). The trade in ornamental plants and pets, and other enterprises that rely on non-native taxa, continue to introduce new species into the city, many remaining undetected and/or unregulated. Some invasive plant species pose serious risks to humans; for example, invasive pines and wattles increase the severity of wildfires near residential areas (van Wilgen and Scott 2001). Invasive non-native species also pose a serious risk to the native vegetation in conservation areas within the city and outside of the city as the endemic fynbos vegetation is highly susceptible to plant invasions (Gaertner et al. 2016).

The city spends 70% of its invasive species funds on controlling terrestrial invasive plants in nature reserves, targeting mainly Australian tree species (e.g. *Acacia* and *Eucalyptus* trees). Priority is given to nature reserves, followed by areas where invasive plants are a fire hazard or provide shelter for criminals. However, control efforts are often compromised because of significant conflicts over the management of invasive species. For example, conservationists and citizen science groups support the control of invasive species (mainly to protect biodiversity and to decrease negative impacts on ecosystem services), but control operations are met with strong resistance from commercial foresters, in the case of pine plantations, and also from citizens who have concerns over loss of forested space in urban environments (van Wilgen 2012).

Experience in managing invasive plant species in Cape Town has shown that decision frameworks adapted to the special situations in urban areas need to be used to decide on a course of action when any new invasive species becomes a potential target for control (Gaertner et al. 2017b). In order to make logical and defensible choices, it is further essential to gauge the views of stakeholders. Where significant numbers of people are likely to oppose management, the potential political consequences of different courses of action must be explicitly considered. It also has to be accepted that conventional goals (such as eradication or reductions in populations) will need to be replaced with partial or complete tolerance of invasive species in some cases. However, given Cape Town's high conservation significance due its location in the Cape floristic region, we caution that where some form of cohabitation is agreed upon, approaches should be sought for minimizing potential negative impacts of retaining areas with invasive species.

Berlin, Germany

Berlin is Germany's largest city, with 3.7 million inhabitants in 2020 in a total area of 892 km^2. Berlin is a historical city with different urban nuclei from the medieval age. Due to Berlin's designation as the German capital in 1871 and its importance as an industrial hotspot, the population grew rapidly from circa 424,000 inhabitants in 1849 to 4 million inhabitants in 1925. Today's Berlin was founded in 1920 as a union of eight cities and 59 villages. The resulting polycentric urban structure is closely intertwined with historical landscape remnants and large green areas. Green and blue spaces cover 41% of the area, including forests (18%), lakes and rivers (6%), parks (6%), allotment gardens (5%), fields (5%), and meadows (1%) (SDUDH 2016). Berlin's climate is temperate, with an annual average temperature of 9.9°C and a mean annual precipitation of 576 mm (observation period 1981–2010; Cubasch and Kadow 2011).

Berlin is located within the central European broadleaf forest biome and thus not in a global biodiversity hotspot. Nevertheless, nature conservation plays an important role in urban development. Berlin's protected area system covers about 24% of the city's surface, including 15 Fauna-Flora-Habitat areas protected under European law, 43 nature reserves, and other landscape protection areas (SenUVK 2021). These conservation areas include remnants of the natural landscape (deciduous forests, lakes, rivers, wetlands), the preindustrial cultural landscape (anthropogenic grasslands, forest plantations), and some historical landscape parks. In addition, some of the novel urban wilderness areas that emerged after World War II through the natural revegetation of vacant land are now also designated as protected areas (Kowarik 2018). Beyond conservation areas, Berlin's biodiversity strategy (SenSU 2012) attempts to foster approaches towards a biodiversity friendly development of each land use type.

Thanks to a long tradition of natural history studies over a period of >300 years and a wealth of studies by the so-called "Berlin School of Urban Ecology" since the 1970s, Berlin's biodiversity and its responses to urbanization are well-known (Sukopp 1987, 1990; Kowarik 2020). Red Lists of extinct and endangered plant and animal species in Berlin exist since the 1980s and support urban conservation policies. The recent Red List for plants in Berlin (Seitz et al. 2018) reports a total number of 1,220 native species (including archaeophytes) and 307 established non-native species (neophytes). About half of the total flora (46.4%) is red listed.

Previous studies have revealed a high share of non-natives plants in the vegetation of anthropogenic sites in Berlin (Kowarik 2008). Immature novel ecosystems harbor more non-native species than expected – in contrast to natural remnants where native plant species dominate (Kowarik and von der Lippe 2018). Conflicts with biodiversity conservation exist but are much less relevant than other risk factors. Most prominent were efforts to manage the North American tree, *Prunus serotina*, in Berlin's forest, where it had been previously deliberately planted by foresters (Starfinger et al. 2003). In some conservation areas, herbaceous species from the genera *Solidago* and *Fallopia* and the tree *Robinia pseudoacacia* are managed if conflicts with biodiversity conservation are obvious. Compared to other efforts in biodiversity conservation, the management of non-native plants is much less relevant since other threats to biodiversity are more pressing.

Given the co-occurrence of introduced and native species in the Berlin region for several hundreds of years, non-native species are generally accepted as a component of Berlin's nature. Accordingly, Berlin's Biodiversity Strategy (SenSU 2012) formulates differentiated targets concerning non-natives species, including three major components: (1) non-native species have become characteristic components of many urban habitats and are generally accepted here; (2) invasive non-native species are monitored, but only regulated if they significantly affect biodiversity and appropriate measures promise sustainable success; (3) the focus is on approaches to prevent negative impacts of non-native species.

Urban nature conservation policy in Berlin therefore relies on a context-specific approach to non-native species. Action is only taken if a species poses a concrete threat to biodiversity at a particular site. As a preventive measure, non-native species can only be planted outside the built-up area if permission has been granted. An example of a threefold differentiated strategy, also with regard to invasive non-native plants, is the handling of *R. pseudoacacia*, an invasive species in large parts of Europe (Cierjacks et al. 2013; Vitková et al. 2017). When the species invades valuable dry grassland in Berlin, mostly in conservation areas, it is managed vigorously. At the same time, *Robinia* is a valuable species for tree plantings along streets that are usually out of reach of susceptible habitats of conservation concern. Finally, emerging forests dominated by *Robinia* are part of an urban nature reserve on a derelict freight yard in Berlin. Here, one conservation goal is to allow the undisturbed adaptation of novel communities of native and non-native species to urban conditions (Kowarik and Langer 2005).

Synthesis of the comparison

A comparison of the challenges and opportunities related to non-native plant species in Cape Town and Berlin reveals fundamental differences between the two cities. These can be traced back to the history of the cities and the different biogeographical contexts. Berlin is embedded in a cultural landscape with ecosystems that have been shaped by agriculture and forestry for centuries. The city retains important natural remnants of temperate deciduous forest, as well as many water bodies and wetlands. Both cultural and natural remnants harbor some non-native plant species, but serious threats to biodiversity are rare. In contrast, the intersection of urban and wilderness is much more pronounced in Cape Town, where residential areas often extend directly adjacent to natural ecosystems. Here, fynbos vegetation is highly susceptible to invasion and is often under pressure from invasive trees. Unlike in Berlin, the "tree" life form is (largely) absent from Cape Town's natural fynbos vegetation. An invasive tree here represents a novel life form with significant ecological consequences, whereas the natural vegetation in Berlin is dominated by trees.

In both cities, non-native plants provide a wide range of cultural and regulating ecosystem services. However, planting of non-native ornamental plants is much more common in Cape Town within the reach of vulnerable ecosystems. Important differences also exist in the social realm. Many less privileged people in Cape Town benefit directly from non-native plants in their daily lives, for example in the use of timber from invasive *Acacia* populations. In Berlin, non-native plants are also foraged as edible plants, but the socioeconomic dimension of people's use of non-native plant species is much less important than in Cape Town. This has implications for the acceptance of management measures. While control of alien species is much more urgent in Cape Town than in Berlin, it is much more difficult in South Africa to gain support in the neighborhoods concerned.

Conclusions

Biological invasions are closely linked to urbanization processes as the spread and establishment of non-native species is promoted by a variety of anthropogenic activities. Therefore, non-native species already play a significant role in urban regions, and this role will continue to increase as urbanization accelerates worldwide. The invasion of non-native plants in urban areas is associated with both risks and opportunities. For example, some non-native plant species can affect native species and trigger ecosystem disservices. On the other hand, introduced species are often well adapted to novel urban conditions and can also support a variety of ecosystem services in cities. The contrasting examples of Cape Town and Berlin shows that the impact and management of non-native species in cities cannot be generalized. In this respect, it is a crucial task of future research to explore urban invasions and their impacts in different contexts to better inform management approaches. Important factors that modulate the extent and consequences of urban invasions include the following:

- The biogeographical context of a city
- The cultural history of a city in regard to the diversity, strength, and duration of pathways of plant introductions
- The adjacency of urban environments to natural surroundings or ecosystems of conservation concern
- The socioeconomic relevance of invasions in terms of ecosystem services or disserves and how these are being addressed in urban environmental policies
- The availability of knowledge about invasions and their consequences
- The availability of funding for closing invasion-related knowledge gaps and for developing and implementing management strategies in urban environmental policies

Key companion papers

Didham, R.K., J.M. Tylianakis, M.A. Hutchison, R.M. Ewers, N.J. Gemmell. 2005. Are invasive species the drivers of ecological change? *Trends in Ecology & Evolution* 20(9):470–474.

Kowarik, I., L.K. Fischer. 2020. Alien plants in cities: Human-driven patterns, risks and benefits. Pages 472–482 in I. Douglas, et al., editors. *The Routledge Handbook of Urban Ecology*. Routledge, London.

Potgieter, L.J., et al. 2022. Prioritization and thresholds for managing biological invasions in urban ecosystems. *Urban Ecosystems* 25:253–271.

Schlaepfer, M.A., B.P. Guinaudeau, P. Martin, N. Wyler. 2020. Quantifying the contributions of native and non-native trees to a city's biodiversity and ecosystem services. *Urban Forestry & Urban Greening* 56:126861.

References

Aronson, M.F.J., S.N. Handel, I.P. La Puma, S.E. Clemants. 2015. Urbanization promotes non-native woody species and diverse plant assemblages in the New York metropolitan region. *Urban Ecosystems* 18(1):31–45.

Aronson, M.F.J., F.A. La Sorte, C.H. Nilon, M. Katti, M.A. Goddard, C.A. Lepczyk, P.S. Warren, N.S.G. Williams, S. Cilliers, B. Clarkson, C. Dobbs, R. Dolan, M. Hedblom, S. Klotz, J.L. Kooijmans, I. Kühn, I. MacGregor-Fors, M. McDonnell, U. Mörtberg, P. Pysek, S. Siebert, J. Sushinsky, P. Werner, M. Winter. 2014. A global analysis of the impacts of urbanization on bird and plant diversity reveals key anthropogenic drivers. *Proceedings of the Royal Society B: Biological Sciences* 281(1780):20133330.

Arrington, A. 2021. Urban foraging of five non-native plants in NYC: Balancing ecosystem services and invasive species management. *Urban Forestry & Urban Greening* 58:126896.

Bacher, S., T.M. Blackburn, F. Essl, P. Genovesi, J. Heikkilä, J.M. Jeschke, G. Jones, R. Keller, M. Kenis, C. Kueffer, A.F. Martinou, W. Nentwig, J. Pergl, P. Pyšek, W. Rabitsch, D.M. Richardson, H.E. Roy, W.-C. Saul, R. Scalera, M. Vilà, J.R.U. Wilson, S. Kumschick. 2018. Socio-economic impact classification of alien taxa (SEICAT). *Methods in Ecology and Evolution* 9:159–168.

Bartz, R., U. Heink, I. Kowarik. 2010. Proposed definition of environmental damage illustrated by the cases of genetically modified crops and invasive species. *Conservation Biology* 24(3):675–681.

Bartz, R., I. Kowarik. 2019. Assessing the environmental impacts of invasive alien plants: A review of assessment approaches. *NeoBiota* 43:69–99.

Bastian, O., D. Haase, K. Grunewald. 2012. Ecosystem properties, potentials and services – The EPPS conceptual framework and an urban application example. *Ecological Indicators* 21:7–16.

Bímová, K., B. Mandák, P. Pyšek. 2003. Experimental study of vegetative regeneration in four invasive *Reynoutria taxa* (Polygonaceae). *Plant Ecology* 166(1):1–11.

Blackburn, T.M., P. Pyšek, S. Bacher, J.T. Carlton, R.P. Duncan, V.J. Jarošík, J.R.U. Wilson, D.M. Richardson. 2011. A proposed unified framework for biological invasions. *Trends in Ecology & Evolution* 26(7):333–339.

Boraine, A., O. Crankshaw, C. Engelbrecht, G. Gotz, S. Mbanga, M. Narsoo, S. Parnell. 2006. The state of South African cities a decade after democracy. *Urban Studies* 43(2):259–284.

Born, W., F. Rauschmayer, I. Bräuer. 2005. Economic evaluation of biological invasions – A survey. *Ecological Economics* 55(3):321–336.

Botham, M.S., P. Rothery, P.E. Hulme, M.O. Hill, C.D. Preston, D.B. Roy. 2009. Do urban areas act as foci for the spread of alien plant species? An assessment of temporal trends in the UK. *Diversity and Distributions* 15(2):338–345.

Burton, M.L., L.J. Samuelson, S. Pan, S. 2005. Riparian woody plant diversity and forest structure along an urban-rural gradient. *Urban Ecosystems* 8(1):93–106.

Cadotte, M.W., S.L.E. Yasui, S. Livingstone, J.S. MacIvor. 2017. Are urban systems beneficial, detrimental, or indifferent for biological invasion? *Biological Invasions* 19(12):3489–3503.

Carruthers, J., L. Robin, J.P. Hattingh, C.A. Kull, H. Rangan, B.W. van Wilgen. 2011. A native at home and abroad: The history, politics, ethics and aesthetics of acacias. *Diversity & Distributions* 17(5):810–821.

Casella, F., M. Vurro. 2013. *Ailanthus altissima* (Ailanthus): Spread and harmfulness in a case-study urban area. *Arboricultural Journal: The International Journal of Urban Forestry* 35(3):172–181.

Castro-Díez, P., Á. Alonso, A. Saldaña-López, E. Granda. 2021. Effects of widespread non-native trees on regulating ecosystem services. *Science of the Total Environment* 778:146141.

Celesti-Grapow, L., C. Ricotta. 2020. Plant invasion as an emerging challenge for the conservation of heritage sites: The spread of ornamental trees on ancient monuments in Rome, Italy. *Biol Invasions* 23:1191–1206.

Čeplová, N., V. Kalusová, Z. Lososová. 2017. Effects of settlement size, urban heat island and habitat type on urban plant biodiversity. *Landscape and Urban Planning* 159:15–22.

Čeplová, N., Z. Lososova, D. Zelený, M. Chytrý, J. Danihelka, K. Fajmon, D. Láníková, Z. Preislerová, V. Řehořek, L. Tichý, L. 2015. Phylogenetic diversity of central-European urban plant communities: Effects of alien species and habitat types. *Preslia* 87(1):1–16.

Chocholoušková, Z., P. Pyšek. 2003. Changes in composition and structure of urban flora over 120 years: A case study of the city of Plzeň. *Flora-Morphology, Distribution, Functional Ecology of Plants* 198(5):366–376.

Cierjacks, A., I. Kowarik, J. Joshi, S. Hempel, M. Ristow, M. von der Lippe, E. Weber. 2013. Biological flora of the British Isles: *Robinia pseudoacacia*. *Journal of Ecology* 101(6):1623–1640.

Cowling, R.M., P.W. Rundel, B.B. Lamont, M.K. Arroyo, M. Arianoutsou. 1996. Plant diversity in Mediterranean-climate regions. *Trends in Ecology and Evolution* 11(9):362–366.

Crowley, S.L., S. Hinchliffe, R.A. McDonald. 2017. Conflict in invasive species management. *Frontiers in Ecology and the Environment* 15(3):133–141.

Cubasch, U., C. Kadow. 2011. Global climate change and aspects of regional climate change in the Berlin-Brandenburg region. *Die Erde* 142(1–2):3–20.

D'Amato, G., F.T.M. Spieksma, G. Liccardi, S. Jäger, M. Russo, K. Kontou-Fili, H. Nikkels, B. Wüthrich, S. Bonini. 1998. Pollen-related allergy in Europe. *Allergy* 53:567–578.

Davoren, E., C.M. Shackleton. 2021. Urban ecosystem disservices in the Global South. Pages 265–292 in C.M. Shackleton, S.S. Cilliers, E. Davore, M.J. du Toit, editors. *Urban Ecology in the Global South.* Springer, Cham, Switzerland.

Dehnen-Schmutz, K., J. Touza, C. Perrings, M. Williamson. 2007. A century of the ornamental plant trade and its impact on invasion success. *Diversity and Distributions* 13(5):527–534.

de Wit, M.P., D.J. Crookes, B.W. van Wilgen. 2001. Conflicts of interest in environmental management: Estimating the costs and benefits of a tree invasion. *Biological Invasions* 3:167–178.

Dickie, I.A., B.M. Bennett, L.E. Burrows, M.A. Nuñez, D.A. Peltzer, A. Porté, D.M. Richardson, M. Rejmánek, P.W. Rundel, B.W. wan Wilgen. 2014. Conflicting values: Ecosystem services and invasive tree management. *Biological Invasions* 16:705–719.

Didham, R.K., J.M. Tylianakis, M.A. Hutchison, R.M. Ewers, N.J. Gemmell. 2005. Are invasive species the drivers of ecological change? *Trends in Ecology & Evolution* 20(9):470–474.

Dobbs, C., F.J. Escobedo, W.C. Zipperer. 2011. A framework for developing urban forest ecosystem services and goods indicators. *Landscape Urban Planning* 99:196–206.

Essl, F., S. Bacher, P. Genovesi, P.E. Hulme, J.M. Jeschke, S. Katsanevakis, I. Kowarik, I. Kühn, P. Pyšek, W. Rabitsch, S. Schindler, M. van Kleunen, M. Vilà, J.R.U. Wilson, D.M. Richardson. 2018. Which taxa are alien? Criteria, applications, and uncertainties. *BioScience* 68(7):496–509.

Essl, F., K. Biró, D. Brandes, O. Broennimann, J.M. Bullock, D.S. Chapman, B. Chauvel, S. Dullinger, B. Fumanal, A. Guisan, G. Karrer, G. Kazinczi, C. Kueffer, B. Laitung, C. Lavoie, M. Leitner, T. Mang, D. Moser, H. Müller-Schärer, B. Petitpierre, R. Richter, U. Schaffner, M. Smith, U. Starfinger, R. Vautard, G. Vogl, M. von der Lippe, S. Follak. 2015. Biological flora of the British Isles: *Ambrosia artemisiifolia. Journal of Ecology* 103(4):1069–1098.

Estévez, R.A., C.B. Anderson, J.C. Pizarro, M.A. Burgman. 2015. Clarifying values, risk perceptions, and attitudes to resolve or avoid social conflicts in invasive species management. *Conservation Biology* 29(1):19–30.

Fischer, L.K., I. Kowarik. 2020. Connecting people to biodiversity in cities of tomorrow: Is urban foraging a powerful tool? *Ecological Indicators* 112:106087.

Fischer, L.K., V. Rodorff, M. von der Lippe, I. Kowarik. 2016. Drivers of biodiversity patterns in parks of a growing South American megacity. *Urban Ecosystems* 19(3):1231–1249.

Francis, R.A, M.A. Chadwick. 2015. Urban invasions: Non-native and invasive species in cities. *Geography* 100:144–151.

Gaertner, M., R. Biggs, M. Te Beest, C. Hui, J. Molofsky, D.M. Richardson. 2014. Invasive plants as drivers of regime shifts: Identifying high-priority invaders that alter feedback relationships. *Diversity and Distributions* 20:733–744.

Gaertner, M., B.M.H. Larson, U.M. Irlich, P.M. Holmes, L. Stafford, B.W. van Wilgen, D.M. Richardson. 2016. Managing invasive species in cities: A framework from Cape Town, South Africa. *Landscape and Urban Planning* 151:1–9.

Gaertner, M., A. Novoa, J. Fried, D.M. Richardson. 2017b. Managing invasive species in cities: A decision support framework applied to Cape Town. *Biological Invasions* 19:3707–3723.

Gaertner, M., J.R.U. Wilson, M.W. Cadotte, J.S. MacIvor, R.D. Zenni, D.M. Richardson. 2017a. Non-native species in urban environments: Patterns, processes, impacts and challenges. *Biological Invasions* 19:3461–3469.

Gavier-Pizarro, G.I., V.C. Radeloff, S.I. Stewart, C.D. Huebner, N.S. Keuler. 2010. Housing is positively associated with invasive exotic plant species richness in New England, USA. *Ecological Applications* 20(7):1913–1925.

Géron, C., J.J. Lembrechts, J. Borgelt, J. Lenoir, R. Hamdi, G. Mahy, I. Nijs, A. Monty. 2021. Urban alien plants in temperate oceanic regions of Europe originate from warmer native ranges. *Biological Invasions* 23:1765–1779.

Grimm, N.B., S.H. Faeth, N.E. Golubiewski, C.L. Redman, J. Wu, X. Bai, J.M. Briggs. 2008. Global change and the ecology of cities. *Science* 319(5864):756–760.

Haase, D., N. Larondelle, E. Andersson, M. Artmann, S. Borgström, J. Breuste, E. Gomez-Baggethun, Å. Gren, Z. Hamstead, R. Hansen, N. Kabisch, P. Kremer, J. Langemeyer, E.L. Rall, T. McPhearson, S. Pauleit, S. Qureshi, N. Schwarz, A. Voigt, D. Wurster, T. Elmqvist. 2014. A quantitative review of urban ecosystem service assessments: Concepts, models, and implementation. *AMBIO* 43(4):413–433.

Hahs, A.K., M.J. McDonnell. 2016. Moving beyond biotic homogenization: Searching for new insights into vegetation dynamics. *Journal of Vegetation Science* 27(3):439–440.

Hobbs, R.J., S. Arico, J. Aronson, J.S. Baron, P. Bridgewater, V.A. Cramer, P.R. Epstein, J.J. Ewel, C.A. Klink, A.E. Lugo, D. Norton, D. Ojima, D.M. Richardson, E.W. Sanderson, F. Valladares, M. Vilá, R. Zamora, M. Zobel. 2006. Novel ecosystems: Theoretical and management aspects of the new ecological world order. *Global Ecology and Biogeography* 15(1):1–7.

Hulme, P.E. 2009. Trade, transport and trouble: Managing invasive species pathways in an era of globalization. *Journal of Applied Ecology* 46(1):10–18.

Ignatieva, M., K. Ahrné, J. Wissman, T. Eriksson, P. Tidåker, M. Hedblom, T. Kätterer, H. Marstorp, P. Berg, T. Eriksson, J. Bengtsson. 2015. Lawn as a cultural and ecological phenomenon: A conceptual framework for transdisciplinary research. *Urban Forestry & Urban Greening* 14(2):383–387.

Ives, C.D., P.E. Lentini, C.G. Threlfall, K. Ikin, D.F. Shanahan, G.E. Garrard, S.A. Bekessy, R.A. Fuller, L. Mumaw, L. Rayner, R. Rowe, L.E. Valentine, D. Kendal. 2016. Cities are hotspots for threatened species. *Global Ecology and Biogeography* 25(1):117–126.

Jim, C.Y., W.Y. Chen. 2009. Ecosystem services and valuation of urban forests in China. *Cities* 26(4):187–194.

Kim, G. 2016. Assessing urban forest structure, ecosystem services, and economic benefits on vacant land. *Sustainability* 8(7):679.

Knapp, S., I. Kühn, J. Stolle, S. Klotz. 2010. Changes in the functional composition of a Central European urban flora over three centuries. *Perspectives in Plant Ecology, Evolution and Systematics* 12(3):235–244.

Kowarik, I. 2003. Human agency in biological invasions: Secondary releases foster naturalisation and population expansion of alien plant species. *Biological Invasions* 5:293–312.

Kowarik, I. 2008. On the role of alien species in urban flora and vegetation. Pages 321–338 in J.M. Marzluff, E. Shulenberger, W. Endlicher, M. Alberti, G. Bradley, C. Ryan, U. Simon, C. ZumBrunnen, editors. *Urban Ecology*. Springer, Boston, MA.

Kowarik, I. 2011. Novel urban ecosystems, biodiversity, and conservation. *Environmental Pollution* 159(8–9):1974–1983.

Kowarik, I. 2018. Urban wilderness: Supply, demand, and access. *Urban Forestry & Urban Greening* 29:336–347.

Kowarik, I. 2020. Herbert Sukopp – An inspiring pioneer in the field of urban ecology. *Urban Ecosystems* 23(3):445–455.

Kowarik, I., L.K. Fischer. 2020. Alien plants in cities: Human-driven patterns, risks and benefits. Pages 472–482 in I. Douglas, P.M. Anderson, D. Goode, M.C. Houck, D. Maddox, H. Nagendra, T.P. Yok, editors. *The Routledge Handbook of Urban Ecology*. Routledge, Oxon, Oxfordshire, UK.

Kowarik, I., A. Hiller, G. Planchuelo, B. Seitz, B., M. von der Lippe, S. Buchholz. 2019. Emerging urban forests: Opportunities for promoting the wild side of the urban green infrastructure. *Sustainability* 11(22):6318.

Kowarik, I., A. Langer. 2005. Natur-Park Südgelände: Linking conservation and recreation in an abandoned railyard in Berlin. Pages 288–299 in I. Kowarik, S. Körner, editors. *Wild Urban Woodlands*. Springer, Berlin, Heidelberg, Germany.

Kowarik, I., P. Pyšek. 2012. The first steps towards unifying concepts in invasion ecology were made one hundred years ago: Revisiting the work of the Swiss botanist Albert Thellung. *Diversity and Distributions* 18(12):1243–1252.

Kowarik, I., I. Säumel. 2007. Biological flora of central Europe: *Ailanthus altissima* (Mill.) Swingle. *Perspectives in Plant Ecology, Evolution and Systematics* 8(4):207–237.

Kowarik, I., T.M. Straka, M. Lehmann, R. Studnitzky, L.K. Fischer. 2021. Between approval and disapproval: Citizens' views on the invasive tree *Ailanthus altissima* and its management. *NeoBiota* 66:1–30.

Kowarik, I., M. von der Lippe. 2018. Plant population success across urban ecosystems: A framework to inform biodiversity conservation in cities. *Journal of Applied Ecology* 55(5):2354–2361.

Kueffer, C. 2013. Integrating natural and social sciences for understanding and managing plant invasions. Pages 71–96 in S. Larrue, editor. *Biodiversity and Societies in the Pacific Islands*. Presses Universitaires de Provence, Marseille, France & ANU ePress, Canberra, Australia.

Kueffer, C., C.A. Kull. 2017. Non-native species and the aesthetics of nature. Pages 311–324 in M. Vilà, P.E. Hulme, editors. *Impact of Biological Invasions on Ecosystem Services*. Springer, Berlin, Germany.

Kühn, I., R. Brandl, S. Klotz. 2004. The flora of German cities is naturally species rich. *Evolutionary Ecology Research* 6(5):749–764.

Kühn, I., S. Klotz. 2006. Urbanization and homogenization – Comparing the floras of urban and rural areas in Germany. *Biological Conservation* 127(3):292–300.

Kühn, I., J. Wolf, A. Schneider. 2017. Is there an urban effect in alien plant invasions? *Biological Invasions* 19:3505–3513.

Kull, C.A., C.M. Shackleton, P.J. Cunningham, C. Ducatillon, J.M. Dufour-Dror, K.J. Esler, J.B. Friday, A.C. Gouveia, A.R. Griffin, E. Marchante, S.J. Midgley, A. Pauchard, H. Rangan, D.M. Richardson, T. Rinaudo, J. Tassin, L.S. Urgenson, G.P. von Maltitz, R.D. Zenni, M.J. Zylstra. 2011. Adoption, use and perception of Australian acacias around the world. *Diversity & Distributions* 17(5):822–836.

Kunick, W. 1974. Veränderungen von Flora und Vegetation einer Großstadt dargestellt am Beispiel von Berlin (West). *Dissertation Technische Universität Berlin* 83:1–472.

La Sorte, F.A., M.F.J. Aronson, N.S.G. Williams, L. Celesti-Grapow, S. Cilliers, B.D. Clarkson, R.W. Dolan, A. Hipp, S. Klotz, I. Kühn, P. Pyšek, S. Siebert, M. Winter. 2014. Beta diversity of urban floras among European and non-European cities. *Global Ecology and Biogeography* 23(7):769–779.

Le Maitre, D.C., M. Gaertner, E. Marchante, E.-J. Ens, P.M. Holmes, A. Pauchard, P.J. O'Farrell, A.M. Rogers, R. Blanchard, J. Blignaut, D.M. Richardson. 2011. Impacts of invasive Australian acacias: Implications for management and restoration. *Diversity and Distributions* 17:1015–1029.

Lindemann-Matthies, P. 2016. Beasts or beauties? Lay people's perception of invasive alien plant species in Switzerland and attitudes towards their management. *NeoBiota* 29:15–33.

Lososová, Z., M. Chytrý, L. Tichý, J. Danihelka, K. Fajmon, O. Hájek, K. Kintrová, I. Kühn, D. Láníková, Z. Otýpková, V. Řehořek. 2012. Native and alien floras in urban habitats: A comparison across 32 cities of central Europe. *Global Ecology and Biogeography* 21(5):545–555.

McKinney, M.L. 2006. Urbanization as a major cause of biotic homogenization. *Biological Conservation* 127(3):247–260.

McLean, P., L. Gallien, J.R.U. Wilson, M. Gaertner, D.M. Richardson. 2017. Small urban centres as launching sites for plant invasions in natural areas: Insights from South Africa. *Biological Invasions* 19:3541–3555.

McPherson, E.G. 2003. A benefit-cost analysis of ten street tree species in Modesto, California, U.S. *Journal of Arboriculture* 29(1):1–8.

McPherson, E.G., J.R. Simpson, Q. Xiao, C. Wu. 2011. Million trees Los Angeles canopy cover and benefit assessment. *Landscape and Urban Planning* 99:40–50.

Novoa, A., K. Dehnen-Schmutz, J. Fried, G. Vimercati. 2017. Does public awareness increase support for invasive species management? Promising evidence across taxa and landscape types. *Biological Invasions* 19:3691–3705.

Novoa, A., R. Shackleton, S. Canavan, C. Cybèle, S.J. Davies, K. Dehnen-Schmutz, J. Fried, M. Gaertner, S. Geerts, C.L. Griffiths, H. Kaplan, S. Kumschick, D.C. Le Maitre, G.J. Measey, A.L. Nunes, D.M. Richardson, T.B. Robinson, J. Touza, J.R.U. Wilson. 2018. A framework for engaging stakeholders on the management of alien species. *Journal of Environmental Management* 205:286–297.

Padayachee, A.L., U.M. Irlich, K.T. Faulkner, M. Gaertner, Ş. Procheş, J.R.U. Wilson, M. Rouget. 2017. How do invasive species travel to and through urban environments? *Biological Invasions* 19:3557–3570.

Parepa, M., M. Fischer, C. Krebs, O. Bossdorf. 2014. Hybridization increases invasive knotweed success. *Evolutionary Applications* 7(3):413–420.

Pooley, S. 2014. *Burning Table Mountain: An Environmental History of Fire on the Cape Peninsula*. Palgrave Macmillian, London.

Potgieter, L.J., M. Gaertner, C. Kueffer, B.M.H. Larson, S.W. Livingstone, P.J. O'Farrell, D.M. Richardson. 2017. Alien plants as mediators of ecosystem services and disservices in urban systems: A global review. *Biological Invasions* 19:3571–3588.

Potgieter, L.J., M. Gaertner, P.J. O'Farrell, D.M. Richardson. 2019a. Perceptions of impact: Invasive alien plants in the urban environment. *Journal of Environmental Management* 229:76–87.

Potgieter, L.J., G. Gaertner, P.J. O'Farrell, D.M. Richardson. 2019b. Does vegetation structure influence criminal activity? Insights from Cape Town, South Africa. *Frontiers of Biogeography* 11(1):e42035.

Potgieter, L.J., M. Gaertner, P.J. O'Farrell, D.M. Richardson. 2019c. A fine-scale assessment of the ecosystem service-disservice dichotomy in the context of urban ecosystems affected by alien plant invasions. *Forest Ecosystems* 6:e46.

Pyšek, P. 1998. Alien and native species in central European urban floras: A quantitative comparison. *Journal of Biogeography* 25(1):155–163.

Pyšek, P., M. Křivánek, V. Jarošík. 2009. Planting intensity, residence time, and species traits determine invasion success of alien woody species. *Ecology* 90(10):2734–744.

Pyšek, P., D.M. Richardson. 2010. Invasive species, environmental change and management, and health. *Annual Review of Environmental Resources* 35:25–55.

Qiu, J. 2015. A global synthesis of the effects of biological invasions on greenhouse gas emissions. *Global Ecology & Biogeography* 24(11):1351–1362.

Reinhardt, F., M. Herle, F. Bastiansen, B. Streit. 2003. *Economic Impact of the Spread of Alien Species in Germany*. Federal Ministry of the Environment, Nature, Conservation and Nuclear Safety, Berlin, Germany.

Richardson, D.M., B.W. van Wilgen. 2004. Invasive alien plants in South Africa: How well do we understand the ecological impacts? *South African Journal of Science* 100(1–2):45–52.

Riley, C.B., D.A. Herms, M.M. Gardiner. 2018. Exotic trees contribute to urban forest diversity and ecosystem services in inner-city Cleveland, OH. *Urban Forestry & Urban Greening* 29:367–376.

Roloff, A., S. Gillner, R. Kniesel, D. Zhang. 2018. Interesting and new street tree species for European cities. *Journal of Forest and Landscape Research* 1:1–7.

Roy, S., J. Byrne, C. Pickering. 2012. A systematic quantitative review of urban tree benefits, costs, and assessment methods across cities in different climatic zones. *Urban Forestry & Urban Greening* 11:351–363.

Rusterholz, H.P., D. Wirz, B. Baur. 2012. Garden waste deposits as a source for non-native plants in mixed deciduous forests. *Applied Vegetation Science* 15(3):329–337.

Sádlo, J., M. Vítková, J. Pergl, P. Pyšek. 2017. Towards site-specific management of invasive alien trees based on the assessment of their impacts: The case of *Robinia pseudoacacia*. *NeoBiota* 35:1–34.

Schlaepfer, M.A., B.P. Guinaudeau, P. Martin, N. Wyler. 2020. Quantifying the contributions of native and non-native trees to a city's biodiversity and ecosystem services. *Urban Forestry & Urban Greening* 56:126861.

SDUDH. 2016. *Senate Department for Urban Development and Housing*. Berlin Environmental Atlas. 06.02.1 Actual Use and Vegetation Cover. www.stadtentwicklung.berlin.de/umwelt/umweltatlas/eke601.htm

Seitz, B., W. Ristow, J. Meißner, B. Machatzi, H. Sukopp. 2018. *Rote Liste und Gesamtartenliste der etablierten Farn-und Blütenpflanzen von Berlin*. Universitätsverlag der TU, Berlin, Germany.

SenSU [Senatsverwaltung für Stadtentwicklung und Umwelt]. 2012. *Berliner Strategie zur Biologischen Vielfalt*. Die Senatsverwaltung für Stadtentwicklung und Umwelt, Berlin, Germany.

SenUVK. 2021. *Senatsverwaltung für Umwelt, Verkehr und Klimaschutz*. Schutzgebiete. www.berlin.de/sen/uvk/natur-und-gruen/naturschutz/schutzgebiete/

Shackleton, C.M., P.J. Mograbi. 2020. Meeting a diversity of needs through a diversity of species: Urban residents' favourite and disliked tree species across eleven towns in South Africa and Zimbabwe. *Urban Forestry & Urban Greening* 48:e126507.

Shackleton, C.M., S. Ruwanza, G.K. Sinasson Sanni, S. Bennett, P. De Lacy, R. Modip, N. Mtati, M. Sachikonye, G. Thondhlana. 2016. Unpacking Pandora's box: Understanding and categorising ecosystem disservices for environmental management and human wellbeing. *Ecosystems* 19(4):587–600.

Shackleton, R.T., B.M.H. Larson, A. Novoa, D.M. Richardson, C.A. Kull. 2019. The human and social dimensions of invasion science and management. *Journal of Environmental Management* 229:1–9.

Shackleton, R.T., D.C. Le Maitre, N.M. Pasiecznik, D.M. Richardson. 2014. *Prosopis*: A global assessment of the biogeography, benefits, impacts and management of one of the world's worst woody invasive plant taxa. *AoB Plants* 6:plu027.

Šilc, U. 2010. Synanthropic vegetation: Pattern of various disturbances on life history traits. *Acta Botanica Croatica* 69(2):215–225.

Sladonja, B., M. Sušek, J. Guillermic. 2015. Review on invasive Ailanthus (*Ailanthus altissima* (Mill.) Swingle) conflicting values: Assessment of its ecosystem services and potential biological threat. *Environmental Management* 56(4):1009–1034.

Štajerová, K., P. Šmilauer, J. Brůna, P. Pyšek. 2017. Distribution of invasive plants in urban environments is strongly spatially structured. *Landscape Ecology* 32:681–692.

Starfinger, U., I. Kowarik, M. Rode, H. Schepker. 2003. From desirable ornamental plant to pest to accepted addition to the flora? The perception of an alien tree species through the centuries. *Biological Invasions* 5(4):323–335.

Stohlgren, T.J., D.T. Barnett, J.T. Kartesz. 2003. The rich get richer: Patterns of plant invasions in the United States. *Frontiers in Ecology and the Environment* 1:11–14.

Strayer, D.L., V.T. Eviner, J.M. Jeschke, M.L. Pace. 2006. Understanding the long-term effects of species invasions. *Trends in Ecology & Evolution* 21(11):645–651.

Sukopp, H. 1987. On the history of plant geography and plant ecology in Berlin. *Englera* 7:85–103.

Sukopp, H. 1990. *Stadtökologie: Das Beispiel Berlin*. D. Reimer, Berlin, Germany.

Sukopp, H., P. Werner. 1983. Urban environments and vegetation. Pages 247–260 in W. Holzner, M.J.A. Werger, I. Ikusima, editors. *Man's Impact on Vegetation*. Dr. W Junk Publishers, The Hague, Netherlands.

Sullivan, J.J., S.M. Timmins, P.A. Williams. 2005. Movement of exotic plants into coastal native forests from gardens in northern New Zealand. *New Zealand Journal of Ecology* 29(1):1–10.

Tait, C.J., C.B. Daniels, R.S. Hill. 2005. Changes in species assemblages within the Adelaide metropolitan area, Australia, 1836–2002. *Ecological Applications* 15(1):346–359.

Taylor, K., T. Brummer, M.L. Taper, A. Wing, L.J. Rew. 2012. Human-mediated long-distance dispersal: An empirical evaluation of seed dispersal by vehicles. *Diversity and Distributions* 18(9):942–951.

Thellung, A. 1912. La flore adventice de Montpellier. *Mémoires de la Société Nationale des Sciences, Naturelles et Mathématiques de Cherbourg* 38:57–728.

Trentanovi, G., M. von der Lippe, T. Sitzia, U. Ziechmann, I. Kowarik, A. Cierjacks. 2013. Biotic homogenization at the community scale: Disentangling the roles of urbanization and plant invasion. *Diversity and Distributions* 19(7):738–748.

van Wilgen, B.W. 2012. Evidence, perceptions, and trade-offs associated with invasive alien plant control in the Table Mountain National Park, South Africa. *Ecology and Society* 17(2):1–23.

van Wilgen, B.W., D.B. Richardson. 2014. Challenges and trade-offs in the management of invasive alien trees. *Biological Invasions* 16:721–734.

van Wilgen, B.W., D.F. Scott. 2001. Managing fires on the Cape Peninsula: Dealing with the inevitable. *Journal of Mediterranean Ecology* 2:197–208.

Vaz, A.S., P. Castro-Díez, O. Godoy, Á. Alonso, M. Vilà, A. Saldaña, H. Marchante, Á. Bayón, J.S. Silva, J.R. Vicente, J.P. Honrado. 2018. An indicator-based approach to analyse the effects of non-native tree species on multiple cultural ecosystem services. *Ecological Indicators* 85:48–56.

Vaz, A.S., C. Kueffer, C.A. Kull, D.M. Richardson, S. Schindler, A.J. Muñoz-Pajares, J.R. Vicente, J. Martins, C. Hui, I. Kühn, J.P. Honrado. 2017b. The progress of interdisciplinarity in invasion science. *Ambio* 46:428–442.

Vaz, A.S., C. Kueffer, C.A. Kull, D.M. Richardson, J.R. Vicente, I. Kühn, M. Schröter, J. Hauck, A. Bonn, J.P. Honrado. 2017a. Integrating ecosystem services and disservices: Insights from plant invasions. *Ecosystem Services* 23:94–107.

Vicente, J.R., A.T. Pinto, M.B. Araújo, P.H. Verburg, A. Lomba, C.F. Randin, A. Guisan, J.P. Honrado. 2013. Using life strategies to explore the vulnerability of ecosystem services to invasion by alien plants. *Ecosystems* 16:678–693.

Vítková, M., J. Müllerová, J. Sádlo, J. Pergl, P. Pyšek. 2017. Black locust (*Robinia pseudoacacia*) beloved and despised: A story of an invasive tree in central Europe. *Forest Ecology and Management* 384:287–302.

von der Lippe, M., I. Kowarik. 2007. Long-distance dispersal of plants by vehicles as a driver of plant invasions. *Conservation Biology* 21(4):986–996.

von Döhren, P., D. Haase. 2015. Ecosystem disservices research: A review of the state of the art with a focus on cities. *Ecological Indicators* 52:490–497.

von Döhren, P., D. Haase. 2019. Risk assessment concerning urban ecosystem disservices: The example of street trees in Berlin, Germany. *Ecosystem Services* 40:101031.

Williamson, M., A. Fitter. 1996. The varying success of invaders. *Ecology* 77(6):1661–1666.

Wolf, J., D. Haase, I. Kühn. 2020. The functional composition of the neophytic flora changes in response to environmental conditions along a rural-urban gradient. *NeoBiota* 54:23–47.

Yelenik, S.G., W.D. Stock, D.M. Richardson. 2007. Functional group identity does not predict invader impacts: Differential effects of nitrogen-fixing exotic plants on ecosystem function. *Biological Invasions* 9:117–125.

Zeeman, B.J., M.J. McDonnell, D. Kendal, J.W. Morgan. 2017. Biotic homogenization in an increasingly urbanized temperate grassland ecosystem. *Journal of Vegetation Science* 28(3):550–561.

14

THE IMPORTANCE OF CITIZEN SCIENCE[1] FOR UNDERSTANDING URBAN BIODIVERSITY

Christopher A. Lepczyk and Timothy Vargo

Introduction

Citizen science has gained widespread recognition in the 20th century as both a legitimate form of science and as a way to broaden the scientific enterprise. While exacting definitions vary, it can be described as the "scientific work undertaken by members of the general public, often in collaboration with or under the direction of professional scientists and scientific institutions" (OED). Essentially, then, most iterations of citizen science involve some degree of collaboration between professional and non-professionals in scientific investigations – asking questions, collecting data, or interpreting results.

Historically, a large portion of citizen science was carried out by expert amateurs who conducted scientific investigations independently of professional scientists, or in collaborations as peers. Even today, there remains many expert amateur scientists who make important contributions to many fields, especially taxonomy and natural history (Costello et al. 2013). However, citizen science has now greatly expanded from its historic roots and been of significant importance in ecology, conservation, and natural resource management.

In the case of urban biodiversity, citizen science has played an important role in our understanding of urban ecosystems and offers great promise to continue to do so in the future. In fact, citizen science dovetails exceptionally well with urban biodiversity research because cities are where the majority of people live and have the opportunity to engage with biodiversity. Moreover, some of the oldest citizen science programs in the United States, such as Audubon's Christmas Bird Count and the National Weather Service's Cooperative Observer Programs, which both started over 100 years ago, have received some of their strongest support and participation from urban areas.

How citizen science works

In academic research a best practice is to identify overarching goals and objectives (Tweddle et al. 2020). Citizen science projects (or research in general), however, are often led by methods, study

1 Note that for the purposes of this chapter, we use *citizen science* rather than *community science*. Recent discussions within the citizen science community have raised issues surrounding the use of the word *citizen* and what it connotes. While we understand the evolving nature of language, at present there is a professional society, peer-reviewed journal, and books that operate under the term *citizen science*.

DOI: 10.4324/9781003016120-16

site, or data, rather than the main questions or goals. From an academic perspective, this approach can thereby limit the quality and value of the research as it may appear to be aimless, even if that is not the case. Thus, professionals should first consider whether a citizen science framework is necessary or useful depending upon what the project is seeking to address. For details on the complete process of conceptualizing through implementing a citizen science project from a practitioner's point of view, see the *Handbook of Citizen Science in Ecology and Conservation* (Lepczyk et al. 2020).

Often, citizen science seeks to interact and train the public in assisting with the scientific enterprise. In practice, however, there are many forms that citizen science can take, and the approach used depends on the type and context of the project (Shirk et al. 2012). While most citizen science projects involve some degree of collaboration between professional and community scientists, the degree and type of participation on each side differs substantially from one project to the next. As a result, citizen science projects are often categorized. For example, Shirk et al. (2012) divided citizen science projects into five models based on the degree of participation as follows: contractual projects, contributory projects, collaborative projects, co-created projects, and collegial contributions (Table 14.1). Aside from these five types of models, citizen science projects can also be categorized based on the types of activities in which the participants are involved, such as data collectors, data interpreters, or as full partners (Bonney et al. 2016). In the case of urban biodiversity research, all these models and categories have value and can be used. However, most historic citizen science projects and many present projects tend to be contributory projects or projects where individuals serve as data collectors for professional scientists, and that is the type of model we focus on throughout the chapter.

After goals and objectives have been established and the type of citizen science model chosen, the next step typically is building the team. Team building often requires recruitment, which can vary from simple online invitations to networking at places of work to socializing or setting up event-based recruitment stations (Prysby 2020). Furthermore, when recruiting for a citizen science project, it is important to keep in mind that individuals have various motivations for participating. For instance, a primary reason people participate in any activity is simply because they were asked (Freeman 1997). However, in the case of ecological, conservation, and management related projects, a primary reason for participating is a desire to help the environment (Frensley et al. 2017).

Following recruitment is training. Depending on the type or project, the amount of time and effort required to train, participants can vary markedly. For instance, identifying a specific species

Table 14.1 Models of citizen science based on Shirk et al. (2012)

Model Name	*Focus*
Contractual projects	where communities ask professional researchers to conduct a specific scientific investigation and report on the results
Contributory projects	which are generally designed by scientists and for which members of the public primarily contribute data
Collaborative projects	which are generally designed by scientists and for which members of the public contribute data but also help to refine project design, analyze data, or disseminate findings
Co-created projects	which are designed by scientists and members of the public working together and for which at least some of the public participants are actively involved in most or all aspects of the research process
Collegial contributions	where non-credentialed individuals conduct research independently with varying degrees of expected recognition by institutionalized science or professionals.

of plant or animal will likely require less training than how to identify a variety of species and then measure attributes about them. As a result, incorporating enough time on the planning side of the project to include training and the opportunity to learn from mistakes is essential for participants to carry out the work. Once trained and ready, participants can engage in the project.

Because many citizen science projects are long-term or multi-year in nature, one of the largest challenges is retaining participants. As projects are designed, it should be noted that not all participants will continue over time as people lose interest, move, change careers, or simply have other obligations in their lives. Thus, it is valuable to keep participants engaged on the project and understand their motivations for participating in the work. At the same time, keep in mind that a certain percentage of individuals are likely to leave each season or year and that long-term research will require ongoing or renewed recruitment. Furthermore, there is no single approach that is used to retain participants, and thus multiple methods should be considered, including regular contact with participants and group meetings where participants can interact with one another.

From conception through implementation, citizen science research requires the same amount of commitment, time, and energy as any other ecological research project, if not a considerable amount more. In fact, the added components of managing a large number of participants as well as the data they collect and/or analyze is not trivial. As a result, individuals and practitioners interested in developing an urban biodiversity citizen science project are encouraged to spend time learning the intricate details of each step of the process that we have outlined here as well as the steps that we have left out for brevity.

Strengths and weaknesses of citizen science

Citizen science is a powerful tool for urban biodiversity research (e.g. Callaghan et al. 2020; Vereecken et al. 2021). However, it is critical to assess whether citizen science is an appropriate method for your research before you initiate it. In particular, it is important for scientists and managers to determine whether or not the project's goals and objectives are better served and addressed through collaborations with citizen scientists.

For urban biodiversity research, citizen science provides many benefits (see Shirk and Bonney 2020 for detailed discussion of benefits). First, citizen scientists can cover vastly more areas or larger geographic spaces because of the sheer number of individuals involved in a project and can repeat such coverage over extended periods of time. Hence, citizen science allows for larger and longer spatiotemporal scales than traditional individual-based or academic research. This spatiotemporal coverage is particularly relevant for monitoring-based projects. Second, citizen science can increase the resolution of data collection through targeting efforts on specific locations, land use or property types, or ecological contexts. For instance, private lands form an important component of urban biodiversity (e.g. Lepczyk et al. 2004; Lepczyk 2005; Goddard et al. 2010; Narango et al. 2018) and access to them can be greatly enhanced through collaborations with citizen science landowners and homeowners, such as Neighborhood Nestwatch (https://nationalzoo.si.edu/migratory-birds/neighborhood-nestwatch). Third, citizen science can increase the likelihood of detection. Having a large number of participants involved in a project can increase the likelihood of observing phenological events (Schwartz et al. 2012), species (Vereecken et al. 2021), and attributes of the environment that can affect species (e.g. pollution). For instance, Alabama Water Watch (https://aaes.auburn.edu/alabamawaterwatch/) is a citizen science water monitoring program whose tools are being widely used to evaluate water quality in urban areas (Cendejas et al. 2021). Fourth, citizen science can aid in data and image analyses (Willi et al. 2019). While artificial intelligence (AI) and machine learning have increased markedly in recent years, humans still have important skills in recognizing species and detecting patterns as well as extracting information from existing data. Notably, coupling AI and machine learning with citizen science is particularly valuable for advancing ecological monitoring (McClure et al. 2020). Fifth, citizen scientists may have knowledge and perspectives about local environments and species that can complement or improve professional scientists' interpretations of data and results (Lepczyk 2005; Theobald et al. 2015). Sixth, citizen scientists can help refine or expand research questions as they may have unique perspectives and insight or are affected by and observe local natural resources and the environment in their daily lives. Finally, citizen science can facilitate a shared understanding among managers, scientists, regulators, policymakers, volunteers, and others of the complex social dimensions of the natural systems where people live (McGreavy et al. 2016).

Citizen science helps professional researchers appreciate connections between humans and their environment and is well suited for interdisciplinary collaboration, particularly for projects that include both natural and social dimensions. Natural resource and environmental managers are increasingly tasked with addressing the social aspects of difficult ecological issues, such as managing wildfires in the wildland-urban interface. By engaging local community members, citizen science can facilitate a shared understanding among managers, scientists, regulators, policymakers, volunteers, and others of the complex social dimensions of the natural systems where people live.

While there are numerous benefits of citizen science for urban biodiversity research, there are also instances where this approach might be avoided (Pocock et al. 2014; Shirk and Bonney 2020). For example, if research will be used to inform policy or management decisions and the citizen science partners have a stake in the outcomes, a citizen science approach might not be the best one if there could be strong potential for conflict of interest or bias. Such potential for bias does not necessarily preclude the use of citizen scientists, but there might be need for much more stringent measures in place regarding experimental design and data stewardship. Hence, in such a situation, it might be better to perform the research with a team of professional scientists, assuming they also do not have a stake in the outcome. Another consideration is to first gauge the interest of potential citizen scientist partners. Not all research generates community interest or may occur in locations where there are simply too few individuals available to participate. Thus, projects involving species, systems, or topics that are not widely known or are less charismatic may struggle to find a critical mass of participants. A third consideration is risk. Will the project put the participants, species, habitat, or subjects at risk? Field-based research has inherent risks (Sasse 2003), but there are also many species and habitats, even in urban areas, that are sensitive to human interactions or could pose safety risks to people or species. A fourth point is that it may not be appropriate to work with citizen scientists if sampling frequency is very high or very low. For instance, projects requiring repeat or frequent sampling or that have sampling very rarely (e.g. annually) may make it difficult to maintain an active and engaging research environment. Likewise, projects that require individuals to work on certain days of the week or time of day may be more difficult (e.g. owl surveys at night). A final consideration is the skill level of citizen scientists. If data collection requires a high degree of skill or a specific skill, the amount of time and effort in training might not make the project feasible when compared with hiring technicians or professionals.

Beyond the potential weaknesses of citizen science, please keep in mind the following considerations and caveats. First, citizen science is not a replacement for professional science. There will always be aspects of urban biodiversity research and management requiring highly trained and experienced professionals, just as we still need professional doctors and dentists even though many citizens can perform basic first aid. Training and experience are particularly important in applied science contexts, such as conservation biology and natural resource management, where professionals bring skills that can help steer outcomes away from long-term or permanent environmental problems. Second, citizen science is not a license to use unpaid volunteers to collect data (Fournier and Bond 2015). When considering an urban biodiversity project, it is important to avoid implementing a citizen science approach simply to reduce costs or because a grant did not provide the desired level of funding. Citizen science can help with these things, but collaborating with citizen scientists also brings a responsibility to respect the value of the partnership and the opportunities to engage the public in a meaningful way. Therefore, it is important for nonprofit organizations, government agencies, and academics not to turn to citizen science simply because of a lean budget year or as a means to reduce costs. In other words, citizen science should not be viewed as free labor but as an opportunity to strengthen the power and applications of your research. Seeing and appreciating the value of collaborating with citizen scientists can also strengthen trust in science and scientists from an often skeptical public.

Contributions of citizen science to urban biodiversity science and conservation

Citizen science programs are widespread and continue to grow regarding urban biodiversity, particularly when considering projects that collect species data as part of their approach. Geographically, projects range from single city (e.g. Hamilton, Canada; Cartwright et al. 2015) to global (e.g. eBird; Callaghan et al. 2021; Roadkill Project; Peer et al. 2021). Some projects are long-term or permanent types of monitoring (e.g. variations on the national lilac monitoring program; Schwartz et al. 2012) or educational programs (e.g. Budburst; https://budburst.org/), whereas others are one-time events like a bioblitz. For instance, the Bee Conservancy (https://thebeeconservancy.org/) runs beeblitizes where individuals record all bee species found in a designated area over a short period of time. This wide variation in projects is reflective of the various goals set up by the organization or group responsible for leading the project as there is no singular type of program set up solely for the sake of collecting urban biodiversity data.

What have we learned about urban biodiversity?

Our understanding of which species live in cities, where they are distributed, their abundance, and how biodiversity is related to the urban environment have been greatly informed through citizen science. Moreover, citizen science has provided key data on how species richness and abundance and ecological processes have changed over time in cities around the world. In fact, citizen science has been a major contributor to urban ecological research because by default it has been used for monitoring.

Some of the earliest measures of urban biodiversity in birds came from citizen science via the National Audubon Society's Christmas Bird Count, which has been collecting data annually for well over a century in cities around North America and the world (Table 14.2). Today, birds remain a large focus of urban citizen science research, with projects focused on individual cities (e.g. Tucson Bird Count; www.tucsonbirds.org/) through national and international scales that include urban components (e.g. eBird). This focus on birds likely has been driven in large part because birds are common in urban ecosystems, easy to observe, readily differentiated from one another, and can be used as proxy for other taxa. Furthermore, there is a long history of amateur ornithology (Birkhead and Charmantier 2009; Mayfield 1979) and bird clubs around the world. However, birds are by no means the sole focus of urban citizen science research.

Much of the focus of urban biodiversity citizen science has been to describe the number of individuals of a species (abundance) and the number of unique species (richness) at either specific

Table 14.2 Example long-term citizen science projects that focus or include urban biodiversity

Program	*Focus*	*Website/Citation*
National Audubon Society's Christmas Bird Count	Non-breeding birds in and around cities	www.audubon.org/conservation/science/christmas-bird-count
Great Backyard Bird Count	Birds around homeowner's yards	www.birdcount.org/
Budburst	Monitoring plant life cycles near where people live	https://budburst.org/
Roadkill	Understanding where roadkills occur to affect conservation	https://roadkill.at/en/
URBAN (Urban-Rural Biomonitoring and Assessment Network)	URBAN volunteers have tracked the health of more than 23 wetlands and 8 streams in the Hamilton region. Includes measures of birds and amphibians.	www.urbanmonitoring.ca/

habitat types (e.g. gardens; Pendl et al. 2022) and locations (e.g. urban green spaces), or in general across a city. From these basic measures, a number of additional diversity metrics can be calculated, including, but not limited to, evenness, relative abundance, and indices such as Shannon-Weiner. In addition, because many urban citizen science projects are ongoing, their data can show how population indices and biodiversity change over time. Moreover, consistent recording of species over time has provided important information on phenology of species in urban areas relative to rural locations and how phenology has been changing over time because of climate change. Collectively, urban citizen science is providing a needed form of ecological monitoring that is at the heart of most biodiversity research.

Citizen science projects have also been established to address how specific attributes of the city affect aspects of biodiversity. For instance, the Fledermausforscher Berlin (Bat researchers in Berlin; www.fledermausforscher-berlin.de/) project specifically sought to understand landscape attributes that related to the presence of bats in urban areas as a means to aid in their conservation and management (Lewanzik et al. 2022). As we seek to understand how the physiognomy of cities influences biodiversity, citizen science projects can be developed or revised to include such components directly into their research questions.

New species

Citizen science in urban areas has yielded a wealth of new species data. For instance, the Natural History Museum of Los Angeles County's NHM Biodiversity Science: City and Nature (NHM BioSCAN) project identified 30 new species of dipteran flies (genus *Megaselia*) in Los Angeles (Hartop et al. 2015). Importantly, because most of the global biodiversity that has been fully described are terrestrial vertebrates and higher plants, citizen science provides a critical approach for discovering new native species. Of equal, if not more important, citizen science is also increasingly being used at describing non-native and invasive species (Encarnação et al. 2021). In fact, the use of citizen science is becoming increasingly valuable as a rapid response approach to early detection of potential invaders. Ultimately, the discovery of new species is of particular importance to urban biodiversity as it allows practitioners to develop appropriate management and planning guidelines.

Taxonomic focus

Urban biodiversity research traditionally has been dominated by work on plants, birds, and more charismatic species. Today, plants and birds still dominate the focus of research, but there have been notable increases in expanding the taxonomic focus to be more inclusive of bats and insects as well as several other invertebrate groups (Rega-Brodsky et al. 2022). Notably, work on land snails in urban areas remains understudied (Rega-Brodsky et al. 2022). Furthermore, considering the number of known insect species, the numbers described in most urban ecosystems are still relatively low. Ultimately, it is difficult to fully assess the biodiversity of urban ecosystems as we lack taxonomic representativeness. As a result, we have relatively robust data on some taxonomic groups that allow for evaluating change over time questions (e.g. community turnover, effects of pollution), differences between cities, and to in-depth questions, such as functional diversity and evolutionary distinctness (e.g. La Sorte et al. 2018). However, because of the gaps in taxonomic data, we have relatively low ability to fully evaluate food webs or compare them across cities. Aside from basic biodiversity questions, the lack of taxonomic knowledge about invertebrates and amphibians is also a limit in our ability to understand environmental problems as these groups serve as important environmental sentinels and will hopefully be the focus of future research.

Comparative urban ecology

One of the main frontiers in urban ecology remains the comparative approach (e.g. Aronson et al. 2014). Though citizen science projections range from local to global in extent, all types are increasingly being used to conduct comparative urban ecology research. For example, a variety of studies have used existing citizen science projects to test whether there is support for the occurrence of biological homogenization in cities (Leong and Trautwein 2019). Most of the comparative urban ecology work has focused on birds and plants, but other taxonomic groups are increasingly being studied to the point where more complete comparisons and types of analyses can be carried out, such as evaluating changes in species richness and abundance over time.

Importance to conservation and management

Cities are valuable locations for conserving biodiversity (Soanes and Lentini 2019; Lepczyk et al. 2023), including rare, threatened, endangered, and species of conservation concern (Ives et al. 2016). Likewise, cities encompass a wide variety of urban green spaces that house these species and are of particular importance for urban biodiversity conservation (Aronson et al. 2017; Lepczyk et al. 2017). Citizen science has provided much of the data from which our knowledge of species and habitats has been derived. Ultimately, understanding where species occur, the abundance of these species, and how they interact as part of the ecosystem are essential components of conserving and managing species as well as maintaining ecosystem functions. Thus, citizen science often provides the monitoring that is necessary to both develop conservation and management plans as well as evaluate their success over time.

Shortcomings of urban biodiversity citizen science research

While citizen science has provided a wealth of data that have added greatly to our knowledge of urban biodiversity, it also has several shortcomings that are in need of attention. One shortcoming is simply the lack of representative sampling. As with any strong scientific study, a well-thought-out sampling approach is essential to the success of the work. Haphazardly collected data and poor experimental designs can ultimately doom a project and waste the human capital that was invested in it. Furthermore, if we seek to design and carry out long-term and spatially expansive understanding of urban biodiversity, including global comparative studies, we need citizen science projects that share standardized designs and sampling methods.

A second common shortcoming is the lack of participant representation. As a whole, participation in citizen science research has often been skewed towards certain demographics. Specifically, many citizen science projects are biased towards older Caucasian individuals, many of whom are financially stable or well off (Merenlender et al. 2016; Frensley et al. 2017). Because of these and other known biases, there have been increasing discussions and efforts to diversify participation (Paleco et al. 2021). That said, if we seek to democratize science and engage communities, citizen science needs to increase participant representation. Citizen science can be a means to democratize data, and as such, there is a particular need to increase the diversity of participants if we seek to diversify our understanding of urban biodiversity.

On the biodiversity side, even in cities, a notable challenge that remains is taxonomic bias. Globally, we have excess knowledge of most vertebrate species and some plants (e.g. woody species) but have far less knowledge about most insects, arachnids, and fungi (Troudet et al. 2017). While people seem to have innate preferences towards charismatic species, there have been notable increases in the breadth of species representation (Rega-Brodsky et al. 2022). Thus, while the disparity in taxonomic representativeness narrows, there remain gaps (e.g. land snails) that need attention. These gaps are particularly important as we seek to evaluate biodiversity metrics beyond richness and abundance and aim towards a more complete systems understanding.

Future research needs

Our collective knowledge of urban biodiversity and the socioecological processes responsible for shaping it have greatly increased over the past two decades (Rega-Brodsky et al. 2022). This increase has, in part, been because of citizen science projects that have focused on species monitoring and bioblitzes. In looking toward the future, such programs will continue to be of critical importance. There is also great value in research that focuses on specific needs and questions. The following questions and topics are of particular importance to both citizen science and urban biodiversity:

1 *Increasing representative participation.* One of the most valuable aspects of citizen science programs is the involvement of non-scientists in the scientific enterprise. Not only does this involvement help address scientific questions, but can aid in building ecological knowledge and scientific literacy (Trumbull et al. 2000; Jordan et al. 2012). Thus, it is essential that to increase the value of the citizen science process, we need to continue working towards building a diverse network of participants. Currently we know very little about participants, aside from a few specific demographic groups.

2 *Advancing understanding of urban biodiversity to the community.* Biodiversity underpins many ecosystem services that are essential to both nature and society. An important means to demonstrate the value of biodiversity to the public is through citizen science. While a wide variety of citizen science programs already exist, they are a powerful tool to directly connect people to nature, particularly within urban areas. Furthermore, citizen science can increase the understanding and relevance of how urban biodiversity positively relates to human wellbeing and livelihoods.

3 *Biodiversity on private lands.* Urban ecosystems around the world are predominantly composed of private lands. Because private lands are often inaccessible to the scientific community, they

represent a large segment of the ecosystem for which we have limited ecological information on. Although citizen science programs that target homeowners (e.g. Great Backyard Bird Count; www.birdcount.org/) exist, they often are not strictly centered on cities or urban areas. Hence, what is needed are citizen science projects, or expansion of projects, that focus on collecting species (and other ecological data) from yards, residential neighborhoods, and other private properties. Such data would complement projects from public lands in cities (e.g. parks) and aid in answering a variety of basic questions, including how (a) urban biodiversity relates to land use practices and socioecological processes, (b) biodiversity compares on private lands across cities, and (c) similar or different private lands and public lands are for maintaining urban biodiversity.

4 *Taxonomic representativeness*. Although urban biodiversity research has greatly expanded, placing additional resources and attention on less well-studied taxa is particularly needed. Ultimately, by having a goal of increasing taxonomic representativeness and full inventories of species, we can begin to answer more systems level questions, such as how urban food webs may be similar or different across cities around the world.

5 *Expanding beyond richness and diversity*. Urban biodiversity research continues to focus greatly on simple metrics, such as richness, evenness, and abundance. The next steps are to move beyond basic metrics into more advanced considerations of biodiversity so as to address such questions as how functional and evolutionary diversity changes over time and among cities and how biodiversity is related to public health and ecosystem resilience within and among cites.

Key companion papers

Deguines, N., R. Julliard, M. de Flores, C. Fontaine. 2012. The whereabouts of flower visitors: Contrasting land-use preferences revealed by a country-wide survey based on citizen science. *PLoS ONE* 7(9):e45822.

Lepczyk, C.A., O. Boyle, T. Vargo, editors. 2020. *Handbook of Citizen Science in Ecology and Conservation*. University of California Press, Berkeley, CA.

Shirk, J., et al. 2012. Public participation in scientific research: A framework for deliberate design. *Ecology and Society* 17(2):29.

Spear, D.M., G.B. Pauly, K. Kaiser. 2017. Citizen science as a tool for augmenting museum collection data from urban areas. *Frontiers in Ecology and Evolution* 5. https://doi.org/10.3389/fevo.2017.00086

Strohbach, M.W., S.B. Lerman, P.S. Warren. 2013. Are small greening areas enhancing bird diversity? Insights from community-driven greening projects in Boston. *Landscape and Urban Planning* 11:69–79.

References

Aronson, M.F.J., F.A. La Sorte, C.H. Nilon, M. Katti, M.A. Goddard, C.A. Lepczyk, P.S. Warren, N.S.G. Williams, S. Cilliers, B. Clarkson, C. Dobbs, R. Dolan, M. Hedblom, S. Klotz, J. Louwe Kooijmans, I. Kühn, I. MacGregor-Fors, M. McDonnell, U. Mortberg, P. Pyšek, S. Siebert, J. Sushinsky, P. Werner, M. Winter. 2014. A global analysis of the impacts of urbanization on bird and plant diversity reveals key anthropogenic drivers. *Proceedings of Royal Society of London B* 281:20133330.

Aronson, M.F.J., C.A. Lepczyk, K.L. Evans, M.A. Goddard, S.B. Lerman, J.S. MacIvor, C.H. Nilon, T. Vargo. 2017. Biodiversity in the city: Research priorities and knowledge gaps for urban green space planning and management. *Frontiers in Ecology and the Environment* 15(4):189–196.

Birkhead, T.R., I. Charmantier. 2009. History of ornithology. In *Encyclopedia of Life Sciences (ELS)*. John Wiley & Sons, Chichester, UK.

Bonney, R., T.B. Phillips, H.L. Ballard, J.W. Enck. 2016. Can citizen science enhance public understanding of science? *Public Understanding of Science* 25(1):2–16.

Callaghan, C.T., W.K. Cornwell, A.G.B. Poore, Y. Benedetti, F. Morelli. 2021. Urban tolerance of birds changes throughout the full annual cycle. *Journal of Biogeography* 48(6):1503–1517.

Callaghan, C.T., I. Ozeroff, C. Hitchcock, M. Chandler. 2020. Capitalizing on opportunistic citizen science data to monitor urban biodiversity: A multi-taxa framework. *Biological Conservation* 251:108753.

Cartwright, L.A., M. Cvetkovic, S. Graham, D. Tozer, P. Chow-Fraser. 2015. URBAN: Development of a citizen science biomonitoring program based in Hamilton, Ontario, Canada. *International Journal of Science Education, Part B* 5(2):93–113.

Cendejas, J.C., L.M. Ramirez, J.R. Zierold, J.D. Valenzuela, M.M. Ibarra, S.M.S. de Tagle, A.C. Tellez. 2021. Evaluation of the impacts of land use in water quality and the role of nature-based solutions: A citizen science-based study. *Sustainability* 13:10519.

Costello, M.J., R.M. May, N.E. Stork. 2013. Can we name Earth's species before they go extinct? *Science* 339(6118):413–416.

Encarnação, J., V. Baptista, M.A. Teodósio, P. Morais. 2021. Low-cost citizen science effectively monitors the rapid expansion of a marine invasive species. *Frontiers in Environmental Science* 9:752705.

Fournier, A.M.V., A.L. Bond. 2015. Volunteer field technicians are bad for wildlife ecology. *Wildlife Society Bulletin* 39(4):819–821.

Freeman, R.B. 1997. Working for nothing: The supply of volunteer labor. *Journal of Labor Economics* 15(1):Part 2:S140–S166.

Frensley, T., A. Crall, M. Stern, R. Jordan, S. Gray, M. Prysby, J. Huang. 2017. Bridging the benefits of online and community supported citizen science: A case study on motivation and retention with conservation-oriented volunteers. *Citizen Science: Theory and Practice* 2(1):4.

Goddard, M.A., A.J. Dougill, T.G. Benton. 2010. Scaling up from gardens: Biodiversity conservation in urban environments. *Trends Ecology and Evolution* 25(2):90–98.

Hartop, E.A., B.V. Brown, R.H.L. Disney. 2015. Opportunity in our ignorance: Urban biodiversity study reveals 30 new species and one new Nearctic record for *Megaselia* (Diptera: Phoridae) in Los Angeles (California, USA). *Zootaxa* 3941(4):451–484.

Ives, C.D., P.E. Lentini, C.G. Threlfall, K. Ikin, D.F. Shanahan, G.E. Garrard, S.A. Bekessy, R.A. Fuller, L. Mumaw, L. Rayner, R. Rowe, L.E. Valentine, D. Kendal. 2016. The importance of cities for threatened species. *Global Ecology and Biogeography* 25(1):117–126.

Jordan, R., J.G. Ehrenfeld, S.A. Gray, W.R. Brooks, D.V. How, C.E. Hmelo-Sliver. 2012. Cognitive considerations in the development of citizen science projects. Pages 167–178 in J.L. Dickinson, R. Bonney, editors. *Citizen Science: Public Participation in Environmental Research.* Cornell University Press, Ithaca, NY.

La Sorte, F.A., C.A. Lepczyk, M.F.J. Aronson, M.A. Goddard, M. Katti, I. MacGregor-Fors, U. Mörtberg, C.H. Nilon, P.S. Warren, N.S.G. Williams, J. Yang. 2018. The phylogenetic and functional diversity of regional breeding bird assemblages is reduced and constricted through urbanization. *Diversity and Distributions* 24(7):928–938.

Leong, M., M. Trautwein. 2019. A citizen science approach to evaluating US cities for biotic homogenization. *PeerJ* 7:e6879.

Lepczyk, C.A. 2005. Integrating published data and citizen science to describe bird diversity across a landscape. *Journal of Applied Ecology* 42(4):672–677.

Lepczyk, C.A., M.F.J. Aronson, K.L. Evans, M.A. Goddard, S.B. Lerman, J.S. MacIvor. 2017. Biodiversity in the city: Fundamental questions for understanding the ecology of urban green spaces for biodiversity conservation. *BioScience* 67(9):799–807.

Lepczyk, C.A., M.F.J. Aronson, F.A. La Sorte. 2023. Cities as sanctuaries. *Frontiers in Ecology and the Environment.* 21: 251–259.

Lepczyk, C.A., O. Boyle, T.L.V. Vargo, editors. 2020. *Handbook of Citizen Science in Ecology and Conservation.* University of California Press, Berkeley, CA.

Lepczyk, C.A., A.G. Mertig, J. Liu. 2004. Assessing landowner activities that influence birds across rural-to-urban landscapes. *Environmental Management* 33(1):110–125.

Lewanzik, D., T.M. Straka, J. Lorenz, L. Marggraf, S. Voigt-Heucke, A. Schumann, M. Brandt, C.C. Voigt. 2022. Evaluating the potential of urban areas for bat conservation with citizen science data. *Environmental Pollution* 15(297):118785.

Mayfield, H.F. 1979. The amateur in ornithology. *The Auk* 96:168–171.

McClure, E.C., M. Sievers, C.J. Brown, C.A. Buelow, E.M. Ditria, M.A. Hayes, R.M. Pearson, V.J.D. Tulloch, R.K.F. Unsworth, R.M. Connolly. 2020. Artificial intelligence meets citizen science to supercharge ecological monitoring. *Patterns* 1(7):100109.

McGreavy, B., A.J.K. Calhoun, J. Jansujwicz, V. Levesque. 2016. Citizen science and natural resource governance: Program design for vernal pool policy innovation. *Ecology and Society* 21(2):48.

Merenlender, A., A.W. Crall, S. Drill, M. Prysby, H.L. Ballard. 2016. Evaluating environmental education, citizen science, and stewardship through naturalist programs. *Conservation Biology* 30(6):1255–1265.

Narango, D.L., D.W. Tallamy, P.P. Marra. 2018. Nonnative plants reduce population growth of an insectivorous bird. *Proceedings of the National Academy of Sciences* 115(45):11549–11554.

Paleco, C., S. García Peter, N. Salas Seoane, J. Kaufmann, P. Argyri. 2021. Inclusiveness and diversity in citizen science. Pages 261–81 in K. Vohland, A. Land-Zandstra, L. Ceccaroni, R. Lemmens, J. Perelló, M. Ponti, R. Samson, K. Wagenknecht, editors. *The Science of Citizen Science*. Springer, Cham.

Peer, M., D. Dörler, J.G. Zaller, H. Scheifinger, S. Schweiger, G. Laaha, G. Neuwirth, T. Hübner, F. Heigl. 2021. Predicting spring migration of two European amphibian species with plant phenology using citizen science data. *Scientific Reports* 11:21611.

Pendl, M., R.I. Hussain, D. Moser, T. Frank, T. Drapela. 2022. Influences of landscape structure on butterfly diversity in urban private gardens using a citizen science approach. *Urban Ecosystems* 25:477–486.

Pocock, M.J.O., D.S. Chapman, L.J. Sheppard, H.E. Roy. 2014. *Choosing and Using Citizen Science: A Guide to When and How to Use Citizen Science to Monitor Biodiversity and the Environment.* Centre for Ecology & Hydrology, Wallingford, Oxfordshire, UK.

Prysby, M.D. 2020. Recruitment and building the team. Pages 73–86 in C.A. Lepczyk, O.D. Boyle, T.L.V. Vargo, editors. *Handbook of Citizen Science in Conservation and Ecology*. University of California Press, Oakland, CA.

Rega-Brodsky, C.C., M.F.J. Aronson, M.R. Piana, E.-S. Carpenter, A.K. Hahs, A. Herrera-Montes, S. Knapp, D.J. Kotze, C.A. Lepczyk, M. Moretti, A.B. Salisbury, N.S.G. Williams, K. Jung, M. Katti, I. MacGregor-Fors, J.S. MacIvor, F.A. La Sorte, V. Sheel, C.G. Threlfall, C.H. Nilon. 2022. Urban biodiversity: State of the science and future directions. *Urban Ecosystems* 25:1083–1096.

Sasse, D.B. 2003. Job-related mortality of wildlife workers in the United States, 1937–2000. *Wildlife Society Bulletin* 31(4):1000–1003.

Schwartz, M.D., J.L. Betancourt, J.F. Weltzin. 2012. From Caprio's lilacs to the USA National Phenology Network. *Frontiers in Ecology and the Environment* 10(6):324–327.

Shirk, J.L., H.L. Ballard, C.C. Wilderman, T. Phillips, A. Wiggins, R. Jordan, E. McCallie, M. Minarchek, B. Lewenstein, M. Krasny, R. Bonney. 2012. Public participation in scientific research: A framework for deliberate design. *Ecology and Society* 17(2):29.

Shirk, J.L., R. Bonney. 2020. What is citizen science? Pages 7–15 in C.A. Lepczyk, O.D. Boyle, T.L.V. Vargo, editors. *Handbook of Citizen Science in Ecology and Conservation.* University of California Press, Berkeley, CA.

Soanes, K., P.E. Lentini. 2019. When cities are the last chance for saving species. *Frontiers in Ecology and the Environment* 17(4):225–231.

Theobald, E.J., A.K. Ettinger, H.K. Burgess, L.B. DeBey, N.R. Schmidt, H.E. Froehlich, C. Wagner, J. HilleRisLambers, J. Tewksbury, M.A. Harsch, J.K. Parrish. 2015. Global change and local solutions: Tapping the unrealized potential of citizen science for biodiversity research. *Biological Conservation* 181:236–244.

Troudet, J., P. Grandcolas, A. Bliz, R. Vignes-Lebbe, F. Legendre. 2017. Taxonomic bias in biodiversity data and societal preferences. *Scientific Reports* 7(1):9132.

Trumbull, D.J., R. Bonney, D. Bascom, A. Cabral. 2000. Thinking scientifically during participation in a citizen-science project. *Science Education* 84(2):265–275.

Tweddle, J.C., H.E. Roy, L.D. Robinson, M.J.O. Pocock. 2020. Project planning and design. Pages 33–50 in C.A. Lepczyk, O.D. Boyle, T.L.V. Vargo, editors. *Handbook of Citizen Science in Conservation and Ecology*. University of California Press, Oakland, CA.

Vereecken, N.J., T. Weekers, L. Marshall, J. D'Haeseleer, A. Cuypers, A. Pauly, B. Pasau, N. Leclercq, A. Tshibungu, J.-M. Molenberg, S. De Greef. 2021. Five years of citizen science and standardised field surveys in an informal urban green space reveal a threatened Eden for wild bees in Brussels, Belgium. *Insect Conservation and Diversity* 14:868–876.

Willi, M., R.T. Pitman, A.W. Cardoso, C. Locke, A. Swanson, A. Boyer, M. Veldthuis, L. Fortson. 2019. Identifying animal species in camera trap images using deep learning and citizen science. *Methods in Ecology and Evolution* 10(1):80–91.

PART 3

Population and community ecology of key urban taxa

15

PATTERNS AND DRIVERS OF PLANT BIODIVERSITY IN URBAN AND URBANIZING LANDSCAPES

Sonja Knapp and Wayne Zipperer

Introduction

Global biodiversity is in decline (IPBES 2019), while urbanization is on the rise (Chen et al. 2020). While safeguarding biodiversity will not solely be decided in urban areas, their relevance for biodiversity conservation increases (Seto et al. 2012). To plan, design, and manage urban areas in ways that benefit both humans and biodiversity, it is necessary to understand drivers and patterns of urban biodiversity and underlying mechanisms driving species populations and community dynamics (McDonnell and Hahs 2013; Lepczyk et al. 2017; Kowarik and von der Lippe 2018; Piana et al. 2019).

Plant species as primary producers are key elements of ecosystems. High plant diversity supports the diversity of many other taxa (Bräuniger et al. 2010). Understanding how urban environments promote plant diversity will thus benefit other taxa as well. Moreover, plant diversity is key to the delivery of many ecosystem services (Schwarz et al. 2017). Maintaining and promoting urban plant diversity should thus be a core interest of urban stakeholders.

This chapter overviews patterns of plant biodiversity across urban and urbanizing landscapes and compares these patterns to those in non-urban areas. Specifically, we will examine four attributes of diversity – taxonomic, phylogenetic, genetic, and functional (Table 15.1) – and their responses to urban environments. We will also examine the social-ecological drivers influencing the composition, abundance, and distribution of plant biodiversity. Finally, we will make recommendations on the conservation of plant biodiversity in urban areas.

The chapter focuses principally on vascular plants, but lichen and mosses are also considered. We classify vascular plants into those occurring spontaneously (wildly) and those cultivated by humans. Both groups comprise native and non-native species. These differentiations are necessary because spontaneous and cultivated as well as native and non-native species respond differently to urban influences.

Patterns and drivers of spontaneous vascular plant diversity in and across urban areas

Taxonomic diversity

Urban areas often have a greater species richness (total number of species) than adjacent non-urban regions (e.g. Pyšek 1993; Hope et al. 2003). This richness results from a number of factors,

 DOI: 10.4324/9781003016120-18

Table 15.1 Measures of biodiversity with definitions

Measure of biodiversity	*Definition*
Taxonomic diversity	Number, abundance or evenness of species, genera, or other taxonomic units
Phylogenetic diversity	Wealth of evolutionary information inherent in a species assemblage. While the term "phylogenetic diversity" has been used by Faith (1992) for measuring the sum of branch lengths in an evolutionary/phylogenetic tree, it is often used more broadly for describing phylogenetic richness (sum of accumulated phylogenetic differences among taxa), divergence (mean phylogenetic relatedness among taxa), and regularity (variance in differences among taxa) – cf. Tucker et al. (2017).
Genetic diversity	Measure of within-species diversity quantifying the diversity of alleles present within a species population
Functional diversity	Number, range, or abundance (or combinations of these) of functional traits represented within an assemblage. Different measures exist covering aspects of richness, abundance, and divergence (Villéger et al. 2008).

including urban areas being located in biological hotspots, high degree of spatiotemporal heterogeneity, introduction of non-native species, and age and size of the urban area.

Urban areas often occur in biological hotspots (Seto et al. 2012). These hotspots correspond to regions of high productivity such as deltas and estuaries, high substrate diversity, the confluence of ecotones, and moderate climates. By settling in these regions, humans capitalized on the natural resources, such as timber, game, and fisheries, productive soils for agriculture, and the confluence of rivers for transportation. Consequently, today's urban areas are located in regions of high plant species richness (Kühn et al. 2004).

Urban areas have a high degree of spatial heterogeneity when compared to agricultural landscapes (Sukopp 1998). They consist of a range of land use/cover configurations of which each has a unique social-ecological environment (Cadenasso et al. 2007) which influences not only the abundance and distribution of species but also ecosystem functions. Sites with vegetation (Kowarik 1992) include remnants of natural landscapes (e.g. alluvial forest), patches of agrarian landscapes (e.g. hay meadows), designed urban green spaces (e.g. gardens, parks), and novel urban ecosystems (e.g. vacant lots, railways, road verges).

Non-native species contribute significantly to urban biodiversity (Pyšek 1998). Most of these have been and continue to be introduced with trade and transport (Seebens et al. 2015), either as ornamental or crop plants or unintentional introductions (Reichard and White 2001; Lambdon et al. 2008). Consequently, non-native species richness increases with urbanization (Tait et al. 2005). In addition, cities host many native generalist species as well as a number of rare and threatened species (Ives et al. 2016). Some of the latter, however, only exist in today's urban areas due to extinction debts (Hahs et al. 2009), and on the long term, they might not be able to form viable populations (Kowarik and von der Lippe 2018). The increase in widespread non-native species and the loss of rare species can increase the similarity of urban floras across the globe (McKinney 2006; Winter et al. 2009), but the compositional similarity of urban forests among the world's cities is scale dependent (Yang et al. 2015).

Species richness also increases with age and size of urban areas (Klotz 1990; Aronson et al. 2014). Regardless of size and age, native species richness declines, while non-native species richness increases as one moves from the rural fringe to the urban core (Lososová et al. 2012; Aronson et al. 2015).

Phylogenetic diversity

Studies focusing on species richness provide little insight into how species are related to each other. Similarly, species richness may not be the best indicator how urban environments influence biodiversity. Phylogenetic diversity is a measure of evolutionary relationships and is based on the phylogenetic similarities and differences among species (Table 15.1). In urban areas, species are more closely related to each other than species in the countryside (Knapp et al. 2008a; Ricotta et al. 2009). Across land use/cover types within urban areas, plant communities are phylogenetically clustered (i.e. less diverse than expected by chance; Čeplová et al. 2015). Possible underlying reasons are the high share of non-native species that are closely related to each other and to native species and restricted growing conditions for species mal-adapted to urban environments (Knapp et al. 2008a; Čeplová et al. 2015; Knapp et al. 2017).

Genetic diversity

Broad-scale analyses of genetic diversity across urban and urbanizing landscapes are limited, and there appears to be no general pattern emerging of genetic responses to the urban environment. Nonetheless, the urban landscape can be hostile to gene flow because the habitats remaining within the built-up matrix are highly isolated, which can promote inbreeding and genetic drift, thus decreasing genetic diversity and reproductive success within populations (Van Rossum 2008; Bartlewicz et al. 2015). Yakub and Tiffin (2017) also observed that the individuals of an urban plant population can be more closely related to individuals from other urban centers than to individuals in adjacent non-urban areas. It seems that long-distance transportation of species (which especially applies to non-native species) connects plant populations across cities, thus potentially increasing global genetic homogenization.

The urban landscape can also promote genetic diversity. Scholz (2007) describes a group of species, anecophytes, which are only found in urban landscapes, thus, potentially enhancing genetic diversity. Similarly, Johnson et al. (2018) report that genetic diversity within *Trifolium repens* increases with urbanization. The conservation of genetic diversity within urban landscapes can be achieved by enhancing genetic flow through connecting habitat (Van Rossum 2009) and improving opportunities for pollinators (Culley et al. 2007; Krishnan and Borges 2018; but see Diniz et al. 2019).

Functional diversity

Functional plant traits are anatomical, biochemical, morphological, phenological, physiological, and structural characteristics of species reflecting adaptation to environmental conditions (Kattge et al. 2011). Changes in environmental conditions can cause shifts in the trait composition of species assemblages. With urbanization increasing local temperatures, air pollutants, and nutrient deposition, and lowering soil moisture, for example (Sukopp 1998), urban and non-urban plant assemblages often differ in functional trait composition. In addition to environmental conditions, an urban area's size, history, and geographical location as well as habitat types can influence functional trait composition. Consequently, the direction and strength of trait shifts are not the same globally. Nonetheless, some general patterns are observed (Williams et al. 2015; Palma et al. 2017) including increases in

- short-lived species that reproduce by seeds and colonize disturbed sites;
- woody species, fostered by plantings of shrubs and trees in parks, gardens, or along streets, that escape cultivation and become part of spontaneous vegetation;

- tall growing species and heavier-seeded plants – related but not restricted to the presence of woody species.

By comparison, geophytes (e.g. bulbs, rhizomes, corms, tubers) as well as wind-dispersed species decrease in occurrence in urban as compared to non-urban areas. The latter might relate to land cover. For instance, Cheptou et al. (2008) observed that *Crepis sancta*, which grows in small patches often surrounded by impervious surfaces, had a higher degree of germination success for non-dispersed seeds as compared to dispersed seeds. When seeds dropped from a plant (non-dispersed), they often fall on soil and are able to germinate. In comparison, when seeds are dispersed by wind, they often fall on impervious surfaces and are unable to germinate (Cheptou et al. 2008). In contrast, Kowarik (2011) observed that forested vacant lots were dominated by wind-dispersed species. Seeds dispersed by animals and/or humans (such as those wrapped in fleshy fruits) were shown to benefit plant occurrence in urban areas (Knapp et al. 2008b) and to increase over time (Aronson et al. 2007). These contrasting examples illustrate the complexity of shifting functional traits in urban landscapes, with differences among habitats and across time (e.g. succession on vacant lots) and the drivers influencing that complexity (Aronson et al. 2016).

Urban-induced trait-shifts exist across genera (interspecific) and within species (intraspecific). The latter point towards urbanization being a driver of species evolution (Donihue and Lambert 2014). Examples of intraspecific trait shifts with increasing urban land use/cover comprise shifts from long- to short-distance dispersal (Cheptou et al. 2008), reduced production of antiherbivore defense chemicals (Johnson et al. 2018), or changes in leaf physiology (Alberti 2015).

Measures of functional diversity apart from functional composition have rarely been investigated in urban areas. Lososová et al. (2016), in a study of 32 European cities, showed that across a range of habitat types, plant communities were functionally convergent (i.e. less functional variation than expected for random communities). This low functional diversity is in contrast to high species richness found in the very same communities. It seems that urban habitats allow for a high number of species but restrict the range of functional traits that can occur. Strong environmental filtering and/or competitive exclusion among species might be reasons for this (Lososová et al. 2016).

Patterns and drivers of cultivated vascular plant diversity in designed urban green spaces

In the previous section, we examined diversity attributes for spontaneous vegetation that occurs across all four types of urban nature (Kowarik 1992). Another element of the green infrastructure in urban landscapes is the pool of cultivated species, which is primarily restricted to designed green space. Designed urban green spaces include but are not limited to household and community gardens, stormwater management sites, parks, and street plantings. They are significant land cover types in the world's cities. For instance, domestic gardens alone make up 16% of the total area of Stockholm, Sweden (Colding et al. 2006); 22% to 27% of UK cities (Loram et al. 2007); and 36% of Dunedin, New Zealand (Mathieu et al. 2007). Cultivated species are usually highly managed for aesthetics, leisure, food, or medicine. Social decisions are a much stronger influence on the selection and establishment of cultivated vegetation than on spontaneous vegetation (Swan et al. 2021).

Taxonomic diversity

Designed green spaces are biologically diverse, including both cultivated and spontaneous vegetation. Numbers of cultivated plant species can exceed numbers in natural areas (Norfolk et al. 2013;

Pearse et al. 2018). This reflects the diversity of people's preferences (Kendal et al. 2012a) and, related to that, exceptionally high proportions of non-native plant species available to select for plantings. Unfortunately, ornamental plantings are one of the most important pathways of biological invasions (Reichard and White 2001). In Europe, approximately 60% (Pyšek et al. 2009) and in Australia, 67% (Moss and Walmsley 2005) of all non-native vascular plants are escaped ornamental or horticultural species. Moreover, designed urban green spaces plantings introduce cultivars and hybrids of both native and non-native species (Galluzzi et al. 2010), which can significantly affect ecosystem structure and function (Whelan et al. 2006).

Overall, cultivated plant diversity depends on current and past availability in nurseries (Kendal et al. 2012b; Avolio et al. 2018), homeowner preferences (Cavender-Bares et al. 2020), social factors such as income (Hope et al. 2003; Leong et al. 2018), and prevailing policies (as in the case of public greenery; Kendal et al. 2012b). These relationships exemplify the strong impact of human decisions on cultivated plant diversity. In addition, management plays a significant role in the success of cultivated plants through weeding, irrigating, and applying fertilizers and biocides (Niinemets and Peñuelas 2008). These actions often lessen the effects of the urban environment (Kendal et al. 2012b) and physical attributes of the garden (Cavender-Bares et al. 2020).

Phylogenetic diversity

Less is known about the phylogenetic diversity of cultivated plant species. Studies from domestic gardens in the US indicate that homeowner preferences and economic status affect phylogenetic biodiversity. For instance, in Minnesotan gardens, phylogenetic diversity decreased with both the frequency of fertilizer application and preference for ease of maintenance but increased with preference for orderliness (Cavender-Bares et al. 2020). By comparison, Pearse et al. (2018) observed that cultivated plantings in domestic gardens across seven US cities had higher phylogenetic diversity than natural areas, reflecting the wide range of species available in the horticultural trade.

Genetic diversity

Similarly, genetic diversity across designed urban green spaces can be quite high owing to the availability of cultivars, hybrids, non-native, and native species for planting (Galluzzi et al. 2010). On the other hand, the selection of specific species, cultivars, or clones, such as for street tree plantings, can significantly decrease genetic diversity (Makeeva et al. 2018) and increase the susceptibility to insects and diseases over time (Vanden Broeck et al. 2018).

Functional diversity

Studies on the functional richness of cultivated plants in designed urban green spaces are scarce but case studies indicate that it can exceed that of natural vegetation (Norfolk et al. 2013). The functional composition of cultivated species in designed urban green spaces highly depends on human preferences (Kendal et al. 2012a). For example, aesthetic qualities are often preferred and related to plant growth form, leaf traits (e.g. foliage color, leaf width), flowering traits (e.g. flower size), and tree size (Kendal et al. 2012a; Pataki et al. 2013; Goodness et al. 2016). Similarly, species are selected for ease of maintenance, water use, and providing shade or fruits (Kendal et al. 2012a; Pataki et al. 2013). Accordingly, functional traits can be purposefully used to select those species that are adapted to local environmental conditions and that promote the provision of ecosystem services and the increase in urban functional diversity (Pataki et al. 2013; Lundholm et al. 2015; Filazzola et al. 2019).

Lichen and mosses

Taxonomic diversity

As with vascular plants, the substrate heterogeneity of urban areas promotes the diversity of lichens and mosses. For example, gardens, cemeteries, and tree-dominated habitats were shown to support high species numbers of lichens and mosses (e.g. Smith et al. 2010; Sérgio et al. 2016; Vogt-Schilb et al. 2018; Oishi 2019). Lichens can establish on different anthropogenic substrates such as brick, mortar, or tarmac (Seaward 1982; Gombert et al. 2004). Lichens and mosses can even grow on roofs because of their ability to survive extreme drought (Van Mechelen et al. 2015; Schröder and Kiehl 2020). Still, moss cover, diversity, and abundance are higher in sheltered than sun-exposed areas of roofs (Van Mechelen et al. 2015), exemplifying habitat preferences. Consequently, the composition of moss assemblages differs across urban green spaces, depending on microclimatic conditions (Zepeda-Gómez et al. 2014). Also, lichen diversity in urban woodlands increases with woodland age (Vogt-Schilb et al. 2018), indicating preferences for specific substrates and micro-climatic conditions.

Air pollutant concentrations – especially SO_2, being toxic to lichens – are key to lichen occurrence within urban areas. In fact "lichen deserts" exist in many cities around the world. Nonetheless, lichens can recolonize areas after SO_2 concentrations decreased significantly because of clean air policies (Seaward 1982; Ranta 2001; Gombert et al. 2004; Sérgio et al. 2016). Still, other pollutants, such as traffic-related NO_2 and dust deposition, as well as extremes of temperatures and water supply can reduce species occurrence, abundance, and distribution (Seaward 1982; Sérgio et al. 2016). Consequently, lichen cover and diversity are often low in high-traffic areas (Coffey and Fahrig 2012; Llop et al. 2012). A study in Ottawa, Canada, however, reported that lichen species richness was more related to moisture and colonization sources and less to vehicle pollution (Coffey and Fahrig 2012). Similarly, in Almada, Portugal, climate, and not air pollution, was the principal driver of lichen diversity (Munzi et al. 2014).

In Germany, Knapp et al. (2008a) reported lower lichen richness in urban than non-urban protected areas but no difference in moss richness, whereas Concepción et al. (2016) found no response of moss richness to urban sprawl across Switzerland. These studies indicate that additional research is needed to quantify lichen and moss occurrence, abundance, and distributions across urban landscapes.

Other measures of diversity

Phylogenetic and genetic diversity of lichens and mosses have not been investigated in response to urban land use. Few studies on functional composition exist. For lichens, terricolous species (i.e. those growing on bare soil) tend to be less common in urban environments than saxicolous (on stone) or epiphytic species (on bark) because appropriate open soil is scarce (Seaward 1982). Oligotrophic, hygrophytic, and acidophilous lichen are sensitive to traffic-induced pollution, while eutrophic, xerophytic, basophilous, and nitrophilous lichen are more common in traffic-dominated areas (Llop et al. 2012; Sérgio et al. 2016). Overall, lichen species that reach high abundances in urban areas are often generalists (Seaward 1982), resembling a pattern that is common across many taxonomic groups (Concepción et al. 2015; Deguines et al. 2016).

For moss species, those common in urban areas are often species associated with disturbance (Landis and Leopold 2014). In comparison to non-urban habitats more dendroid (i.e. resembling the form of a tree) and thallose (without leaf-like structures) but less plagiotropic (growing more or less horizontally) moss species were found in urban sites of the Bolzano region, Italy (Spitale et al. 2020).

Mechanisms driving urban plant biodiversity

In summary, both anthropogenic and environmental drivers affect urban plant biodiversity. These can be related to (1) regional climate, biogeography, and land use; (2) human mediated biotic interchange; (3) urban form and development history; (4) socioeconomic and cultural influences; (5) local human facilitation; and (6) species interactions (Table 15.2; Aronson et al. 2016). Still, the complexity of the interactions between drivers and patterns of urban biodiversity and underlying mechanisms driving plant populations and communities dynamics remain largely unanswered (McDonnell and Hahs 2013) but are the focus of emerging research (Piana et al. 2019). One key question in urban plant biodiversity research is, of which type, size, density, connectivity, and quality (e.g. in terms of management) urban green spaces should be maintained or created to enhance long-term population viability, overall plant diversity, and the diversity of specific groups

Table 15.2 Groups of drivers of urban plant biodiversity according to Aronson et al. (2016) with non-exhaustive list of examples of drivers and mechanisms

Group	*Drivers*	*Mechanisms*
Biogeography, land use, and regional climate	Urban heat island	Urban heat is one reason behind rural-urban shifts in plant phenology, but other drivers likely interact in ways pending to be identified (Zohner 2019). It acts across local and regional scales.
	Soils	In response to urban vs. potting soil, native species mainly showed trait divergence, while non-native species mainly showed preadaptation. Strategies for living in urban areas thus comprise phenotypic plasticity and/or genetic-based differences for native but not non-native plants (Borowy and Swan 2020).
Human mediated biotic interchange	Traffic	Long-distance dispersal by vehicles is more common in non-native than native plant species, promoting plant invasions (von der Lippe and Kowarik 2007).
	(Horticultural) trade	Higher market frequency and cheap prices of seeds in horticultural trade promote invasion success (Dehnen-Schmutz et al. 2007).
Urban form and development history	Size of urban area	The larger a city, the more species it hosts (Klotz 1990), pointing towards local extinction and colonization mechanisms (cf. MacArthur and Wilson 1967) and the importance of habitat heterogeneity, which correlates to size (Knapp et al. 2008a).
Socioeconomic and cultural influences	Socioeconomic status	A series of interacting mechanisms relates affluence to plant biodiversity. Particularly in arid regions, irrigation promotes plant diversity, with affluent households able to afford irrigation (Leong et al. 2018).
Local human facilitation	Management and human selection (e.g. irrigation, weeding)	See the previous example (Leong et al. 2018) illustrating that drivers of different groups can interact in affecting urban plant diversity.
Species interactions	Plant-pollinator interactions	Insect visitation rates and Hymenoptera community diversity drive higher urban than non-urban seed set in *Trifolium pratense* and in turn are affected by patch edge density and proportion of arable land (Theodorou et al. 2020).

Table 15.3 Major approaches in the study of urban plant biodiversity

Major approach	*Examples*
Monitoring and observation	• "Urban biotope mapping" in Germany that provides basic knowledge for planning and policy (Sukopp and Weiler 1988). • Citizen/community science programs enable the collection of monitoring data across urban areas (e.g. BioBlitz – see www.nationalgeographic.org/projects/bioblitz/).
Measurements	• Lundholm et al. (2015) measured traits of green roof plant species and linked them to ecosystem services.
Macroecological analysis	• In a global analysis, Aronson et al. (2014) revealed that most plant species occurring in urban areas are native species, few are cosmopolitan, and the number of plant species per km^2 declined with urbanization but increased with city age.
Review and meta-analysis	• Williams et al. (2015) by analyzing publications across the globe identified both broadly valid plant trait patterns in response to urban environmental conditions and inconsistencies pointing towards context-dependency of trait patterns.
Experimental approaches	• Borowy and Swan (2020) used soil treatment with urban soil vs. potting soil in order to test for species' preadaptation to urban conditions. • In a social-ecological experiment, Shwartz et al. (2014) investigated urban dwellers' perception of plant diversity.

such as endangered species (Lepczyk et al. 2017; Kowarik and von der Lippe 2018)? Answering this question can inform urban biodiversity conservation and thus aid in protecting global biodiversity. A range of approaches exist for analyzing and understanding urban plant biodiversity. Some of which are summarized in Table 15.3.

Conservation and management

All four types of urban nature (Kowarik 1992) – remnants of natural landscapes, patches of agrarian landscapes, designed urban green spaces, and novel urban ecosystems – with their variety of habitats together can be home to a diversity of plant species. Consequently, all of them need to be included into measures of biodiversity conservation.

Being the core instrument of biodiversity conservation, protected areas do exist in many urban areas, with natural remnants, local natural heritage, rare species, and ecosystem functions key to human well-being (such as oxygen production) being their main target of conservation. For example, in the United States, natural remnants occupy 84% of municipal parkland (Harnik et al. 2017). Maintaining protected areas and extending their number and size should be one core strategy of urban biodiversity conservation.

Still, the amount of urban land that can be protected is restricted. Thus, in addition to legal protection, other parts of urban nature should be used in a sustainable, multifunctional way that benefits both human health and well-being and biodiversity (an approach termed "urban land sharing"; Lin and Fuller 2013). For example, designed urban parks will host higher levels of plant diversity if hedges, groups of trees, small forests, water bodies, and extensively mown meadows complement short-cut lawns. At the same time, such multifunctional green spaces will provide opportunity for a range of leisure activities as well as more ecosystem services (Lundholm 2015) such as temperature regulation and air purification.

Even novel urban ecosystems, which often host large numbers of non-native species, can be of value for biodiversity conservation. They can be found along unused railway tracks, brownfields, and other typical urban-industrial habitats. An example are emerging forests that gain rare and endangered species the older they grow, thus increasingly benefitting biodiversity (Kowarik et al. 2019).

In summary, across urban-rural landscapes, a dense network of larger and smaller green spaces of all types of urban nature will help plant species to form viable populations (Kowarik et al. 2019). To complement such networks, it is even possible to re-establish habitats of conservation interest within urban settings, such as shown by Fischer et al. (2013), who created novel urban grasslands by reintroducing native plant species into urban wasteland vegetation. The protection of plant species within urban and urbanizing regions will even support the protection of other taxa (Bräuniger et al. 2010). With urban areas growing worldwide, identifying how urban landscapes and their collective urban nature are planned, designed, and managed in order to protect global biodiversity is a research priority.

Key Companion Papers

Aronson, M.F.J., et al. 2016. Hierarchical filters determine community assembly of urban species pools. *Ecology* 97(11):2952–2963.

Hahs, A.K., et al. 2009. A global synthesis of plant extinction rates in urban areas. *Ecology Letters* 12(11):1165–1173.

Kowarik, I. 2011. Novel urban ecosystems, biodiversity, and conservation. *Environmental Pollution* 159(8–9):1974–1983.

Lososová, Z., et al. 2012. Native and alien floras in urban habitats: A comparison across 32 cities of central Europe. *Global Ecology and Biogeography* 21(5):545–555.

Ricotta, C., et al. 2009. Phyloecology of urban alien floras. *Journal of Ecology* 97(6):1243–1251.

Williams, N.S.G., A.K. Hahs, P.A. Vesk. 2015. Urbanisation, plant traits and the composition of urban floras. *Perspectives in Plant Ecology Evolution and* Systematics 17(1):78–86.

References

Alberti, M. 2015. Eco-evolutionary dynamics in an urbanizing planet. *Trends in Ecology & Evolution* 30(2):114–126.

Aronson, M.F.J., S.N. Handel, S.E. Clemants. 2007. Fruit type, life form and origin determine the success of woody plant invaders in an urban landscape. *Biological Invasions* 9(4):465–475.

Aronson, M.F.J., S.N. Handel, I.P. La Puma, S.E. Clemants. 2015. Urbanization promotes non-native woody species and diverse plant assemblages in the New York metropolitan region. *Urban Ecosystems* 18:31–45.

Aronson, M.F.J., F.A. La Sorte, C.H. Nilon, M. Katti, M.A. Goddard, C.A. Lepczyk, P.S. Warren, N.S.G. Williams, S. Cilliers, B. Clarkson, C. Dobbs, R. Dolan, M. Hedblom, S. Klotz, J.L. Kooijmans, I. Kühn, I. MacGregor-Fors, M. McDonnell, U. Mortberg, P. Pyšek, S. Siebert, J. Sushinsky, P. Werner, M. Winter. 2014. A global analysis of the impacts of urbanization on bird and plant diversity reveals key anthropogenic drivers. *Proceedings of the Royal Society B-Biological Sciences* 281(1780):20133330.

Aronson, M.F.J., C.H. Nilon, C.A. Lepczyk, T.S. Parker, P.S. Warren, S.S. Cilliers, M.A. Goddard, A.K. Hahs, C. Herzog, M. Katti, F.A. La Sorte, N.S.G. Williams, W. Zipperer. 2016. Hierarchical filters determine community assembly of urban species pools. *Ecology* 97(11):2952–2963.

Avolio, M.L., D.E. Pataki, T.L.E. Trammell, J. Endter-Wada. 2018. Biodiverse cities: The nursery industry, homeowners, and neighborhood differences drive urban tree composition. *Ecological Monographs* 88(2):259–276.

Bartlewicz, J., K. Vandepitte, H. Jacquemyn, O. Honnay. 2015. Population genetic diversity of the clonal self-incompatible herbaceous plant Linaria vulgaris along an urbanization gradient. *Biological Journal of the Linnean Society* 116(3):603–613.

Borowy, D., C.M. Swan. 2020. A multi-trait comparison of an urban plant species pool reveals the importance of intraspecific trait variation and its influence on distinct functional responses to soil quality. *Frontiers in Ecology and Evolution* 8.

Bräuniger, C., S. Knapp, I. Kühn, S. Klotz. 2010. Testing taxonomic and landscape surrogates for biodiversity in an urban setting. *Landscape and Urban Planning* 97(4):283–295.

Cadenasso, M.L., S.T.A. Pickett, K. Schwarz. 2007. Spatial heterogeneity in urban ecosystems: Reconceptualizing land cover and a framework for classification. *Frontiers in Ecology and the Environment* 5(2):80–88.

Cavender-Bares, J., J. Padullés Cubino, W.D. Pearse, S.E. Hobbie, A.J. Lange, S. Knapp, K.C. Nelson. 2020. Horticultural availability and homeowner preferences drive plant diversity and composition in urban yards. *Ecological Applications* 30(4):e02082.

Čeplová, N., Z. Lososová, D. Zelený, M. Chytrý, J. Danihelka, K. Fajmon, D. Láníková, Z. Preislerová, V. Řehořek, L. Tichý. 2015. Phylogenetic diversity of central-European urban plant communities: Effects of alien species and habitat types. *Preslia* 87:1–16.

Chen, G., X. Li, X. Liu, Y. Chen, X. Liang, J. Leng, X. Xu, W. Liao, Y. Qiu, Q. Wu, K. Huang. 2020. Global projections of future urban land expansion under shared socioeconomic pathways. *Nature Communications* 11:537.

Cheptou, P.O., O. Carrue, S. Rouifed, A. Cantarel. 2008. Rapid evolution of seed dispersal in an urban environment in the weed *Crepis sancta*. *Proceedings of the National Academy of Sciences* 105:3796–3799.

Coffey, H.M.P., L. Fahrig. 2012. Relative effects of vehicle pollution, moisture and colonization sources on urban lichens. *Journal of Applied Ecology* 49(6):1467–1474.

Colding, J., J. Lundberg, C. Folke. 2006. Incorporating green-area user groups in urban ecosystem management. *Ambio* 35(5):237–244.

Concepción, E.D., M. Moretti, F. Altermatt, M.P. Nobis, M.K. Obrist. 2015. Impacts of urbanisation on biodiversity: The role of species mobility, degree of specialisation and spatial scale. *Oikos* 124(12):1571–1582.

Concepción, E.D., M.K. Obrist, M. Moretti, F. Altermatt, B. Baur, M.P. Nobis. 2016. Impacts of urban sprawl on species richness of plants, butterflies, gastropods and birds: Not only built-up area matters. *Urban Ecosystems* 19:225–242.

Culley, T.M., S.J. Sbita, A. Wick. 2007. Population genetic effects of urban habitat fragmentation in the perennial herb *Viola pubescens* (Violaceae) using ISSR markers. *Annals of Botany* 100(1):91–100.

Deguines, N., R. Julliard, M. de Flores, C. Fontaine. 2016. Functional homogenization of flower visitor communities with urbanization. *Ecology and Evolution* 6(7):1967–1976.

Dehnen-Schmutz, K., J. Touza, C. Perrings, M. Williamson. 2007. The horticultural trade and ornamental plant invasions in Britain. *Conservation Biology* 21:224–231.

Diniz, U.M., S.A. Lima, I.C.S. Machado. 2019. Short-distance pollen dispersal by bats in an urban setting: Monitoring the movement of a vertebrate pollinator through fluorescent dyes. *Urban Ecosystems* 22:281–291.

Donihue, C.M., M.R. Lambert. 2014. Adaptive evolution in urban ecosystems. *Ambio* 44:194–203.

Faith, D.P. 1992. Conservation evaluation and phylogenetic diversity. *Biological Conservation* 61:1–10.

Filazzola, A., N. Shrestha, J.S. MacIvor. 2019. The contribution of constructed green infrastructure to urban biodiversity: A synthesis and meta-analysis. *Journal of Applied Ecology* 56(9):2131–2143.

Fischer, L.K., M. von der Lippe, M.C. Rillig, I. Kowarik. 2013. Creating novel urban grasslands by reintroducing native species in wasteland vegetation. *Biological Conservation* 159:119–126.

Galluzzi, G., P. Eyzaguirre, V. Negri. 2010. Home gardens: Neglected hotspots of agro-biodiversity and cultural diversity. *Biodiversity and Conservation* 19(13):3635–3654.

Gombert, S., J. Asta, M.R.D. Seaward. 2004. Assessment of lichen diversity by index of atmospheric purity (IAP), index of human impact (IHI) and other environmental factors in an urban area (Grenoble, southeast France). *Science of the Total Environment* 324(1–3):183–199.

Goodness, J., E. Andersson, P.M.L. Anderson, T. Elmqvist. 2016. Exploring the links between functional traits and cultural ecosystem services to enhance urban ecosystem management. *Ecological Indicators* 70:597–605.

Hahs, A.K., M.J. McDonnell, M.A. McCarthy, P.A. Vesk, R.T. Corlett, B.A. Norton, S.E. Clemants, R.P. Duncan, K. Thompson, M.W. Schwartz, N.S.G. Williams. 2009. A global synthesis of plant extinction rates in urban areas. *Ecology Letters* 12:1165–1173.

Harnik, P., C. McCabe, A. Hiple. 2017. *City Park Facts*. The Trust for Public Lands. www.tpl.org/sites/default/files/files_upload/CityParkFacts_2017.4_7_17.FIN_.LO_.pdf

Hope, D., C. Gries, W.X. Zhu, W.F. Fagan, C.L. Redman, N.B. Grimm, A.L. Nelson, C. Martin, A. Kinzig. 2003. Socioeconomics drive urban plant diversity. *Proceedings of the National Academy of Sciences of the United States of America* 100(15):8788–8792.

IPBES. 2019. *Global Assessment Report on Biodiversity and Ecosystem Services of the Intergovernmental Science- Policy Platform on Biodiversity and Ecosystem Services*. E.S. Brondizio, J. Settele, S. Díaz, H.T. Ngo, editors. IPBES Secretariat, Bonn, Germany.

Ives, C.D., P.E. Lentini, C.G. Threlfall, K. Ikin, D.F. Shanahan, G.E. Garrard, S.A. Bekessy, R.A. Fuller, L. Mumaw, L. Rayner, R. Rowe, L.E. Valentine, D. Kendal. 2016. Cities are hotspots for threatened species. *Global Ecology and Biogeography* 25(1):117–126.

Johnson, M.T.J., C.M. Prashad, M. Lavoignat, H.S. Saini. 2018. Contrasting the effects of natural selection, genetic drift and gene flow on urban evolution in white clover (*Trifolium repens*). *Proceedings of the Royal Society B: Biological Sciences* 285(1883):20181019.

Kattge, J., S. Díaz, S. Lavorel, I.C. Prentice, P. Leadley, G. Bönisch, E. Garnier, M. Westoby, P.B. Reich, I.J. Wright, I.H.C. Cornelissen, C. Violle, S.P. Harrison, P.M. Van Bodegom, M. Reichstein, B.J. Enquist, N.A. Soudzilovskaia, D.D. Ackerly, M. Anand, O. Atkin, M. Bahn, T.R. Baker, D. Baldocchi, R. Bekker, C.C. Blanco, B. Blonder, W.J. Bond, R. Bradstock, D.E. Bunker, F. Casanoves, J. Cavender-Bares, J.Q. Chambers, F.S. Chapin III, J. Chave, D. Coomes, W.K. Cornwell, J.M. Craine, B.H. Dobrin, l. Duarte, W. Durka, J. Elser, G. Esser, M. Estiarte, W.F. Fagan, J. Fang, F. Fernández-Méndez, A. Fidelis, B. Finegan, O. Flores, H. Ford, D. Frank, G.T. Freschet, N.M. Fyllas, R.V. Gallagher, W.A. Green, A.G. Gutierrez, T. Hickler, S.I. Higgins, J.G. Hodgson, A. Jalili, S. Jansen, C.A. Joly, A.J. Kerkhoff, D. Kirkup, K. Kitajima, M. Kleyer, S. Klotz, J.M.H. Knops, K. Kramer, I. Kühn, H. Kurokawa, D. Laughlin, T.D. Lee, M. Leishman, F. Lens, T. Lenz, S.L. Lewis, J. Lloyd, J. Llusià, F. Louault, S. Ma, M.D. Mahecha, P. Manning, T. Massad, B.E. Medlyn, J. Messier, A.T. Moles, S.C. Müller, K. Nadrowski, S. Naeem, U. Niinemets, S. Nöllert, A. Nüske, R. Ogaya, J. Oleksyn, V.G. Onipchenko, Y. Onoda, J. Ordoñez, G. Overbeck, W.A. Ozinga, S. Patiño, S. Paula, J.G. Pausas, J. Peñuelas, O.L. Phillips, V. Pillar, H. Poorter, L. Poorter, P. Poschlod, A. Prinzing, R. Proulx, A. Rammig, S. Reinsch, B. Reu, L. Sack, B. Salgado-Negret, J. Sardans, S. Shiodera, B. Shipley, A. Siefert, E. Sosinski, J.-F. Soussana, E. Swaine, N. Swenson, K. Thompson, P. Thornton, M. Waldram, E. Weiher, M. White, S. White, S.J. Wright, B. Yguel, S. Zaehle, A.E. Zanne, C. Wirth. 2011. TRY – A global database of plant traits. *Global Change Biology* 17(9):2905–2935.

Kendal, D., K.J.H. Williams, N.S.G. Williams. 2012a. Plant traits link people's plant preferences to the composition of their gardens. *Landscape and Urban Planning* 105(1–2):34–42.

Kendal, D., N.S.G. Williams, K.J.H. Williams. 2012b. A cultivated environment: Exploring the global distribution of plants in gardens, parks and streetscapes. *Urban Ecosystems* 15:637–652.

Klotz, S. 1990. Species/area and species/inhabitants relations in European cities. In H. Sukopp, S. Hejný, I. Kowarik, editors. *Urban Ecology: Plants and Plant Communities in Urban Environment*. SPB Academic Publishing, The Hague, Netherlands.

Knapp, S., I. Kühn, O. Schweiger, S. Klotz. 2008a. Challenging urban species diversity: Contrasting phylogenetic patterns across plant functional groups in Germany. *Ecology Letters* 11:1054–1064.

Knapp, S., I. Kühn, R. Wittig, W.A. Ozinga, P. Poschlod, S. Klotz. 2008b. Urbanization causes shifts in species' trait state frequencies. *Preslia* 80:375–388.

Knapp, S., M. Winter, S. Klotz. 2017. Increasing species richness but decreasing phylogenetic richness and divergence over a 320-year period of urbanization. *Journal of Applied Ecology* 54:1152–1160.

Kowarik, I. 1992. Das Besondere der städtischen Vegetation. *Schriftenreihe des Deutschen Rates für Landespflege* 61:33–47.

Kowarik, I. 2011. Novel urban ecosystems, biodiversity, and conservation. *Environmental Pollution* 159:1974–1983.

Kowarik, I., A. Hiller, G. Planchuelo, B. Seitz, M. von der Lippe, S. Buchholz. 2019. Emerging urban forests: Opportunities for promoting the wild side of the urban green infrastructure. *Sustainability* 11:6318.

Kowarik, I., M. von der Lippe. 2018. Plant population success across urban ecosystems: A framework to inform biodiversity conservation in cities. *Journal of Applied Ecology* 55:2354–2361.

Krishnan, A., R.M. Borges. 2018. A fig tree in a concrete jungle: Fine-scale population genetic structure of the cluster fig *Ficus racemosa* in an urban environment. *Urban Ecosystems* 21:171–181.

Kühn, I., R. Brandl, S. Klotz. 2004. The flora of German cities is naturally species rich. *Evolutionary Ecology Research* 6:749–764.

Lambdon, P.W., P. Pyšek, C. Basnou, M. Hejda, M. Arianoutsou, F. Essl, V. Jarosik, J. Pergl, M. Winter, P. Anastasiu, P. Andriopoulos, I. Bazos, G. Brundu, L. Celesti-Grapow, P. Chassot, P. Delipetrou, M. Josefsson, S. Kark, S. Klotz, Y. Kokkoris, I. Kühn, H. Marchante, I. Perglova, J. Pino, M. Vila, A. Zikos, D. Roy, P.E. Hulme. 2008. Alien flora of Europe: Species diversity, temporal trends, geographical patterns and research needs. *Preslia* 80:101–149.

Landis, C.L., D.J. Leopold. 2014. Natural plant establishment along an urban stream, Onondaga Creek, New York. *Northeastern Naturalist* 21:303–322.
Leong, M., R.R. Dunn, M.D. Trautwein. 2018. Biodiversity and socioeconomics in the city: A review of the luxury effect. *Biology Letters* 14:20180082.
Lepczyk, C.A., M.F.J. Aronson, K.L. Evans, M.A. Goddard, S.B. Lerman, J.S. MacIvor. 2017. Biodiversity in the city: Fundamental questions for understanding the ecology of urban green spaces for biodiversity conservation. *BioScience* 67(9):799–807.
Lin, B.B., R.A. Fuller. 2013. Sharing or sparing? How should we grow the world's cities? *Journal of Applied Ecology* 50:1161–1168.
Llop, E., P. Pinho, P. Matos, M.J. Pereira, C. Branquinho. 2012. The use of lichen functional groups as indicators of air quality in a Mediterranean urban environment. *Ecological Indicators* 13:215–221.
Loram, A., J. Tratalos, P.H. Warren, K.J. Gaston. 2007. Urban domestic gardens (X): The extent & structure of the resource in five major cities. *Landscape Ecology* 22:601–615.
Lososová, Z., N. Čeplová, M. Chytrý, L. Tichý, J. Danihelka, K. Fajmon, D. Láníková, Z. Preislerová, V. Řehořek. 2016. Is phylogenetic diversity a good proxy for functional diversity of plant communities? A case study from urban habitats. *Journal of Vegetation Science* 27(5):1036–1046.
Lososová, Z., M. Chytrý, L. Tichý, J. Danihelka, K. Fajmon, O. Hájek, K. Kintrová, I. Kühn, D. Láníková, Z. Otýpková, V. Řehořek. 2012. Native and alien floras in urban habitats: A comparison across 32 cities of central Europe. *Global Ecology and Biogeography* 21(5):545–555.
Lundholm, J.T. 2015. Green roof plant species diversity improves ecosystem multifunctionality. *Journal of Applied Ecology* 52(3):726–734.
Lundholm, J.T., S. Tran, L. Gebert. 2015. Plant functional traits predict green roof ecosystem services. *Environmental Science and Technology* 49(4):2366–2374.
MacArthur, R.H., E.O. Wilson. 1967. *The Theory of Island Biogeography*. Princeton University Press, Princeton, NJ.
Makeeva, V.M., A.V. Smurov, D.V. Politov, M.M. Belokon, Y.S. Belokon, E.G. Suslova. 2018. Comparative assessment of the gene pool and the viability of forest plantations from Moscow and natural populations from the Moscow region by example of Norway Spruce (*Picea abies* (L.) Karst.). *Russian Journal of Genetics* 54:1040–1049.
Mathieu, R., C. Freeman, J. Aryal. 2007. Mapping private gardens in urban areas using object-oriented techniques and very high-resolution satellite imagery. *Landscape and Urban Planning* 81:179–192.
McDonnell, M.J., A.K. Hahs. 2013. The future of urban biodiversity research: Moving beyond the 'low-hanging fruit'. *Urban Ecosystems* 16:397–409.
McKinney, M.L. 2006. Urbanization as a major cause of biotic homogenization. *Biological Conservation* 127:247–260.
Moss, W., R. Walmsley. 2005. *Controlling the Sale of Invasive Garden Plants: Why Voluntary Measures Alone Fail.* WWF-Australia Discussion Paper. WWF-Australia, Sydney.
Munzi, S., O. Correia, P. Silva, N. Lopes, C. Freitas, C. Branquinho, P. Pinho. 2014. Lichens as ecological indicators in urban areas: Beyond the effects of pollutants. *Journal of Applied Ecology* 51(6):1750–1757.
Niinemets, Ü., J. Peñuelas. 2008. Gardening and urban landscaping: Significant players in global change. *Trends in Plant Science* 13(2):60–65.
Norfolk, O., M.P. Eichhorn, F. Gilbert. 2013. Traditional agricultural gardens conserve wild plants and functional richness in arid South Sinai. *Basic and Applied Ecology* 14:659–669.
Oishi, Y. 2019. The influence of microclimate on bryophyte diversity in an urban Japanese garden landscape. *Landscape and Ecological Engineering* 15:167–176.
Palma, E., J.A. Catford, R.T. Corlett, R.P. Duncan, A.K. Hahs, M.A. McCarthy, M.J. McDonnell, K. Thompson, N.S.G. Williams, P.A. Vesk. 2017. Functional trait changes in the floras of 11 cities across the globe in response to urbanization. *Ecography* 40(7):875–886.
Pataki, D.E., H.R. McCarthy, T. Gillespie, G.D. Jenerette, S. Pincetl. 2013. A trait-based ecology of the Los Angeles urban forest. *Ecosphere* 4(6):1–20.
Pearse, W.D., J. Cavender-Bares, S.E. Hobbie, M.L. Avolio, N. Bettez, R. Roy Chowdhury, L.E. Darling, P.M. Groffman, J.M. Grove, S.J. Hall, J.B. Heffernan, J. Learned, C. Neill, K.C. Nelson, D.E. Pataki, B.L. Ruddell, M.K. Steele, T.L.E. Trammell. 2018. Homogenization of plant diversity, composition, and structure in North American urban yards. *Ecosphere* 9:e02105.
Piana, M.R., M.F.J. Aronson, S.T.A. Pickett, S.N. Handel. 2019. Plants in the city: Understanding recruitment dynamics in urban landscapes. *Frontiers in Ecology and the Environment* 17(8):455–463.

Pyšek, P. 1993. Factors affecting the diversity of flora and vegetation in central European settlements. *Vegetatio* 106:89–100.
Pyšek, P. 1998. Alien and native species in Central European urban floras: A quantitative comparison. *Journal of Biogeography* 25(1):155–163.
Pyšek, P., P.W. Lambdon, M. Arianoutsou, I. Kühn, J. Pino, M. Winter, J.A. Drake. 2009. Alien vascular plants of Europe. Pages 43–61 in DAISIE, editor. *Handbook of Alien Species in Europe*. Springer, Dordrecht, Netherlands.
Ranta, P. 2001. Changes in urban lichen diversity after a fall in sulphur dioxide levels in the city of Tampere, SW Finland. *Annales Botanici Fennici* 38:295–304.
Reichard, S.H., P. White. 2001. Horticulture as a pathway of invasive plant introductions in the United States. *Bioscience* 51:103–113.
Ricotta, C., F.A. La Sorte, P. Pysek, G.L. Rapson, L. Celesti-Grapow, K. Thompson. 2009. Phyloecology of urban alien floras. *Journal of Ecology* 97(6):1243–1251.
Scholz, H. 2007. Questions about indigenous plants and anecophytes. *Taxon* 56(4):1255–1260.
Schröder, R., K. Kiehl. 2020. Extensive roof greening with native sandy dry grassland species: Effects of different greening methods on vegetation development over four years. *Ecological Engineering* 145:105728.
Schwarz, N., M. Moretti, M.N. Bugalho, Z.G. Davies, D. Haase, J. Hack, A. Hof, Y. Melero, T.J. Pett, S. Knapp. 2017. Understanding biodiversity-ecosystem service relationships in urban areas: A comprehensive literature review. *Ecosystem Services* 27:161–171.
Seaward, M.R.D. 1982. Lichen ecology of changing urban environments. Pages 181–189 in R. Bornkamm, J.A. Lee, M.R.D. Seaward, editors. *Urban Ecology*. Blackwell Scientific Publications. Oxford.
Seebens, H., F. Essl, W. Dawson, N. Fuentes, D. Moser, J. Pergl, P. Pysek, M. van Kleunen, E. Weber, M. Winter, B. Blasius. 2015. Global trade will accelerate plant invasions in emerging economies under climate change. *Global Change Biology* 21:4128–4140.
Sérgio, C., P. Carvalho, C.A. Garcia, E. Almeida, V. Novais, M. Sim-Sim, H. Jordão, A.J. Sousa. 2016. Floristic changes of epiphytic flora in the Metropolitan Lisbon area between 1980–1981 and 2010–2011 related to urban air quality. *Ecological Indicators* 67:839–852.
Seto, K.C., B. Güneralp, L.R. Hutyra. 2012. Global forecasts of urban expansion to 2030 and direct impacts on biodiversity and carbon pools. *Proceedings of the National Academy of Sciences of the United States of America* 109(40):16083–16088.
Shwartz, A., A. Turbé, L. Simon, R. Julliard. 2014. Enhancing urban biodiversity and its influence on city-dwellers: An experiment. *Biological Conservation* 171:82–90.
Smith, R.M., K. Thompson, P.H. Warren, K.J. Gaston. 2010. Urban domestic gardens (XIII): Composition of the bryophyte and lichen floras, and determinants of species richness. *Biological Conservation* 143(4):873–882.
Spitale, D., P. Mair, J. Nascimbene. 2020. Patterns of bryophyte life-forms are predictable across land cover types. *Ecological Indicators* 109:105799.
Sukopp, H. 1998. Urban ecology – Scientific and practical aspects. Pages 3–16 in J. Breuste, H. Feldmann, O. Uhlmann, editors. *Urban Ecology*. Springer Verlag, Berlin, Heidelberg, Germany.
Sukopp, H., S. Weiler. 1988. Biotope mapping and nature conservation strategies in urban areas of the Federal Republic of Germany. *Landscape and Urban Planning* 15:39–58.
Swan, C.M., B.V. Brown, D. Borowy, J. Cavender-Bares, A. Jeliazkov, S. Knapp, Z. Lososová, J. Padullés Cubino, S. Pavoine, C. Ricotta, D. Sol. 2021. A framework for understanding how biodiversity patterns unfold across multiple spatial scales in urban ecosystems. *Ecosphere* 12(7):e03650.
Tait, C.J., C.B. Daniels, R.S. Hill. 2005. Changes in species assemblages within the Adelaide Metropolitan Area, Australia, 1836–2002. *Ecological Applications* 15(1):346–359.
Theodorou, P., R. Radzevičiūtė, G. Lentendu, B. Kahnt, M. Husemann, C. Bleidorn, J. Settele, O. Schweiger, I. Grosse, T. Wubet, T.E. Murray, R.J. Paxton. 2020. Urban areas as hotspots for bees and pollination but not a panacea for all insects. *Nature Communications* 11:576.
Tucker, C.M., M.W. Cadotte, S.B. Carvalho, T.J. Davies, S. Ferrier, S.A. Fritz, R. Grenyer, M.R. Helmus, L.S. Jin, A.O. Mooers, S. Pavoine, O. Purschke, D.W. Redding, D.F. Rosauer, M. Winter, F. Mazel. 2017. A guide to phylogenetic metrics for conservation, community ecology and macroecology. *Biological Reviews* 92:698–715.
Vanden Broeck, A., K. Cox, I. Melosik, B. Maes, K. Smets. 2018. Genetic diversity loss and homogenization in urban trees: The case of *Tilia × europaea* in Belgium and the Netherlands. *Biodiversity and Conservation* 27:3777–3792.

Van Mechelen, C., T. Dutoit, M. Hermy. 2015. Vegetation development on different extensive green roof types in a Mediterranean and temperate maritime climate. *Ecological Engineering* 82:571–582.
Van Rossum, F. 2008. Conservation of long-lived perennial forest herbs in an urban context: *Primula elatior* as study case. *Conservation Genetics* 9(1):119–128.
Van Rossum, F. 2009. Conservation of wet meadow species in an urban context: *Lychnis Flos-Cuculi* as study case. *Belgian Journal of Botany* 142(2):98–110.
Villéger, S., N.W.H. Mason, D. Mouillot. 2008. New multidimensional functional diversity indices for a multifaceted framework in functional ecology. *Ecology* 89:2290–2301.
Vogt-Schilb, H., P.A. Moreau, J.C. Malaval, B. Schatz, F. Richard. 2018. Effects of long-term landscape dynamics and city growth on biodiversity in the Mediterranean: The case study of Montpellier, France. *Urban Ecosystems* 21(5):921–932.
von der Lippe, M., I. Kowarik. 2007. Long-distance dispersal of plants by vehicles as a driver of plant invasions. *Conservation Biology* 21:986–996.
Whelan, R.J., D.G. Roberts, P.R. England, D.J. Ayre. 2006. The potential for genetic contamination vs. augmentation by native plants in urban gardens. *Biological Conservation* 128(4):493–500.
Williams, N.S.G., A.K. Hahs, P.A. Vesk. 2015. Urbanisation, plant traits and the composition of urban floras. *Perspectives in Plant Ecology, Evolution and Systematics* 17:78–86.
Winter, M., O. Schweiger, S. Klotz, W. Nentwig, P. Andriopoulos, M. Arianoutsou, C. Basnou, P. Delipetrou, V. Didžiulis, M. Hejda, P.E. Hulme, P.W. Lambdon, J. Pergl, P. Pyšek, D.B. Roy, I. Kühn. 2009. Plant extinctions and introductions lead to phylogenetic and taxonomic homogenization of the European flora. *Proceedings of the National Academy of Sciences of the United States of America* 106:21721–21725.
Yakub, M., P. Tiffin. 2017. Living in the city: Urban environments shape the evolution of a native annual plant. *Global Change Biology* 23:2082–2089.
Yang, J., F.A. La Sorte, P. Pyšek, P. Yan, D. Nowak, J. McBride. 2015. The compositional similarity of urban forests among the world's cities is scale dependent. *Global Ecology and Biogeography* 24(12):1413–1423.
Zepeda-Gómez, C., P. Ávila-Pérez, U.S. Díaz-García, Y. Alanís-Martínez, G. Zarazúa-Ortega, A. Amaya-Chávez. 2014. Diversidad de musgos epifitos de la zona metropolitana del valle de Toluca, México. *Revista Mexicana de Biodiversidad* 85:108–124.
Zohner, C.M. 2019. Phenology and the city. *Nature Ecology & Evolution* 3:1618–1619.

16

URBANIZATION SHAPES INSECT DIVERSITY

Emily K. Meineke, Elsa Youngsteadt, Mia K. Lippey, and Katherine C.R. Baldock

Introduction

As the most diverse and massive group of animals on the planet, insects play vital ecological roles in all terrestrial ecosystems, including cities (Bar-On et al. 2018; Stork 2018; Wagner 2020). Viewed from a human perspective, insects provide both ecosystem services and disservices. They pollinate plants, decompose organic materials, and aerate soils in cities. They also serve as food for vertebrates (Wagner 2020) and provide biological control of their own pestiferous relatives. Disservices include disease transmission, excessive insect herbivore damage in gardens and green infrastructure, structural damage to buildings, and indoor nuisances, such as ants in the sugar. While indoor insects are a critical topic in human well-being, this chapter focuses on the insects that participate in outdoor food webs built on plants and detritus.

Cities arose only in the past 5,000 years (Redman 2011), a recent development in the 500-million-year evolutionary history of insects (Misof et al. 2014). Built environments present novel challenges and opportunities, through changes in dispersal, nutrition, communication, and metabolism. An insect in the city is faced with a patchwork of habitats; some may resemble the native ecosystem, while others are transformed by managed vegetation and built structures, such as pavement and buildings. Cities also manufacture localized climates – often hotter and drier than native ecosystems (Oke et al. 2017) – with direct effects on the timing of insect life history events, metabolism, and hydration. Urban pollutants and chemical contamination of air, water, and soil may also directly interfere with insect communication, navigation, and biochemistry.

Many studies show that native insect diversity is lower in urbanized areas compared to less disturbed native habitats (Fenoglio et al. 2020). Urbanization can reduce insect diversity by isolating habitats from potential colonists and by imposing strong filters that exclude species adapted to local, non-urban environments. Moreover, urbanization may have contributed to global declines in insect abundance and biomass, also known as "the insect apocalypse" (van Klink et al. 2020). However, the effects of urbanization are not uniformly negative for insects. Like other drivers of global change, urbanization creates "winners" and "losers", wherein some species and genotypes survive and even thrive in urban areas, while others decline or simply cannot penetrate the urban-rural interface (Dornelas et al. 2019).

The diversity of responses to urbanization among insects is perhaps not surprising because they are themselves phenomenally diverse. In particular, functional groups vary in their responses. For instance, herbivorous insects may be more sensitive to pollutant-driven changes in plant quality

 DOI: 10.4324/9781003016120-19

than are their predators – which may be more sensitive to urban fragmentation. Similarly, insects with small size, weak flight, or specialized diets may experience an urban habitat as more fragmented than do large, strong-flying generalists.

This chapter highlights how urban environments affect insect diversity, typically measured as species-level richness. We focus on several key urban environmental variables: fragmentation, climate, pollution, land use, vegetation complexity, urban area age, and affluence (Figure 16.1). We highlight effects of these drivers on five insect functional groups that shape the world around us (Box 16.1). Occasionally, one functional group will appear within another group's section for context, and in some cases, we include non-insect arthropods, such as arachnids, that participate in insect food webs. The chapter concludes by summarizing areas of consensus and highlighting knowledge gaps for future research.

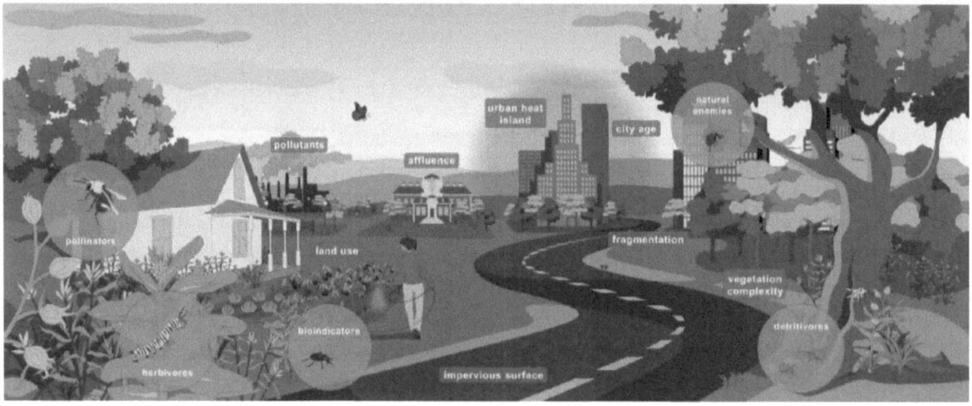

Figure 16.1 Illustration of the key drivers of urban insect diversity and insect functional groups.

Box 16.1 Functional groups

Herbivores feed on live plants and include species from almost all insect orders. Those most often studied in cities are butterflies and moths, with a strong bias toward adult butterflies.

Pollinators include all species that facilitate plant sexual reproduction by transferring pollen. We use this term for flower-visiting insects that have the *potential* to fill this role. Pollinators are taxonomically diverse, including certain lineages of bees, flies, butterflies, moths, wasps, and beetles (Rader et al. 2016).

Detritivores eat decaying matter. They aid in the critical ecosystem service of decomposition and include ants, cockroaches, beetles, and non-insect arthropods such as springtails.

Natural enemies are insects that feed on other insects. Predators and parasitoids of herbivorous insects are the natural enemies most often addressed in urban ecology because they aid in biological control. Controlling herbivores without pesticides, and with the help of natural enemies, is often a goal in urban landscapes.

Bioindicator groups. Bioindicators are taxa that are thought to be particularly sensitive to environmental changes. The bioindicators most commonly addressed in urban ecology are ants and carabid beetles. Bioindicator taxa often include detritivores and natural enemies and are thus combined into one section where appropriate.

Fragmentation

Fragmentation is the process by which a once-continuous habitat is divided into patches of land uses (Laurance 2010). A core concept in fragmentation research is the equilibrium theory of island biogeography, which famously predicted that island size and isolation would determine the diversity of species on the island – thus invoking both species-area relationships and dispersal limitation (MacArthur and Wilson 1963). Studies of fragmentation have similarly predicted that smaller, more isolated habitat fragments would support fewer species than larger fragments near continuous habitat.

Effects of fragmentation may extend beyond simple island-biogeographic relationships. In urban environments, the matrix surrounding a focal habitat may affect its isolation and local habitat quality. Fragment shape and size further determine the dominance of edge effects in driving the diversity and species composition in a fragment. Most predicted effects of urban fragmentation are similar across functional groups:

- Smaller habitat fragments will harbor fewer species; those with large or specialized resource demands are most likely to be lost (Holt et al. 1999).
- More isolated fragments will lose species with poor dispersal ability – often those that cannot fly or have smaller bodies. Thus, fragment size and isolation may have opposing effects on body size at the community level.
- Edges of woodlands will favor more dry-adapted species, and species that utilize resources restricted to edge habitats. The opposite pattern is expected in arid biomes, where urban edges are wetter and may support more wet-adapted species.
- Predators and parasitoids are expected to be more sensitive to fragmentation since they rely on pre-existence of their hosts in a fragment, and their populations are inherently smaller than those of the host and thus more subject to stochastic extinction (Kruess and Tscharntke 1994; but see Holt et al. 1999).

Herbivores

Studies of herbivores in urban habitat fragments have focused largely on butterflies, which are moderately mobile and thus able to disperse from unsuitable to suitable habitats. As a result, the species composition of urban butterflies is driven more strongly by patch size than isolation (Krauss et al. 2003; Soga and Koike 2012; Soga et al. 2015). Other taxonomic groups also support the prediction that larger habitat patches support higher herbivore abundance and diversity – whether a "patch" is a single tree or a nature reserve (Wood and Pullin 2002; Angold et al. 2006; Herrmann et al. 2012; Turrini and Knop 2015; Sing et al. 2016). A few studies have identified specific drivers of higher butterfly diversity in larger patches. For instance, in Tokyo, Japan, butterfly diversity in woodland habitat fragments increased with fragment size and higher abundance of herbaceous, nectar-producing plants – which likely represent resources for adults rather than herbivorous larvae (Soga and Koike 2012).

Effects of the amount of edge habitat on insect diversity are taxon-specific and probably depend on whether herbivores use resources provided by edges (e.g. flowering herbaceous plants) and can survive the environmental conditions of forest edges, which can often be hotter and drier than forest interiors (but wetter than arid habitats). Edge habitats can support herbivore communities with different feeding habits than those of interior habitats (Christie et al. 2010), though the specific mechanism is unclear. For other taxa, edges may be inappropriate habitats, such that species maintain higher abundances in patch interiors, as is the case for woodland moths in central Scotland (Lintott et al. 2014).

Pollinators

Effects of natural habitat fragmentation on flower-visiting insects have been investigated primarily in bees and butterflies. Most studies that have tested for a species-area relationship in these groups have found one, in habitats including desert, grassland, and temperate and tropical forest in an urban matrix (e.g. Cane et al. 2006; Hinners et al. 2012; Soga and Koike 2012; Stewart et al. 2018). This pattern is not, however, universal. Work in Canadian oak-scrub fragments did not recover overall species-area relationships and instead found that individual pollinator guilds (such as aboveground nesters) responded differently to habitat area and local habitat features (Neame et al. 2013; Wray and Elle 2015). Moreover, some studies find greater species richness in large fragments (generally >1 to tens of ha, depending on the study) than in continuous natural habitat, suggesting that access to urban edge or matrix habitats can increase pollinator diversity (e.g. Cane et al. 2006; Hinners et al. 2012).

Pollinator community assembly in fragments is not a neutral process of island biogeography; instead, there is strong evidence for environmental or biotic filtering. For example, pollinator communities in sage scrub fragments (3–117 ha) in San Diego, California, USA, lost more functional diversity than expected by chance, a signature of environmental filtering (Hung et al. 2019). Other studies have found specific guilds, particularly aboveground nesting bees, to be increasingly overrepresented as fragments shrunk below one hectare (Cane et al. 2006; Hinners et al. 2012). In Halle, Germany, fragmentation may have reduced pollinator diversity by reducing plant diversity – a biotic filter (Theodorou et al. 2020). Effects of fragmentation on body size are inconsistent across studies, perhaps reflecting different local trade-offs between the mobility and resource demands of a large body (Cane et al. 2006; Neame et al. 2013; Merckx et al. 2018a).

Whether pollinator communities in fragments are filtered more strongly by local habitat features or by the landscape matrix remains unclear. This question is often examined by applying model selection to many possible predictors, which often identifies drivers of diversity at both scales. Locally, floral abundance, floral diversity, or nesting resources are often important, while landscape factors such as percent impervious surface and percent green space at various radii are also selected (e.g. Soga and Koike 2012; Wray and Elle 2015; Wenzel et al. 2020). This approach may miss cross-scale interactions and biologically relevant mechanisms. For example, communities in small or low-quality fragments may interact more with the surrounding matrix than do those in large or high-quality fragments, such that the effect of habitat at one scale is contingent upon conditions at another scale (Dunning et al. 1992). This idea was recently supported in an urban landscape in Texas, USA, where pollinator diversity in poor habitats depended on the amount of semi-natural habitat in the surrounding landscape, while diversity in native grassland fragments did not (Ballare et al. 2019).

Finally, attempts to evaluate fragment isolation or matrix quality based on Euclidean distances are likely flawed because urban pollinators may not disperse or forage uniformly within a fixed radius. For example, a study in Stockholm, Sweden considered two predictors of bee and wasp diversity: the total amount of foraging habitat in a fixed radius of the sampling point or the amount of habitat accessible without crossing putative barriers such as roads (Johansson et al. 2018). Only the latter metric was a significant predictor of diversity.

Detritivores

Research on the effects of fragmentation on detritivorous insects is limited. Carrion beetles (Coleoptera: Silphidae) contribute to nutrient cycling and soil turnover in temperate forests by burying dead vertebrates as a food source for larvae. Along an urban-rural gradient of forest fragmentation in central New York State, USA, carrion beetle species richness declined by one-third, and abundance by two-thirds, in fragmented forests compared to continuous forests, with patterns more pronounced when the matrix was urban rather than rural (Gibbs and Stanton 2001). In contrast, muscoid flies, the beetles' primary competitors, were more abundant in fragments compared to continuous forests. Although forest fragmentation has a profound impact on carrion beetle communities, the underlying mechanisms are unclear. In a study of forest fragmentation and decomposition of vertebrate carcasses in an urban-agricultural matrix, burial rates by beetles increased in larger fragments, whilst carcass removal rates by vertebrates decreased (Sugiura et al. 2013). Fragmentation may have a greater effect on carrion beetles as they are less mobile than flies and vertebrates.

Natural enemies and bioindicator groups

Studies of ants, carabids, wasps, and broader groups of arthropods in urban habitat fragments almost always yield the expected species-area relationships (e.g. Faeth and Kane 1978; Bolger et al. 2000), while occasional counter examples remain unexplained (Gibb and Hochuli 2002).

Rarely have studies of bioindicator groups or natural enemies taken advantage of their variation in body size or trophic level to query more specific hypotheses about effects of fragment area and isolation. Results from carabid beetles are consistent with the prediction that less mobile species should be lost from isolated fragments; highly urbanized (and presumably fragmented) sites lack flightless or weak-flying species (Martinson and Raupp 2013; Piano et al. 2017).

The prediction that predators would decline more steeply than herbivores has not generally been supported. In San Diego, California, USA, the richness of spiders increased in smaller fragments (while most other arthropods declined); these predators may have defied expectations by capitalizing on abundant, invasive Argentine ants that also dominated small fragments (Bolger et al. 2000), or by escaping predation by scorpions present in less disturbed fragments (Longcore 2003). In Sydney, Australia, arthropods trapped in pitfall and yellow pan traps did not differ in overall richness between large and small fragments, although compositional changes among predators were stronger than those among lower trophic levels (Gibb and Hochuli 2002). However, studies rarely assess the effects of fragmentation on explicit predator-prey interaction networks and instead focus on taxonomic or functional group classifications in generalized sampling.

In rare cases, data can be gathered on predators alone that can indicate to what extent prey availability drives their diversity. For instance, larger forest fragments across two cities in Japan supported more diverse web-building spider assemblages (Miyashita et al. 1998). Larger fragments also tended to support spider species with larger body sizes and, within species, larger individuals. Because spider body size is often driven by prey availability, the putative explanation was that larger forest patches may support more prey.

Finally, studies on ants have documented profound edge effects in urban habitat fragments. In arid California, USA, for example, urban fragment edges are moister than fragment interiors or natural areas, likely due to irrigation and runoff. Strong circumstantial evidence suggests that Argentine ants rely on these high-moisture environments, where they suppress diversity of native ants (Suarez et al. 1998; Bolger et al. 2000).

Conclusions

Species-area relationships are a consistent feature of ecosystems worldwide, and urban habitat fragments rarely depart from this pattern for any functional group. Nevertheless, for mobile animals who use different, patchy resources for foraging, nesting, or egg-laying, an island biogeography model of fragmentation in cities is too simple. The reviewed studies showed that fragmentation can reduce insect diversity not only through dispersal limitation (as evidenced by community-level shifts in body size or mobility) but also environmental and biotic filtering. Fragment edges have distinct communities of herbivores and bioindicators, with other groups often neglected. Few studies have compared fragmentation effects across trophic or functional groups, and this literature may benefit by examining species interaction networks rather than generalized samples collected in traps. Future work will benefit by considering semi-natural fragments as one of many land uses in the urban mosaic, by examining cross-scale interactions between local and landscape features, and by defining habitat isolation in terms of not only distance but also barriers to movement, among habitat patches.

Climate

Cities dramatically alter the texture, thermal, and radiative properties of the Earth's surface – and thus the way it interacts with solar energy and precipitation to produce climate. Two of the best documented features of urban climates are the urban heat island (UHI) and the urban dry island (UDI) (Moriwaki et al. 2013; Oke et al. 2017). The UHI describes the elevated temperature of urban air, surface, and soil compared to those outside the city; the UDI describes the reduced humidity in cities. Although these phenomena are less pronounced in desert or tropical cities than in temperate, mesic ones, the UHI and UDI are consistent features of many cities.

Insects are physiologically adapted to their local climate or microclimate (Sunday et al. 2011; García-Robledo et al. 2016), and altered urban climate may impose strong environmental filters upon this fauna. As ectotherms, insects respond directly to environmental temperatures, which alter their development, activity, and fitness; as small organisms with a large surface area, they are also at risk of desiccation (Harrison et al. 2012).

Predicted effects of urban climate on arthropods include:

- Urban communities will over-represent taxa that are tolerant of heat and desiccation, as indicated by physiological traits or habitat-of-origin (Menke et al. 2011; Merckx et al. 2018b).
- Urban warming may cause steeper declines in higher trophic levels than in their prey (Chown and Duffy 2015).
- Effects of urban warming and desiccation on diversity and community structure should depend on background climate. Warming may benefit insects in cool climates but reduce diversity in already-hot climates (Diamond et al. 2015).
- Urban climates may drive diversity loss through phenological mismatch between interacting organisms (Renner and Zohner 2018).

Herbivores

Most literature addressing effects of urban heat on insects focuses on how urban warming affects insect herbivore abundance, not diversity. If an individual species becomes hyper-abundant in response to urban warming (e.g. Meineke et al. 2013; Dale and Frank 2014), it may outcompete other species with overlapping niche space. Thus, urban heat may create "winners" that outcompete "losers", reducing species evenness and diversity within functional groups. Several studies have addressed how arthropod communities that *include* herbivores respond to urban warming. For example, malaise traps deployed in backyards across the Los Angeles Basin captured phorid flies (mixed functional groups), butterflies (herbivores), and bees, among other taxa. Warmer, drier sites that were climatically more similar to local native habitats housed the most diverse assemblages (Adams et al. 2020). Urban warming can also drive changes in host plant water stress that can benefit or hamper herbivorous arthropod fitness in cities, though the relationship between plant water stress regimes and insect herbivores appears to be guild-specific (Gely et al. 2020).

Pollinators

Bees require high thoracic temperatures to fly and may generate body heat (at an energetic cost) to meet this requirement in cool habitats (Willmer and Stone 2004). In hot habitats, however, flight can generate dangerous heat, limiting activity to cooler times of day or causing bees to fall from the air when they exceed their heat tolerance (Chappell 1984). The effects of urban warming on bees may therefore depend on the season and region.

Urban climate does appear to filter bee communities based on physiological tolerance to heat and drought. For example, in Raleigh, North Carolina, USA, species with the lowest physiological heat tolerance also declined most steeply in abundance along an urban warming gradient (Hamblin et al. 2017). These changes altered community composition but not richness (Hamblin et al. 2018). In Toledo, Ohio, USA, bumblebees and sweat bees had hotter body temperatures in more urban locations, while honey bees became less hydrated (Burdine and McCluney 2019a). The idea that cities favor heat- and desiccation-tolerant species is supported by an analysis of the bee fauna in Poznan, Poland, where 16% of species in the city, but only 9% in a nature reserve, had geographic ranges extending to hotter, drier locations (Banaszak-Cibicka 2014).

Urban warming causes flowers to bloom earlier in cities (Neil and Wu 2006). If pollinator activity does not shift accordingly, reduced access to floral resources could cause extinction (Memmott et al. 2007). Few studies have examined this problem, and existing results are mixed. In some cases, urban flowers experience an earlier spring, while bees do not (Fisogni et al. 2020); in others, abundance of both bees and flowers peak later in urban than natural sites (Wray and Elle 2015). Many studies have examined urban bee phenology without corresponding floral data and often find that pollinator diversity or abundance peak later in urban than natural habitats due to dominance of late-flying, generalist species (e.g. Hung et al. 2017; Harrison et al. 2018a; Luder et al. 2018). More work is needed to evaluate phenological mismatch in cities and to disentangle the contributions of phenotypic plasticity versus altered community composition. Because floral abundance can affect pollinator detectability (Packer and Darla-West 2021), observing phenologies of pollinator nesting activity and emergence will be valuable.

Natural enemies

Studies addressing the effects of urban climates on natural enemies tended to focus on phenology, abundance, and/or efficacy in controlling herbivorous insects rather than diversity, *per se*. In one case, hotter versus cooler urban trees in Raleigh, North Carolina, USA hosted different spider communities, in large part because ghost spiders were common at cooler sites but absent in the hottest trees (Meineke et al. 2017). The mechanism driving this pattern is unknown but could result from arboreal temperatures approaching or exceeding ghost spider thermal maxima.

Detritivores and bioindicator groups

Studies in urban ants and carabids have used geographic range or habitat of origin as a proxy for species' tolerances of hot or desiccating environments. For example, ant species in Raleigh, North Carolina, USA, and carabids in urban sites in northern Belgium originated from hotter, drier geographic ranges than those in neighboring forest or non-urban habitat (Menke et al. 2011; Piano et al. 2017), while ants in irrigated parks in arid Phoenix, Arizona, USA, showed the opposite trend (Miguelena and Baker 2019).

In ectotherms, body size is generally expected to shrink in hotter environments due to metabolic costs of fueling a large body at high temperatures (Sheridan and Bickford 2011). Recent work has therefore investigated whether invertebrate communities exposed to urban warming comprise smaller species than those in reference habitats (Magura et al. 2020). In northern Belgium, for example, the expected pattern of reduced community-average body size was found in five of ten taxa considered, two did not shift size distribution, and three increased (Merckx et al. 2018b). The taxa whose urban communities increased in body size were all flying insects for whom larger size corresponds to higher mobility, suggesting a trade-off between coping with climate and fragmentation. Because body size responds to many factors other than temperature, work is still needed to isolate the contribution of climate to this pattern, and the extent to which urban warming drives immigration of small species, selective loss of large species, or both.

Conclusions

Pollinators, ants, and carabids all show evidence of environmental filtering by urban climate. Most studies have used geographic origin as a proxy for physiological tolerance. Herbivores and detritivores are understudied in this regard, but given their impact on green infrastructure and carbon cycling, links between urban climate, herbivore and detritivores diversity, and function deserve attention. As with fragmentation, direct comparison of diversity loss across interacting trophic levels is lacking. Phenological mismatch is often suggested as a consequence of warming, but results remain few and mixed, with specific examples in parasitoids and evidence of diffuse mismatch in pollinators. To understand the effects of urban climate on diversity, additional work is needed within documented interaction networks, including herbivore-host, parasitoid-host, and pollinator-plant interactions.

Nutrients and pollutants

In cities, nutrients accumulate due to fertilizer application, pet excretion, and fossil fuel combustion (Baker et al. 2001). Urban industry and transportation produce an urban "CO_2 dome" along with other aerial pollutants (Idso et al. 2001). In this section, we summarize the direct and indirect effects of such chemicals and nutrients on insect diversity. We do not discuss pesticides, which are often specifically designed to cause insect mortality.

Because macronutrients, including nitrogen and carbon, often limit plant growth or reproduction, altered nutrient cycles shift plant species diversity, productivity, and nutritional quality, with consequences for higher trophic levels (Robinson et al. 2012; Nijssen et al. 2017; David et al. 2019). For example, long-term experimental nitrogen addition in a (non-urban) grassland reduced species richness of herbivorous and predatory insects while increasing richness of detritivores (Haddad et al. 2000). However, if both carbon and nitrogen are elevated in urban systems, the combined effects on nutritional quality of plants are difficult to anticipate. In addition, certain pollutants tend to travel to rural areas, where they remain in higher concentrations than in urban areas, notably ozone (e.g. Gregg et al. 2003).

Predicted effects of chemical landscapes in urban areas on arthropods include:

- Effects of elevated nitrogen and CO_2 on insects will be indirect via plants and should be most pronounced among herbivores that directly consume those plants (including pollen, nectar, and litter).

- Because nitrogen is a limiting nutrient in many habitats, urban nitrogen subsidies should increase the abundance of certain herbivorous insects, with unpredictable effects on overall insect diversity.
- Effects of pollutants on plants and herbivores are likely to cascade, affecting diversity, abundance, and distributions of organisms at higher trophic levels.

Herbivores

Inputs of nitrogen generally enhance the fitness of common insect herbivores on ornamental plants (Raupp et al. 2010). However, we recovered no studies assessing how urban nutrient additions affect diversity of herbivores. Most studies of urban pollutants and their effects on herbivores have occurred in more natural ecosystems, where pollutants waft in from cities. In a study of three plant species in the Los Angeles Basin, herbivores were sampled to determine differences in diversity and community composition across a range of rural sites with different intensities of urban pollution. Neither herbivore diversity nor abundance varied consistently across the pollution gradient, but insect communities at more polluted sites consistently comprised more chewing herbivores than other feeding groups, potentially because nitrogen and ozone pollution enhanced leaf nutritional quality (Jones and Paine 2006).

In certain cases, nitrogen deposition from nearby urban areas may enhance the invasion potential of non-native grasses that out-compete native plants in rural areas with historically nutrient-poor soils. Moths and butterflies that rely on host plants from low nitrogen environments are particularly at risk (Weiss 1999; Fox 2013). While these two studies display the potential strong effects of aerial pollutants on herbivores, we did not recover studies investigating the specific effects of pollutants on herbivores within cities where many pollutants are present in elevated concentrations.

Pollinators

If nitrogen deposition shifts plant communities from flowering herbs to grasses, it may reduce flower cover and food availability to pollinators (Burkle and Irwin 2010) and eliminate host plants of specialist bees (David et al. 2019). Nitrogen and CO_2 enrichment can also alter the nutritional quality of pollen and nectar (David et al. 2019). Bees, which rely on floral resources at all life stages, are likely to be particularly sensitive to these changes (Vaudo et al. 2015), but nectar quality may also affect the fitness of non-bee pollinators who consume nectar only as adults (Mevi-Schütz and Erhardt 2005; Hoover et al. 2012). Finally, urban atmospheric chemistry, including CO_2, nitrogen, and other pollutants such as ozone, can change the production or persistence of floral scent, with potential to alter pollinators' foraging preferences or abilities (Jamieson et al. 2017; David et al. 2019). Despite the well-documented mechanisms by which urban pollutants could alter plant composition and floral quality, these effects on pollinators in an urban context have not been examined.

Natural enemies, detritivores and bioindicator groups

Few studies have examined the effects of urban pollutants on these functional groups. A study of soil arthropod communities, primarily detritivores, associated with urban and suburban holm oak (*Quercus ilex*) in Italy found no effect of pollution on arthropod abundance and diversity (Rota et al. 2015). Heavy metal contamination influenced the establishment and functional diversity of ground-dwelling beetle communities in vacant lots in Cleveland, Ohio, USA (Perry et al. 2020). Finally, a comparison of terrestrial invertebrates (ants, beetles, mites, and springtails) living on three streambanks in Manaus, Brazil, found that the banks of sewage-polluted streams hosted less diverse communities than did an unpolluted reference stream (Couceiro et al. 2007). Further research is needed to understand the types of pollutants that most affect these functional groups, and how they move through food webs to alter the quality of prey, detritus, or plant-insect interactions. Natural enemies that use plant volatiles to locate prey may be particularly sensitive to pollutant-driven changes in plant signaling.

Conclusions

We identified mechanisms by which pollutant effects on plant quality could alter interactions, fitness, and diversity of each functional group. Research on these effects in an urban setting remains underdeveloped, and uneven across groups. Herbivore and pollinator work focuses on airborne pollutants, while detritivore and bioindicator studies focus on soil or water pollution. Ultimately, in an environment where humans directly manipulate plant community composition, the importance of pollutants as a further modulator of plant quality and plant-insect interactions remains unclear.

Land use

An outstanding feature of urban landscapes is their heterogeneity. Parks, lawns, gardens, vacant land, industrial areas, and natural habitat remnants form a dense patchwork of land uses. The transitions between land uses are patterned on a finer spatial scale than outside cities, where landscape heterogeneity often include larger areas composed of forests, prairies, deserts, or agriculture. Effects of land use on insect diversity stem not from the land use itself but from the emergent qualities associated with land use, many of which are described in separate sections in this chapter. For instance, some land uses may support higher plant species diversity than others and thus support more herbivore species. Though land use is often not a proximate driver of insect diversity, it is a useful categorization that facilitates management decisions and comparisons of the effects of land use types across cities.

From studies along urban-rural gradients and among land use classes, the concept of biotic homogenization has arisen. Cities share certain environmental features worldwide, which may favor a subset of functionally or taxonomically similar organisms (Groffman et al. 2014). Pigeons and house sparrows exemplify the intuitive appeal of this idea. If cities drive biotic homogenization, then urban insect communities should be more similar to one another – from one site to the next, or one city to the next – than are non-urban insect communities. Even if an individual urban site harbors as many species as an individual comparably sized natural site, biotic homogenization suggests that the urban sites would all harbor the same species, while each natural site would host some unique species, thus accumulating an overall larger species pool in the natural areas.

Predicted effects of urban land use on arthropods include:

- Land uses that closely match natural habitats (e.g. parks and yards with native plants) will support more native species diversity.
- Land uses that are considered more "urban" – with more impervious surfaces and less vegetation – will support lower insect diversity and will favor a subset of native and non-native species.
- As a result of elevated non-native species abundances and reduced native insect diversities, cities will be more homogenized worldwide than natural habits are.

Herbivores

Few studies explicitly tested how herbivore diversity varies among urban land use types. In one exception, Mata et al. (2017) found that golf courses support higher diversity of heteropteran herbivores than gardens and parks in Melbourne, Australia. Most studies of urban herbivores quantify diversity in urban and rural areas, or across gradients of impervious surface cover, rather than for specific land use types. Most of these studies focused on transect counts of adult butterflies and showed that diversity declines at high densities of impervious surface cover (Clark et al. 2007; Dallimer et al. 2012). Across many studies, species that are locally rare and host plant specific are most vulnerable to the development of high-density urban areas (Clark et al. 2007). In a rare study of multiple herbivore taxa in Switzerland, the biotic homogenization hypothesis was supported, with evidence that urban areas were more similar in terms of specialist species composition on birch trees than rural areas (Knop 2016).

Suburban habitats, in contrast, can support diverse butterfly assemblages. In the Philippines, butterfly richness was similar in rural, suburban, and forested areas, suggesting that low-level development may not filter out as many species as dense urban development (Posa and Sodhi 2006). Suburban habitats may be populated by vagrants flying in from surrounding forests and may not themselves support viable populations. However, it is also possible that more nectar-producing or caterpillar host plants are planted and/or recruited naturally in suburban habitats (Clark et al. 2007). Suburban areas may also support more habitat diversity than urban areas (Dallimer et al. 2012). Several studies from around the world suggest that butterfly richness peaks in suburban areas that experience intermediate levels of human disturbance (Hogsden and Hutchinson 2004; Clark et al. 2007).

Pollinators

Pollinator responses to urbanization are mixed. Pollinator diversities in urban areas are often lower than in natural areas but greater than agricultural areas (e.g. Tonietto et al. 2011; Baldock et al. 2015; reviewed by Wenzel et al. 2020). Pollinator taxa often respond differently to urbanization. Several studies have shown positive impacts on bees but negative effects on hoverflies (e.g. Verboven et al. 2014; Baldock et al. 2015). Homogenization of urban pollinator communities compared to those in non-urban landscapes has been noted in the UK, France, Poland, and the USA (Baldock et al. 2015; Deguines et al. 2016; Harrison et al. 2018b; Banaszak-Cibicka and Zmihorski 2020). Few studies have examined pollinators across urban-rural gradients in the tropics, although in a study of West African cities, bee abundance and richness were not affected by urbanization, whilst wasps and beetles declined (Guenat et al. 2019).

Impervious surfaces tend to diminish pollinator communities. A recent review suggests that areas with more than 50% impervious surface are linked with declines in pollinators and pollination services (Wenzel et al. 2020). Landscape-level imperviousness can reduce diversity even within green space sites. For example, declines in bee abundance and species richness in allotments (also known as community gardens) and city parks in Toledo, Ohio, USA, were correlated with increases in the area of nearby impervious surface (Burdine and McCluney 2019b). However, high floral resource availability seemed to counteract the negative impacts of impervious surface proximity on bee richness. The effects appear to differ among bee taxa, with ground-nesting bees more sensitive to impervious surface cover than cavity-nesting or long-tongued species (Fortel et al. 2014; Geslin et al. 2016; Bennett and Lovell 2019).

Urban land uses differ in their suitability for pollinators, and those managed by individual gardeners may be better for pollinators. In the UK, a comparison of nine urban land uses found that residential and allotment gardens were important for pollinators (Baldock et al. 2019), perhaps because they produced significantly more nectar per m^2 than the majority of public green space sites and land uses dominated by impervious surfaces (Tew et al. 2021). Proximity to residential gardens was also important for pollinator diversity in central France (Leve et al. 2019). Bee communities in allotments in Montreal and Quebec, Canada, had higher functional trait diversity than those in nearby cemeteries, although lower functional trait diversity than for parks (Normandin et al. 2017).

Detritivores

Few studies have compared urban detritivore communities to those in other landscapes. In Riyadh, Saudi Arabia, carrion beetle communities were similar between desert and urban habitats but different from agriculture habitats (Mashaly 2017). In Italy, the proportion of detritivores in soil arthropod communities was similar across forest, agricultural, and urban forest habitats (Maisto et al. 2017). A study that examined nest-dwelling arthropods in great tit (*Parus major*) nests in Belgium found that, although urbanization reduced the abundances of saprophagous beetles found in nests, there was no effect on species richness or diversity (Baardsen et al. 2021). Finally, a study in Phoenix, Arizona, USA, compared ground arthropods among the four most abundant types of land use (residential, industrial, agricultural, desert remnant) (McIntyre et al. 2001). The most dominant taxa, both detritivorous groups, were Collembola (springtails) and Formicidae (ants). Overall, detritivores were most abundant in agricultural sites, although their high overall numbers indicate that they are tolerant of urbanization. These studies suggest that some groups of insect detritivores do well in urban landscapes, but there is a need for mechanistic work to explain and predict variation among groups.

Natural enemies

Densely urbanized areas generally support lower diversities of natural enemies than nearby rural or natural areas (e.g. Burks and Philpott 2017). For instance, in a study of parasitoids captured on yellow sticky cards across an urbanization gradient in Wisconsin, total diversity declined 10% in the most highly urbanized areas. In the same study system, natural pest suppression increased with natural enemy diversity, highlighting that heavy urbanization may reduce pest suppression (Bennett and Gratton 2012).

Though most research points to low natural enemy diversity at developed sites, spiders show more variable responses that might be due to differences in particular cities' development regimes (i.e. habitats surrounded by 30% impervious surface might look much different in Finland than in Brazil) or to sampling methods specific to particular studies. For instance, foliage-dwelling spiders in Argentina captured in aspirators tended to be more diverse at urban than rural sites, whereas studies from pitfall traps find various effects of urbanization, from negative (Egerer et al. 2017) to no effect (Alaruikka et al. 2002) to peak abundance at intermediate levels of development (Vergnes et al. 2014). Individually, these studies are informative, especially for localized conservation efforts. However, to understand why results vary so widely, more coordinated research is needed, with multiple standardized collection procedures (pitfall traps, foliage sampling) in a variety of green space types across multiple cities.

Natural enemy diversity has been assessed across specific land use classifications, though these studies are rare. In Phoenix, Arizona, ground-dwelling spiders were sampled using pitfall traps across a variety of land use categories that spanned the rural to urban gradient, including desert parks, agricultural fields, desert remnant habitats, xeric yards, industrial areas, and mesic yards. Because of water inputs, mesic yards and agricultural fields were most productive in terms of vegetation and were characterized by high abundances, but low diversities, of spiders. Specifically, wolf spiders were particularly common in these wetter habitats. This study documents an example of biotic homogenization, not driven by the urbanization gradient, but instead by anthropogenic water inputs (Shochat et al. 2008).

Bioindicator groups

Ant and carabid richness often, but not always, decline at more urbanized sites (e.g. Martinson and Raupp 2013; Perez and Diamond 2019). To disentangle effects of urbanization at different spatial scales, a comprehensive study in Belgium used a unique study design that examined high, medium, and low levels of local urbanization (200 m scale) blocked within landscapes that themselves represent high, medium, or low urbanization (3 km scale; Piano et al. 2020). Terrestrial arthropods examined in this framework included carabids, butterflies, moths, and spiders. For most groups, species richness declined with urbanization at the landscape scale – with the most striking losses among moths – and often also at the local scale. However, groups differed in the relative contributions of local versus landscape effects and whether changes in overall diversity arose from site-level losses versus spatial homogenization. This heterogeneity implies that a single measure of urbanization does not mean the same thing to all taxa, and relevant measures of habitat quality and connectivity may be needed to predict the responses of different functional groups to urbanization.

The frequent loss of species with urban intensification underlies a trade-off between land sparing and land sharing approaches to urban development (Lin and Fuller 2013). To accommodate an equal number of humans, a dense development takes up less space, potentially leaving more land undisturbed. In contrast, a low-density development might harbor more diversity but leaves less undisturbed land. Which approach maximizes the number of species conserved overall? In Tokyo, Japan, and Ottawa and Gatineau, Canada, carabid diversity was maximized by restricting development to the maximum density and smallest footprint (Gagné and Fahrig 2010; Soga et al. 2014). In contrast, open-habitat butterflies in Tokyo benefited from a land-sharing approach (Soga et al. 2014), introducing the need to prioritize functional groups or address a complicated multigroup optimization problem.

Most bioindicator studies have used broad measures of urbanization, but a few have compared diversity among distinct land use types. For example, in New York City, ant diversity was highest in urban forests, intermediate in parks, and lowest in street medians – a result that could represent combined effects of habitat size and complexity (Savage et al. 2015). Diversity also differed among land-use types for ants in Raleigh, North Carolina, USA (Menke et al. 2011) and ground arthropods in Sydney, Australia (Leonard et al. 2018). In these cases, land use category was a more important predictor of species richness than were continuous metrics of urbanization surrounding the sites or vegetation complexity within the sites.

Conclusions

Studies vary in how they quantify urbanization and in how they classify land uses within cities. They additionally vary in whether organisms of interest are sampled within semi-natural habitats along an urbanization gradient or within the urban matrix in street trees, gardens, or parks. These differences in focal habitat type could explain a lot of variation among studies, but this is difficult to parse given that these studies also differ in their approaches and their locations around the world. More studies are needed across cities that look at landscape and local scale effects of land use and land cover. Many studies only use one or, at best, two years of data. Taken together, studies on arthropod diversity and land use that we reviewed underscore the need for standardized, long-term biodiversity monitoring across land use types within and outside of cities. Additionally, including variables in models describing past land uses may reveal any legacy effects of past development regimes (e.g. Duan et al. 2019).

Even within land uses, there can be high variability in suitability of sites for invertebrates. The addition of man-made features, such as green roofs to buildings, can create habitat in locations often lacking in green space. Although we did not exhaustively review literature on green roofs, bioswales, and other highly engineered green spaces, a growing body of research shows that they can host diverse insect communities and have great potential to retrofit densely built areas with functional green space (Spring 2017; Domínguez et al. 2020).

The issue of land sparing versus sharing is one with practical relevance to urban planning, but it has been directly examined in only a few taxa (carabids and butterflies). Given that effects of land sparing and sharing can be modeled from data collected on any gradient of urban development intensity, a reanalysis of existing data in many functional groups could provide important insights.

Biotic homogenization is commonly detected in urban studies that focus on a single land use or site type (Holway and Suarez 2006; Deguines et al. 2016; Knop 2016). However, arthropod communities clearly vary among different land use types, and this variation is an important source of overall diversity in cities (Fournier et al. 2020). In terms of maintaining spatial variation in diversity across cities, it remains to be seen whether the community variation among land uses is enough to counteract the homogenization detected within each individual land use.

Vegetation complexity and diversity

Plants are foundation species in most food chains in natural and urban habitats. In cities, the total amount of vegetation, measured as vegetated area or volume, is a consistent predictor of insect diversity across multiple functional guilds (see "Fragmentation" section). However, the quality of vegetation can additionally affect the number of insect species that are supported by urban habitats. For instance, insect herbivores are often specialized to plants in certain families or genera (Forister et al. 2015) and require these hosts to survive within the urban matrix. Another consequence of host plant specialization is that higher plant diversity usually results in higher herbivore diversity (Haddad et al. 2009). In some cases, host plant diversity may also scale up to increase natural enemy diversity and/or abundance (Dinnage et al. 2012).

However, the diversity of higher trophic levels might not simply be a function of the number of plant species. Native plants (Burghardt et al. 2009) and non-native plants that are closely related to natives (Pearse and Hipp 2009) host more insect species and/or individuals than distantly related non-native plants (Salisbury et al. 2015; Salisbury et al. 2017). Such studies have led to a movement toward planting native plant species in residential yards and commercial landscapes. However, non-native plants can also support herbivores that switch hosts in urban areas and can provide important food sources (larvae and nectar) for adult insects.

Another important aspect of vegetation is its complexity, that is, the diversity in height and texture within a given vegetated area. For instance, a golf course that contains just turf and trees could be considered "simple", while another golf course that has turf, shrubs, herbaceous and/or flowering plants, and trees would be rated as more "complex". Habitat complexity may promote insect diversity, though it may do so through diverse mechanisms. For instance, more complex landscapes likely provide a larger diversity of microclimates, host plant species, and areas from which natural enemies can hide and hunt.

Predicted effects of vegetation complexity and diversity on arthropods include:

- Areas with more native plant diversity will have higher insect herbivore, natural enemy, and pollinator diversity.
- Areas with more complex vegetation will have higher natural enemy diversity

- Species that are adapted to cooler habitats (i.e. species from higher latitudes) may be more likely to occur in complex than simple urban habitats due to the availability of microclimates.
- Species that can effectively thermoregulate and/or disperse might be more likely to occur in simple habitats.

Herbivores

Herbivorous insect diversity generally increases as the area of vegetation increases, but effects of plant diversity, as opposed to area, might be more taxon-specific. In Melbourne, Australia, diversity of native herbivorous heteropterans increased in response to higher local vegetation volumes, not plant diversity (Mata et al. 2017), whereas diversity of butterflies and moths does typically increase with plant diversity (Nagase et al. 2019).

Increasing evidence points to the importance for of native plants, which have co-evolved with native insects, for supporting insect herbivore species in urban habitats. For instance, in a study that quantified caterpillar diversity, yards landscaped with just native plants supported more species than yards with a mix of native and non-native plant species (Burghardt et al. 2009). In another study, a suite of native tree species was experimentally planted within a forest that was previously dominated by a non-native plant species. The planted natives supported significantly more caterpillar diversity than the non-native plant species (Grunzweig et al. 2015). However, even within native species, a small number of keystone plant genera support the vast majority of caterpillar species, with *Quercus*, *Salix*, and *Prunus* supporting the most total species within the USA (Narango et al. 2020).

While non-native urban plant species are generally thought to be poor resources for native herbivores compared to native plants, non-native plants can also serve as critical habitat and food plants, especially in cases where an herbivore's co-evolved host plant is not available (Graves and Shapiro 2003). In particular, non-native plant species that are more closely related to native species might be more likely to support native herbivores. For instance, in a common garden in Davis, California, USA, native oak (*Quercus*) individuals had the highest levels of insect herbivory, and non-native oaks had more leaf damage if they were closely related to native oaks (Pearse and Hipp 2009). Though this study did not explicitly test for herbivore diversity, the results point to more herbivores being supported on non-native oaks closely related to native oaks.

The process of breeding plants for cultivation in urban settings by definition favors a subset of the total range of plant traits genotypes. Though it has not been studied until recently, the process of breeding may additionally affect herbivore diversity via the choice of plant traits represented in urban cultivars. For instance, more diverse caterpillar assemblages were documented on variegated cultivars (i.e. those with variously colored leaves) than cultivars of just one color (Baisden et al. 2018). However, another study found no difference in Hemipteran herbivore species richness across four paired cultivar-wild type species (Poythress and Affolter 2018). More work is needed to elucidate why some studies find effects of breeding and others do not.

Compared to vegetation diversity, the effects of vegetation complexity on herbivore diversity are harder to predict and less well studied. Because more complex habitats support more and more diverse predator communities, herbivore abundance could be lower in more complex habitats, which might promote a higher diversity of herbivores through reduced competition and the more varied niches available in complex habitats. It is also possible that the effects of vegetation complexity on herbivore diversity hinge on plant diversity and/or identity, such that more complex habits support more herbivore species but only if plant species diversity is high or plants that support diverse herbivores are present. However, these hypotheses remain untested to our knowledge.

Pollinators

Pollinator-focused studies typically examine the relationships between flowering plants and pollinator richness and abundance, as all taxa rely on flowers for at least part of their life cycle. A global review concluded that while pollinator diversity benefits from urban green space area at a landscape scale, it also increases with the local availability of nesting resources and flowering plants (Wenzel et al. 2020). Both flowering plant species richness and floral abundance are likely to be important for pollinator communities. A more diverse flowering plant community has the potential to support a more diverse pollinator community, and increased numbers of flowers will provide more nectar and pollen resources for pollinators. For example, UK towns and cities hosted more flowering plant species than nearby agricultural areas, and this pattern was driven by high numbers of non-native species in urban areas (Baldock et al. 2015).

Multiple studies have demonstrated beneficial effects of increased flowering plant diversity or floral abundance on pollinating insects in urban landscapes (e.g. Smith et al. 2006; Matteson and Langellotto 2010; Pardee and Philpott 2014; Hulsmann et al. 2015; Banaszak-Cibicka et al. 2016; Quistberg et al. 2016; Ballare et al. 2019). Studies have shown differing effects of native and non-native flowers on pollinator communities. For example, increases in native plantings increased bee richness and abundance in backyard gardens in Ohio, USA (Pardee and Philpott 2014). In contrast, adding native plant species to community gardens in New York did not increase the species richness of bees, butterflies, or predatory wasps, as butterflies and megachilid (leaf-cutter) bees frequently foraged on non-native species (Matteson and Langellotto 2011). A recent review concluded that positive effects of adding flowers were independent of the plant's origin, which suggests that from a pollinator's perspective, urban land uses with more native flowers are not necessarily better than those dominated by non-natives (Wenzel et al. 2020).

Detritivores

Limited research addresses how vegetation diversity and complexity affect insect detritivores in urban landscapes. A study in the metropolitan area of Tokyo, Japan, found that arthropod communities varied among grassland types (Hironaka and Koike 2013). In more artificial urban grasslands, insect detritivores, as well as small carnivores and omnivores, were abundant; whereas larger predators were abundant in rural, semi-natural grasslands. However, it is difficult to disentangle whether these findings are due to differences in vegetation complexity, landscape context, or a combination of these factors, as vegetation also varies along the urbanization gradient. A study in Melbourne, Australia, examined the effects of habitat complexity (including vegetation) of low- and high-complexity urban parks and high-complexity woodland remnants on detritivore assemblages (Ossola et al. 2016). While the abundance and species richness of macrofauna detritivores (Isopoda, Diplopoda, Amphipoda) were higher in more complex habitats, those of mesofauna detritivores (mainly Collembola) were not. A replicated experiment in the UK examined how arthropod assemblages differed among horticultural plant communities with different geographical origins: native to the UK, near-native (Northern Hemisphere), and non-native (Southern Hemisphere; Salisbury et al. 2017). Detritivores (mainly Collembola) made up a large proportion of the sampled arthropods, and higher numbers were associated with native plants compared to non-native plants. Together, these studies suggest more research is needed to understand the role of vegetation diversity and complexity in structuring detritivore communities.

Natural enemies

While herbivorous insects have clear evolutionary associations with plants, and most are specialized to plant genera or families (Forister et al. 2015), natural enemies vary more in their specialization; many species eat a wide variety of insects that feed on different plants (e.g. spiders and lady beetles). However, natural enemy species may tend to be attracted more to some plants than others. For instance, among ten plant species native to Maryland, USA, the composition of natural enemy taxa found on individual plant species varied (Frank et al. 2008). Thus, we may be able to tailor plant species selection to target focal insect pests. Similarly, the diversity of predaceous Heteropteran bug diversity increased with plant diversity, but vegetation volume might be an even better predictor (Mata et al. 2017). However, localized native plantings intended to enhance natural enemy abundance and/or diversity may fail to do so if barriers to natural enemy dispersal exist at the landscape scale (Matteson and Langellotto 2011).

Effects of plant origin on natural enemy diversity in urban systems is similarly underexplored, and results for the few studies that exist are mixed. For example, in one case, non-native plantings supported lower natural enemy richness than native congeners, although the effect was small and inconsistent across plant genera (Parsons et al. 2020). In contrast, native and non-native oak and maple street trees supported similar natural enemy communities, even though natives hosted more abundant herbivores (Frank et al. 2019). The relative benefits of planting native versus non-native taxa in yards for enhancing natural enemy diversity remains an area for future research.

Strong evidence links vegetation complexity and natural enemy abundance or pest control (e.g. Shrewsbury and Raupp 2000), but fewer studies address effects of complexity on natural enemy diversity. More research is needed to elucidate whether vegetation complexity benefits a small subset of predators, or results in greater overall diversity of natural enemies. Further, the specific mechanisms driving effects of complexity on natural enemies may include some combination of increased alternative prey, pollen and nectar resources, and more varied microclimates, and future studies could investigate which of these specific mechanisms is most important to support conservation efforts in urban landscapes.

Bioindicator groups

Most ants and carabids are not herbivores, so vegetation complexity and diversity most likely affect these groups by enhancing overall habitat volume and heterogeneity or by indirect effects via lower trophic levels. Accordingly, effects of vegetation complexity on ant and carabid diversity are mixed. In Sydney, Australia, golf courses, ant diversity increased with vegetation complexity (represented by more layers of vegetation including groundcover, shrubs, and trees; Nooten et al. 2019). Similarly, pitfall traps in Durban, South Africa, detected more arthropod species in complex traffic islands than those planted with lawn alone (Whitmore et al. 2002).

Contrasting results are also common. For example, studies in Melbourne golf courses and reserves (Ossola et al. 2015), Sydney parks, road verges, and median strips (Leonard et al. 2018), and Perth coastal habitat remnants (Cross et al. 2016), detected neutral or negative effects of vegetation complexity or diversity on ant and arthropod species richness. Each study suggested an alternative driver that may have overpowered the predicted effects of vegetation – such as the presence of an invasive ant species, effects of land use, or surrounding matrix.

These mixed results suggest that, compared to other urban drivers, vegetation complexity and diversity may exert weak effects on bioindicator groups. Methodological caution is also in order. Ants and carabids are often sampled using pitfall traps, which may work better in simple habitats (Lang 2000).

Conclusions

Studies addressing effects of urban vegetation on insect diversity commonly focus on butterflies, but it is unclear to what extent these species are dispersing into the urban matrix or using urban plants as host plants in larval stages. Thus, there is a need to disentangle effects on herbivorous versus flower-visiting life stages and, accordingly, which life stage represents the demographically limiting step for urban populations. More studies along these lines would provide better information on how vegetation diversity and complexity might serve all insect life stages.

As predicted, effects of vegetation diversity and complexity appear to be stronger, or at least better studied, in herbivores and pollinators than in other groups. The roles of native and non-native plants on these groups are nuanced. While caterpillars may be most diverse on natives, non-natives from the same hemisphere or clade may also support herbivores, and an even broader suite of plants can support nectar feeders and natural enemies.

City or fragment age

All of the drivers discussed previously can operate over time and space. For instance, habitats within cities can become more or less fragmented over time, and land use can change in the span of weeks or years. However, the drivers of insect biodiversity have been studied much more over space than over time.

Overall effects of city age on biodiversity may represent tension between extinction debt and colonization by native and non-native species. At a local scale, biodiversity in older fragments of natural or semi-natural habitats should be driven by the same processes. As cities expand, and as areas within cities are demolished and rebuilt, parcels within cities can vary greatly in time since development, which complicates predictions about how biodiversity may respond to overall aging of a city. Underlying plant diversity – which may drive insect diversity, particularly for herbivores – tends to decline within cities over time (Tait et al. 2005; Hahs et al. 2009). We do not provide predictions for this section, as we think it is equally likely that city age may increase or reduce insect diversity.

All groups

Few studies have assessed the role of time since urbanization on diversity of insects. One notable exception monitored ant diversity before, during, and after conversion of an abandoned orchard to a housing development in West Lafayette, Indiana, USA (Buczkowski and Richmond 2012). Most species disappeared during construction, and recolonization was still incomplete at 10-year-old houses. A study of carabids and spiders in Hubei province, China, suggested that community composition (not diversity) in agricultural habitats changed between 1995 and 2013 as the amount of surrounding urbanization increased (Duan et al. 2019). Finally, one study examined a reverse situation – the "rewilding" of brownfields that had been abandoned for 2 to 20 years in the West Midlands region in the UK (Small et al. 2002). Carabid richness peaked at early successional sites. However, since succession proceeded at different rates on different sites, there was not a clear relationship with time per se.

Conclusions

As urban and suburban areas continue to expand, there is tremendous potential to examine temporal trends in biodiversity and species composition by collecting high-resolution time-series data while habitats are converted or fragmented. Such studies will reveal the temporal scale of eradication, recolonization, or realization of extinction debt, as well as effects of urbanization on patterns of succession and community assembly.

Affluence

Urban biodiversity often increases in more affluent neighborhoods, an important pattern from the perspective of social justice. Sometimes termed the "luxury effect", this pattern was originally identified in plants (Hope et al. 2008) and has since been found in numerous taxa worldwide (Leong et al. 2018; Kuras et al. 2020). Insects, however, remain heavily underrepresented.

Predicted effects of affluence on insect diversity include:

- If affluence increases plant diversity, the diversity of insect herbivores, natural enemies, and pollinators should also increase.
- If affluence increases total vegetation cover or complexity, it may also affect the abundance and diversity of other functional groups.
- However, landscaping practices associated with affluence may depress the value of the added plant diversity. If increased vegetation diversity comes from non-native plants, cultivars that provide few resources, or is associated with heavy chemical inputs and frequent disturbance, affluence may break the links between plant diversity and higher trophic levels.

All groups

We recovered no studies that related affluence to diversity of herbivores, natural enemies, detritivores, or arthropod bioindicator groups. Only for pollinators has the literature begun to link affluence to garden management, floral resource diversity, and insect diversity. Although Matteson et al. (2013) found no relationship between income and floral richness in New York City, USA, other studies have supported links between affluence and diversity. Lowenstein and Minor (2016) found that household income, race, and ethnicity were related to flowering plant community composition in Chicago, Illinois, USA, with more weedy species in lower-income neighborhoods. In UK residential gardens, higher household income was linked to a greater abundance and diversity of flowering plants, as well as greater pollinator abundance (Baldock et al. 2019). Residential gardens play an important role in floral resource provision in UK cities, providing up to 85% of available nectar, predominantly from non-native plants (Tew et al. 2021). Therefore, householder choice of horticultural plants likely plays an important role in shaping pollinator communities both in gardens and in the wider urban landscape. A more diverse flowering plant community could be expected to result in a greater range of floral morphologies, accessible to a wider range of pollinator species, as well as potentially longer flowering seasons if there are early-flowering and later-flowering species within the communities.

Conclusions

Although effects of affluence on biodiversity are frequently documented, work in insects lags behind other groups, and thus, insect ecologists have much to contribute to the growing literature on the hidden effects of socioeconomic status on human quality of life (Schell et al. 2020). Little information is available to support or refute our predictions related to affluence and insect diversity, and future work should test these predictions, examine the underlying social mechanisms, and address the spatial scale at which affluence acts on different functional groups – that is, whether effects are dominated by management choices in individual yards, neighborhoods, or larger-scale, publicly managed resources such as street trees.

Future directions: research and conservation

More long-term studies with standardized, biologically relevant protocols across networks of cities are needed to improve our understanding of the spatiotemporal mechanisms driving insect diversity. For instance, many studies measure landscape properties within multiple Euclidean radii of the focal site, then conduct model selection to identify the relevant spatial scale for their study system. Although this approach is appropriate for exploratory studies, the field is ripe for a shift to predictive, rather than exploratory, work. Exploratory results cannot be generalized to other cities or taxa, and Euclidean radii may not represent the most biologically relevant variable (accessible foraging area) for insects. Instead, we need to develop a mechanistic understanding of which urban landscape features promote or reduce functional connectivity among habitat patches. Ideally, these and other studies would span multiple years to decades to reveal to what extent cities support insect diversity long-term, and in what circumstances they act as sinks wherein diversity declines over time. Most urban studies use space-for-time substitutions to understand effects of urban drivers, but temporal studies are needed to validate the assumption that space can substitute for time.

Many urban studies correlate diversity to environmental conditions – occasionally, biotic environmental conditions (such as vegetation) – but almost none look at coexistence mechanisms among organisms. We found evidence that urban habitat patches lose diversity because they are smaller, more isolated, and are climatically and structurally altered from their natural counterparts. These features may directly filter out species, but they might also break coexistence mechanisms among species, by disrupting niche partitioning, tradeoffs, or competitive networks. More research along these lines is needed to better understand the mechanisms underlying diversity patterns and to facilitate effective management.

Our review identified several other broad knowledge gaps in urban insect diversity. There is a strong need for studies that focus on the effects of variables with demonstrated consequences for certain insects in the limited studies that have been done, notably urban pollution (light, noise, and chemical), urban heat, and urban moisture deficits. Tropical regions (Baldock 2020; Silva et al. 2020; Wenzel et al. 2020) and insect larval stages are additionally underrepresented in urban research. In particular, integrated assessments are needed across life cycles for holometabolous groups that may be most sensitive to different environmental drivers during each life stage. Belowground insects, in particular detritivores, are sorely underrepresented in the urban insect diversity literature, which is perhaps surprising given the critical service they provide.

Overall, the most consistent driver of urban insect diversity in the studies we reviewed is patch size. Thus, the more urban areas can be rewilded, the more likely they are to support diverse insect populations into the future. Although human preferences for urban vegetation are complex (Aronson et al. 2017), growing evidence suggests that people do prefer more densely vegetated or diverse landscapes in parks and yards when these are visibly maintained (Peterson et al. 2012; Harris et al. 2018). Thus, complex, diverse vegetation may benefit both biodiversity and human experience, and replacing urban lawns with diverse, layered landscapes may be a win-win scenario.

Key companion papers

Burghardt, K.T., D.W. Tallamy, W.G. Shriver. 2009. Impact of native plants on bird and butterfly biodiversity in suburban landscapes. *Conservation Biology* 23(1):219–224.

Fenoglio, M.S., M.R. Rossetti, M. Videla. 2020. Negative effects of urbanization on terrestrial arthropod communities: A meta-analysis. *Global Ecology and Biogeography* 29(8):1412–1429.

Piano, E., et al. 2017. Urbanization drives community shifts towards thermophilic and dispersive species at local and landscape scales. *Global Change Biology* 23(7):2554–2564.

Wenzel, A., I. Grass, V.V. Belavadi, T. Tscharntke. 2020. How urbanization is driving pollinator diversity and pollination – A systematic review. *Biological Conservation* 241:108321.

Acknowledgements

EKM and EY contributed equally to this manuscript. This work was supported by the USDA National Institute of Food and Agriculture, Hatch project 1018689 to EY.

References

Adams, B.J., E. Li, C.A. Bahlai, E.K. Meineke, T.P. McGlynn, B.V. Brown. 2020. Local-and landscape-scale variables shape insect diversity in an urban biodiversity hot spot. *Ecological Applications* 30(4):e02089.

Alaruikka, D., D.J. Kotze, K. Matveinen, J. Niemelä. 2002. Carabid beetle and spider assemblages along a forested urban – Rural gradient in southern Finland. *Journal of Insect Conservation* 6:195–206.

Angold, P.G., J.P. Sadler, M.O. Hill, A. Pullin, S. Rushton, K. Austin, E. Small, B. Wood, R. Wadsworth, R. Sanderson, K. Thompson. 2006. Biodiversity in urban habitat patches. *Science of the Total Environment* 360(1–3):196–204.

Aronson, M.F.J., C.A. Lepczyk, K.L. Evans, M.A. Goddard, S.B. Lerman, J.S. MacIvor, C.H. Nilon, T. Vargo. 2017. Biodiversity in the city: Key challenges for urban green space management. *Frontiers in Ecology and the Environment* 15(4):189–196.

Baardsen, L.F., L. De Bruyn, F. Adriaensen, J. Elst, D. Strubbe, D. Heylen, E. Matthysen. 2021. No overall effect of urbanization on nest-dwelling arthropods of great tits (*Parus major*). *Urban Ecosystems* 24(4):959–972.

Baisden, E.C., D.W. Tallamy, D.L. Narango, E. Boyle. 2018. Do cultivars of native plants support insect herbivores? *HortTechnology* 28(5):596–606.

Baker, L.A., D. Hope, Y. Xu, J. Edmonds, L. Lauver. 2001. Nitrogen balance for the Central Arizona – Phoenix (CAP) ecosystem. *Ecosystems* 4:582–602.

Baldock, K.C.R. 2020. Opportunities and threats for pollinator conservation in global towns and cities. *Current Opinion in Insect Science* 38:63–71.

Baldock, K.C.R., M.A. Goddard, D.M. Hicks, W.E. Kunin, N. Mitschunas, H. Morse, L.M. Osgathorpe, S.G. Potts, K.M. Robertson, A.V. Scott, P.P.A. Staniczenko, G.N. Stone, I.P. Vaughan, J. Memmott. 2019. A systems approach reveals urban pollinator hotspots and conservation opportunities. *Nature Ecology & Evolution* 3(3):363–373.

Baldock, K.C.R., M.A. Goddard, D.M. Hicks, W.E. Kunin, N. Mitschunas, L.M. Osgathorpe, S.G. Potts, K.M. Robertson, A.V. Scott, G.N. Stone, I.P. Vaughan, J. Memmott. 2015. Where is the UK's pollinator biodiversity? The importance of urban areas for flower-visiting insects. *Proceedings of the Royal Society of London Series B: Biological Sciences* 282(1803):20142849.

Ballare, K.M., J.L. Neff, R. Ruppel, S. Jha. 2019. Multi-scalar drivers of biodiversity: Local management mediates wild bee community response to regional urbanization. *Ecological Applications* 29(3):e01869.

Banaszak-Cibicka, W. 2014. Are urban areas suitable for thermophilic and xerothermic bee species (Hymenoptera: Apoidea: Apiformes)? *Apidologie* 45:145–155.

Banaszak-Cibicka, W., H. Ratyńska, Ł. Dylewskia. 2016. Features of urban green space favourable for large and diverse bee populations (Hymenoptera: Apoidea: Apiformes). *Urban Forestry & Urban Greening* 20:448–452.

Banaszak-Cibicka, W., M. Żmihorski. 2020. Are cities hotspots for bees? Local and regional diversity patterns lead to different conclusions. *Urban Ecosystems* 23:713–722.

Bar-On, Y.M., R. Phillips, R. Milo. 2018. The biomass distribution on Earth. *Proceedings of the National Academy of Sciences* 115(25):6506–6511.

Bennett, A.B., C. Gratton. 2012. Local and landscape scale variables impact parasitoid assemblages across an urbanization gradient. *Landscape and Urban Planning* 104(1):26–33.

Bennett, A.B., S. Lovell. 2019. Landscape and local site variables differentially influence pollinators and pollination services in urban agricultural sites. *PLoS ONE* 14(2):e0212034.

Bolger, D.T., A.V. Suarez, K.R. Crooks, S.A. Morrison, T.J. Case. 2000. Arthropods in urban habitat fragments in southern California: Area, age, and edge effects. *Ecological Applications* 10(4):1230–1248.

Buczkowski, G., D.S. Richmond. 2012. The effect of urbanization on ant abundance and diversity: A temporal examination of factors affecting biodiversity. *PLoS ONE* 7(8):e41729.

Burdine, J.D., K.E. McCluney. 2019a. Differential sensitivity of bees to urbanization-driven changes in body temperature and water content. *Scientific Reports* 9:1643.

Burdine, J.D., K.E. McCluney. 2019b. Interactive effects of urbanization and local habitat characteristics influence bee communities and flower visitation rates. *Oecologia* 190(4):715–723.

Burghardt, K.T., D.W. Tallamy, W. Gregory Shriver. 2009. Impact of native plants on bird and butterfly biodiversity in suburban landscapes. *Conservation Biology* 23(1):219–224.

Burkle, L.A., R.E. Irwin. 2010. Beyond biomass: Measuring the effects of community-level nitrogen enrichment on floral traits, pollinator visitation and plant reproduction. *Journal of Ecology* 98(3):705–717.

Burks, J.M., S.M. Philpott. 2017. Local and landscape drivers of parasitoid abundance, richness, and composition in urban gardens. *Environmental Entomology* 46(2):201–209.

Cane, J.H., R.L. Minckley, L.J. Kervin, T.H. Toulston, N.M. Williams. 2006. Complex responses within a desert bee guild (Hymenoptera: Apiformes) to urban habitat fragmentation. *Ecological Applications* 16(2):632–644.

Chappell, M.A. 1984. Temperature regulation and energetics of the solitary bee *Centris pallida* during foraging and intermale mate competition. *Physiological Zoology* 57(2):215–225.

Chown, S.L., G.A. Duffy. 2015. Thermal physiology and urbanization: Perspectives on exit, entry and transformation rules. *Functional Ecology* 29(2):902–912.

Christie, F.J., G. Cassis, D.F. Hochuli. 2010. Urbanization affects the trophic structure of arboreal arthropod communities. *Urban Ecosystems* 13(2):169–180.

Clark, P.J., J.M. Reed, F.S. Chew. 2007. Effects of urbanization on butterfly species richness, guild structure, and rarity. *Urban Ecosystems* 10:321–337.

Couceiro, S.R.M., N. Hamada, R.L.M. Ferreira, B.R. Forsberg, J.O. da Silva. 2007. Domestic sewage and oil spills in streams: Effects on edaphic invertebrates in flooded forest, Manaus, Amazonas, Brazil. *Water, Air, & Soil Pollution* 180:249–259.

Cross, S.L., A.T. Cross, D.J. Merritt, K.W. Dixon, A.N. Andersen. 2016. Biodiversity responses to vegetation structure in a fragmented landscape: Ant communities in a peri-urban coastal dune system. *Journal of Insect Conservation* 20(3):485–495.

Dale, A.G., S.D. Frank. 2014. Urban warming trumps natural enemy regulation of herbivorous pests. *Ecological Applications* 24(7):1596–1607.

Dallimer, M., J.R. Rouquette, A.M.J. Skinner, P.R. Armsworth, L.M. Maltby, P.H. Warren, K.J. Gaston. 2012. Contrasting patterns in species richness of birds, butterflies and plants along riparian corridors in an urban landscape. *Diversity and Distributions* 18(8):742–753.

David, T.I., J. Storkey, C.J. Stevens. 2019. Understanding how changing soil nitrogen affects plant – pollinator interactions. *Arthropod-Plant Interactions* 13:671–684.

Deguines, N., R. Julliard, M. de Flores, C. Fontaine. 2016. Functional homogenization of flower visitor communities with urbanization. *Ecology and Evolution* 6(7):1967–1976.

Diamond, S.E., R.R. Dunn, S.D. Frank, N.M. Haddad, R.A. Martin. 2015. Shared and unique responses of insects to the interaction of urbanization and background climate. *Current Opinion in Insect Science* 11:71–77.

Dinnage, R., M.W. Cadotte, N.M. Haddad, G.M. Crutsinger, D. Tilman. 2012. Diversity of plant evolutionary lineages promotes arthropod diversity. *Ecology Letters* 15(11):1308–1317.

Domínguez, M.V.S., E. González, D. Fabián, A. Salvo, M.S. Fenoglio. 2020. Arthropod diversity and ecological processes on green roofs in a semi-rural area of Argentina: Similarity to neighbor ground habitats and landscape effects. *Landscape and Urban Planning* 199:103816.

Dornelas, M., N.J. Gotelli, H. Shimadzu, F. Moyes, A.E. Magurran, B.J. McGill. 2019. A balance of winners and losers in the Anthropocene. *Ecology Letters* 22(5):847–854.

Duan, M., W. Hu, Y. Liu, Z. Yu, X. Li, P. Wu, F. Zhang, H. Shi, J. Baudry. 2019. The influence of landscape alterations on changes in ground beetle (Carabidae) and spider (Araneae) functional groups between 1995 and 2013 in an urban fringe of China. *Science of the Total Environment* 689:516–525.

Dunning, J.B., B.J. Danielson, H.R. Pulliam. 1992. Ecological processes that affect populations in complex landscapes. *Oikos* 65(1):169–175.

Egerer, M.H., C. Arel, M.D. Otoshi, R.D. Quistberg, P. Bichier, S.M. Philpott. 2017. Urban arthropods respond variably to changes in landscape context and spatial scale. *Journal of Urban Ecology* 3(1):jux001.

Faeth, S.H., T.C. Kane. 1978. Urban biogeography: City parks as islands for Diptera and Coleoptera. *Oecologia* 32(1):127–133.

Fenoglio, M.S., M.R. Rossetti, M. Videla. 2020. Negative effects of urbanization on terrestrial arthropod communities: A meta-analysis. *Global Ecology and Biogeography* 29(8):1412–1429.

Fisogni, A., N. Hautekèete, Y. Piquot, M. Brun, C. Vanappelghem, D. Michez, F. Massol. 2020. Urbanization drives an early spring for plants but not for pollinators. *Oikos* 129(11):1681–1691.

Forister, M.L., V. Novotny, A.K. Panorska, L. Baje, Y. Basset, P.T. Butterill, L. Cizek, P.D. Coley, F. Dem, I.R. Diniz, P. Drozd, M. Fox, A.E. Glassmire, R. Hazen, J. Hrcek, J.P. Jahner, O. Kaman, T.J. Kozubowski, T.A. Kursar, O.T. Lewis, J. Lill, R.J. Marquis, S.E. Miller, H.C. Morais, M. Murakami, H. Nickel, N.A. Pardikes, R.E. Ricklefs, M.S. Singer, A.M. Smilanich, J.O. Stireman, S. Villamarin-Cortez, S. Vodka, M. Volf, D.L. Wagner, T. Walla, G.D. Weiblen, L.A. Dyer. 2015. The global distribution of diet breadth in insect herbivores. *Proceedings of the National Academy of Sciences* 112(2):442–447.

Fortel, L., M. Henry, L. Guilbaud, A.L. Guirao, M. Kuhlmann, H. Mouret, O. Rollin, B.E. Vaissière. 2014. Decreasing abundance, increasing diversity and changing structure of the wild bee community (Hymenoptera: Anthophila) along an urbanization gradient. *PLoS ONE* 9(8):e104679.

Fournier, B., D. Frey, M. Moretti. 2020. The origin of urban communities: From the regional species pool to community assemblages in city. *Journal of Biogeography* 47(3):615–629.

Fox, R. 2013. The decline of moths in Great Britain: A review of possible causes. *Insect Conservation and Diversity* 6(1):5–19.

Frank, S.D., K.M. Backe, C. McDaniel, M. Green, S. Widney, R.R. Dunn. 2019. Exotic urban trees conserve similar natural enemy communities to native congeners but have fewer pests. *PeerJ* 7:e6531.

Frank, S.D., P.M. Shrewsbury, O. Esiekpe. 2008. Spatial and temporal variation in natural enemy assemblages on Maryland native plant species. *Environmental Entomology* 37(2):478–486.

Gagné, S.A., L. Fahrig. 2010. The trade-off between housing density and sprawl area: Minimizing impacts to carabid beetles (Coleoptera: Carabidae). *Ecology and Society* 15(4):12.

García-Robledo, C., E.K. Kuprewicz, C.L. Staines, T.L. Erwin, W.J. Kress. 2016. Limited tolerance by insects to high temperatures across tropical elevational gradients and the implications of global warming for extinction. *Proceedings of the National Academy of Sciences* 113(3):680–685.

Gely, C., S.G.W. Laurance, N.E. Stork. 2020. How do herbivorous insects respond to drought stress in trees? *Biological Reviews* 95(2):434–448.

Geslin, B., V. Le Feon, M. Folschweiller, F. Flacher, D. Carmignac, E. Motard, S. Perret, I. Dajoz. 2016. The proportion of impervious surfaces at the landscape scale structures wild bee assemblages in a densely populated region. *Ecology and Evolution* 6(18):6599–6615.

Gibb, H., D.F. Hochuli. 2002. Habitat fragmentation in an urban environment: Large and small fragments support different arthropod assemblages. *Biological Conservation* 106(1):91–100.

Gibbs, J.P., E.J. Stanton. 2001. Habitat fragmentation and arthropod community change: Carrion beetles, phoretic mites, and flies. *Ecological Applications* 11(1):79–85.

Graves, S.D., A.M. Shapiro. 2003. Exotics as host plants of the California butterfly fauna. *Biological Conservation* 110:413–433.

Gregg, J.W., C.G. Jones, T.E. Dawson. 2003. Urbanization effects on tree growth in the vicinity of New York City. *Nature* 424(6945):183–187.

Groffman, P.M., J. Cavender-Bares, N.D. Bettez, J.M. Grove, S.J. Hall, J.B. Heffernan, S.E. Hobbie, K.L. Larson, J.L. Morse, C. Neill, K. Nelson, J. O'Neil-Dunne, L. Ogden, D.E. Pataki, C. Polsky, R.R. Chowdhury, M.K. Steele. 2014. Ecological homogenization of urban USA. *Frontiers in Ecology and the Environment* 12(1):74–81.

Grunzweig, L., D.J. Spiering, A. Labatore, R.J. Warren. 2015. Non-native plant invader renders suitable habitat unsuitable. *Arthropod-Plant Interactions* 9(6):577–583.

Guenat, S., W.E. Kunin, A.J. Dougill, M. Dallimer. 2019. Effects of urbanisation and management practices on pollinators in tropical Africa. *Journal of Applied Ecology* 56(1):214–224.

Haddad, N.M., G.M. Crutsinger, K. Gross, J. Haarstad, J.M.H. Knops, D. Tilman. 2009. Plant species loss decreases arthropod diversity and shifts trophic structure. *Ecology Letters* 12(10):1029–1039.

Haddad, N.M., J. Haarstad, D. Tilman. 2000. The effects of long-term nitrogen loading on grassland insect communities. *Oecologia* 124:73–84.

Hahs, A.K., M.J. McDonnell, M.A. McCarthy, P.A. Vesk, R.T. Corlett, B.A. Norton, S.E. Clemants, R.P. Duncan, K. Thompson, M.W. Schwartz, N.S.G. Williams. 2009. A global synthesis of plant extinction rates in urban areas. *Ecology Letters* 12(11):1165–1173.

Hamblin, A.L., E. Youngsteadt, S.D. Frank. 2018. Wild bee abundance declines with urban warming, regardless of floral density. *Urban Ecosystems* 21(3):419–428.

Hamblin, A.L., E. Youngsteadt, M.M. López-Uribe, S.D. Frank. 2017. Physiological thermal limits predict differential responses of bees to urban heat-island effects. *Biology Letters* 13(6):20170125.

Harris, V., D. Kendal, A.K. Hahs, C.G. Threlfall. 2018. Green space context and vegetation complexity shape people's preferences for urban public parks and residential gardens. *Landscape Research* 43(1):150–162.

Harrison, J.F., H.A. Woods, S.P. Roberts, editors. 2012. *Ecological and Environmental Physiology of Insects.* Oxford University Press, Oxford.

Harrison, T., J. Gibbs, R. Winfree. 2018a. Forest bees are replaced in agricultural and urban landscapes by native species with different phenologies and life-history traits. *Global Change Biology* 24(1):287–296.

Harrison, T., J. Gibbs, R. Winfree. 2018b. Phylogenetic homogenization of bee communities across ecoregions. *Global Ecology and Biogeography* 27(12):1457–1466.

Herrmann, D.L., I.S. Pearse, J.H. Baty. 2012. Drivers of specialist herbivore diversity across 10 cities. *Landscape and Urban Planning* 108(2–4):123–130.

Hinners, S.J., C.A. Kearns, C.A. Wessman. 2012. Roles of scale, matrix, and native habitat in supporting a diverse suburban pollinator assemblage. *Ecological Applications* 22(2):1923–1935.

Hironaka, Y., F. Koike. 2013. Guild structure in the food web of grassland arthropod communities along an urban-rural landscape gradient. *Ecoscience* 20(2):148–160.

Hogsden, K.L., T. Hutchinson. 2004. Butterfly assemblages along a human disturbance gradient in Ontario, Canada. *Canadian Journal of Zoology* 82(5):739–748.

Holt, R.D., J.H. Lawton, G.A. Polis, N.D. Martinez. 1999. Trophic rank and the species – area relationship. *Ecology* 80(5):1495–1504.

Holway, D.A., A.V. Suarez. 2006. Homogenization of ant communities in mediterranean California: The effects of urbanization and invasion. *Biological Conservation* 127(3):319–326.

Hoover, S.E.R., J.J. Ladley, A.A. Shchepetkina, M. Tisch, S.P. Gieseg, J.M. Tylianakis. 2012. Warming, CO_2, and nitrogen deposition interactively affect a plant-pollinator mutualism. *Ecology Letters* 15(3):227–234.

Hope, D., C. Gries, W. Zhu, W.F. Fagan, C.L. Redman, N.B. Grimm, A.L. Nelson, C. Martin, A. Kinzig. 2008. Socioeconomics drive urban plant diversity. *Proceedings of the National Academy of Sciences of the United States of America* 100(15):8788–8792.

Hulsmann, M., H. von Wehrden, A.-M. Klein, S.D. Leonhardt. 2015. Plant diversity and composition compensate for negative effects of urbanization on foraging bumble bees. *Apidologie* 46:760–770.

Hung, K.L.J., J.S. Ascher, J.A. Davids, D.A. Holway. 2019. Ecological filtering in scrub fragments restructures the taxonomic and functional composition of native bee assemblages. *Ecology* 100(5):e02654.

Hung, K.L.J., J.S. Ascher, D.A. Holway. 2017. Urbanization-induced habitat fragmentation erodes multiple components of temporal diversity in a Southern California native bee assemblage. *PLoS ONE* 12(8):e0184136.

Idso, C.D., S.B. Idso, R.C. Balling. 2001. An intensive two-week study of an urban CO_2 dome in Phoenix, Arizona, USA. *Atmospheric Environment* 35:995–1000.

Jamieson, M.A., L.A. Burkle, J.S. Manson, J.B. Runyon, A.M. Trowbridge, J. Zientek. 2017. Global change effects on plant – insect interactions: The role of phytochemistry. *Current Opinion in Insect Science* 23:70–80.

Johansson, V., A. Koffman, M. Hedblom, G. Deboni, P. Andersson. 2018. Estimates of accessible food resources for pollinators in urban landscapes should take landscape friction into account. *Ecosphere* 9(10):e02486.

Jones, M.E., T.D. Paine. 2006. Detecting changes in insect herbivore communities along a pollution gradient. *Environmental Pollution* 143(3):377–387.

Knop, E. 2016. Biotic homogenization of three insect groups due to urbanization. *Global Change Biology* 22(1):228–236.

Krauss, J., I. Steffan-Dewenter, T. Tscharntke. 2003. How does landscape context contribute to effects of habitat fragmentation on diversity and population density of butterflies? *Journal of Biogeography* 30(6):889–900.

Kruess, A., T. Tscharntke. 1994. Habitat fragmentation, species loss, and biological control. *Science* 264:1581–1584.

Kuras, E.R., P.S. Warren, J.A. Zinda, M.F.J. Aronson, S. Cilliers, M.A. Goddard, C.H. Nilon, R. Winkler. 2020. Urban socioeconomic inequality and biodiversity often converge, but not always: A global meta-analysis. *Landscape and Urban Planning* 198:103799.

Lang, A. 2000. The pitfalls of pitfalls: A comparison of pitfall trap catches and absolute density estimates of epigeal invertebrate predators in arable land. *Journal of Pest Science* 73:99–106.

Laurance, W.F. 2010. Beyond island biogeography theory: Understanding habitat fragmentation in the real world. Pages 214 in R.E.R. Jonathan, B. Losos, editors. *The Theory of Island Biogeography Revisited.* Princeton University Press, Princeton, NJ.

Leonard, R.J., C. McArthur, D.F. Hochuli. 2018. Habitat complexity does not affect arthropod community composition in roadside greenspaces. *Urban Forestry & Urban Greening* 30:108–114.

Leong, M., R.R. Dunn, M.D. Trautwein. 2018. Biodiversity and socioeconomics in the city: A review of the luxury effect. *Biology Letters* 14(5):20180082.

Leve, M., E. Baudry, C. Bessa-Gomes. 2019. Domestic gardens as favorable pollinator habitats in impervious landscapes. *Science of the Total Environment* 647:420–430.
Lin, B.B., R.A. Fuller. 2013. Sharing or sparing? How should we grow the world's cities? *Journal of Applied Ecology* 50(5):1161–1168.
Lintott, P.R., N. Bunnefeld, E. Fuentes-Montemayor, J. Minderman, L.M. Blackmore, D. Goulson, K.J. Park. 2014. Moth species richness, abundance and diversity in fragmented urban woodlands: Implications for conservation and management strategies. *Biodiversity and Conservation* 23(11):2875–2901.
Longcore, T. 2003. Terrestrial arthropods as indicators of ecological restoration success in coastal sage scrub (California, USA). *Restoration Ecology* 11(4):397–409.
Lowenstein, D.M., E.S. Minor. 2016. Diversity in flowering plants and their characteristics: Integrating humans as a driver of urban floral resources. *Urban Ecosystems* 19:1735–1748.
Luder, K., E. Knop, M.H.M. Menz. 2018. Contrasting responses in community structure and phenology of migratory and non-migratory pollinators to urbanization. *Diversity and Distributions* 24(7):919–927.
MacArthur, R.H., E.O. Wilson. 1963. An equilibrium theory of insular zoogeography. *Evolution* 17(4):373–387.
Magura, T., M. Ferrante, G.L. Lövei. 2020. Only habitat specialists become smaller with advancing urbanization. *Global Ecology and Biogeography* 29(11):1978–1987.
Maisto, G., V. Milano, L. Santorufo. 2017. Relationships among site characteristics, taxonomical structure and functional trait distribution of arthropods in forest, urban and agricultural soils of Southern Italy. *Ecological Research* 32(4):511–521.
Martinson, H.M., M.J. Raupp. 2013. A meta-analysis of the effects of urbanization on ground beetle communities. *Ecosphere* 4(5):1–24.
Mashaly, A.M.A. 2017. Carrion beetles succession in three different habitats in Riyadh, Saudi Arabia. *Saudi Journal of Biological Sciences* 24(2):430–435.
Mata, L., C.G. Threlfall, N.S.G. Williams, A.K. Hahs, M. Malipatil, N.E. Stork, S.J. Livesley. 2017. Conserving herbivorous and predatory insects in urban green spaces. *Scientific Reports* 7:1–12.
Matteson, K.C., J.B. Grace, E.S. Minor. 2013. Direct and indirect effects of land use on floral resources and flower-visiting insects across an urban landscape. *Oikos* 122(5):682–694.
Matteson, K.C., G.A. Langellotto. 2010. Determinates of inner city butterfly and bee species richness. *Urban Ecosystems* 13:333–347.
Matteson, K.C., G.A. Langellotto. 2011. Small scale additions of native plants fail to increase beneficial insect richness in urban gardens. *Insect Conservation and Diversity* 4(2):89–98.
McIntyre, N., J. Rango, W. Fagan, S. Faeth. 2001. Ground arthropod community structure in a heterogeneous urban environment. *Landscape and Urban Planning* 52(4):257–274.
Meineke, E.K., R.R. Dunn, J.O. Sexton, S.D. Frank. 2013. Urban warming drives insect pest abundance on street trees. *PLoS ONE* 8(3):e59687.
Meineke, E.K., A.J. Holmquist, G.M. Wimp, S.D. Frank. 2017. Changes in spider community composition are associated with urban temperature, not herbivore abundance. *Journal of Urban Ecology* 3:juw010.
Memmott, J., P.G. Craze, N.M. Waser, M.V. Price. 2007. Global warming and the disruption of plant-pollinator interactions. *Ecology Letters* 10(8):710–717.
Menke, S.B., B. Guenard, J.O. Sexton, M.D. Weiser, R.R. Dunn, J. Silverman. 2011. Urban areas may serve as habitat and corridors for dry-adapted, heat tolerant species; an example from ants. *Urban Ecosystems* 14(2):135–163.
Merckx, T., A. Kaiser, H. Van Dyck. 2018a. Increased body size along urbanization gradients at both community and intraspecific level in macro-moths. *Global Change Biology* 24(8):3837–3848.
Merckx, T., C. Souffreau, A. Kaiser, L.F. Baardsen, T. Backeljau, D. Bonte, K.I. Brans, M. Cours, M. Dahirel, N. Debortoli, K. De Wolf, J.M.T. Engelen, D. Fontaneto, A.T. Gianuca, L. Govaert, F. Hendrickx, J. Higuti, L. Lens, K. Martens, H. Matheve, E. Matthysen, E. Piano, R. Sablon, I. Schön, K. Van Doninck, L. De Meester, H. Van Dyck. 2018b. Body-size shifts in aquatic and terrestrial urban communities. *Nature* 558(7708):113–116.
Mevi-Schütz, J., A. Erhardt. 2005. Amino acids in nectar enhance butterfly fecundity: A long-awaited link. *The American Naturalist* 165(4):411–419.
Miguelena, J.G., P.B. Baker. 2019. Effects of urbanization on the diversity, abundance, and composition of ant assemblages in an arid city. *Environmental Entomology* 48(4):836–846.
Misof, B., S. Liu, K. Meusemann, R.S. Peters, A. Donath, C. Mayer, P.B. Frandsen, J. Ware, T. Flouri, R.G. Beutel, O. Niehuis, M. Petersen, F. Izquierdo-Carrasco, T. Wappler, J. Rust, A.J. Aberer, U. Aspöck, H. Aspöck, D. Bartel, A. Blanke, S. Berger, A. Böhm, T.R. Buckley, B. Calcott, J. Chen, F. Friedrich, M. Fukui, M. Fujita, C. Greve, P. Grobe, S. Gu, Y. Huang, L.S. Jermiin, A.Y. Kawahara, L. Krogmann, M. Kubiak, R. Lanfear, H. Letsch, Y. Li, Z. Li, J. Li, H. Lu, R. Machida, Y. Mashimo, P. Kapli, D.D. McKenna, G. Meng, Y.

Nakagaki, J.L. Navarrete-Heredia, M. Ott, Y. Ou, G. Pass, L. Podsiadlowski, H. Pohl, B.M. von Reumont, K. Schütte, K. Sekiya, S. Shimizu, A. Slipinski, A. Stamatakis, W. Song, X. Su, N.U. Szucsich, M. Tan, X. Tan, M. Tang, J. Tang, G. Timelthaler, S. Tomizuka, M. Trautwein, X. Tong, T. Uchifune, M.G. Walzl, B.M. Wiegmann, J. Wilbrandt, B. Wipfler, T.K.F. Wong, Q. Wu, G. Wu, Y. Xie, S. Yang, Q. Yang, D.K. Yeates, K. Yoshizawa, Q. Zhang, R. Zhang, W. Zhang, Y. Zhang, J. Zhao, C. Zhou, L. Zhou, T. Ziesmann, S. Zou, Y. Li, X. Xu, Y. Zhang, H. Yang, J. Wang, J. Wang, K.M. Kjer, X. Zhou. 2014. Phylogenomics resolves the timing and pattern of insect evolution. *Science* 346:763–767.

Miyashita, T., A. Shinkai, T. Chida. 1998. The effects of forest fragmentation on web spider communities in urban areas. *Biological Conservation* 86(3):357–364.

Moriwaki, R., K. Watanabe, K. Morimoto. 2013. Urban dry island phenomenon and its impact on cloud base level. *Journal of JSCE* 1(1):521–529.

Nagase, A., M. Kurashina, M. Nomura, J.S. MacIvor. 2019. Patterns in urban butterflies and spontaneous plants across a University campus in Japan. *The Pan-Pacific Entomologist* 94(4):195–215.

Narango, D.L., D.W. Tallamy, K.J. Shropshire. 2020. Few keystone plant genera support the majority of Lepidoptera species. *Nature Communications* 11:1–8.

Neame, L.A., T. Griswold, E. Elle. 2013. Pollinator nesting guilds respond differently to urban habitat fragmentation in an oak-savannah ecosystem. *Insect Conservation and Diversity* 6(1):57–66.

Neil, K., J. Wu. 2006. Effects of urbanization on plant flowering phenology: A review. *Urban Ecosystems* 9:243–257.

Nijssen, M.E., M.F. WallisDeVries, H. Siepel. 2017. Pathways for the effects of increased nitrogen deposition on fauna. *Biological Conservation* 212(B):423–431.

Nooten, S., P. Schultheiss, R.C. Rowe, S.L. Facey, J.M. Cook. 2019. Habitat complexity affects functional traits and diversity of ant assemblages in urban green spaces (Hymenoptera: Formicidae). *Myrmecological News* 29:67–77.

Normandin, E., N.J. Vereecken, C.M. Buddle, V. Fournier. 2017. Taxonomic and functional trait diversity of wild bees in different urban settings. *PeerJ* 5:e3051.

Oke, T.R., G.M. Mills, J.A. Voogt, editors. 2017. *Urban Climates*. Cambridge University Press, Cambridge.

Ossola, A., A.K. Hahs, M.A. Nash, S.J. Livesley. 2016. Habitat complexity enhances comminution and decomposition processes in urban ecosystems. *Ecosystems* 19:927–941.

Ossola, A., M.A. Nash, F.J. Christie, A.K. Hahs, S.J. Livesley. 2015. Urban habitat complexity affects species richness but not environmental filtering of morphologically-diverse ants. *PeerJ* 3:e1356.

Packer, L., G. Darla-West. 2021. Bees: How and why to sample them. Pages 55–83 in J.C. Santos, G.W. Fernandes, editors. *Measuring Arthropod Biodiversity: A Handbook of Sampling Methods*. Springer, Cham, Switzerland.

Pardee, G.L., S.M. Philpott. 2014. Native plants are the bee's knees: Local and landscape predictors of bee richness and abundance in backyard gardens. *Urban Ecosystems* 17:641–659.

Parsons, S.E., L.M. Kerner, S.D. Frank. 2020. Effects of native and exotic congeners on diversity of invertebrate natural enemies, available spider biomass, and pest control services in residential landscapes. *Biodiversity and Conservation* 29:1241–1262.

Pearse, I.S., A.L. Hipp. 2009. Phylogenetic and trait similarity to a native species predict herbivory on non-native oaks. *Proceedings of the National Academy of Sciences* 106(43):18097–18102.

Perez, A., S.E. Diamond. 2019. Idiosyncrasies in cities: Evaluating patterns and drivers of ant biodiversity along urbanization gradients. *Journal of Urban Ecology* 5(1):juz017.

Perry, K.I., N.C. Hoekstra, Y.A. Delgado de la flor, M.M. Gardiner. 2020. Disentangling landscape and local drivers of ground-dwelling beetle community assembly in an urban ecosystem. *Ecological Applications* 30(8):e02191.

Peterson, M.N., B. Thurmond, M. Mchale, S. Rodriguez, H.D. Bondell, M. Cook. 2012. Predicting native plant landscaping preferences in urban areas. *Sustainable Cities and Society* 5:70–76.

Piano, E., K. De Wolf, F. Bona, D. Bonte, D.E. Bowler, M. Isaia, L. Lens, T. Merckx, D. Mertens, M. Van Kerckvoorde, L. De Meester, F. Hendrickx. 2017. Urbanization drives community shifts towards thermophilic and dispersive species at local and landscape scales. *Global Change Biology* 23(7):2554–2564.

Piano, E., C. Souffreau, T. Merckx, L.F. Baardsen, T. Backeljau, D. Bonte, K.I. Brans, M. Cours, M. Dahirel, N. Debortoli, E. Decaestecker, K. De Wolf, J.M.T. Engelen, D. Fontaneto, A.T. Gianuca, L. Govaert, F.T.T. Hanashiro, J. Higuti, L. Lens, K. Martens, H. Matheve, E. Matthysen, E. Pinseel, R. Sablon, I. Schon, R. Stoks, K. Van Doninck, H. Van Dyck, P. Vanormelingen, J. Van Wichelen, W. Vyverman, L. De Meester, F. Hendrickx. 2020. Urbanization drives cross-taxon declines in abundance and diversity at multiple spatial scales. *Global Change Biology* 26(3):1196–1211.

Posa, M.R.C., Sodhi, N.S., 2006. Effects of anthropogenic land use on forest birds and butterflies in Subic Bay, Philippines. *Biological Conservation* 129:256–270.
Poythress, J.C., J.M. Affolter. 2018. Ecological value of native plant cultivars versus wild-type native plants for promoting hemipteran diversity in suburban areas. *Environmental Entomology* 47(4):890–901.
Quistberg, R.D., P. Bichier, S.M. Philpott. 2016. Landscape and local correlates of bee abundance and species richness in urban gardens. *Environmental Entomology* 45(3):592–601.
Rader, R., I. Bartomeus, L.A. Garibaldi, M.P.D. Garratt, B.G. Howlett, R. Winfree, S.A. Cunningham, M.M. Mayfield, A.D. Arthur, G.K.S. Andersson, R. Bommarco, C. Brittain, L.G. Carvalheiro, N.P. Chacoff, M.H. Entling, B. Foully, B.M. Freitas, B. Gemmill-Herren, J. Ghazoul, S.R. Griffin, C.L. Gross, L. Herbertsson, F. Herzog, J. Hipolito, S. Jaggar, F. Jauker, A-M. Klein, D. Kleijn, S. Krishnan, C.Q. Lemos, S.A.M. Lindstrom, Y. mandelik, V.M. Monteiro, W. Nelson, L. Nilsson, D.E. Pattemore, N. de O. Pereira, G. Pisanty, S.G. Potts, M. Reemer, M. Rundlof, C.S. Sheffield, J. Scheper, C. Schuepp, H.G. Smith, P.A. Stanley, J.C. Stout, H. Szentgyorgyi, H. Taki, C.H. Vergara, B.F. Viana, M. Woyciechowski. 2016. Non-bee insects are important contributors to global crop pollination. *Proceedings of the National Academy of Sciences* 113(1):146–151.
Raupp, M.J., P.M. Shrewsbury, D.A. Herms. 2010. Ecology of herbivorous arthropods in urban landscapes. *Annual Review of Entomology* 55:19–38.
Redman, C. 2011. Social-ecological transformations in urban landscapes: A historical perspective. Pages 206–212 in J. Niemelä, J.H. Breuste, T. Elmqvist, G. Guntenspergen, P. James, N.E. McIntyre, editors. *Urban Ecology: Patterns, Processes, and Applications*. Oxford University Press, Oxford.
Renner, S.S., C.M. Zohner. 2018. Climate change and phenological mismatch in trophic interactions among plants, insects, and vertebrates. *Annual Review of Ecology, Evolution, and Systematics* 49:165–182.
Robinson, E.A., G.D. Ryan, J.A. Newman. 2012. A meta-analytical review of the effects of elevated CO2 on plant – arthropod interactions highlights the importance of interacting environmental and biological variables. *New Phytologist* 194(2):321–336.
Rota, E., T. Caruso, M. Migliorini, F. Monaci, V. Agamennone, G. Biagini, R. Bargagli. 2015. Diversity and abundance of soil arthropods in urban and suburban holm oak stands. *Urban Ecosystems* 18:715–728.
Salisbury, A., S. Al-Beidh, J. Armitage, S. Bird, H. Bostock, A. Platoni, M. Tatchell, K. Thompson, J. Perry. 2017. Enhancing gardens as habitats for plant-associated invertebrates: Should we plant native or exotic species? *Biodiversity and Conservation* 26:2657–2673.
Salisbury, A., J. Armitage, H. Bostock, J. Perry, M. Tatchell, K. Thompson. 2015. Enhancing gardens as habitats for flower-visiting aerial insects (pollinators): Should we plant native or exotic species? *Journal of Applied Ecology* 52(5):1156–1164.
Savage, A.M., B. Hackett, B. Guénard, E.K. Youngsteadt, R.R. Dunn. 2015. Fine-scale heterogeneity across Manhattan's urban habitat mosaic is associated with variation in ant composition and richness. *Insect Conservation and Diversity* 8:216–228.
Schell, C.J., K. Dyson, T.L. Fuentes, S. Des Roches, N.C. Harris, D.S. Miller, C.A. Woelfle-Erskine, M.R. Lambert. 2020. The ecological and evolutionary consequences of systemic racism in urban environments. *Science* 369(6510):eaay4497.
Sheridan, J.A., D. Bickford. 2011. Shrinking body size as an ecological response to climate change. *Nature Climate Change* 1:401–406.
Shochat, E., W.L. Stefanov, M.E.A. Whitehouse, S.H. Faeth. 2008. Urbanization and spider diversity: Influences of human modification of habitat structure and productivity. Pages 455–472 in J.M. Marzluff, E. Schulenberger, W. Endlicher, M. Alberti, G. Bradley, C. Ryan, U. SImon, C. ZumBrunnen, editors. *Urban Ecology*. Springer, Boston, MA.
Shrewsbury, P.M., M.J. Raupp. 2000. Evaluation of components of vegetational texture for predicting azalea lace bug, *Stephanitis pyrioides* (Heteroptera: Tingidae), abundance in managed landscapes. *Environmental Entomology* 29(5):919–926.
Silva, J.L.S., M.T.P. de Oliveira, O. Cruz-Neto, M. Tabarelli, A.V. Lopes. 2020. Plant – pollinator interactions in urban ecosystems worldwide: A comprehensive review including research funding and policy actions. *Ambio* 50(4):884–890.
Sing, K.-W., H. Dong, W.-Z. Wang, J.-J. Wilson. 2016. Can butterflies cope with city life? Butterfly diversity in a young megacity in southern China. *Genome* 59(9):751–761.
Small, E.C., J.P. Sadler, M.G. Telfer. 2002. Carabid beetle assemblages on urban derelict sites in Birmingham, UK. *Journal of Insect Conservation* 6:233–246.

Smith, R.M., P.H. Warren, K. Thompson, K.J. Gaston. 2006. Urban domestic gardens (VI): Environmental correlates of invertebrate species richness. *Biodiversity & Conservation* 15:2415–2438.

Soga, M., T. Kawahara, K. Fukuyama, K. Sayama, T. Kato, M. Shimomura, T. Itoh, T. Yoshida, K. Ozaki. 2015. Landscape versus local factors shaping butterfly communities in fragmented landscapes: Does host plant diversity matter? *Journal of Insect Conservation* 19(4):781–790.

Soga, M., S. Koike. 2012. Relative importance of quantity, quality and isolation of patches for butterfly diversity in fragmented urban forests. *Ecological Research* 27(2):265–271.

Soga, M., Y. Yamaura, S. Koike, K.J. Gaston. 2014. Land sharing vs. land sparing: Does the compact city reconcile urban development and biodiversity conservation? *Journal of Applied Ecology* 51(5):1378–1386.

Spring, M.R. 2017. *Impacts of Urban Greenspace Management on Beneficial Insect Communities.* MS Thesis, The Ohio State University, Columbus, OH.

Stewart, A.B., T. Sritongchuay, P. Teartisup, S. Kaewsomboon, S. Bumrungsri. 2018. Habitat and landscape factors influence pollinators in a tropical megacity, Bangkok, Thailand. *PeerJ* 6:e5335.

Stork, N.E. 2018. How many species of insects and other terrestrial arthropods are there on Earth? *Annual Review of Entomology* 63:31–45.

Suarez, A.V., D.T. Bolger, T.J. Case. 1998. Effects of fragmentation and invasion on native ant communities in coastal southern California. *Ecology* 79(6):2041–2056.

Sugiura, S., R. Tanaka, H. Taki, N. Kanzaki. 2013. Differential responses of scavenging arthropods and vertebrates to forest loss maintain ecosystem function in a heterogeneous landscape. *Biological Conservation* 159:206–213.

Sunday, J.M., A.E. Bates, N.K. Dulvy. 2011. Global analysis of thermal tolerance and latitude in ectotherms. *Proceedings of the Royal Society of London Series B: Biological Sciences* 278(1713):1823–1830.

Tait, C.J., C.B. Daniels, R.S. Hill. 2005. Changes in species assemblages within the Adelaide metropolitan area, Australia, 1836–2002. *Ecological Applications* 15(1):346–359.

Tew, N.E., J. Memmott, I.P. Vaughan, S. Bird, G.N. Stone, S.G. Potts, K.C.R. Baldock. 2021. Quantifying nectar production by flowering plants in urban and rural landscapes. *Journal of Ecology* 109(4):1747–1757.

Theodorou, P., S.-C. Herbst, B. Kahnt, P. Landaverde-González, L.M. Baltz, J. Osterman, R.J. Paxton. 2020. Urban fragmentation leads to lower floral diversity, with knock-on impacts on bee biodiversity. *Scientific Reports* 10:1–11.

Tonietto, R., J. Fant, J. Ascher, K. Ellis, D. Larkin. 2011. A comparison of bee communities of Chicago green roofs, parks and prairies. *Landscape and Urban Planning* 103:102–108.

Turrini, T., E. Knop. 2015. A landscape ecology approach identifies important drivers of urban biodiversity. *Global Change Biology* 21(4):1652–1667.

van Klink, R., D.E. Bowler, K.B. Gongalsky, A.B. Swengel, A. Gentile, J.M. Chase. 2020. Meta-analysis reveals declines in terrestrial but increases in freshwater insect abundances. *Science* 368(6489):417–420.

Vaudo, A.D., J.F. Tooker, C.M. Grozinger, H.M. Patch. 2015. Bee nutrition and floral resource restoration. *Current Opinion in Insect Science* 10:133–141.

Verboven, H.A., R. Uyttenbroeck, R. Brys, M. Hermy. 2014. Different responses of bees and hoverflies to land use in an urban – Rural gradient show the importance of the nature of the rural land use. *Landscape and Urban Planning* 126:31–41.

Vergnes, A., V. Pellissier, G. Lemperiere, C. Rollard, P. Clergeau. 2014. Urban densification causes the decline of ground-dwelling arthropods. *Biodiversity and Conservation* 23(8):1859–1877.

Wagner, D.L. 2020. Insect declines in the Anthropocene. *Annual Review of Entomology* 65:457–480.

Weiss, S.B. 1999. Cars, cows, and checkerspot butterflies: Nitrogen deposition and management of nutrient-poor grasslands for a threatened species. *Conservation Biology* 13(6):1476–1486.

Wenzel, A., I. Grass, V.V. Belavadi, T. Tscharntke. 2020. How urbanization is driving pollinator diversity and pollination – A systematic review. *Biological Conservation* 241:108321.

Whitmore, C., T. Crouch, R. Slotow. 2002. Conservation of biodiversity in urban environments: Invertebrates on structurally enhanced road islands. *African Entomology* 10(1):113–126.

Willmer, P.G., G.N. Stone. 2004. Behavioral, ecological, and physiological determinants of the activity patterns of bees. *Advances in the Study of Behavior* 34:347–466.

Wood, B.C., A.S. Pullin. 2002. Persistence of species in a fragmented urban landscape: The importance of dispersal ability and habitat availability for grassland butterflies. *Biodiversity & Conservation* 11:1451–1468.

Wray, J.C., E. Elle. 2015. Flowering phenology and nesting resources influence pollinator community composition in a fragmented ecosystem. *Landscape Ecology* 30(2):261–272.

17

BIRDS IN AN URBAN WORLD

Community patterns across cities

Christine Rega-Brodsky and Ian MacGregor-Fors

Introduction

Birds have long been among the best studied organisms in urban settings (Marzluff et al. 2001; McKinney 2008). There are several traits that have made them appropriate and appealing to study in cities, including that they are highly diverse, conspicuous, relatively easy to survey with proper training, and form complex assemblages across urban landscapes, including heavily urbanized sites (Chase and Walsh 2006; MacGregor-Fors et al. 2009). Thus, since the 1970s, ornithologists and ecologists have used birds as model taxa to understand how urbanization affects fauna and their abilities to adapt, and even thrive, within cities. As with other branches of urban ecology, interest in avian biological and ecological phenomena that occur within cities boosted in the 1990s, when the consolidation of the discipline occurred (McDonnell and MacGregor-Fors 2016). Several reviews have synthesized the knowledge since 2001, when John Marzluff and colleagues (2001) compiled a book on bird ecology and conservation in urban systems. The opening chapter of that book provides a historical perspective, identifying major drawbacks and knowledge gaps. Marzluff et al. (2001) found that most studies were one- or two-year correlational assessments of the breeding birds of focal sites in urban settings from developed upland temperate regions, limiting our understanding of the broader picture. They identified an important bias toward avian community structure and composition focused on relative abundance quantification, with a pressing need to assess avian demography in urban bird studies.

Five years later, Chase and Walsh (2006) reviewed a much larger set of papers, as the publication rate increased While they open their review stating that "The effect of urbanization can be immense, yet our understanding is rudimentary" (p. 46), they identified additional enlightening patterns and provided future directions on the study of urban birds that have been a guideline to date. Among the most relevant findings, they summarize: (1) bird community composition in heavily developed urban sites are more similar among them than when contrasted with less-developed urban sites; (2) omnivore, granivore, and cavity nesting species are well represented within urban settings; (3) avian species richness tends to decrease with urbanization, while abundances tend to increase; (4) given the large extent of the home range of urban dwelling raptors, their ecological requirements do not entirely depend on urban systems, which are important food sources

 DOI: 10.4324/9781003016120-20

for them; (5) native vegetation is positively associated with native avian communities; (6) fecundity of urban birds is related to species-specific adaptation processes; (7) urbanite activities have negative effects on urban birds; and (8) urban hazards such as bird-window collisions affect avian survivorship in urban settings. Chase and Walsh (2006) clearly state the importance of understanding the processes behind the summarized population and community responses and associations and the need to address the lack of data from tropical regions. Finally, they pinpoint the importance of generating local knowledge before implementing management policies and actions to reduce the impacts of urbanization on birds.

Evans et al. (2009) reviewed the factors associated with avian assemblages together with an analysis of a common bird monitoring scheme to assess some of the reviewed topics. Their main findings were that bird assemblages (1) respond more to local factors than to regional ones; (2) are driven by habitat fragmentation, with patch size being more important than patch isolation; and (3) respond positively to increasing vegetation structural complexity. Most recently, Marzluff (2016) performed an update to the aforementioned reviews after recognizing the urban bird studies had populated the literature exponentially. Marzluff (2016) detected some novel trends in bird studies, such as the following: (1) the shift of studies from the assemblage/community level to demographic responses in the Global North; (2) studies focused on assemblage/community level patterns remain in the Global South; (3) global studies are revealing evolutionary adaptations of urban birds (associated with phenotypic plasticity); and (4) the field remains rarely experimental, although the number of experimental studies is growing.

Recent meta-analyses and literature reviews have also highlighted the broad impacts of urbanization on bird assemblages, particularly focusing on the general disagreements with their findings and trends. Even though Chase and Walsh (2006) reported the aforementioned negative trend in bird communities along the urbanization gradient, more recent literature has found differing relationships between urbanization and bird abundances (Saari et al. 2016) and diversity (Batáry et al. 2018) worldwide. These variable trends may be the result of a number of different study methods or variables used to characterize urban systems (Padilla and Sutherland 2019). In particular, Péter Batáry and colleagues (2018) found that study design may greatly impact community patterns along gradients, as the gradient length and land uses studied vastly impact whether the abundance and richness trends illustrate a linear or intermediate peak along the gradient. Variations in methods and gradient classifications may consequently yield inconsistent generalizations and may also discount the wide variation of urban characteristics worldwide.

While the aforementioned reviews have made a laudable attempt in addressing the general patterns across the globe, they have overlooked some aspects of the state of the knowledge of urban bird studies, especially from outside the Global North (e.g. MacGregor-Fors and Escobar-Ibáñez 2017). In this chapter, we addressed a particular topic with the widest knowledge from studies from across the globe: bird assemblage and community patterns in urban settings. Although there has been extensive, pivotal research on bird population trends across the urbanization gradient, our goal was to detail urbanization impacts on the broad community rather than focus on one particular species or guild. For this, we performed a thorough literature review. Based on a set of retrieved publications, we provide a bibliometric summary, describe the main frameworks used to tackle research questions, identify drivers of avian response in urban settings, and provide suggested areas for future management actions, research topics, and directions.

Literature review

To gather a representative sample of the existing knowledge regarding urban bird communities from across the world, we performed a search in the Web of Science using the following advanced Boolean operator string: TS=((*urban* NOT disturbance* OR city OR cities OR town) AND (bird OR avian) AND (communit* OR assemblage* OR biodiversity OR "species richness" OR richness OR diversity OR composition)). This search, completed on 28 May 2020, retrieved a total of 2,596 publications. Once obtaining these papers, we screened-out publications that were not in the focus of this chapter following three main criteria: (a) urban focus, (b) avian community/assemblage approach, and (c) original primary peer-reviewed documents. Regarding the latter, we excluded all reviews and meta-analyses to avoid replicating any studies in our analyses. Additionally, three papers were removed from the search findings because they were either a duplicate of a paper within our retrieved publications (n = 1) or a reprint that was unavailable after an exhaustive search (n = 2). The final set of publications included 405 documents, which we used for further analysis from two perspectives: (a) bibliometrics and (b) reported methodological framework and findings.

As has happened with most urban ecology studies, the number of publications in our dataset was relatively low until the 1990s. The set of reviewed publications shows that the number of urban bird community studies rose steadily for over a decade, finally peaking after 2004 (Figure 17.1a). Most urban bird community studies were published in journals with a predominantly urban focus (e.g. *Landscape and Urban Planning*, 14%; and *Urban Ecosystems*, 11%), rather than taxa-specific journals (e.g. *Condor*, 2%; *Ornitología Neotropical*, 1.7%) or high-impact journals for broader audiences (e.g. *Global Change Biology*, 0.7%; *Nature Communications*, 0.2%).

The retrieved set of studies included results from hundreds of urban centers across 78 countries (and territories), led by the USA (90 publications), Australia (30), Brazil (23), Mexico (23), and Spain (19). Interestingly, among the ten most studied countries in terms of publications, half are from the Global South (i.e. Brazil, Mexico, Argentina, China, India). Regarding regional patterns, most countries in the American continent have at least one study regarding this topic, which also applies

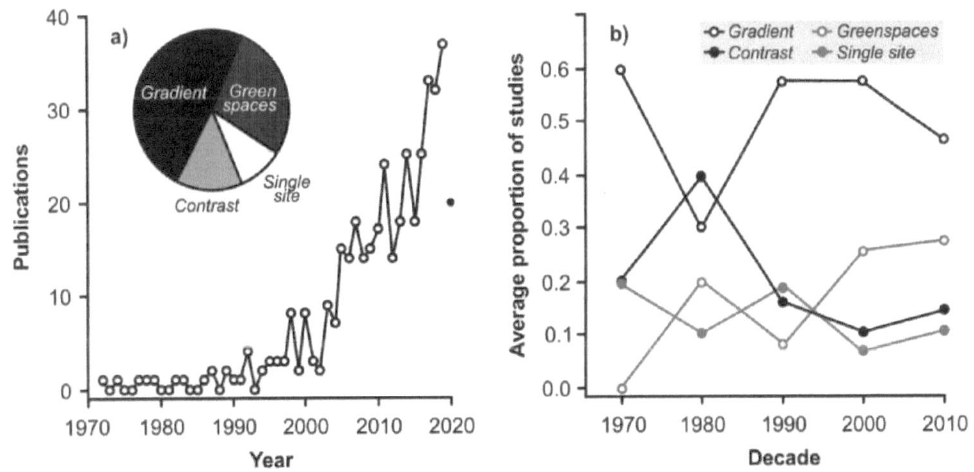

Figure 17.1 Trends in publications by time and focus between 1970 and May 2020.

Figure 17.2 The top 500 terms used within urban bird literature titles (without bird/avian, community, urban). Word cloud in the shape of a bird.

to Europe (with a higher value for Western Europe) (MacGregor-Fors et al. 2020). Regarding focal cities, those with five or more studies were Phoenix (USA), Madrid (Spain), Rovaniemi (Finland), Singapore (Singapore), Mar del Plata (Argentina), Prague (Czech Republic), Paris (France), Melbourne (Australia), Rome (Italy), and Seattle (USA).

Most studies have been framed under the urban-rural gradient approach, followed by studies performed uniquely within urban green spaces, those focused under a contrast approach (e.g. urban vs. forest), and single site studies (Figure 17.1a). Temporally, studies focused on urban-rural gradients have always comprised an important proportion of the total publications by decade. Regarding green spaces, the proportion of studies per decade has increased, while single site studies have decreased. In the case of contrast studies, they showed a similar pattern to that of single sites, but with an important peak in the 1980s (Figure 17.1b). Regarding the topics assessed in the retrieved set of publications, we found studies shifted in topic and complexity, always with a strong focus on a diversity approach (Figure 17.2). As can be seen in Figure 17.3, an analysis of the thematic evolution of publication titles shows a basic focus of studies before the 1990s, which increased during the 1990s and peaked in the 2000s, where more comprehensive approaches of studies were detected (Figure 17.3).

Major approaches

Survey methods

Most studies utilized standard survey methods (Bibby et al. 2000) to assess bird communities, either through point count (52%) or transect (29%) surveys. Community science gained more

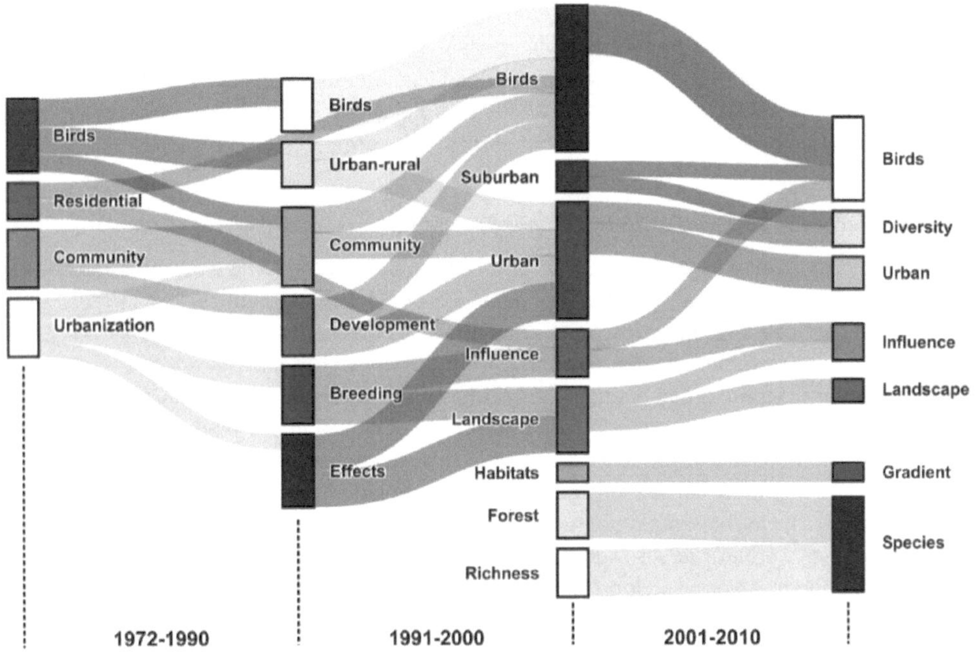

Figure 17.3 Thematic evolution of publication titles following an Inclusion Index weighted by word occurrences (max number of words 250, min. cluster frequency = 2; time slices: 1990, 2000, 2010).

popularity over the survey period, with 6% of studies obtaining their data either through the North American Breeding Bird Survey, eBird, Christmas Bird Count, and other community science efforts (e.g. Neighborhood Nestwatch, French Breeding Bird Survey, Bird Society of Japan). It is interesting to note that 8% of studies utilized multiple survey methods, usually in conjunction with point counts or transects, to widen their scope (e.g. capture/banding programs, acoustic data collection, territory mapping).

Half of the retrieved studies (52%) were performed in ≤ 6-month periods, with 2-month studies being the most frequent, similar to findings in published reviews (Marzluff et al. 2001). Notably, 12-month studies were not uncommon (9%). Studies lasting > 5 years were generally scarce (2%), while extremely few studies utilized long-term datasets spanning 20 years or more (0.7%; e.g. Pidgeon et al. 2014; Strohbach et al. 2014; Plummer et al. 2019). Most studies occurred exclusively during the breeding season (47%), followed by year-round studies (23%), studies combining multiple seasons (i.e. breeding and non-breeding samples; 11%), or studies conducted solely within the non-breeding (winter) season (11%).

Although most of the retrieved publications focused mainly on resident or breeding landbirds, fewer studies (< 5%) focused solely on aquatic birds, migrants, introduced/exotics, cavity nesters, raptors, birds at feeders, and woodpeckers. Not surprisingly, the methods used to analyze bird communities have shifted from the simplistic species composition, density, richness, and Shannon's Diversity Index in the 1970s and 1980s to more modern choices in describing and analyzing community trends, especially at the landscape scale (Figure 17.3). Urban bird community assemblages were first analyzed in the 1990s, linking composition and relative abundances via multivariate analyses (e.g. Nilon et al. 1995). In the mid-2000s, the urban bird literature started to explore functional traits and analyze the community within a functional group framework (e.g. Bishop and Myers 2005; Blair and Johnson 2008; Filippi-Codaccioni et al. 2009). This initiated a stream of new metrics, applying the traditional methods of urban bird community analyses with new, innovative ways to describe how communities function, exploring their spatial and temporal turnover (e.g. Husté and Boulinier 2007; Catterall et al. 2010), phylogenetic diversity (e.g. Palacio et al. 2018; Weideman et al. 2020), evolutionary distinctiveness (e.g. La Sorte et al. 2018), and more.

Urban gradient

The vast majority of papers studying urban bird communities were conducted across an urbanization gradient (Figure 17.1a), a method popularized by McDonnell and Pickett (1990; Figure 17.1b). Gradients can be defined by their endpoints (i.e. "urban" vs. "rural") or assess trends across multiple land use/land covers, spanning from the most built and "gray" urban centers to native habitat far from the urban boundary. Locations for sampling bird communities along these gradients were either selected *a priori* for their urbanization ranking or placed in a randomized pattern within and outside of cities to assess their urbanization values in parallel with the bird community (e.g. pollution, noise, and artificial light sampled at randomly selected plots; Ciach and Fröhlich 2017). On the same note, the number of urbanization levels selected varied greatly, from the simplistic "urban" versus "suburban" versus "rural" to hundreds of samples in between. The most common means to develop these urbanization gradients was to quantify variables such as built cover (and contrastingly, vegetation cover), distance from city center, housing density, human activity (e.g. pedestrian abundance), development age, and human population density. Other studies have used multivariate approaches to synthesize the complexity of urbanization through ordination procedures (e.g. Padilla and Sutherland 2019), qualitatively estimated urbanization via land use and cover categories of the sampled area or surrounding matrix, or described gradients of human modifications (e.g. levels of riparian zone engineering; Banville et al. 2017).

Of the papers that reported trends in species richness or diversity for the entire bird community across urbanization gradients (n = 159 of 199 gradient papers), approximately 19% reported an intermediate peak pattern, while 69% reported a direct relationship. Of those that reported direct relationships, all but five resulted in a negative relationship with increasing urbanization (Hohtola 1978; Leveau and Leveau 2005; Pautasso and Dinetti 2009; Naumkin 2010; Weideman et al. 2020). These five studies reported an increase in bird species richness with human population size (Pautasso and Dinetti 2009), urban and matrix land uses (Weideman et al. 2020), and suburban and peri-urban versus rural land cover (Leveau and Leveau 2005); while bird diversity increased with decreasing plant volume (Hohtola 1978) and more urbanized pine forests (Naumkin 2010). Three of these five studies occurred within smaller cities (i.e. Kuopio, Finland; Kungur, Russia; and Phalaborwa, South Africa), while Carlos M. Leveau and Lucas M. Leveau's (2005) study in Mar del Plata, Argentina, only sampled suburban, peri-urban, and rural land uses, which may have resulted in such a trend without an adequately dense urban area to anchor the trend. Nine studies reported a lack of relationship between urbanization and bird species richness or diversity; there were no similarities across these nine studies in regard to their location, study duration, or sampling season.

Typically, it is challenging to determine a true urbanization value based on one metric, as many "urban" variables can be specific to the particular city studied. Smaller cities, even if considered "urban", may influence the bird community in a much different way compared to communities sampled within a "megacity". Similarly, surrounding land use types may deeply alter community trends if the non-urban points are located within unfragmented habitats versus rural agricultural settings. Papers that integrate multiple urbanization variables may offer a more accurate and broader generalization of the study's findings. For example, in their review on the subject, Padilla and Sutherland (2019) recommended a standardized landscape gradient framework to encourage a consistent definition of the urbanization gradient. This may be achieved through multivariate gradients (e.g. Alberti et al. 2001; Padilla and Sutherland 2019) or the use of scaled urbanization indices, such as the "wildness value" and "bioscore" utilized in Müller et al. (2018) that integrated local biodiversity values, vegetative cover, and a wildness metric into one representative urban value. Efforts to standardize urbanization by merging variables in such means would be beneficial for the discipline as a whole to make urban biodiversity studies directly comparable; however, we do recognize the challenges related to its consistent implementation at a global scale.

Green spaces

Research in urban green spaces has become more common over time (Figure 17.1b), especially as these remaining habitat fragments are often the highlight of conservation efforts in cities. Urban bird studies overwhelmingly focus on communities within forested land cover (39%), while few studied birds within grasslands, wetlands, lakes, desert, or mangrove habitats (all < 5%), skewing generalities of urban bird communities toward forest species. Even so, the majority of the most frequently studied cities are within forested biomes, save for Phoenix (USA), increasing the attention of non-forest ecosystems such as deserts. As for the land uses highlighted within urban bird community studies, most research occurred in urban parks (42%) and remnant habitats (37%), which tend to be larger fragments with the potential to test various questions concerning the impacts of habitat edge, fragment size, and vegetation communities on birds, all within public lands. Increasing rates of studies located within vacant lots (e.g. Villaseñor et al. 2020), residential yards (e.g. Lerman and Warren 2011), and other small green spaces (e.g. Carbó-Ramírez and Zuria 2011) are a new, welcomed approach to determine the value of small habitat fragments within the core of most cities.

Situating urban bird studies within green spaces is beneficial to test conservation actions or management plans (at the city administration or individual homeowner level) within land cover types that formerly dominated the city's landscape (e.g. habitat fragments) or new vegetation communities (e.g. lawn habitats). As many of the cities studied within this literature search are located within forested biomes, studying the response of the bird community to green space variables lends itself well to generalizable findings. One particular generality studied across most of these green space studies tested green space size, with the majority of papers highlighting the positive relationship between size and bird species richness (e.g. La Sorte 2020).

Urban contrasts

About a quarter of all sampled studies from our search framed their research question by contrasting bird communities across two locations that differed in their urbanization classification. This approach was diverse in its undertaking as some studies focused solely on "gray" (i.e. highly developed urban sites) versus "green" (i.e. well-vegetated, often urban) features within city boundaries (32% contrast studies), while others developed their sampling design to reflect a contrast between urban features within city boundaries versus the non-urban features beyond the city (68%). Contrasts were most often described by their location in and outside city boundaries, land cover (e.g. built vs. vegetated), or land use (e.g. residential vs. commercial), while some studies uniquely categorized their urban contrast by noise level (e.g. Fontana et al. 2011) or native versus exotic streetscapes (e.g. White et al. 2005).

All but two waterfowl studies reflected broad study designs to capture differences spanning the entire bird community, yielding contrasting patterns in the more traditional metrics of species richness, abundance, diversity, similarity, and assemblages. More modern study designs contrasted urban bird communities from their rural or green counterparts by their evolutionary distinctiveness (e.g. Ibáñez-Álamo et al. 2020), functional similarity (e.g. Ortega-Álvarez and MacGregor-Fors 2011), and functional dispersion indices (e.g. Leveau and Leveau 2020).

Contrasting urban bird communities against non-urban counterparts yields similar challenges as discussed in gradient studies. It is often difficult to generalize these study findings across cities due to the urban versus non-urban contrasts. For example, some studies considered residential areas as "gray" locations within a city (Ortega-Álvarez and MacGregor-Fors 2009), while others considered them "green" (Taylor et al. 2013). Similarly, how agricultural land uses are classified varied widely in the same respect for both contrast and gradient studies, either serving as the most rural sampling location (Leveau et al. 2018) or as an intermediate urbanization value (Shochat et al. 2010). Rather than relying upon ill-defined land use categories, especially those with indeterminate cutoffs of continuous variables (e.g. high vs. low housing density), we found that successful studies contrasted regional species pools against urban bird communities to assess non-urban community characteristics (e.g. Ferenc et al. 2014).

Single site

Literature reporting findings from a single site were rare among the literature (9%). The majority of these studies were completed at nearby university campuses (26%), habitat reserves (16%), and urban parks (16%). These sites either were sampled due to convenience (e.g. university campus), their unique features lacked a suitable nearby replicate (e.g. airport, zoo, military fort), or unique management actions taken at the one location. Findings from these studies, although informative regarding local urban bird community composition and unique urban land uses, may not be generalizable across different geographic areas. Even so, single site studies have their particular significance within the literature. Compared to other papers analyzed in this study, single site papers included the longest studies conducted, with data collection spanning a 35.6-month average compared to 11.6-month average for all studies. This is particularly informative for documenting changes in community composition due to climate change and shifts in landscape composition. Additionally, focusing research efforts on one location may also allow for greater uses of unique sampling methods for urban areas (e.g. banding programs) and studies that incorporate greater diversity of sampling to explore beyond relative abundance of each species (e.g. nesting patterns, migration counts, acoustic data collection; Inzunza and Rodriguez 2009).

Drivers of urban bird communities

There was not one common pattern throughout all reported studies regarding urbanization impacts on bird communities; however, some similarities were apparent. The most species rich and diverse urban bird communities occur in association with (1) increasing habitat patch size, age, and connectivity; (2) food resources, such as fruiting plants or bird feeders; (3) perennial water features; (4) vegetative cover (tree cover reported most often in the literature) and diversity in plant height; (5) less dense residential areas, specifically those with single-family units; and (6) protected areas and habitat reserves within cities. Features detrimental to urban bird richness and diversity reported in our surveyed literature included high rates of (1) built cover; (2) building density; (3) noise; (4) commercial and industrial land uses; (5) vehicle and pedestrian traffic; (6) presence of domestic cats and dogs; (7) areas of high housing density, such as dense apartment complexes; and (8) locations within the city center. Other variables such as agricultural land cover, altitude, golf course land uses, and socioeconomic status have variable impacts on the bird community reported throughout this sampled literature. While many of these variables are well-studied, this does not reduce the value of future research on these variables since, as discussed previously, there is still a lack of consensus of bird community trends (i.e. linear or intermediate peak pattern) and uneven sampling of many of these drivers across cities globally. This variability in results highlights the need to broaden our methodological scope, exploring new and more comprehensive variables used to quantify urbanization, new study locations (i.e. city locations and size, climates, gradients, or contrasting end points), guilds (e.g. inverse relationships between generalist vs. specialist species richness across gradients), and/or seasons sampled.

Management and conservation

The vast majority of studies focused on urban bird communities or assemblages suggest or request that their findings be used in conservation, management, and/or planning strategies and actions, yet many have remained in publications as a "dead letter". Among the most common suggestions to act upon in favor of birds in urban settings are green space management and planning and the study and control of nuisance (often exotic and/or invasive) species. These suggestions emphasize topics such as the importance of education plans and moving forward toward bird-friendly legislation and policies (MacGregor-Fors et al. 2020). Some studies have focused and developed studies with management and/or conservation perspective from the local to the landscape scales, depending on the desired results (Fernández-Juricic and Jokimäki 2001; Murgui and Hedblom 2017).

Given the heterogeneity of urban systems and the contrasts in habitat quality for most birds among highly-developed urban sites and large urban green spaces, conservation, management, and planning strategies are heavily focused on green spaces. While we agree that enhancing the ecological quality of the green space network of cities is crucial for their avian communities, more attention needs to be paid to the harsher urbanized matrix of which they are surrounded (Hostetler et al. 2011). Studies focused on the land sharing versus land sparing contrast in urban systems have provided important insight on the importance of biodiversity conservation of urban green space networks, but also the enhancement of human life quality in greener urban matrices, which also tends to benefit an important proportion of species (Soga et al. 2014; Jokimäki et al. 2019; Ibáñez-Álamo et al. 2020).

There is no doubt that using evidence-based knowledge to develop urban conservation, management, and planning strategies and actions in favor of birds would not only benefit them but other taxa as well, as their habitat requirements often make them surrogate bioindicators (Moreno et al. 2007). If materialized, the potential of enhancing biodiversity in urban systems with proper management and planning is vast. Translating and extending knowledge to action relies on a sum of stakeholder willingness, which is the cornerstone of bridging the gap on our race toward building more resilient and biodiverse cities (McDonnell and MacGregor-Fors 2016).

Future directions

Studies on bird communities are increasingly exploring new cities, becoming more geographically diverse over time, and utilizing modern methods. Even so, we would like to highlight a few recommendations and examples as to how this field can transcend the already robust publication resources currently available. Specifically, we suggest enhancing the field of urban ornithology to increase the publications that sample beyond landbirds, explore broader geographic and temporal time scales, investigate blue spaces and the urban core, reclassify and homogenize urbanization values, and refine community ecology metrics.

The taxa studied in the papers we reviewed were predominantly landbirds; however, this broad sampling method, while informative, also obscures the guilds and unique species found within cities. Specifically, studies on waterfowl, nocturnal species, and soaring species are understudied, perhaps due to the limited studies conducted around waterways, challenges associated with personal safety for evening surveys, and sampling independence for soaring species. Even so, publications that focused on a particular bird guild such as woodpeckers (e.g. Morrison and Chapman 2005; McClure et al. 2015), waterfowl (e.g. Rosa et al. 2003; Chen et al. 2020), and egrets (e.g. Zhang et al. 2018) offer important trends for these species' management and conservation. Another area to explore is the link between urban birds and other taxa. Most papers within our literature search surveyed only bird and vegetation variables, missing the relationships between the urban bird community and other taxa (e.g. insects, mammals, gastropods). Studies that successfully integrated a multi-taxonomic approach provide an ecosystem-level understanding to assess resources and competition for the bird community, information which is desperately needed for our knowledge of urban ecosystems (Concepción et al. 2016; MacGregor-Fors et al. 2016).

Another future direction to broaden beyond taxonomy is the temporal and geographic scales at which studies are conducted. We have been aware of the limited temporal scope of most urban bird studies since John Marzluff's review in 2001. While some studies have heeded his advice, more multi-season and multi-year studies are still necessary to obtain a wider view of how bird communities interact with urban ecosystems, as urbanization may impact ecosystem functions and communities over time (Palacio et al. 2018) and seasons (Leveau and Leveau 2016). Additionally, urban bird studies have been predominantly performed in the Global North (MacGregor-Fors et al. 2020) and large cities, such as Phoenix, Madrid, and Singapore. Studies should venture beyond these borders, particularly further into Africa, the Middle East, small urban areas (e.g. Gorosito and Cueto 2020), and multicity studies (e.g. Clucas and Marzluff 2015; Leveau et al. 2017). Studying bird communities long-term over wide geographic areas is a major feat; however, utilization of inclusive community science data (e.g. eBird) or collaborations with researchers and existing databases in multiple cities, countries, or regions (e.g. Global Biodiversity Information Facility (GBIF), Urban Biodiversity Research Coordination Network (UrBioNet)) may alleviate challenges associated with these study designs. Many research networks have successfully been able to conduct these types of studies, such as the Baltimore and Phoenix Long Term Ecological Research (LTER) collaborations (e.g. Shochat et al. 2010) and collaborator groups throughout the European continent (e.g. Ferenc et al. 2014; Morelli et al. 2020).

While focusing on green spaces within cities lends itself well to bird sampling methods (e.g. timed area searches within defined habitat fragments) and key priority areas of the city (e.g. rainforest reserve habitats; Tonetti et al. 2017), it is shortsighted to not extend our views beyond these fragments. Among the less known systems within cities, ecologically, are the heavily built ones, which are those that need more management and planning attention. Species common to the urban core have adapted well to these conditions without proper management or planning, utilizing chimneys (e.g. chimney swifts, *Chaetura pelagica*), building ledges (e.g. peregrine falcons, *Falco peregrinus*), and other city features that are worth investigating as their own habitat type. Urban blue spaces are also not well represented in the literature concerning urban bird communities, as only 13% of these studies centered their efforts on sampling urban rivers, lakes, wetlands, irrigation, and waste-stabilization ponds. Bird communities within these gray and blue spaces often are unique within cities and should be explored in future work to determine how these areas can be best managed for bird diversity.

As discussed throughout this chapter, most urban bird community studies utilized the gradient or contrast framework to design their studies, albeit without consistent metrics of how urbanization is defined nor its endpoints. Urbanization terms need to be standardized before we can progress further within this field. Terms need to be efficient, generalizable, and well-suited for a variety of urban ecosystems, whether we are utilizing quantitative or qualitative terms to describe what it means to be urban or rural (Moll et al. 2019; MacGregor-Fors and Vázquez 2020). Without such terms, results of these studies may continue to be variable and location-dependent. The integration of both local and landscape features (e.g. Concepción et al. 2016; Ferenc et al. 2016), socioeconomic factors (e.g. Lerman and Warren 2011), land use density classifications (e.g. Pidgeon et al. 2014), and noise and pollution impacts on urban ecosystems (e.g. Ciach and Fröhlich 2017) are just some ways that we can begin to move toward a unified definition of what it means to be urban.

Our last recommendation for studying urban bird communities is to refine and explore new experimental and analytical methods to quantify urban bird communities. Large amounts of bird community publications utilize the terms "diversity" and "richness" (Figure 17.2); however, in some of these papers, those two terms are used interchangeably. Authors should be specific in regard to how communities are described and also move towards new means of evaluating urbanization's impacts on birds. New diversity metrics (e.g. functional diversity, phylogenetic diversity), while introduced in the early 2000s, are still relatively rare in the reviewed urban bird literature. Additionally, most studies (1) lack the incorporation of detection models for bird surveys, which is particularly detrimental in assessing communities in noisy urban environments; (2) use observed species richness for analyses rather than richness estimations; (3) conduct natural rather than manipulative experiments to test process-driven mechanisms; and (4) utilize linear models for determining relationships between the community and urbanization variables, which may not be accurate for communities with an intermediate peak along a gradient. Statistical packages to solve these three issues are abundant and free, allowing urban ecologists to design and analyze urban bird communities with greater certainty and confidence.

In this chapter, we have described how far urban bird ecology has come since its consolidation in the early 1990s. We recognize and applaud the feats taken on by the global community to expand and evolve the field, providing current and invaluable information for urban bird management and conservation. We hope that this chapter helps in synthesizing the current status of urban bird community literature while offering potential guidance and suggestions as to the future of this area of research. By expanding our global networks and continuing to push experimental and analytical boundaries, our understanding of bird communities will be a strong contributor to urban biodiversity as a whole.

Key companion papers

Chase, J.F., J.J. Walsh. 2006. Urban effects on native avifauna: A review. *Landscape and Urban Planning* 20:202–226.

Clucas, B., J.M. Marzluff. 2015. A cross-continental look at the patterns of avian species diversity and composition across an urbanisation gradient. *Wildlife Research* 42:554–562.

Ferenc, M., et al. 2014. Are cities different? Patterns of species richness and beta diversity of urban bird communities and regional species assemblages in Europe. *Global Ecology and Biogeography* 23:479–489.

Jokimäki, J., et al. 2019. Land-sharing vs. land-sparing urban development modulate predator – prey interactions in Europe. *Ecological Applications* 30:e02049.

Palacio, F.X., L.M. Ibañez, R.E. Maragliano, D. Montalti. 2018. Urbanization as a driver of taxonomic, functional, and phylogenetic diversity losses in bird communities. *Canadian Journal of Zoology* 96:1114–1121.

Shochat, E., et al. 2010. Invasion, competition, and biodiversity loss in urban ecosystems. *BioScience* 60:199–208.

References

Alberti, M., E. Botsford, A. Cohen. 2001. Quantifying the urban gradient: Linking urban planning and ecology. Pages 89–115 in J.M. Marzluff, R. Bowman, R. Donnelly, editors. *Avian Ecology and Conservation in an Urbanizing World*. Springer, Boston, MA.

Banville, M.J., H.L. Bateman, S.R. Earl, P.S. Warren. 2017. Decadal declines in bird abundance and diversity in urban riparian zones. *Landscape and Urban Planning* 159:48–61.

Batáry, P., K. Kurucz, M. Suarez-Rubio, D.E. Chamberlain. 2018. Non-linearities in bird responses across urbanization gradients: A meta-analysis. *Global Change Biology* 24:1046–1054.

Bibby, C.J., N.D. Burgess, D.A. Hill. 2000. *Bird Census Techniques*, 2nd edition. Academic Press Limited, Cambridge.

Bishop, J.A., W.L. Myers. 2005. Associations between avian functional guild response and regional landscape properties for conservation planning. *Ecological Indicators* 5:33–48.

Blair, R.B., E.M. Johnson. 2008. Suburban habitats and their role for birds in the urban – Rural habitat network: Points of local invasion and extinction? *Landscape Ecology* 23:1157–1169.

Carbó-Ramírez, P., I. Zuria. 2011. The value of small urban greenspaces for birds in a Mexican city. *Landscape and Urban Planning* 100:213–222.

Catterall, C.P., J.A. Cousin, S. Piper, G. Johnson. 2010. Long-term dynamics of bird diversity in forest and suburb: Decay, turnover or homogenization? *Diversity and Distributions* 16:559–570.

Chase, J.F., J.J. Walsh. 2006. Urban effects on native avifauna: A review. *Landscape and Urban Planning* 20:202–226.

Chen, T., L. Bao, L. Bao Zhu, Y. Tian, Q. Xu, Y. Hu. 2020. The diversity of birds in typical urban lake-wetlands and its response to the landscape heterogeneity in the buffer zone based on GIS and field investigation in Daqing, China. *European Journal of Remote Sensing* 53:1–9.

Ciach, M., A. Fröhlich. 2017. Habitat type, food resources, noise and light pollution explain the species composition, abundance and stability of a winter bird assemblage in an urban environment. *Urban Ecosystems* 20:547–559.

Clucas, B., J.M. Marzluff. 2015. A cross-continental look at the patterns of avian species diversity and composition across an urbanisation gradient. *Wildlife Research* 42:554–562.

Concepción, E.D., M.K. Obrist, M. Moretti, F. Altermatt, B. Baur, M.P. Nobis. 2016. Impacts of urban sprawl on species richness of plants, butterflies, gastropods and birds: Not only built-up area matters. *Urban Ecosystems* 19:225–242.

Evans, K.L., K.J. Gaston, S.P. Sharp, A. McGowan, B.J. Hatchwell. 2009. The effects of urbanization on avian morphology and latitudinal gradients in body size. *Oikos* 118:251–259.

Ferenc, M., O. Sedláček, J. Mourková, A. Exnerová, J. Škopek, J. Formánek, R. Fuchs. 2016. Disentangling the influences of habitat availability, heterogeneity and spatial position on the species richness and rarity of urban bird communities in a central European city. *Urban Ecosystems* 19:1265–1281.

Ferenc, M., O. Sedláček, R. Fuchs, M. Dinetti, M. Fraissinet, D. Storch. 2014. Are cities different? Patterns of species richness and beta diversity of urban bird communities and regional species assemblages in Europe. *Global Ecology and Biogeography* 23:479–489.

Fernández-Juricic, E., J. Jokimäki. 2001. A habitat island approach to conserving birds in urban landscapes: Case studies from southern and northern Europe. *Biodiversity and Conservation* 10:2023–2043.

Filippi-Codaccioni, O., J. Clobert, R. Jilliard 2009. Urbanisation effects on the functional diversity of avian agricultural communities. *Acta Oecologica* 35:705–710.

Fontana, S., T. Sattler, F. Bontadina, M. Moretti. 2011. How to manage the urban green to improve bird diversity and community structure. *Landscape and Urban Planning* 101:278–285.

Gorosito, C.A., V.R. Cueto. 2020. Do small cities affect bird assemblages? An evaluation from Patagonia. *Urban Ecosystems* 23:289–300.

Hohtola, E. 1978. Differential changes in bird community structure with urbanisation: A study in central Finland. *Ornis Scandinavica* 9:94–100.

Hostetler, M., W. Allen, C. Meurk. 2011. Conserving urban biodiversity? Creating green infrastructure is only the first step. *Landscape and Urban Planning* 100:369–371.

Husté, A., T. Boulinier. 2007. Determinants of local extinctions and turnover rates in urban bird communities. *Ecological Applications* 17:168–180.

Ibáñez-Álamo, J.D., F. Morelli, Y. Benedetti, E. Rubio, J. Jokimäki, T. Pérez-Contreras, P. Sprau, J. Suhonen, P. Tryjanowski, M.-L. Kaisanlahti-Jokimäki, A.P. Møller, M. Díaz. 2020. Biodiversity within the city: Effects of land sharing and land sparing urban development on avian diversity. *Science of The Total Environment* 707:135477.

Inzunza, E.R., S.H.A. Rodriguez. 2009. The urban avifauna of the Macuiltépetl Ecological Park in Xalapa, Veracruz, Mexico. *Ornitologia Neotropical* 21:87–103.

Jokimäki, J., J. Suhonen, Y. Benedetti, M. Diaz, M-L. Kaisanlahti-Jokimäki, F. Morelli, T. Pérez-Contreras, E. Rubio, P. Sprau, P. Tryjanowski, J.D. Ibánez-Álamo. 2019. Land-sharing vs. land-sparing urban development modulate predator – prey interactions in Europe. *Ecological Applications* 30:e02049.

La Sorte, F.A., M.F.J. Aronson, C.A. Lepczyk, K.G. Horton. 2020. Area is the primary correlate of annual and seasonal patterns of avian species richness in urban green spaces. *Landscape and Urban Planning* 203:103892.

La Sorte, F.A., C.A. Lepczyk, M.F.J. Aronson, M.A. Goddard, M. Hedblom, M. Katti, I. MacGregor-Fors, U. Mörtberg, C.H. Nilon, P.S. Warren, N.S.G. Williams, J. Yang. 2018. The phylogenetic and functional diversity of regional breeding bird assemblages is reduced and constricted through urbanization. *Diversity and Distributions* 24:928–938.

Lerman, S.B., P.S. Warren. 2011. The conservation value of residential yards: Linking birds to people. *Ecological Applications* 21:1327–1339.

Leveau, C.M., L.M. Leveau. 2005. Avian community response to urbanization in the Pampean Region, Argentina. *Ornitologia Neotropical* 16:503–510.

Leveau, L.M., F.I. Isla, M.I. Bellocq. 2018. Predicting the seasonal dynamics of bird communities along an urban-rural gradient using NDVI. *Landscape and Urban Planning* 177:103–113.

Leveau, L.M., C.M. Leveau. 2016. Does urbanization affect the seasonal dynamics of bird communities in urban parks? *Urban Ecosystems* 19:631–647.

Leveau, L.M., C.M. Leveau. 2020. Street design in suburban areas and its impact on bird communities: Considering different diversity facets over the year. *Urban Forestry & Urban Greening* 48:126578.

Leveau, L.M., C.M. Leveau, M. Villegas, J.A. Cursach, C.G. Suazo. 2017. Bird communities along urbanization gradients: Comparative analysis among three Neotropical cities. *Ornitologia Neotropical* 28:77–87.

MacGregor-Fors, I., F. Escobar, R. Rueda-Hernández, S. Avendaño-Reyes, M.L. Baena, V.M. Bandala, S. Chacón-Zapata, A. Guillén-Servent, F. González-García, F. Lorea-Hernández, E. Montes de Oca, L. Montoya, E. Pineda, L. Ramírez-Restrepo, E. Rivera-García, E. Utrera-Barrillas. 2016. City 'green' contributions: The role of urban greenspaces as reservoirs for biodiversity. *Forests* 7:146.

MacGregor-Fors, I., J.F. Escobar-Ibáñez, editors. 2017. *Avian Ecology in Latin American Cityscapes*. Springer International Publishing, Cham, Switzerland.

MacGregor-Fors, I., M.A. Gómez-Martínez, M. García-Arroyo, C.A. Chávez-Zichinelli. 2020. A dead letter? Urban conservation, management, and planning strategies from the Mexican urban bird literature. *Urban Ecosystems* 23:1107–1115.

MacGregor-Fors, I., R. Ortega-Álvarez, J.E. Schondube. 2009. On the ecological quality of urban systems: An ornithological perspective. Pages 5–66 in D.S. Graber, K.A. Birmingham, editors. *Urban Planning in the 21st Century*. Nova Science Publishers, New York.

MacGregor-Fors, I., C.C. Rega-Brodsky, M. Garcia-Arroyo, M.A. Gomez-Martinez, L-B. Vazquez. 2020. Urban bird ecologists cite more publications from the Global North: Why? *Journal of Urban Ecology* 6:juaa027.

MacGregor-Fors, I., L-B. Vázquez. 2020. Revisiting 'rural'. *Science of the Total Environment* 741:132789.

Marzluff, J.M. 2016. A decadal review of urban ornithology and a prospectus for the future. *Ibis* 159:1–13.

Marzluff, J.M., R. Bowman, R. Donnelly. 2001. A historical perspective on urban bird research: Trends, terms, and approaches. Pages 1–17 in J.M. Marzluff, R. Bowman, R. Donnelly, editors. *Avian Ecology and Conservation in an Urbanizing World*. Springer US, Boston, MA.

McClure, C.J.W., A.C. Korte, J.A. Heath, J.R. Barber. 2015. Pavement and riparian forest shape the bird community along an urban river corridor. *Global Ecology and Conservation* 4:291–310.

McDonnell, M.J., I. MacGregor-Fors. 2016. The ecological future of cities. *Science* 352:936–938.

McDonnell, M.J., S.T.A. Pickett. 1990. Ecosystem structure and function along urban-rural gradients: An unexploited opportunity for ecology. *Ecology* 71:1232–1237.

McKinney, M.L. 2008. Effects of urbanization on species richness: A review of plants and animals. *Urban Ecosystems* 11:161–176.
Moll, R.J., J.D. Cepek, P.D. Lorch, P.M. Dennis, E. Tans, T. Robison, J.J. Millspaugh, R.A. Montgomery. 2019. What does urbanization actually mean? A framework for urban metrics in wildlife research. *Journal of Applied Ecology* 56:1289–1300.
Morelli, F., Y. Benedetti, J.D. Ibáñez-Álamo, P. Tryjanowski, J. Jokimäki, M.-L. Kaisanlahti-Jokimäki, T. Pérez-Contreras, P. Sprau, J. Suhonen, R. Yosef, M. Díaz, A.P. Møller. 2020. Insurance for the future? Potential avian community resilience in cities across Europe. *Climatic Change* 159:195–214.
Moreno, C.E., G. Sanchez-Rojas, E. Pineda, F. Escobar. 2007. Shortcuts for biodiversity evaluation: A review of terminology and recommendations for the use of target groups, bioindicators and surrogates. *International Journal of Environment and Health* 1:71–86.
Morrison, J.L., W.C. Chapman. 2005. Can urban parks provide habitat for woodpeckers? *Northeastern Naturalist* 12:253–262.
Müller, A., P.K. Bøcher, C. Fischer, J.-C. Svenning. 2018. 'Wild' in the city context: Do relative wild areas offer opportunities for urban biodiversity? *Landscape and Urban Planning* 170:256–265.
Murgui, E., M. Hedblom. 2017. Improving research towards conservation objectives. Pages 491–507 in E. Murgui, M. Hedblom, editors. *Ecology and Conservation of Birds in Urban Environments*. Springer International Publishing, Cham, Switzerland.
Naumkin, D.V. 2010. Bird population in the pine and birch forests of Kungurskii Raion (Perm'; Krai) under anthropogenic load of various levels. *Contemporary Problems of Ecology* 3:76–80.
Nilon, C.H., C.N. Long, W.C. Zipperer. 1995. Effects of wildland development on forest bird communities. *Landscape and Urban Planning* 32:81–92.
Ortega-Álvarez, R., I. MacGregor-Fors. 2009. Living in the big city: Effects of urban land-use on bird community structure, diversity, and composition. *Landscape and Urban Planning* 90:189–195.
Ortega-Álvarez, R., I. MacGregor-Fors. 2011. Dusting-off the file: A review of knowledge on urban ornithology in Latin America. *Landscape and Urban Planning* 101:1–10.
Padilla, B.J., C. Sutherland. 2019. A framework for transparent quantification of urban landscape gradients. *Landscape Ecology* 34:1219–1229.
Palacio, F.X., L.M. Ibañez, R.E. Maragliano, D. Montalti. 2018. Urbanization as a driver of taxonomic, functional, and phylogenetic diversity losses in bird communities. *Canadian Journal of Zoology* 96:1114–1121.
Pautasso, M., M. Dinetti. 2009. Avian species richness, human population and protected areas across Italy's regions. *Environmental Conservation* 36:22–31.
Pidgeon, A.M., C.H. Flather, V.C. Radeloff, C.A. Lepczyk, N.S. Keuler, E.M. Wood, S.I. Stewart, R.B. Hammer. 2014. Systematic temporal patterns in the relationship between housing development and forest bird biodiversity. *Conservation Biology* 28:1291–1301.
Plummer, K.E., K. Risely, M.P. Toms, G.M. Siriwardena. 2019. The composition of British bird communities is associated with long-term garden bird feeding. *Nature Communications* 10:2088.
Rosa, S., J.M. Palmerim, F. Moreira. 2003. Factors affecting waterbird abundance and species richness in an increasingly urbanized area of the Tagus Estuary in Portugal. *Waterbirds* 26:226–232.
Saari, S., S. Richter, M. Higgins, M. Oberhofer, A. Jennings, S.H. Faeth. 2016. Urbanization is not associated with increased abundance or decreased richness of terrestrial animals – dissecting the literature through meta-analysis. *Urban Ecosystems* 19:1251–1264.
Shochat, E., S.B. Lerman, J.M. Anderies, P.S. Warren, S.H. Faeth, C.H. Nilon. 2010. Invasion, competition, and biodiversity loss in urban ecosystems. *BioScience* 60:199–208.
Soga, M., Y. Yamaura, S. Koike, K.J. Gaston. 2014. Land sharing vs. land sparing: Does the compact city reconcile urban development and biodiversity conservation? *Journal of Applied Ecology* 51:1378–1386.
Strohbach, M.W., A. Hrycyna, P.S. Warren. 2014. 150 years of changes in bird life in Cambridge, Massachusetts from 1860 to 2012. *The Wilson Journal of Ornithology* 126:192–206.
Taylor, L., C. Taylor, A. Davis. 2013. The impact of urbanisation on avian species: The inextricable link between people and birds. *Urban Ecosystems* 16:481–498.
Tonetti, V.R., M.A. Rego, A. De Luca, P.F. Develey, F. Schunck, L.F. Silveira. 2017. Historical knowledge, richness and relative representativeness of the avifauna of the largest native urban rainforest in the world. *Zoologia* 34:e13728.

Villaseñor, N.R., L.A. Chiang, H.J. Hernández, M.A.H. Escobar. 2020. Vacant lands as refuges for native birds: An opportunity for biodiversity conservation in cities. *Urban Forestry & Urban Greening* 49:126632.

Weideman, E.A., J.A. Slingsby, R.L. Thomson, B.T.W. Coetzee. 2020. Land cover change homogenizes functional and phylogenetic diversity within and among African savanna bird assemblages. *Landscape Ecology* 35:145–157.

White, J.G., M.J. Antos, J.A. Fitzsimons, G.C. Palmer. 2005. Non-uniform bird assemblages in urban environments: The influence of streetscape vegetation. *Landscape and Urban Planning* 71:123–135.

Zhang, M., Y.-M. Hong, J.G. Castley, F.-S. Zou, Q. Zhang, H.-M. Fan. 2018. Habitat features rather than competition explain the distribution and co-occurrence of Ardeidae in a highly urbanized landscape. *Waterbirds* 41:46–55.

18

A CROSS-CONTINENTAL SYNTHESIS OF URBAN INSECTIVOROUS BAT RESEARCH AND FUTURE DIRECTIONS

Caragh G. Threlfall and Kirsten Jung

Introduction

Among urban wildlife, bats are the most secretive cohabitants of urban areas. They may roost in close proximity to humans, in attics, behind window shields, and in building crevices but often remain unnoticed due to their nocturnal lifestyle. Nevertheless, bats commonly occur in urban areas all over the globe (Russo and Ancillotto 2015; Jung and Threlfall 2016, 2018). Besides taking advantage of roosts in human-made structures (Ancillotto et al. 2015; e.g. Voigt et al. 2016, Figure 18.1A), some bat species are known to readily exploit fruiting trees and nectar sources in urban gardens, insect accumulations around streetlights (e.g. Rydell 1992; Avila-Flores and Fenton 2005; Jung and Kalko 2010; Lewanzik and Voigt 2017), occupy urban parklands (e.g. Fabianek et al. 2011; Threlfall et al. 2016), and forage above urban rooftops (Avila-Flores and Fenton 2005; Jung and Kalko 2011).

There are currently 1,421 species of bats worldwide (Simmons and Cirranello 2020), where they account for about 25% of the living mammalian diversity (Burgin et al. 2018) and have colonised all continental regions of the Earth except extreme latitudes (Simmons 2005). Compared to other similar-sized mammals, bats have a remarkably long lifespan and a slow reproductive cycle typically rearing only one or two young per year, making them slow to recover from population declines (Findley 1993). They can exhibit very complex social systems (McCracken and Wilkinson 2000) and form colonies of just a few or millions of individuals (Kunz and Lumsden 2003; Kerth 2008).

Bats feed on a wide variety of resources, including pollen, fruit, arthropods, and vertebrates, and thus play important ecological roles as pollinators, seed dispersers, and in insect control (Boyles et al. 2011; Kunz et al. 2011). A considerable number of bats undergo significant seasonal movements between habitats (Fleming 2019) and migrate hundreds or even thousands of kilometres each year, avoiding unfavourable conditions during winter or following seasonal food resource gradients (Fleming 2019). Although bats contribute valuable economic benefits to fruit production, forest regeneration, and insect pest control, the whole taxon has recurrently been discredited as airborne vectors for several wildlife-born human diseases, including rabies, Ebola, and Sars-CoV-2 (COVID-19). Consequently, their persistence in rural and urban environments – and in such close proximity to humans – poses additional threats to bat populations around the globe.

DOI: 10.4324/9781003016120-21

Bats are unique among mammals for their abilities of powered flight (Norberg and Rayner 1987) and echolocation (Kunz 1982). Whereas Pteropodids mainly rely on vision as a sensory cue, Microchiroptera rely predominantly on sophisticated echolocation to navigate and forage (Neuweiler 1984; Jones and Holderied 2007). They emit high frequency (ultrasonic) calls and evaluate the incoming echoes to gain information about, distance, size, and direction of targets. Differences in echolocation performance among species are reflected in the high variety to use the aerospace (Schnitzler et al. 2003; Denzinger and Schnitzler 2013). Some species have fast and straight flight and prefer to forage in high altitudes away from vegetation and other obstacles in the open skies, while others are characterised by butterfly-like, slow, and agile flight and are capable of manoeuvring within dense vegetation (Denzinger and Schnitzler 2013).

Within this chapter we focus on insectivorous species because of their wide distribution across many cities and their ability to echolocate, which has led to an expansive literature on insectivorous urban bat ecology mainly in the last two decades, utilising acoustic monitoring techniques. We give a brief insight into the approaches to study bats and then summarise the current knowledge about insectivorous bat ecology with a specific focus on the effect of urbanisation on species richness, habitat use, and species composition. We highlight which habitat elements appear to be important within the urban context (e.g. roost trees, green spaces, water sources) and discuss advantages and

Figure 18.1 (A) *Myotis myotis* roosting in a church attic, Lisa Printz; (B) Acoustic recording of urban winter roost activity of *Nyctalus noctula*, Kirsten Jung; (C) Radiotracking of *Nyctophilus gouldi*, Caragh Threlfall; (D) Urban roost tree in suburban Sydney, Australia, Caragh Threlfall.

risks for bat populations in urban areas. Finally, we outline key research gaps and highlight future directions for the study of bat populations and communities within city landscapes.

Research techniques and tools employed in the study of urban bat ecology

Bats are cryptic, nocturnal mammals and thus difficult to study. Most studies predominantly employ passive acoustic monitoring techniques, radio tracking, or roost surveillance to study the ecology of bats in urban environments.

Use of acoustic methods to examine urban bat ecology

Acoustic monitoring, a method that includes the recording of echolocation signals emitted from free-flying bats (Figure 18.1B), and identifying species based on echolocation signal characteristics (call structure and frequency components) has developed into the most sophisticated method to gain crucial insights about species habitat use, activity patterns, and assemblage composition. Especially in urban areas, acoustic monitoring is a highly valuable tool, as capture techniques such as mist netting or using "harp" traps (special type of large, free-standing trap that captures bats when in flight) is often not a feasible method due to traffic and human interference. Other methods such as hand capture from roosting sites is also often unfeasible due to roost sites being cryptic, non-permanent, or unknown, and research activity at night within some cities may pose safety risks to researchers (Scanlon and Petit 2008).

Early acoustic monitoring studies of bats in urban areas were conducted with hand-held ultrasound recorders (Pettersson, Laar detectors) or custom-made equipment by walking transects or stationary recording (e.g. Gaisler et al. 1998; Jung and Kalko 2011). Most of these earlier studies were limited by battery life of the equipment, data storage capacity, and manpower. Recent technical advances in the last decades however made it possible to place several automated ultrasound recording devices simultaneously in different locations or equip volunteers to participate in citizen science projects (Newson et al. 2015), gaining comparable and thus highly valuable bat activity data to assess the influence of various environmental factors on bats (e.g. Threlfall et al. 2012; Gili et al. 2020).

The biggest bottleneck of acoustic monitoring studies, however, remains the evaluation of the acoustic data. In some parts of the world – especially in tropical countries – knowledge of species-specific echolocation call characteristics and signal variability are still scarce. In addition, analysing acoustic sequences is labour intensive and, although computer software has been developed, is improving constantly, and promises automated species identification, expert knowledge is still needed to verify the software output. Despite the enormous value of acoustic techniques, this passive and noninvasive method however provides limited information on movement of individuals, roosting locations, social interactions, or data related to population ecology such as survival, fitness, and population structure.

Understanding habitat use and population ecology via animal capture, radio tracking, and roost surveillance

Several studies have investigated habitat and roosting site selection (Evelyn et al. 2004), home range size (Popa-Lisseanu et al. 2009), and movement in relation to sex (Patriquin et al. 2019) and reproductive status (Popa-Lisseanu et al. 2009) in rural, suburban, and urban environments by capturing individuals using mist nets or harp traps and radiotracking their movement over several nights and days (Figure 18.1C).

Radiotelemetry is a highly valuable tool to understand habitat use of bats within landscapes. However, radiotelemetry is mostly restricted to a small sample size of individuals and limited in temporal and spatial resolution due to transmitter battery power and range of transmitter signals (Gottwald et al. 2019). Especially in urban areas, ground-based radio telemetry is challenging because radio signals are easily lost in densely built up areas (Kubista and Bruckner 2015).

Some studies have thus examined movement patterns of bats (Popa-Lisseanu et al. 2009) and interactions between individuals using GPS telemetry and onboard ultrasonic microphones (Egert-Berg et al. 2018). However due to the transmitter weight these techniques are currently limited to larger bat species. More recently, lightweight animal-borne sensors have been used (Ripperger et al. 2016; Gottwald et al. 2019) to study fine scale movement of bats and represent a promising tool to evaluate interactions among individuals and investigate close contacts. This technology opens new avenues of research, allowing detailed investigation of events associated with social contact, such as mating behaviour, pathogen transmission, social learning, and resource sharing (Ripperger et al. 2016, 2020).

Although radiotelemetry has its technical limitations, it is critical for detecting foraging habitats and unknown roosting sites of bats in a city, which is important for conservation of urban populations. By enabling the discovery of roosting sites, this method can be the key to future opportunities in bat research, including molecular techniques. Locating roost sites, and thus bat colonies, allows for demographic assessment of populations, assessment of social structure or the health status of colonies via blood or genetic sampling, determining food preferences of individuals and microbiome analysis using guano samples, and identifying preferred prey items or seasonal dietary changes via barcoding.

The response of bats to urbanisation: a summary of research to date

Overall trends in species richness and activity

Using mostly acoustic techniques, in addition to telemetry or capture studies, we have gained a very broad understanding of which factors influence bat species responses to urbanisation. In general, most bats are very sensitive to environmental change and can act as bioindicators of environmental conditions (Jones et al. 2009). Due to their ability to fly, they are thought to respond promptly to unfavourable conditions in their environment and move to more suitable habitats. To this end, several countrywide schemes have been established to monitor distributions of bat species and assess population trends, for example in the UK, France, and US. Overall, it has been found that urbanisation negatively impact many bat species, reducing species richness and diversity (Jung and Threlfall 2016). A recent global meta-analysis quantified the negative consequences of urbanisation on bats and indicated a 34% reduction in the intensity of habitat use by bats in urban compared with non-urban areas (Jung and Threlfall 2018). Studies conducted in large urban areas (e.g. Mexico City; Avila-Flores and Fenton 2005) revealed a stronger negative effect of urbanisation on species richness and activity patterns (Jung and Threlfall 2016) compared to studies conducted in smaller towns (e.g. Furlonger et al. 1987; Luck et al. 2013) and villages (e.g. Rainho 2007; Jung and Kalko 2011).

Studies conducted in relatively large cities have shown that the majority of the bat community is very sensitive to reductions in the cover of trees. For example, in Sydney, Australia, it was found that areas with 10–30% tree cover supported high bat activity and species richness, but below this level, bat species richness significantly declined (Threlfall et al. 2011). Similarly, in Melbourne, Australia, Caryl et al. (2016) found peaks in occupancy of edge and patch species with 20–60%

tree cover; however, this positive effect of trees declined with increasing housing density. Caryl et al. (2016) point out that perhaps this bat assemblage is so sensitive to urban density because of the relatively young age of this city, in comparison to European cities, for example. Therefore, suggesting that landscapes with a longer history of intensive human use (for urban or agricultural activities) may harbour less sensitive assemblages (or individuals).

In contrast to larger cities, smaller urban areas have been shown to harbour a high species richness and high abundances of bats. For example, Luck et al. (2013) surveyed bats across 18 smaller cities and towns in south-eastern Australia (populations ranged from 16,000–78,000 people) and found higher diversity of species and greater bat activity when compared to other Australian studies in larger cities (Brisbane: Hourigan et al. 2010; Sydney: Basham et al. 2011). Luck et al. (2013) sampled within residential areas and urban parklands and found that moderate amounts of vegetation cover in these areas supported greater bat activity and species richness, suggesting that the distance between suitable roosting and foraging areas in small urban centres is short enough to allow bats to utilise much of the landscape.

In the agricultural-dominated areas of the US and Europe, urban areas may provide more roosting and foraging opportunities compared to habitat-poor agricultural landscapes. Within an ongoing study in Germany, bat species richness and activity in urban settlements exceeds semi-natural areas in production forests and agricultural grasslands (Jung unpub.). Several bat species here take advantage from additional roosting sites in stables, barns, and house crevices and feed on insect accumulations drawn from the surrounding semi-natural landscapes to the lit areas along the village roads. The benefits of small urban areas in the broader landscape was also explored in a study by Gili et al. (2020). Here, the authors investigated the effect of urban habitat configuration and composition on bat species occurrence and activity. Their results revealed that continuous urban areas within the landscape are less likely to host bat species, but fragmented urban areas might support the occurrence of some species (e.g. *Myotis* spp.). In addition, Gili et al. (2020) show that the probability of occurrence and activity of bats increased with water surface, in particular lakes, and the availability of discontinuous woodland, while continuous woodland (in this case predominantly coniferous plantations) only benefited some species. Gili et al. (2020) used an extensive dataset to predict the effect of urbanisation under five different scenarios of urban development and showed that sustainable urban development (United Nations 2018 goal) is indeed possible by careful urban planning. They highlighted that expanding existing urban areas in fragments and creating blocks of urban patches, accompanied by a strategy that creates bat-friendly habitat of similar size, such as discontinuous woodland, areas with intermediate tree cover, hedgerows, tree lines, and gardens nearby, could compensate for potential negative effects of urban expansion.

Urban bat community ecology

The composition and structure of ecological communities is known to change in urban areas. Single city studies have previously suggested that bats of some families (e.g. Molossids; Jung and Kalko 2011) are better pre-adapted for life in urban environments compared to others (e.g. Rhinolophids; Stone et al. 2009). However, bat families differ in evolutionary history and geographical distribution (Teeling et al. 2012), which limits generalisations made from a single region. A global analysis revealed that neither phylogenetic relatedness nor functional ecology alone explain different responses of bat species to urbanisation (Jung and Threlfall 2016). Instead, response of bats to urbanisation is very species-specific and related to species morphological and behavioural traits (Jung and Threlfall 2018).

Common to studies of bats in large cities is the finding that certain species can tolerate highly urbanised parts of the landscape and profit from increased insect densities in certain land uses (e.g. Mexico City; Avila-Flores and Fenton 2005). In a study across a medium-density area of the UK (West Midlands metropolitan area, 2.3 million people), Hale et al. (2012) found that most bat species were sensitive to urban land cover; however one species, *Pipistrellus pipistrellus*, was reported to be present up until around 60% urban land cover. The finding that a small handful of species can utilise large portions of the urban landscape also extends to studies in medium-sized cities, for example, in Panama City (1.8 million people) and the nearby small town of Gamboa (Jung and Kalko 2011). In this study, overall bat activity was highest at the forest – town interface and lowest at the urban sites. However, there were strong compositional differences in species assemblages with mainly fast-flying Molossids dominating the urban sites. The small town of Gamboa revealed a similar composition to forest sites, but even here forest species were missing, and the proportional abundance of Molossids increased due to additional roosting sites in houses (Jung and Kalko 2011).

Jung and Threlfall (2018) suggest that urban tolerance in bats is correlated with traits related to greater flexibility in resource requirements and greater mobility, while urban avoidance is correlated with species exhibiting very specialised foraging modes such as perch hunting or gleaning, or specific roosting preferences (e.g. caves). Species that tolerate urban areas are generally open and edge aerial foragers and trawling bats, with flexible roosting strategies and species with higher weight and aspect ratio (narrow-shaped wings) at a given body size. These species might take advantage of increased roosting opportunities such as bat boxes or buildings in urban environments. In addition, the higher weight and aspect ratio in relation to other bats with similar body size likely correlates with higher flight speed and the ability to sustain longer flights. As food resources are likely scattered across a city, requiring longer commutes or longer foraging bouts per night, these flight characteristics may allow these species to maintain or even increase foraging efficiency.

However, even though broad trait-based patterns have been identified, some studies do report differential responses to urbanisation from morphologically similar species (Lintott et al. 2015; Caryl et al. 2016). These studies suggest that other aspects of bat behaviour or ecology may instead explain differences in sensitivity to urbanisation, for example, competitive interactions between bat species (Lintott et al. 2015). In a UK-wide study of two morphologically similar bat species, Lintott et al. (2016) show that even species considered to be common and tolerant of the urban landscape have a negative response to urban land cover at a local scale (1 km), suggesting that conservation efforts should consider schemes that protect and enhance habitat for common species to ensure their long-term persistence in urban areas.

Because urban areas contain complex mosaics of novel habitat conditions, they can support a diversity of bat species. Densely built up commercial and residential areas may neighbour relatively green neighbourhoods, intercepted by parklands and built infrastructure (e.g. roads, rail, industrial areas). Given the complexity of urban environments, it is difficult to evaluate and understand the specific mechanisms (Alberti 2005) by which urbanisation changes the abundance of bats and the structure of bat communities. Several studies have targeted the effect of urban heterogeneity on bats by correlating bat abundance with tree cover and vegetation structure (Suarez-Rubio et al. 2018), different levels of light pollution (Voigt et al. 2020), availability, type, and size of green spaces (Threlfall et al. 2016; Tena et al. 2020), green roofs (Parkins and Clark 2015), and water bodies (Straka et al. 2016). More recently, scientists have attempted to grasp the heterogeneity of a city by using the socioeconomic wealth of human inhabitants in relation to the distribution of bat species (Li and Wilkins 2014). These studies revealed an increase in bat diversity and activity with greater socioeconomic wealth of human inhabitants. Li et al. (2019) showed that certain bat species have greater activity in areas of higher socioeconomic wealth, suggesting this could be related to the increased level of tree cover in these areas, which may support greater populations of insects, in addition to the potential greater access bats have to water (e.g. home swimming pools) in areas of greater socioeconomic wealth.

Urban bat population ecology

Very little is known about the population-level impacts of urban living, including impacts on species fitness. A few studies have focussed on widespread species, such as *Pipistrellus kuhlii* (Ancillotto et al. 2015, 2016) and *Hypsugo savii*, which demonstrate range expansion in association with climate change and urbanisation. One study focused on *Myotis lucifugus* (Coleman and Barclay 2011) and tested whether the increase of habitat complexity linked with urbanisation in North American prairies would benefit this synanthropic bat by making more resources (day-roosts, tree cover, insect food) available, thus improving its fitness. The study targeted urban versus non-urban riparian sites. Bats that did best were located in transition habitat comprised between genuinely urban and rural areas. High-density urban populations exhibited decreased body condition and smaller production of juveniles, perhaps due to the higher competition for food, stress, disease transmission, or acoustic pollution, suggesting that this habitat may represent an ecological trap. This case is instructive as it shows that high population density is not an effective indicator that urban sites constitute population sources for bats (Coleman and Barclay 2011; Russo and Ancillotto 2015).

Urban roosts

Most bats cannot build roosts themselves (Kunz 1982) and thus depend on existing structures as day roosts that protect them from weather and predators. Roost structure, microclimate, and degree of protection from predators critically influence bat survival and reproductive success (Kunz and Lumsden 2003). In nature, roosts may be found in tree cavities, caves, or crevices of rock formations. In urban areas, these natural roosts are limited or non-existent and thus several bat species have taken advantage of human made roosting structures in urban areas (Figure 18.1A). For example, the Mauritian tomb bat, *Taphazous mauritianus*, is a widespread species that roosts in buildings in Lilongwe city, Malawi (O'Malley et al. 2020). This species selects buildings to occupy based on building and landscape features, such as building size and proximity to woodland. Whilst cities have many buildings, the preference for certain types of buildings selected for roosting means this species is vulnerable to roost loss and disturbance, especially given the close contact they have with humans at roost sites (O'Malley et al. 2020). In most areas of the world, this house sharing is not perceived positively by people. Although bats provide vital ecosystem services, people around the globe mostly perceive them as a burden and pest and vectors of virus borne diseases. Thus, loss of roost sites and disturbance at roosts is a direct and very prominent threat to urban bats (Voigt et al. 2016).

Despite the use of buildings by some species, several authors highlight the importance of large and older trees (Evelyn et al. 2004; Kubista and Bruckner 2015) or dead wood (Rhodes and Catterall 2008) for roosting (Figure 18.1D). Evelyn et al. (2004) studied roost selection of *Myotis yumanensis* in suburban residential areas and revealed the necessity for large trees of mean diameter 115 cm. These studies suggest that day roosts are the main limiting factor for urban bat species, highlighting the importance of retaining large trees as roosting habitat for bats within urban landscapes. A study of the roosting ecology of Gould's long-eared bat, *Nyctophilus gouldi*, found that this species roosted in trees across a range of diameters and instead was selecting roost trees that were often in gullies close to water, in areas with high canopy and understorey cover (Threlfall et al. 2013a). Evelyn et al. (2004) also found *Myotis yumanensis* to select roosts close to water with substantial forest cover. Differences in roost preferences across species indicates that not only are large, old hollow bearing trees important in cities, so too are patches of vegetation with adequate canopy coverage in proximity to water sources.

Threats to bats in urban areas

Recent reviews describe a range of key threats to bats in urban areas, including artificial light at night, noise pollution, foraging and roosting habitat loss, disease, heat/thermal limits, and new threats such as modernisations of older buildings in cities, solar panels on roof tops or facades, declines in the diversity and distribution of insect prey, and interactions with invasive or alien species (Russo and Ancillotto 2015; Moretto and Francis 2017). Nearly all of these factors have the potential to threaten populations of urban bats via degrading habitat quality at roost or foraging sites or via direct mortality. For example, seemingly suitable roost sites may lure large numbers of bats and expose them to opportunistic predators, which reach especially high densities in urban environments, including crows, kestrels, seagulls, rats, possums, dogs, and especially domestic cats (Ancillotto et al. 2013; Threlfall et al. 2013b); these factors may potentially lead to the disappearance of entire colonies.

Current key questions and challenges facing urban bat ecology

Despite the increasing number of studies that focus on urban bats, there are several research gaps that remain. These are centred on the topics of spatial and temporal effects of urbanisation, impacts on community and population ecology, and the application of knowledge to inform city planning and management. Key questions include the following:

1 *What are the impacts of urbanisation on populations and communities at different spatial scales?* For example, at large spatial scales such as the scale of a country or entire continent, do urban areas provide refuge habitat for bat species in comparison to non-urban land?
2 *At which spatial extent does urbanisation alter community composition and bat fitness. Is there a threshold level of urbanisation at which impacts are seen?* What drives the apparent differences in responses of bats to urbanisation in large versus small cities?
3 *Within a given urban landscape, at what spatial scale do critical habitats need to co-occur (e.g. foraging and roosting habitat) for bats to be able to establish or maintain a population, and which bat species specifically profit from such habitat complementarity?*
4 *Which traits determine species ability to tolerate urban landscapes?* Do particular species utilise different strategies to cope with the multiple stresses that cities impose? Do species display local adaptation to city conditions, and does this vary across a species range?
5 *How common are nectarivores and frugivorous bats in cities worldwide, and to what extent do they rely on cities or certain habitat types, such as private gardens or parklands within the cities, as refuge habitat?*
6 *How can we better utilise knowledge of bat biology and life history traits to inform conservation ecology, leading to improved city design and management and resolve human-wildlife conflicts especially in developing countries?* Are there certain attributes of urban landscapes acting as a source or sink for bat populations, and if so, how can cities be better designed to benefit bat diversity while improving overall biodiversity and human wellbeing?

To answer these questions, several research challenges must be overcome. Deploying acoustic recording devices or utilising trapping and telemetry techniques across an entire city can prove problematic. For example, certain parts of a city may present greater security issues to researchers or their equipment, leading to sampling bias geared towards areas of higher socioeconomic status. Further, acoustic devices do not account for the number of individual bats being recorded, and hence estimating population sizes or densities of individuals in the landscape are often not feasible (Frick et al. 2020). Use of different types of acoustic technology and data-handling techniques employed in different parts of the world further restricts research synthesis, as some studies record presence/absence of individuals versus others which may record nightly species activity. Human-derived urban sound has also caused many problems for the detection of telemetry transmitter signals that are unlikely to be experienced in other, non-urban settings. Tagged bats can also roost within private and inaccessible inner courtyards and go undetected because buildings shield the transmitter signal to the streets.

Making a difference in urban design to benefit bats

The wealth of knowledge gained from urban bat studies can be usefully applied to urban bat conservation. Many studies report the benefit of small patches of habitat interspersed throughout an urban matrix, supporting the idea that land sparing may benefit urban bat conservation (Caryl et al. 2016; Gili et al. 2020). For example, Caryl et al. (2016) suggest that retaining forest patches in 20% or more of a landscape will substantially improve occupancy for a variety of species. Indeed, habitat loss, fragmentation, and degradation are the major threats to all organisms living in urban environments (McKinney 2002), and attenuating these threats are obvious first choices for making cities more livable for fauna and flora. Similar to the concept of land sparing, Margules and Pressey (2000) suggested the "node, buffer, corridor" principle for landscape-scale conservation. Nodes are high-quality habitats with little to no anthropogenic disturbances. Nodes are surrounded by buffer zones in which human activity decreases with proximity to nodes. Corridors connect the nodes to create a more accessible habitat network. Applying these ideas to the habitats required by bats suggests that preservation of near-natural habitat and woodlands within suburban and urban landscapes would provide many benefits for bats, while protecting overall biodiversity within cities. Connectivity between habitats could be provided by the urban tree network, where studies have found high tree cover can provide commuting routes (Hale et al. 2012) and mitigate the negative impacts of urbanisation for some species by providing protection from artificial light at night (Straka et al. 2019; Haddock et al. 2019).

Whilst urban land in many parts of the world is too costly to set aside for conservation (Soanes et al. 2019), there are critical elements of bat habitat that can be incorporated into the urban fabric. One of the most critical habitats include potential roosting sites (e.g. bat boxes) and roost trees, which for many bat species include larger or older trees, often larger than 40–50 cm diameter (Threlfall et al. 2016). Unfortunately, common management regimes in urban areas involve the removal of these types of trees because they are often perceived as a threat to public safety (Kubista and Bruckner 2015). Particular types of arboricultural practices could be better promoted to reduce the need for large tree removal, including repeated thinning out of the crown of older trees, which can help prevent tree windthrow, thus safely prolonging the lifespan of the tree (Juillerat and Vögeli 2006). Additionally, to secure an older tree, big branches can be supported by ropes and beams, or by ground anchoring (Juillerat and Vögeli 2006). Physical barriers in the form of understorey vegetation and garden beds can also be implemented (Threlfall et al. 2016), to deter pedestrians from entering underneath a dangerous tree (Kubista and Bruckner 2015).

Conclusions

Effects of urbanisation on bats are complex and act at multiple scales. In general, tolerance of bats to urban environments is correlated with species-specific traits related to mobility and roost site flexibility. Larger urban areas have strong negative effects on overall bat occurrence than smaller urban areas, and green spaces providing natural roosting sites in cities are crucial prerequisites for species persistence. However, we still do not know how and at which spatial scale specific components of the heterogeneous urban landscape affect species composition, functional diversity, and population structure. Such knowledge is of great interest not only for bat species conservation but also to incorporate into urban planning for biodiversity in general.

Key companion papers

Frick, W.F., T. Kingston, J. Flanders. 2020. A review of the major threats and challenges to global bat conservation. *Annals of the New York Academy of Sciences* 1469:5–25.

Gallo, T., et al. 2018. Need for multiscale planning for conservation of urban bats. *Conservation Biology* 32:638–647.

Jung, K., C.G. Threlfall. 2018. Trait-dependent tolerance of bats to urbanization: A global meta-analysis. *Proceedings of the Royal Society B: Biological Sciences* 285:20181222.

Russo, D., L. Ancillotto. 2015. Sensitivity of bats to urbanization: A review. *Mammalian Biology* 80:205–212.

Sagot, M., G. Chaverri. 2015. Effects of roost specialization on extinction risk in bats. *Conservation Biology* 29(6):1666–1673.

Wolf, J.M., J.M. Jeschke, C.C. Voigt, Y. Itescu. 2022. Urban affinity and its associated traits: A global analysis of bats. *Global Change Biology* 28:5667–5682.

References

Alberti, M. 2005. The effects of urban patterns on ecosystem function. *International Regional Science Review* 28(2):168–192.

Ancillotto, L., L. Santini, N. Ranc, L. Maiorano, D. Russo. 2016. Extraordinary range expansion in a common bat: The potential roles of climate change and urbanisation. *The Science of Nature* 103(3–4):15.

Ancillotto, L., M.T. Serangeli, D. Russo. 2013. Curiosity killed the bat: Domestic cats as bat predators. *Mammalian Biology* 78:369–373.

Ancillotto, L., A. Tomassini, D. Russo. 2015. The fancy city life: Kuhl's pipistrelle, *Pipistrellus kuhlii*, benefits from urbanization. *Wildlife Research* 42:598–606.

Avila-Flores, R., M.B. Fenton. 2005. Use of spatial features by foraging insectivorous bats in a large urban landscape. *Journal of Mammalogy* 86(6):1193–1204.

Basham, R., B. Law, P. Banks. 2011. Microbats in a 'leafy' Sydney urban landscape: Are they persisting, and what factors influence their presence? *Austral Ecology* 36(6):663–678.

Boyles, J.G., P.M. Cryan, G.F. McCracken, T.H. Kunz. 2011. Economic importance of bats in agriculture. *Science* 332(6025):41–42.

Burgin, C.J., J.P. Colella, P.L. Kahn, N.S. Upham. 2018. How many species of mammals are there? *Journal of Mammalogy* 99(1):1–14.

Caryl, F.M., L.F. Lumsden, R. van der Ree, B.A. Wintle. 2016. Functional responses of insectivorous bats to increasing housing density support 'land-sparing' rather than 'land-sharing' urban growth strategies. *Journal of Applied Ecology* 53(1):191–201.

Coleman, J.L., R.M.R. Barclay. 2011. Influence of urbanization on demography of little brown bats (*Myotis lucifugus*) in the prairies of North America. *PLoS ONE* 6:e20483.

Denzinger, A., H.-U. Schnitzler. 2013. Bat guilds, a concept to classify the highly diverse foraging and echolocation behaviors of microchiropteran bats. *Frontiers in Physiology* 4:164.

Egert-Berg, K., E.R. Hurme, S. Greif, A. Goldstein, L. Harten, L.G. Herrera, M.J.J. Flores-Martínez, A.T. Valdés, D.S. Johnston, O. Eitan, I. Borissov, J.R. Shipley, R.A. Medellin, G.S. Wilkinson, H.R. Goerlitz, Y. Yovel. 2018. Resource ephemerality drives social foraging in bats. *Current Biology* 28:3667–3673.

Evelyn, M.J., D.A. Stiles, R.A. Young. 2004. Conservation of bats in suburban landscapes: Roost selection by *Myotis yumanensis* in a residential area in California. *Biological Conservation* 115:463–473.

Fabianek, F., D. Gagnon, M. Delorme. 2011. Bat distribution and activity in Montréal Island green spaces: Responses to multi-scale habitat effects in a densely urbanized area. *Ecoscience* 18(1):9–17.

Findley, J.S. 1993. *Bats: A Community Perspective*. Cambridge University Press, Cambridge.

Fleming, T.H. 2019. Bat migration. Pages 605–610 in J.C. Choe, editor. *Encyclopedia of Animal Behavior*, 2nd edition. Academic Press, Oxford.

Frick, W.F., T. Kingston, J. Flanders. 2020. A review of the major threats and challenges to global bat conservation. *Annals of the New York Academy of Sciences* 1469(1):5–25.

Furlonger, C.L., H.J. Dewar, B. Fenton. 1987. Habitat use by foraging insectivorous bats. *Canadian Journal of Zoology* 65:284–288.

Gaisler, J., J. Zukal, Z. Rehak, M. Homolka. 1998. Habitat preference and flight activity of bats in a city. *Journal of Zoology* 244:439–445.

Gili, F., S.E. Newson, S. Gillings, D.E. Chamberlain, J.A. Border. 2020. Bats in urbanising landscapes: Habitat selection and recommendations for a sustainable future. *Biological Conservation* 241:08343.

Gottwald, J., R. Zeidler, N. Friess, M. Ludwig, C. Reudenbach, T. Nauss. 2019. Introduction of an automatic and open-source radio-tracking system for small animals. *Methods in Ecology and Evolution* 10(12):2163–2172.

Haddock, J.K., C.G. Threlfall, B. Law, D.F. Hochuli. 2019. Light pollution at the urban forest edge negatively impacts insectivorous bats. *Biological Conservation* 236:17–28.

Hale, J.D., A.J. Fairbrass, T.J. Matthews, J.P. Sadler. 2012. Habitat composition and connectivity predicts bat presence and activity at foraging sites in a large UK conurbation. *PLoS ONE* 7:e33300.

Hourigan, C.L., C.P. Catterall, D. Jones, M. Rhodes. 2010. The diversity of insectivorous bat assemblages among habitats within a subtropical urban landscape. *Austral Ecology* 35(8):849–857.

Jones, G., M.W. Holderied. 2007. Bat echolocation calls: Adaptation and convergent evolution. *Proceedings of the Royal Society B: Biological Sciences* 274(1612):905–912.

Jones, G., D.S. Jacobs, T.H. Kunz, M.R. Willig, P.A. Racey. 2009. Carpe noctem: The importance of bats as bioindicators. *Endangered Species Research* 8:93–115.

Juillerat, L., M. Vögeli. 2006. *Pflege alter Bäume zum Erhalt der Totholzkäfer im Stadtgebiet*. Centre Suisse de Cartographie de la Faune (CSCF) Neuchatel, Switzerland, 1–24.

Jung, K., E.K.V. Kalko. 2010. Where forest meets urbanization: Foraging plasticity of aerial insectivorous bats in an anthropogenically altered environment. *Journal of Mammalogy* 91(1):144–153.

Jung, K., E.K.V. Kalko. 2011. Adaptability and vulnerability of high flying Neotropical aerial insectivorous bats to urbanization. *Diversity and Distributions* 17(2):262–274.

Jung, K., C.G. Threlfall. 2016. Urbanisation and its effects on bats – A global meta-analysis. Pages 13–33 in C.C. Voigt, T. Kingston, editors. *Bats in the Anthropocene: Conservation of Bats in a Changing World*. Springer International Publishing, Cham, Switzerland.

Jung, K., C.G. Threlfall. 2018. Trait-dependent tolerance of bats to urbanization: A global meta-analysis. *Proceedings of the Royal Society B: Biological Sciences* 285:20181222.

Kerth, G. 2008. Causes and consequences of sociality in bats. *BioScience* 58(8):737–746.

Kubista, C.E., A. Bruckner. 2015. Importance of urban trees and buildings as daytime roosts for bats. *Biologia* 70:1545–1552.

Kunz, T.H. 1982. Roosting ecology of bats. Pages 1–55 in T.H. Kunz, editor. *Ecology of Bats*. Plenum Press, New York.

Kunz, T.H., E. Braun de Torrez, D. Bauer, T. Lobova, T.H. Fleming. 2011. Ecosystem services provided by bats. *Annals of the New York Academy of Sciences* 1223(1):1–38.

Kunz, T.H., L.F. Lumsden. 2003. Ecology of cavity and foliage roosting bats. Pages 3–89 in T.H. Kunz, B. Fenton, editors. *Bat Ecology*. The University of Chicago Press, Chicago, IL.

Lewanzik, D., C.C. Voigt. 2017. Transition from conventional to light-emitting diode street lighting changes activity of urban bats. *Journal of Applied Ecology* 54(1):264–271.

Li, H., K.A. Parker, M.C. Kalcounis-Rueppell. 2019. The luxury effect beyond cities: Bats respond to socioeconomic variation across landscapes. *BMC Ecology* 19:46.

Li, H., K.T. Wilkins. 2014. Patch or mosaic: Bat activity responds to fine-scale urban heterogeneity in a medium-sized city in the United States. *Urban Ecosystems* 17(1):1013–1031.

Lintott, P.R., K. Barlow, N. Bunnefeld, P. Briggs, C. Gojas Roig, K. Park. 2016. Differential responses of cryptic bat species to the urban landscape. *Ecology and Evolution* 6(7):2044–2052.

Lintott, P.R., N. Bunnefeld, J. Minderman, E. Fuentes-Montemayor, R.J. Mayhew, L. Olley, K.J. Park. 2015. Differential responses to woodland character and landscape context by cryptic bats in urban environments. *PLoS ONE* 10:e0126850.

Luck, G.W., L. Smallbone, C. Threlfall, B. Law. 2013. Patterns in bat functional guilds across multiple urban centres in south-eastern Australia. *Landscape Ecology* 28:455–469.

Margules, C.R., R.L. Pressey. 2000. Systematic conservation planning. *Nature* 405(6783):243–253.

McCracken, G.F., G.S. Wilkinson. 2000. Bat mating systems. Pages 321–362 in E.G. Crichton, P.H. Krutzsch, editors. *Reproductive Biology of Bats*. Academic Press, London.

McKinney, M.L. 2002. Urbanization, biodiversity, and conservation. *BioScience* 52:883–890.

Moretto, L., C.M. Francis. 2017. What factors limit bat abundance and diversity in temperate, North American urban environments? *Journal of Urban Ecology* 3:16.

Neuweiler, G. 1984. Foraging, echolocation and audition in bats. *Naturwissenschaften* 71:446–455.

Newson, S.E., H.E. Evans, S. Gillings. 2015. A novel citizen science approach for large-scale standardised monitoring of bat activity and distribution, evaluated in eastern England. *Biological Conservation* 191:38–49.

Norberg, U.M., J.M.V. Rayner. 1987. Ecological morphology and flight in bats (Mammalia; Chiroptera): Wing adaptations, flight performance, foraging strategy and echolocation. *Philosophical Transactions of the Royal Society of London. Series B: Biological Sciences* 316:335–427.

O'Malley, K.D., W.E. Kunin, M. Town, W.O. Mgoola, E.L. Stone. 2020. Roost selection by Mauritian tomb bats (Taphozus mauritianus) in Lilongwe city, Malawi – importance of woodland for sustainable urban planning. *PLoS ONE* 15:e0240434.

Parkins, K.L., J.A. Clark. 2015. Green roofs provide habitat for urban bats. *Global Ecology and Conservation* 4:349–357.

Patriquin, K.J., C. Guy, J. Hinds, J.M. Ratcliffe. 2019. Male and female bats differ in their use of a large urban park. *Journal of Urban Ecology* 5:1–13.

Popa-Lisseanu, A.G., F. Bontadina, C. Ibáñez. 2009. Giant noctule bats face conflicting constraints between roosting and foraging in a fragmented and heterogeneous landscape. *Journal of Zoology* 278(2):126–133.

Rainho, A. 2007. Summer foraging habitats of bats in a Mediterranean region of the Iberian Peninsula. *Acta Chiropterologica* 9(1):171–181.

Rhodes, M., C. Catterall. 2008. Spatial foraging behaviour and use of an urban landscape by a fast-flying bat, the Molossid *Tadarida australis*. *Journal of Mammalogy* 89(1):34–42.

Ripperger, S.P., D. Josic, M. Hierold, A. Koelpin, R. Weigel, M. Hartmann, R. Page, F. Mayer. 2016. Automated proximity sensing in small vertebrates: Design of miniaturized sensor nodes and first field tests in bats. *Ecology and Evolution* 6(7):2179–2189.

Ripperger, S.P., S. Stockmaier, G.G. Carter. 2020. Tracking sickness effects on social encounters via continuous proximity sensing in wild vampire bats. *Behavioral Ecology* 31:1296–1302.

Russo, D., L. Ancillotto. 2015. Sensitivity of bats to urbanization: A review. *Mammalian Biology* 80:205–212.

Rydell, J. 1992. Exploitation of insects around streetlamps by bats in Sweden. *Functional Ecology* 6(6):744–750.

Scanlon, A.T., S. Petit. 2008. Effects of site, time, weather and light on urban bat activity and richness: Considerations for survey effort. *Wildlife Research* 35:821–834.

Schnitzler, H.-U., C.F. Moss, A. Denzinger. 2003. From spatial orientation to food acquisition in echolocating bats. *Trends in Ecology & Evolution* 18(8):386–394.

Simmons, N.B. 2005. Order Chiroptera. Pages 312–529 in D.E. Wilson, D.M. Reeder, editors. *Mammal Species of the World: A Taxonomic and Geographic Reference*, 3rd edition. Johns Hopkins University Press, Baltimore, MD.

Simmons, N.B., A.L. Cirranello. 2020. *Bat Species of the World: A Taxonomic and Geographic Database*. Accessed 18/08/2020 https://batnames.org/about.html

Soanes, K., M. Sievers, Y.E. Chee, N.S.G. Williams, M. Bhardwaj, A.J. Marshall, K.M. Parris. 2019. Correcting common misconceptions to inspire conservation action in urban environments. *Conservation Biology* 33(2):300–306.

Stone, E.L., G. Jones, S. Harris. 2009. Street lighting disturbs commuting bats. *Current Biology* 19(13): 1123–1127.

Straka, T.M., P.E. Lentini, L.F. Lumsden, B.A. Wintle, R. van der Ree. 2016. Urban bat communities are affected by wetland size, quality, and pollution levels. *Ecology and Evolution* 6(14):4761–4774.

Straka, T.M., M. Wolf, P. Gras, S. Buchholz, C.C. Voigt. 2019. Tree cover mediates the effect of artificial light on urban bats. *Frontiers in Ecology and Evolution* 7:91.
Suarez-Rubio, M., C. Ille, A. Bruckner. 2018. Insectivorous bats respond to vegetation complexity in urban green spaces. *Ecology and Evolution* 8(6):3240–3253.
Teeling, E.C., S. Dool, M.S. Springer. 2012. Phylogenies, fossils and functional genes: The evolution of echolocation in bats. Pages 1–22 in G.F. Gunnell, N.B. Simmons, editors. *Evolutionary History of Bats.* Cambridge University Press, Cambridge.
Tena, E., G. Fandos, Ó. de Paz, R. de la Peña, J.L. Tellería. 2020. Size does matter: Passive sampling in urban parks of a regional bat assemblage. *Urban Ecosystems* 23:227–234.
Threlfall, C.G., B. Law, P.B. Banks. 2012. Sensitivity of insectivorous bats to urbanization: Implications for suburban conservation planning. *Biological Conservation* 146:41–52.
Threlfall, C.G., B. Law, P.B. Banks. 2013a. Roost selection in suburban bushland by the urban sensitive bat Nyctophilus gouldi. *Journal of Mammalogy* 94:307–319.
Threlfall, C.G., B. Law, P.B. Banks. 2013b. Odour cues influence predation risk at artificial bat roosts in urban bushland. *Biology Letters* 9(3):20121144.
Threlfall, C.G., B. Law, T. Penman, P.B. Banks. 2011. Ecological processes in urban landscapes: Mechanisms influencing the distribution and activity of insectivorous bats. *Ecography* 34(5):814–826.
Threlfall, C.G., N.S.G. Williams, A.K. Hahs, S.J. Livesley. 2016. Approaches to urban vegetation management and the impacts on urban bird and bat assemblages. *Landscape and Urban Planning* 153:28–39.
Voigt, C.C., K.L. Phelps, L.F. Aguirre, M. Corrie Schoeman, J. Vanitharani, A. Zubaid. 2016. Bats and buildings: The conservation of synanthropic bats. Pages 427–462 in C.C. Voigt, T. Kingston, editors. *Bats in the Anthropocene: Conservation of Bats in a Changing World.* Springer International Publishing, Cham, Switzerland.
Voigt, C.C., J.M. Scholl, J. Bauer, T. Teige, Y. Yovel, S. Kramer-Schadt, P. Gras. 2020. Movement responses of common noctule bats to the illuminated urban landscape. *Landscape Ecology* 35:189–201.

19

IMPACTS OF URBANIZATION ON AMPHIBIANS AND OPPORTUNITIES FOR THEIR CONSERVATION IN CITIES

Kirsten M. Parris

Introduction

Amphibians are small, ground- or tree-dwelling vertebrates that form an important part of many ecological communities, from the tropics to the temperate regions of Earth. Amphibians act as prey across all life stages (eggs, larvae, and adults) and as important predators in both freshwater and terrestrial habitats (Wells 2007). Species with a free-swimming larval stage are also an important conduit for the movement of energy and nutrients between the freshwater and terrestrial realms (Fritz and Whiles 2018; Regester et al. 2006). Extant amphibians belong to three orders: Anura (frogs and toads), Caudata (salamanders and newts), and Gymnophiona (caecilians: limbless, wormlike amphibians that live in soil, stream sediments, or leaf litter; Figure 19.1; Wells 2007). According to the IUCN (2022), 41% of 7,296 recognized species are currently at risk of extinction, making amphibians the most threatened vertebrate class of those assessed to date. Of 2,490 amphibian species classified as extinct in the wild, critically endangered, endangered, or vulnerable, 42% are under threat from residential and commercial development (IUCN 2022).

Important pressures faced by amphibians in urban environments include habitat loss and fragmentation, degradation of remaining habitat, barriers to movement across the landscape, and exposure to chemical pollution, noise, disease, and introduced predators (Hamer and Parris 2013; Martel et al. 2013; Heard et al. 2014; Diaz et al. 2020; Jorge et al. 2020; Covarrubias et al. 2021). The species richness of amphibian communities tends to decline in response to urbanization (e.g. Acosta et al. 2005; Shu et al. 2016; Cordier et al. 2021). Yet amphibians are also a viable target for urban conservation, as many species can coexist with humans in cities if suitable provisions are made to protect their habitat and movement requirements, including water quality at breeding sites (e.g. Calhoun et al. 2014; Kruger et al. 2015; Oertli and Parris 2019). This chapter will summarize the current state of knowledge regarding urban amphibians and identify key challenges for conservation of this group in coming years and decades.

What do we know about amphibians in urban areas?

The early years of urban ecology saw a strong focus on studies of birds and vascular plants; however, research on amphibians in urban areas has been increasing steadily over time. A Web

 DOI: 10.4324/9781003016120-22

Figure 19.1 Representatives of the three orders of extant amphibians: Anura (frogs and toads); Gymnophiona (caecilians); and Caudata (salamanders and newts). (A) The western leopard toad *Sclerophrys pantherine*, which is threatened by residential development in Cape Town, South Africa (credit: Serban Proches; CC BY-SA 2.5); (B) the Yunnan caecilian, *Ichthyophis bannanicus*, from Guangdong Province, China (credit: Zichen Wang; CC BY-NC-SA 2.0); and (C) the California tiger salamander, *Ambystoma californiense* (credit: Natalie McNear; CC BY-NC 2.0).

of Science search on urban amphibians reveals 1,885 publications, 85% of which have been published since 2005 (Figure 19.2). The majority of these have focused on anurans (52%), followed by salamanders and newts (31%). Only 13 studies to date (0.7%) have investigated the ecology of caecilians in urban landscapes. The remaining 308 studies (16%) only mention 'amphibians' in general. As is the case for many other taxa, the overall diversity of amphibian communities is highest in the tropics and decreases towards the poles, with this pattern dominated by anurans (Wiens 2007; IUCN 2022). However, approximately half the world's salamander species can be found in the Nearctic and Palearctic realms (IUCN 2022), and the highest diversity of salamanders occurs in the Appalachian region of the eastern United States (Milanovich et al. 2010). While approximately 80% of amphibian species occur in

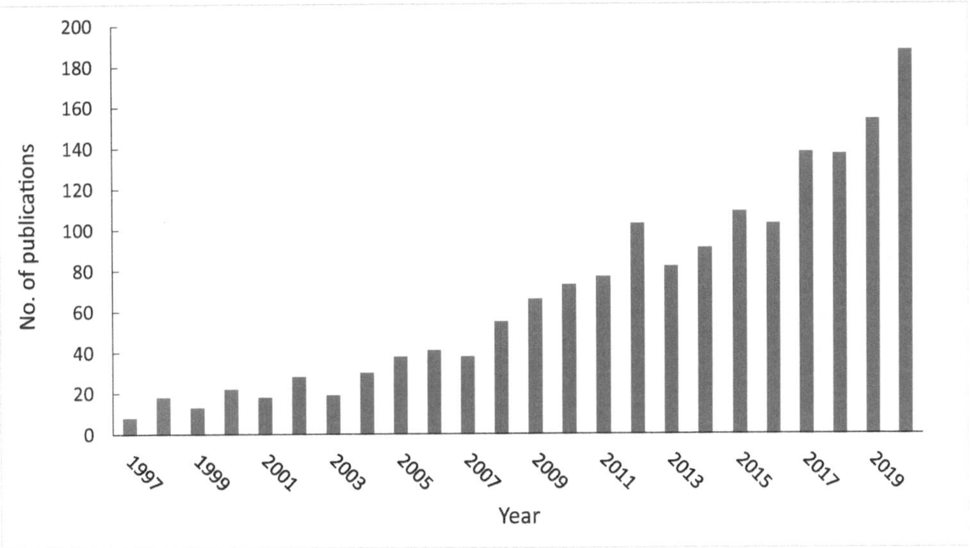

Figure 19.2 Number of publications per year on urban amphibians from 1997 to 2020, as recorded in the Web of Science database (search string: ALL=((amphibia* OR frog* OR toad* OR newt* OR salamander* OR anuran* OR caecilian*) AND (urban* OR city OR suburb*)) and Zoology or Ecology or Environmental Sciences or Biodiversity Conservation or Urban Studies or Marine Freshwater Biology (Web of Science Categories) and Articles or Proceedings Papers or Book Chapters or Review Articles or Books or Data Papers or Letters (Document Types). Retrieved August 2021.

the Afrotropical, Indomalayan, and Neotropical realms (IUCN 2022), studies of urban amphibians are strongly skewed to North America and Europe (Figure 19.3A), and the USA, in particular (Figure 19.3B). This indicates a significant knowledge gap regarding both the impact of urbanization on amphibians in tropical areas and approaches to their conservation in tropical urban landscapes.

Many amphibians retain a biphasic lifestyle that includes both aquatic and terrestrial phases. Consequently, both aquatic and terrestrial habitats, and the opportunity to travel safely between them as required, must be conserved in cities to allow the continued persistence of amphibians. This complexity of habitat requirements sets amphibians apart from many other taxonomic groups. Aquatic habitats are under significant pressure in urban landscapes (Parris 2016; Oertli and Parris 2019). Streams are often altered through engineering works, such as the redirection of first-order (headwater) streams into underground pipes, and the straightening and concreting of stream channels. Further, increased impermeable-surface cover in urban watersheds and efficient stormwater systems combine to deliver large volumes of water to streams following even modest rainfall events (Walsh et al. 2005). This results in the 'urban stream syndrome', characterized by highly channelized streams, scoured banks, low base flows, and turbid and polluted water (Walsh et al. 2005; Booth et al. 2016). Eggs and tadpoles of stream-breeding frogs can be swept away by rapid flows after rain, as can in-stream habitat elements such as aquatic vegetation (Canessa and Parris 2013).

Ephemeral wetlands including vernal pools, small ponds, and soaks are often destroyed during urban development, while permanent lentic wetlands can be highly altered (Parris 2006; Oertli

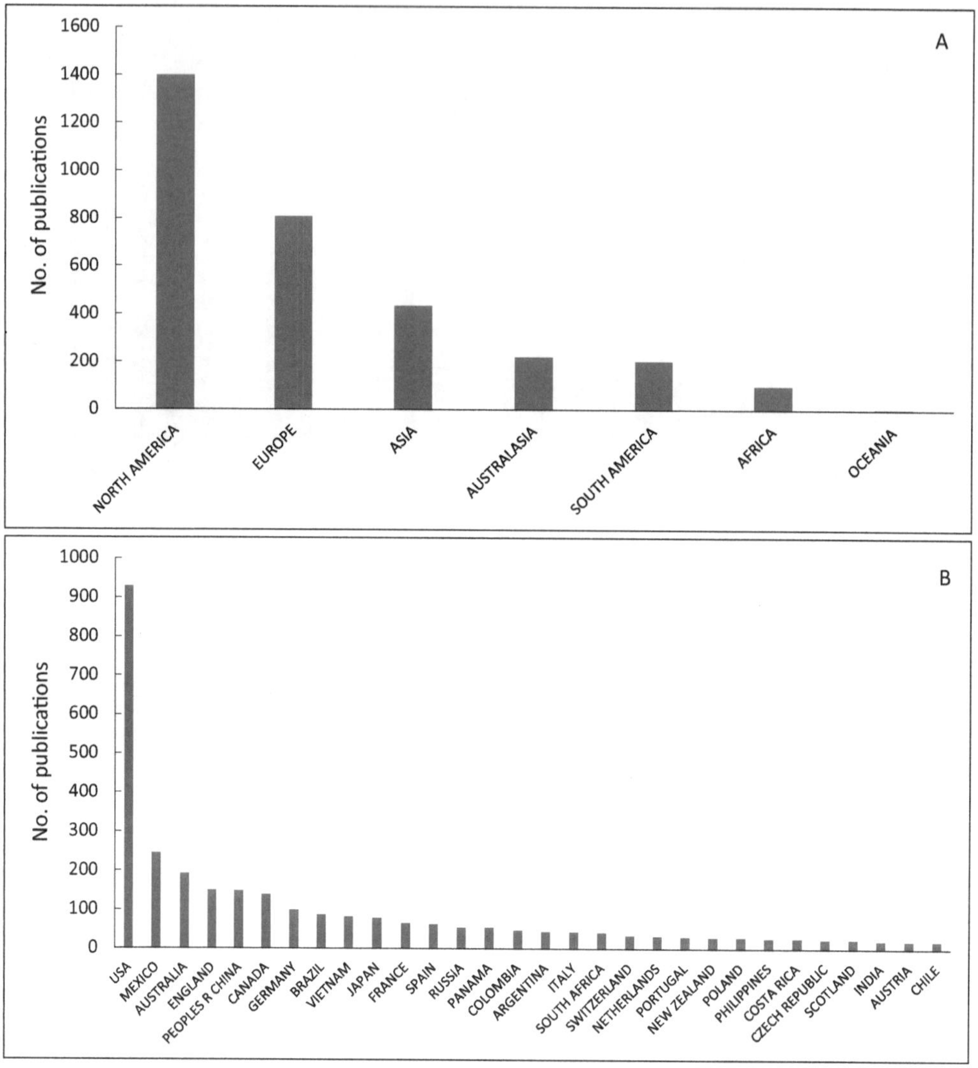

Figure 19.3 Number of publications per year on urban amphibians from 1997 to 2020, as recorded in the Web of Science database: by region (A) and by top 30 countries (B). Note that the total (3,170) exceeds the number of publications in the dataset (1,885), as a single publication can cover more than one country or region.

and Parris 2019). Common changes include a loss of structural-habitat diversity, the replacement of natural edges with ornamental walls, and influxes of nutrients such as nitrogen and phosphorous leading to algal blooms and eutrophication (Paerl et al. 2018; Oertli and Parris 2019). Anthropogenic wetlands constructed to intercept stormwater runoff and protect receiving waters from pollution are also a characteristic feature of many urban landscapes. In some cases, these can provide suitable breeding habitat for amphibians. However, they can also act as ecological traps, attracting animals to substandard habitats where their fitness is compromised (Hale et al. 2015; Clevenot et al. 2018; Sievers et al. 2018a).

Stream-breeding amphibians

A range of studies has investigated the impacts of urbanization on stream-breeding amphibians, particularly frogs, salamanders, and newts (e.g. Canessa and Parris 2013; Barrett and Price 2014; Diaz et al. 2020). These have focused on local site characteristics such as stream morphology, water quality, in-stream and riparian vegetation, as well as landscape- or catchment-scale variables such as the density of roads and the cover of impermeable surfaces or terrestrial vegetation in the watershed. Streams that retain conditions closer to those that existed pre-urbanization tend to support more diverse amphibian assemblages, but different species respond to environmental variables at different scales (Canessa and Parris 2013). Amphibians that breed in first-order streams, such as many salamanders (Barrett and Price 2014), are substantially impacted when a landscape is urbanized; this is due to the common practice of engineering these streams to flow in underground pipes (Elmore and Kaushal 2008; Itsukushima and Ohtsuki 2021). Culverts under roads can also interrupt the natural gradient of streams and prevent the movement of individuals up and down the stream corridor (Barrett and Price 2014).

Changes to water quality and stream hydrology associated with the urbanization of watersheds impact on stream-breeding amphibians both directly and indirectly. Water-borne pollutants such as heavy metals, pesticides, industrial chemicals, and fire retardants are readily absorbed through the permeable skin of amphibian larvae and adults, while benthic-feeding larvae can be further exposed to pollutants bound in sediments. For example, a study in central Texas found increasing concentrations of organic pollutants in the tissues of salamanders (*Eurycea* spp.) as time since urbanization of the stream catchment increased (Diaz et al. 2020). Similarly, concentrations of mercury in the bodies of salamanders and anurans increased along a contamination gradient in the South River in Virginia (Bergeron et al. 2010). This contamination is a legacy from a polluting industry that ceased the use of mercury in the 1950s, approximately 60 years before the study was conducted. Pollution of stream water and sediments associated with urbanization also reduces the diversity of the macroinvertebrate assemblages that form a primary food source for many stream-dwelling amphibians (Barrett and Price 2014; Diaz et al. 2020). Loss of riparian vegetation in urban areas increases water temperatures through loss of shading and reduces the deposition of leaf litter into streams; this, in turn, affects local microhabitat conditions and the abundance of macroinvertebrate prey (Barrett and Price 2014).

In new housing or industrial developments, retention of first-order streams, protection of riparian vegetation, retention of permeable soil and transpiring vegetation across a watershed, and the application of water-sensitive urban design measures (WSUD; also known as low-impact design or green infrastructure; Fletcher et al. 2015) to reduce the speed and volume of stormwater flow to streams following rainfall will reduce the characteristic impacts of urbanization on stream-breeding amphibians and other in-stream biota (Parris et al. 2018). These WSUD measures include the use of water tanks, permeable pavements, green roofs and walls, retarding basins, raingardens, and vegetated swales (Hoban 2019). An example of this approach to urban design can be found in the Kronsberg neighborhood in Hannover, Germany (Codispoti 2021). In existing urban areas, some retrofitting of WSUD measures is possible (Vietz et al. 2016). However, engineering works to restore a concreted channel or 'daylight' a buried stream are very expensive and, despite our best efforts, may not succeed in recreating lost water quality, microhabitats, or ecological communities once a stream catchment has been heavily impacted by urban development (e.g. Violin et al. 2011).

Pond-breeding amphibians

Loss of wetland habitat, modification of remaining wetlands, isolation by roads and other urban infrastructure, chemical pollution of water and sediments, and predation by non-native fish are significant threatening processes for pond-breeding amphibians in urban environments (Parris 2006; Hamer and McDonnell 2008; Hamer and Parris 2013; Sievers et al. 2018b; Hintz and Relyea 2019; Covarrubias et al. 2021). Different species of amphibians are adapted to breed in different conditions along the pond-permanence gradient (Wellborn et al. 1996), from ephemeral ponds that contain few predators (short larval lifespan, few behavioral adaptations for predator avoidance) to ponds that permanently contain water and are therefore likely to support significant populations of fish predators (longer larval lifespan, well-developed strategies for avoiding predation). Thus, in rural environments, hydroperiod is a strong determinant of amphibian community composition (Werner et al. 2007).

As ephemeral ponds are preferentially destroyed in urbanizing landscapes, this skews the remaining ponds towards the permanent end of the gradient. Further, the widespread introduction of non-native fish such as mosquitofish and goldfish in cities means that many amphibians are restricted to habitats containing high densities of piscean predators (e.g. Hamer and Parris 2013). A study of tadpole communities in Melbourne, Australia, found a strong inverse relationship between the density of fish and the density of tadpoles in urban ponds, although this relationship was mediated by aquatic vegetation; well-vegetated ponds provided cover for tadpoles and reduced predation rates (Hamer and Parris 2013). As the name suggests, mosquitofish (*Gambusia* spp.) have been widely introduced in efforts to control mosquito larvae and mosquito-borne diseases in cities (Watters et al. 2018). Interestingly, recent evidence indicates that salamander larvae can eat a comparable number of mosquito larvae to *Gambusia affinis* (DuRant and Hopkins 2008; Watters et al. 2018).

Isolation of ponds by roads and other constructed features of the urban landscape is a recurring problem for amphibian communities in cities throughout the world (Parris 2006, 2018; Covarrubias et al. 2021; Hamer et al. 2021). Isolated populations tend to have lower genetic diversity and are more vulnerable to stochastic events (Keely et al. 2015; Sugawara et al. 2016; Covarrubias et al. 2021). Further, if isolated populations go locally extinct, it's unlikely that new individuals will immigrate successfully to recolonize these habitats (Semlitsch 2000; Scroggie et al. 2019). Amphibians can also suffer significant mortality while attempting to cross roads (Hels and Buchwald 2001; Gibbs and Shriver 2005), with substantial impacts on populations. Road overpasses and underpasses have been trialled as strategies to overcome pond isolation and facilitate the movement of individuals across the urban landscape, with varying degrees of success (Hamer et al. 2014). Like other engineered solutions, these are often difficult and expensive to construct. Reducing the barrier effect of roads through reduced speed limits, seasonal road closures, or 'bucket brigades' may also be successful (Smith et al. 2020), although these approaches require significant community support.

As mentioned previously, constructed wetlands designed to intercept stormwater runoff and filter pollutants are a prominent feature of many urban landscapes. However, heavy metals, pesticides, road de-icing salts, and other chemicals that accumulate in water and sediments can be harmful to amphibians, even at low concentrations (Clotfelter et al. 2004; Snodgrass et al. 2008; Van Meter et al. 2011; Hintz and Relyea 2019; Sievers et al. 2019). Behavioral impacts include changes to feeding and reproductive activity, abnormal swimming patterns, and impacts on the olfactory detection of predators. At higher concentrations, common pollutants in urban wetlands can cause low embryonic survival, developmental delays, developmental and reproductive abnormalities, and even death (Karraker et al. 2008; Lambert et al. 2017; Sievers et al. 2019; Slaby et al. 2019). Further thought must be given to discouraging amphibians from using sub-optimal habitats such as polluted stormwater wetlands, while at the same time offering alternative breeding sites in cities that are fed via rainfall or treated water only to ensure adequate water and sediment quality (Hale et al. 2015).

Amphibians with specialized reproductive strategies

Aquatic deposition of eggs followed by a free-swimming larval stage is considered the ancestral condition for amphibians. While this reproductive strategy remains common, an impressive diversity of specialized breeding strategies has evolved in anurans, salamanders, and caecilians (Wells 2007). For example, in tropical areas with high rainfall and high atmospheric humidity, many frogs lay eggs on vegetation and rocks near waterbodies, or in or above small pools of water that collect in plants such as bromeliads (known as phytotelms; Duellman 1992; Lehtinen et al. 2004). Tropical climbing salamanders (*Bolitoglossa* spp.) and moss salamanders (*Nototriton* spp.) lay their eggs on tree branches or in bromeliads, where they develop directly into juveniles without a free-swimming larval stage (Wells 2007). Terrestrial breeding, in which eggs are laid in nests followed by direct development to juvenile frogs or salamanders, has evolved in both mesic and arid environments (e.g. Plethodontid salamanders in the Americas, *Eleutherodactylus* spp. from the Caribbean and tropical North and Central America, *Breviceps* spp. from Africa, and *Cophixalus* spp. from northern Australia; Hoskin 2004; Wells 2007; Hedges et al. 2008; Du Preez and Carruthers 2009).

Specialized breeding strategies may also be accompanied by extended parental care that involves protection of developing eggs and larvae by, on, or even inside the body of the mother or father (Duellman 1992; Wells 2007). For example, across multiple families, female frogs are known to feed their developing tadpoles with unfertilized eggs, providing nutrition in microhabitats where food availability is otherwise low (Lehtinen et al. 2004; Rowley et al. 2012). Many caecilians also exhibit complex reproductive strategies, ranging from the terrestrial deposition of eggs with parental care followed by a free-swimming larval stage (e.g. the Yunnan caecilian *Ichthyophis bannanicus*; Bei et al. 2012; Figure 19.1A) to extended internal gestation of larvae followed by live birth (Wake 1977).

Globally, urbanization causes the loss and fragmentation of forests – both directly through the removal of native vegetation to make way for urban infrastructure and indirectly through increasing the demand for forest products and the consumption of commodities (Curtis et al. 2018; Hoang and Kanemoto 2021). This is likely to impact a wide range of amphibians with specialized reproductive strategies, by destroying shelter and breeding sites while also changing local microclimatic conditions. Many specialized reproductive strategies rely on high levels of environmental moisture (e.g. in the atmosphere, the leaf-litter, or soil), and the removal of vegetation cover will lead to substantial drying of these elements. Further, the replacement of permeable with impermeable surfaces in cities effectively paves over habitat for ground-dwelling and terrestrial-nesting species, while water extraction for human domestic use can deplete aquifers and lead to the further drying of microhabitats.

For example, 144 of 186 recognized species in the genus *Eleutherodactylus* are considered threatened, with 51 of these under pressure from residential and commercial development (IUCN 2022). One such species is the great peeping frog, *E. grandis*, which has lost much of its habitat to urban development associated with the expansion of Mexico City (IUCN SSC Amphibian Specialist Group 2020). Poison dart frogs (Family Dendrobatidae) from tropical Central and South America display a range of specialized breeding strategies, making them vulnerable to the processes of urbanization. The endangered sky-blue poison dart frog, *Hyloxalus azureiventris*, from Peru inhabits rock piles and caves in tropical lowland rainforest. Females lay a small clutch of eggs in bromeliads, leaf litter, or coconut shells that have filled with water; males guard the eggs then transport the hatched tadpoles to other small waterbodies (Lötters et al. 2000). The habitat of *H. azureiventris* is currently threatened by a range of human activities, including residential development (IUCN SSC Amphibian Specialist Group 2018).

Key challenges for urban amphibian conservation

Maintaining or restoring breeding habitats in cities

The protection of existing breeding habitats for amphibians – and the restoration or creation of new breeding habitats in an effort to replace those that have been lost – remain key challenges in urban environments. It is easier to protect existing soaks, ponds, streams, riparian zones, and phytotelma-supporting vegetation than to recreate conditions that will be suitable for the suite of amphibians in a given area (Violin et al. 2011; Parris et al. 2018). While the creation of new ponds and other wetlands for amphibians can be successful in attracting and maintaining populations of the species they are designed to support (Smith et al. 2020), this is not always the case (e.g. Pyke et al. 2008; Koehler et al. 2015). As an example, isolated wetlands away from stream corridors in Melbourne, Australia, are currently being lost to urban development. Many of these wetlands provide high-quality, fish-free breeding habitat for the threatened growling grass frog *Litoria raniformis* (Heard et al. 2008). While there is a well-considered plan to create new wetlands in riparian zones to partially offset this loss, based on a spatial-optimization approach (Scroggie et al. 2019), there will still be a net reduction in suitable breeding habitat available for this – and other – frog species across the city. New planning approaches that protect existing wetlands on public and private land, including their hydrology and water quality, are required to ensure the persistence of diverse amphibian communities (Calhoun et al. 2014; Parris et al. 2018). One such approach is biodiversity sensitive urban design or BSUD (analogous to WSUD; Garrard et al. 2018), which is currently being trialled in a range of new urban developments across Australia (e.g. Kirk et al. 2021).

Maintaining or restoring ecological connectivity across the urban landscape

The ability of amphibians to disperse away from their natal site, move between breeding and non-breeding habitats on a seasonal basis, and/or make smaller-scale movements for foraging can be severely constrained in urban environments. Maintaining or re-establishing ecological connectivity is therefore another important challenge during the planning and construction of new urban areas, or during the retrofitting of established neighborhoods (Parris et al. 2018; Kirk et al. 2021). Protecting riparian corridors is important for species that breed in streams and adjacent wetlands and soaks, but provision for safe lateral movement between streams and into upland areas is also important for maintaining the viability of populations and metapopulations over time (e.g. Keely et al. 2015).

The provision of habitat corridors to link breeding sites, either through the protection of public or private green space or by combining the benefits of connected, wildlife-friendly gardens, has proven effective in some cities (e.g. Saarikivi et al. 2013). Construction of tunnels or culverts under roads to allow amphibians to move safely from one side to the other have had mixed success (Hamer et al. 2014; Smith et al. 2020). These crossing structures must be designed carefully to ensure they suit the species concerned and do not introduce additional hazards such as flooding, exposure to predators or heavy-metal pollutants (Lesbarreres et al. 2004; van der Grift 2005; Smith et al. 2020). Green bridges built over roads and railway lines are another option for maintaining or reinstating connectivity in the urban landscape; these have been employed to some success in a number of countries (e.g. as part of the landscape-defragmentation program in The Netherlands; Sijtsma et al. 2020). Lastly, community efforts to protect migrating amphibians use a diversity of approaches, including seasonal road closures and 'bucket brigades', in which animals are collected from one side of the road and carried to the other during periods of high migration (Hamer et al. 2015; Sterrett et al. 2019).

Identifying and reducing the impacts of urban noise on acoustic communication in anurans

Urban noise from road and air traffic interferes with acoustic communication in frogs, reducing the distance over which a male's call can be detected and interpreted by females (Bee and Swanson 2007; Parris et al. 2009; Kruger and Du Preez 2016). Interest in the ecological effects of urban noise has increased in recent decades, with a range of reviews considering changes to animal behavior in response to noise and possible population-level impacts across a range of taxonomic groups, including birds, amphibians, mammals, and insects (e.g. Barber et al. 2010; Kight and Swaddle 2011; Shannon et al. 2016). Effects of urban noise on factors such as physiological stress, immune function, and mate choice in frogs are also receiving increased research attention (e.g. Troïanowski et al. 2017; Tennessen et al. 2018; Candolin and Wong 2019). In general, larger frogs with lower-pitched calls will experience greater acoustic interference in urban noise; this relationship holds both within and between species (Parris 2013). However, understanding of the ultimate impacts of urban noise on the individual fitness, breeding success, and population persistence of frogs remains limited. Strategies for protecting the acoustic environment of urban frogs include reducing noise when animals of concern (such as threatened species) are expected to be particularly vulnerable to its effects (e.g. closing roads or reducing traffic speeds during the breeding season; Parris 2015). Sound barriers may also help to protect significant habitats from road-traffic noise, although design considerations need to include the efficacy of noise reduction versus the physical barrier created to movement across the landscape for terrestrial animals (Parris 2015). Protecting the acoustic habitat of frogs from air-traffic noise is more difficult. However, except for locations under the busiest flight paths, this noise is more likely than road-traffic noise to be intermittent – offering quieter windows for signal transmission.

Identifying and minimizing effects of chemical pollutants on urban amphibians

Amphibians in urban environments can be exposed to thousands of different chemical pollutants in water, sediments, and soil, including heavy metals, pesticides, and industrial contaminants (e.g. Sánchez-Bayo et al. 2016; Slaby et al. 2019). As mentioned previously, wetlands designed to intercept stormwater and remove pollutants before they reach receiving waters may contain high concentrations of pollutants (Snodgrass et al. 2008; Sievers et al. 2018a, 2018b), while soil contaminated with chemicals from industry, landfills, or past firefighting efforts is also widespread in cities (Buck et al. 2011). Tens of thousands of unique chemicals are routinely used for industrial applications, but the persistence, environmental behavior, and toxicology of many of these remain unknown. Heavy metals, endocrine disrupters, and persistent organic pollutants such as PFAS can biomagnify through food chains, accumulating in wildlife with significant impacts on fitness (Clotfelter et al. 2004; Bergeron et al. 2010; Giesy et al. 2010; Todd et al. 2011; Ahrens and Bundschuh 2014; Lambert and Skelly 2016). Exposure to urban pollutants may also increase the susceptibility of amphibians to diseases and parasites (Brodkin et al. 2007; Haas et al. 2017).

Due to the complex, biphasic life cycle of many amphibians and their highly permeable egg and skin membranes, the group presents an ideal system for studying the ecological effects of chemical pollutants (Wells 2007). However, much work remains to be done to identify how chemical pollutants affect amphibian eggs, larvae and adults, especially sub-lethal effects on behavior and the effects of multiple chemicals in combination (Sievers et al. 2018b, 2019). Minimizing the negative impacts of chemical pollutants on amphibians as individuals, populations and communities in urban environments will require additional efforts to provide breeding habitats that are not substantially contaminated, as well as increased regulation to limit the use of chemical compounds known to be harmful to the environment, wildlife, and humans. Regulation remains uneven across jurisdictions, with some substances banned in the European Union still in regular use in other parts of the world, including the USA and Australia (e.g. the herbicide atrazine, an endocrine disruptor that causes the feminization of male frogs; Hayes et al. 2011). A more holistic understanding of the benefits of an unpolluted environment for wildlife and people alike will likely spur on efforts to phase out harmful chemical compounds in years to come, but managers must reckon with the legacy of persistent pollutants in urban landscapes.

Key companion papers

Barber, J.R., K.R. Crooks, K.M. Fristrup. 2010. The costs of chronic noise exposure for terrestrial organisms. *Trends in Ecology & Evolution* 25(3):180–189.

Candolin, U., B.B.M. Wong. 2019. Mate choice in a polluted world: Consequences for individuals, populations and communities. *Philosophical Transactions of the Royal Society B: Biological Sciences* 374(1781):20180055.

Covarrubias, S., C. González, Gutiérrez-Rodríguez. 2021. Effects of natural and anthropogenic features on functional connectivity of anurans: A review of landscape genetics studies in temperate, subtropical and tropical species. *Journal of Zoology* 313(3):159–171.

Hamer, A.J., M.J. McDonnell. 2008. Amphibian ecology and conservation in the urbanising world: A review. *Biological Conservation* 141(10):2432–2449.

Parris, K.M., et al. 2018. The seven lamps of planning for biodiversity in the city. *Cities* 83:44–53.

Smith, R.K., H. Meredith, W.J. Sutherland. 2020. Amphibian Conservation. Pages 9–64 in W.J. Sutherland, L.V. Dicks, S.O. Petrovan, R.K. Smith, editors. *What Works in Conservation 2020*. Open Book Publishers, Cambridge.

References

Acosta, R., R. Vera, A. Núñez. 2005. Fauna de anuros en la ciudad de Salta, Argentina. *Revista de Biología Tropical* 53(3–4):569–575.

Ahrens, L., M. Bundschuh. 2014. Fate and effects of poly- and perfluoroalkyl substances in the aquatic environment: A review. *Environmental Toxicology and Chemistry* 33(9):1921–1929.

Barber, J.R., K.R. Crooks, K.M. Fristrup. 2010. The costs of chronic noise exposure for terrestrial organisms. *Trends in Ecology & Evolution* 25(3):180–189.

Barrett, K., S.J. Price. 2014. Urbanization and stream salamanders: A review, conservation options, and research needs. *Freshwater Science* 33(3):927–940.

Bee, M.A., E.M. Swanson. 2007. Auditory masking of anuran advertisement calls by road traffic noise. *Animal Behaviour* 74(6):1765–1776.

Bei, Y., S. Meng, G. Li, W. Xie, J. Li, L. Zhang. 2012. First record of nest site and egg guarding in the caecilian *Ichthyophis bannanicus* (Amphibia: Gymnophiona: Ichthyophiidae). *Journal of Natural History* 46(13):859–865.

Bergeron, C.M., C.M. Bodinof, J.M. Unrine, W.A. Hopkins. 2010. Mercury accumulation along a contamination gradient and nondestructive indices of bioaccumulation in amphibians. *Environmental Toxicology and Chemistry* 29(4):980–988.

Booth, D.B., A.H. Roy, B. Smith, K.A. Capps. 2016. Global perspectives on the urban stream syndrome. *Freshwater Science* 35(1):412–420.

Brodkin, M.A., H. Madhoun, M. Rameswaran, I. Vatnick. 2007. Atrazine is an immune disruptor in adult northern leopard frogs (*Rana pipiens*). *Environmental Toxicology and Chemistry* 26(1):80–84.

Buck, R.C., J. Franklin, U. Berger, J.M. Conder, I.T. Cousins, P. de Voogt, A. Astrup Jensen, K. Kannan, S.A. Mabury, S.P.J. van Leeuwen. 2011. Perfluoroalkyl and polyfluoroalkyl substances in the environment: Terminology, classification, and origins. *Integrated Environmental Assessment and Management* 7(4):513–541.

Calhoun, A.J.K., J.S. Jansujwicz, K.P. Bell, M.L. Hunter, Jr. 2014. Improving management of small natural features on private lands by negotiating the science – Policy boundary for Maine vernal pools. *Proceedings of the Natural Academy of Sciences* 111(30):11002–11006.

Candolin, U., B.B.M. Wong. 2019. Mate choice in a polluted world: Consequences for individuals, populations and communities. *Philosophical Transactions of the Royal Society B: Biological Sciences* 374(1781):20180055.

Canessa, S., K.M. Parris. 2013. Multi-Scale, direct and indirect effects of the urban stream syndrome on amphibian communities in streams. *PLoS ONE* 8(7):e70262.

Clevenot, L., C. Carré, P. Pech. 2018. A review of the factors that determine whether stormwater ponds are ecological traps and/or high-quality breeding sites for amphibians. *Frontiers in Ecology and Evolution* 6:40.

Clotfelter, E.D., A.M. Bell, K.R. Levering. 2004. The role of animal behaviour in the study of endocrine-disrupting chemicals. *Animal Behaviour* 68(4):665–676.

Codispoti, O. 2021. Sustainable urban forms: Eco-neighbourhoods in Europe. *Journal of Urbanism: International Research on Placemaking and Urban Sustainability*:1–26.

Cordier, J.M., R. Aguilar, J.N. Lescano, G.C. Leynaud, A. Bonino, D. Miloch, R. Loyola, J. Nori. 2021. A global assessment of amphibian and reptile responses to land-use changes. *Biological Conservation* 253:108863.

Covarrubias, S., C. González, Gutiérrez-Rodríguez. 2021. Effects of natural and anthropogenic features on functional connectivity of anurans: A review of landscape genetics studies in temperate, subtropical and tropical species. *Journal of Zoology* 313(3):159–171.

Curtis, P.G., C.M. Slay, N.L. Harris, A. Tyukavina, M.C. Hansen. 2018. Classifying drivers of global forest loss. *Science* 361(6407):1108–1111.

Diaz, P.H., E.L. Orsak, F.W. Weckerly, M.A. Montagne, D.A. Alvarez. 2020. Urban stream syndrome and contaminant uptake in salamanders of central Texas. *Journal of Fish and Wildlife Management* 11(1):287–299.

Duellman, W.E. 1992. Reproductive strategies of frogs. *Scientific American* 267:80–87.

Du Preez, L., V. Carruthers, editors. 2009. *A Complete Guide to the Frogs of Southern Africa*. Struik Nature, Cape Town, South Africa.

DuRant, S.E., W.A. Hopkins. 2008. Amphibian predation on larval mosquitoes. *Canadian Journal of Zoology* 86(10):1159–1164.

Elmore, A.J., S.S. Kaushal. 2008. Disappearing headwaters: Patterns of stream burial due to urbanization. *Frontiers in Ecology and the Environment* 6(6):308–312.

Fletcher, T.D., W. Shuster, W.F. Hunt, R. Ashley, D. Butler, S. Arthur, S. Trowsdale, S. Barraud, A. Semadeni-Davies, J.-L. Bertrand-Krajewski, P.S. Mikkelsen, G. Rivard, M. Uhl, D. Dagenais, M. Viklander. 2015. SUDS, LID, BMPs, WSUD and more – The evolution and application of terminology surrounding urban drainage. *Urban Water Journal* 12(7):525–542.

Fritz, K.A., M.R. Whiles. 2018. Amphibian-mediated nutrient fluxes across aquatic – Terrestrial boundaries of temporary wetlands. *Freshwater Biology* 63(10):1250–1259.

Garrard, G.E., N.S.G. Williams, L. Mata, J. Thomas, S.A. Bekessy. 2018. Biodiversity sensitive urban design. *Conservation Letters* 11:e12411.

Gibbs, J.P., W.G. Shriver. 2005. Can road mortality limit populations of pool-breeding amphibians? *Wetlands Ecology and Management* 13(3):281–289.

Giesy, J.P., J.E. Naile, J.S. Khim, P.D. Jones, J.L. Newsted. 2010. Aquatic toxicology of perfluorinated chemicals. Pages 1–52 in D.M. Whitacre, editor. *Reviews of Environmental Contamination and Toxicology. Reviews of Environmental Contamination and Toxicology*, Vol. 202. Springer, New York.

Haas, S.E., M.K. Reeves, A.E. Pinkney, P.T. Johnson. 2017. Continental-extent patterns in amphibian malformations linked to parasites, chemical contaminants, and their interactions. *Global Change Biology* 24(1):1–14.

Hale, R., R. Coleman, V. Pettigrove, S.E. Swearer. 2015. Identifying, preventing and mitigating ecological traps to improve the management of urban aquatic ecosystems. *Journal of Applied Ecology* 52(4):928–939.

Hamer, A.J., B. Barta, A. Bohus, B. Gál, D. Schmera. 2021. Roads reduce amphibian abundance in ponds across a fragmented landscape. *Global Ecology and Conservation* 28:e01663.

Hamer, A.J., T.E.S. Langton, D. Lesbarrères. 2015. Making a safe leap forward: Mitigating road impacts on amphibians. Pages 261–270 in R. van der Ree, C. Grilo, D. Smith, editors. *Handbook of Road Ecology*. Wiley, New York.

Hamer, A.J., M.J. McDonnell. 2008. Amphibian ecology and conservation in the urbanising world: A review. *Biological Conservation* 141(10):2432–2449.

Hamer, A.J., K.M. Parris. 2013. Predation modifies larval amphibian communities in urban ponds. *Wetlands* 33(4):641–652.

Hamer, A.J., R. van der Ree, M.J. Mahony, T. Langton. 2014. Usage rates of an under-road tunnel by three Australian frog species: Implications for road mitigation. *Animal Conservation* 17(4):379–387.

Hayes, T.B., L.L. Anderson, V.R. Beasley, S.R. de Solla, T. Iguchi, H. Ingraham, P. Kestemont, J. Kniewald, Z. Kniewald, V.S. Langlois, E.H. Luque, K.A. McCoy, M. Munoz-de-Toro, T. Oka, C.A. Oliveira, F. Orton, S. Ruby, M. Suzawa, L.E. Tavera-Mendoza, V.L. Trudeau, A.B. Victor-Costa, E. Willingham. 2011. Demasculinization and feminization of male gonads by atrazine: Consistent effects across vertebrate classes. *Journal of Steroid Biochemistry and Molecular Biology* 127(1–2):64–73.

Heard, G.W., P. Robertson, M.P. Scroggie. 2008. Microhabitat preferences of the endangered Growling Grass Frog *Litoria raniformis* in southern Victoria. *Australian Zoologist* 34(3):414–425.

Heard, G.W., M.P. Scroggie, N. Clemann, D.S.L. Ramsey. 2014. Wetland characteristics influence disease risk for a threatened amphibian. *Ecological Applications* 24(4):650–662.

Hedges, S.B., W.E. Duellman, M.P. Heinicke 2008. New World direct-developing frogs (Anura: Terrarana): Molecular phylogeny, classification, biogeography, and conservation. *Zootaxa* 1737:1–182.

Hels, T., E. Buchwald. 2001. The effect of road kills on amphibian populations. *Biological Conservation* 99(3):331–340.

Hintz, W.D., R.A. Relyea. 2019. A review of the species, community, and ecosystem impacts of road salt salinisation in fresh waters. *Freshwater Biology* 64(6):1081–1097.

Hoang, N.T., K. Kanemoto. 2021. Mapping the deforestation footprint of nations reveals growing threat to tropical forests. *Nature Ecology & Evolution* 5:845–853.

Hoban, A. 2019. Water sensitive urban design approaches and their description. Pages. 25–47 in A.K. Sharma, T. Gardner, D. Begbie, editors. *Approaches to Water Sensitive Urban Design*. Woodhead Publishing, Cambridge.

Hoskin, C.J. 2004. Australian microhylid frogs (*Cophixalus* and *Austrochaperina*): Phylogeny, taxonomy, calls, distributions and breeding biology. *Australian Journal of Zoology* 52(3):237–269.

Itsukushima, R., K.A. Ohtsuki. 2021. A century of stream burial due to urbanization in the Tokyo Metropolitan Area. *Environmental Earth Sciences* 80:237.

IUCN. 2022. *IUCN Red List of Threatened Species*. Accessed 08/2021 www.iucnredlist.org

IUCN SSC Amphibian Specialist Group. 2018. *Hyloxalus Azureiventris*. The IUCN Red List of Threatened Species 2018:e.T55169A89200449. https://doi.org/10.2305/IUCN.UK.2018-1.RLTS.T55169A89200449.en

IUCN SSC Amphibian Specialist Group. 2020. *Eleutherodactylus Grandis*. The IUCN Red List of Threatened Species 2020:e.T56629A46775155. https://doi.org/10.2305/IUCN.UK.2020-2.RLTS.T56629A46775155.en

Jorge, R.F., W.E. Magnusson, D.A. Silva, E.M. Polo, A.P. Lima. 2020. Urban growth threatens the lowland Amazonian Manaus harlequin frog, which represents an evolutionarily significant unit within the genus *Atelopus* (Amphibia: Anura: Bufonidae). *Journal of Zoological Systematics and Evolutionary Research* 58(4):1195–1205.

Karraker, N.E., J.P. Gibbs, J.R. Vonesh. 2008. Impacts of road deicing salt on the demography of vernal pool-breeding amphibians. *Ecological Applications* 18(3):724–734.

Keely, C.C., J.M. Hale, G.W. Heard, K.M. Parris, J. Sumner, A.J. Hamer, J. Melville. 2015. Genetic structure and diversity of the endangered growling grass frog in a rapidly urbanizing region. *Royal Society Open Science* 2(8):140255.

Kight, C.R., J.P. Swaddle. 2011. How and why environmental noise impacts animals: An integrative, mechanistic review. *Ecology Letters* 14(10):1052–1061.

Kirk, H., G.E. Garrard, T. Croeser, A. Backstrom, K. Berthon, C. Furlong, J. Hurley, F. Thomas, A. Webb, S.A. Bekessy. 2021. Building biodiversity into the urban fabric: A case study in applying Biodiversity Sensitive Urban Design (BSUD). *Urban Forestry & Urban Greening* 62:127176.

Koehler, S., D. Gilmore, D. Newell. 2015. Translocation of the threatened Growling Grass Frog *Litoria raniformis*: A case study. *Australian Zoologist* 37(3):321–336.

Kruger, D.J.D., L.H. Du Preez. 2016. The effect of airplane noise on frogs: A case study on the critically endangered Pickersgill's reed frog (*Hyperolius pickersgilli*). *Ecological Research* 31(3):393–405.

Kruger, D.J.D., A.J. Hamer, L.H. Du Preez. 2015. Urbanization affects frog communities at multiple scales in a rapidly developing African city. *Urban Ecosystems* 18:1333–1352.

Lambert, M.R., D.K. Skelly. 2016. Diverse sources for endocrine disruption in the wild. *Endocrine Disruptors* 4(1):e1148803.

Lambert, M.R., A.B. Stoler, M.S. Smylie, R.A. Relyea, D.K. Skelly. 2017. Interactive effects of road salt and leaf litter on wood frog sex ratios and sexual size dimorphism. *Canadian Journal of Fisheries and Aquatic Sciences* 74(2):141–146.

Lehtinen, R.M., C.M. Richards, R.A. Nussbaum. 2004. Origin of a complex reproductive trait: Phytotelm breeding in mantelline frogs. *Miscellaneous Publications of the Museum of Zoology, University of Michigan* 193:45–54.
Lesbarreres, D., T. Lode, J. Merila. 2004. Short communication: What type of amphibian tunnel could reduce road kills? *Oryx* 38(2):220–223.
Lötters, S., K.H. Jungfer, A. Widmer. 2000. A new genus of aposematic poison frog (Amphibia: Anura: Dendrobatidae) from the upper Amazon Basin with notes on its reproductive behaviour and tadpole morphology. *Jahreshefte der Gesellschaft für Naturkunde in Wüttemberg* 156:234–243.
Martel, A., A. Spitzen-van der Sluijs, M. Blooi, W. Bert, R. Ducatelle, M.C. Fisher, A. Woeltjes, W. Bosman, K. Chiefs, F. Bossuyt, F. Pasmans. 2013. *Batrachochytrium salamandrivorans* sp. nov. causes lethal chytridiomycosis in amphibians. *Proceedings of the National Academy of Science* 110(38):15325–15329.
Milanovich, J.R., W.E. Peterman, N.P. Nibbelink, J.C. Maerz. 2010. Projected loss of a salamander diversity hotspot as a consequence of projected global climate change. *PLoS ONE* 5(8):e12189.
Oertli, B., K.M. Parris. 2019. Review: Towards management of urban ponds for freshwater biodiversity. *Ecosphere* 10(7):e02810.
Paerl, H.W., T.G. Otten, R. Kudela. 2018. Mitigating the expansion of harmful algal blooms across the freshwater-to-marine continuum. *Environmental Science & Technology* 52(10):5519–5529.
Parris, K.M. 2006. Urban amphibian assemblages as metacommunities. *Journal of Animal Ecology* 75(3):757–764.
Parris, K.M. 2013. Anthropogenic noise constrains acoustic communication in urban-dwelling frogs. *Proceedings of Meetings on Acoustics* 19:010055.
Parris, K.M. 2015. Ecological impacts of road noise and options for mitigation. Pages 151–158 in R. van der Ree, D.J. Smith, C. Grilo, editors. *Handbook of Road Ecology*. Wiley, New York.
Parris, K.M. 2016. *Ecology of Urban Environments*. Wiley Blackwell, Chichester, West Sussex.
Parris, K.M. 2018. Existing ecological theory applies to urban environments. *Landscape and Ecological Engineering* 14:201–208.
Parris, K.M., M. Amati, S.A. Bekessy, D. Dagenais, O. Fryd, A.K. Hahs, D. Hes, S.J. Imberger, S.J. Livesley, A.J. Marshall, J.R. Rhodes, C.G. Threlfall, R. Tingley, R. van der Ree, C.J. Walsh, M.L. Wilkerson, N.S.G Williams. 2018. The seven lamps of planning for biodiversity in the city. *Cities* 83:44–53.
Parris, K.M., M. Velik-Lord, J.M.A. North. 2009. Frogs call at a higher pitch in traffic noise. *Ecology and Society* 14(1):25.
Pyke, G.H., J. Rowley, J. Shoulder, A.W. White. 2008. Attempted reintroduction of the endangered Green and Golden Bell Frog to Long Reef Golf Course: A step towards recovery? *Australian Zoologist* 34(3):361–372.
Regester, K.J., K.R. Lips, M.R. Whiles. 2006. Energy flow and subsidies associated with the complex life cycle of ambystomatid salamanders in ponds and adjacent forest in southern Illinois. *Oecologia* 147:303–314.
Rowley, J.J.L., D.T.A. Tran, D.T.T. Le, H.D. Hoang, R. Altig. 2012. The strangest tadpole: The oophagous, tree-hole dwelling tadpole of *Rhacophorus vampyrus* (Anura: Rhacophoridae) from Vietnam. *Journal of Natural History* 46(47–48):2969–2978.
Saarikivi, J., T. Knopp, A. Granroth, J. Merilä. 2013. The role of golf courses in maintaining genetic connectivity between common frog (*Rana temporaria*) populations in an urban setting. *Conservation Genetics* 14(5):1057–1064.
Sánchez-Bayo, F., K. Goka, D. Hayasaka. 2016. Contamination of the aquatic environment with neonicotinoids and its implication for ecosystems. *Frontiers in Environmental Science* 4:71.
Scroggie, M.P., K. Preece, E. Nicholson, M.A. McCarthy, K.M. Parris, G.W. Heard. 2019. Optimizing habitat management for amphibians: From simple models to complex decisions. *Biological Conservation* 236:60–69.
Semlitsch, R.D. 2000. Principles for management of aquatic-breeding amphibians. *Journal of Wildlife Management* 64(3):615–631.
Shannon, G., M.F. McKenna, L.M. Angeloni, K.R. Crooks, K.M. Fristrup, E. Brown, K.A. Warner, M.D. Nelson, C. White, J. Briggs, S. McFarland, G. Wittemyer. 2016. A synthesis of two decades of research documenting the effects of noise on wildlife. *Biological Reviews of the Cambridge Philosophical Society* 91(4):982–1005.
Shu, X., W. Zhang, B. Li, E. Pei, X. Yuan, T. Wang, Z. Wang. 2016. Major factors affecting the distribution of Anuran communities in the urban, suburban and rural areas of Shanghai, China. *Asian Herpetological Research* 7(4):287–294.
Sievers, M., R. Hale, K.M. Parris, S.D. Melvin, C.M. Lanctôt, S.E. Swearer. 2019. Contaminant-induced behavioural changes in amphibians: A meta-analysis. *Science of the Total Environment* 693:133570.

Sievers, M., R. Hale, S.E. Swearer, K.M. Parris. 2018b. Contaminant mixtures interact to impair predator-avoidance behaviours and survival in a larval amphibian. *Ecotoxicology and Environmental Safety* 161:482–488.
Sievers, M., K.M. Parris, S.E. Swearer, R. Hale. 2018a. Stormwater wetlands can function as ecological traps for urban frogs. *Ecological Applications* 28(4):1106–1115.
Sijtsma, F.J., E. van der Veen, A. van Hinsberg, R. Pouwels, R. Bekker, R.E. van Dijk, M. Grutters, R. Klaassen, M. Krijn, M. Mouissie, E. Wymenga. 2020. Ecological impact and cost-effectiveness of wildlife crossings in a highly fragmented landscape: A multi-method approach. *Landscape Ecology* 35:1701–1720.
Slaby, S., M. Marin, G. Marchand, S. Lemiere. 2019. Exposures to chemical contaminants: What can we learn from reproduction and development endpoints in the amphibian toxicology literature? *Environmental Pollution* 248:478–495.
Smith, R.K., H. Meredith, W.J. Sutherland. 2020. Amphibian conservation. Pages 9–64 in W.J. Sutherland, L.V. Dicks, S.O. Petrovan, R.K. Smith, editors. *What Works in Conservation 2020*. Open Book Publishers, Cambridge.
Snodgrass, J.W., R.E. Casey, D. Joseph, J.A. Simon. 2008. Microcosm investigations of stormwater pond sediment toxicity to embryonic and larval amphibians: Variation in sensitivity among species. *Environmental Pollution* 154(2):291–297.
Sterrett, S.C., R.A. Katz, W.R. Fields, E.H.C. Grant. 2019. The contribution of road-based citizen science to the conservation of pond-breeding amphibians. *Journal of Applied Ecology* 56(4):988–995.
Sugawara, H., T. Kusano, F. Hayashi. 2016. Fine-scale genetic differentiation in a salamander *Hynobius tokyoensis* living in fragmented urban habitats in and around Tokyo, Japan. *Zoological Science* 33(5):476–484.
Tennessen, J.B., S.E. Parks, L. Swierk, L.K. Reinert, W.M. Holden, L.A. Rollins-Smith, K.A. Walsh, T. Langkilde. 2018. Frogs adapt to physiologically costly anthropogenic noise. *Proceedings of the Royal Society B: Biological Sciences* 285(1891):20182194.
Todd, B.D., C.M. Bergeron, M.J. Hepner, W.A. Hopkins. 2011. Aquatic and terrestrial stressors in amphibians: A test of the double jeopardy hypothesis based on maternally and trophically derived contaminants. *Environmental Toxicology and Chemistry* 30(10):2277–2284.
Troïanowski, M., N. Mondy, A. Dumet, C. Arcanjo, T. Lengagne. 2017. Effects of traffic noise on tree frog stress levels, immunity, and color signaling. *Conservation Biology* 31(5):1132–1140.
van der Grift, E.A. 2005. Defragmentation in the Netherlands: A success story? *GAIA* 14(2):144–147.
Van Meter, R.J., C.M. Swan, J.W. Snodgrass. 2011. Salinization alters ecosystem structure in urban stormwater detention ponds. *Urban Ecosystems* 14(4):723–736.
Vietz, G.J., I.D. Rutherfurd, T.D. Fletcher, C.J. Walsh. 2016. Thinking outside the channel: Challenges and opportunities for protection and restoration of stream morphology in urbanizing catchments. *Landscape and Urban Planning* 145:34–44.
Violin, C.R., P. Cada, E.B. Sudduth, B.A. Hassett, D.L. Penrose, E.S. Bernhardt. 2011. Effects of urbanization and urban stream restoration on the physical and biological structure of stream ecosystems. *Ecological Applications* 21:1932–1949.
Wake, M.H. 1977. Fetal maintenance and its evolutionary significance in the Amphibia: Gymnophiona. *Journal of Herpetology* 11(4):379–386.
Walsh, C.J., A.H. Roy, J.W. Feminella, P.D. Cottingham, P.M. Groffman, R.P Morgan. 2005. The urban stream syndrome: Current knowledge and the search for a cure. *Journal of the North American Benthological Society* 24(3):706–723.
Watters, A.M., F.E. Rowland, R.D. Semlitsch. 2018. Larval salamanders are as effective at short-term mosquito predation as mosquitofish. *Canadian Journal of Zoology* 96(10):1165–1169.
Wellborn, G.A., D.K. Skelly, E.E. Werner. 1996. Mechanisms creating community structure across a freshwater habitat gradient. *Annual Review of Ecology and Systematics* 27:337–363.
Wells, K.D., editor. 2007. *The Ecology and Behavior of Amphibians*. University of Chicago Press, Chicago, IL.
Werner, E.E., D.K. Skelly, R.A. Relyea, K.L. Yurewicz. 2007. Amphibian species richness across environmental gradients. *Oikos* 116(10):1697–1712.
Wiens, J.J. 2007. Global patterns of diversification and species richness in amphibians. *The American Naturalist* 170(52):S86–S106.

20
FRESHWATER BIODIVERSITY IN URBAN ECOSYSTEMS

Augustin C. Engman and Allison H. Roy

Introduction

Cities are historically built around aquatic ecosystems because water is essential for human life. For native and indigenous people, proximity to waterways was critical for transportation, food, and water (Kaushal et al. 2015). Energy harnessed from waterways (e.g. mills, hydroelectric power) also fueled communities as they industrialized, while rivers exported human and industrial wastes. As watersheds continue to urbanize, streams and wetlands are filled and piped to make way for infrastructure (Kaushal and Belt 2012). These piped streams contain few aquatic biota and are rarely sampled; thus, this chapter focuses on aboveground, freshwater ecosystems.

The varied aquatic ecosystems of urban, suburban, and exurban landscapes (broadly termed urban aquatic ecosystems, or UAEs) are vital habitat for many aquatic organisms (Figure 20.1). Freshwater streams, lakes, and wetlands in existence prior to city development are now intertwined with and altered by the urban infrastructure in their watersheds. Additionally, wholly anthropogenically constructed UAEs are common in cities. Reservoirs, canals, stormwater ponds, and garden/ornamental ponds are explicitly designed for human needs, while other systems such as ephemeral pools and unintentional wetlands are byproducts of human infrastructure (Palta et al. 2017). Collectively, these anthropogenically altered and constructed UAEs provide essential habitat for some aquatic biota and critical resources for terrestrial and semi-aquatic organisms.

UAEs harbor a subset of the total native fauna that 'made it through the filter' of urban disturbance (see Aronson et al., this volume). Altered hydrology, physical habitat structure, water quality, and habitat connectivity are common and interconnected urban stressors that dictate urban biodiversity. Impervious surfaces and stormwater drainage infrastructure increase runoff and reduce infiltration, typically resulting in flashier streamflow and reduced baseflows, although these responses vary across cities and physiographic setting, and increased hydrologic permanence is common (Hopkins et al. 2015; Bhaskar et al. 2016). Altered flows, combined with land cover change and stream channelization, result in changes to physical habitat structure (e.g. altered bed texture, reduced canopy cover, homogenization) and water quality (e.g. increased nutrients, temperature, toxics; Walsh et al. 2005), causing buildup of pollutants in impoundments and lentic waterbodies. Moreover, reduced flows, channelization, barriers (e.g. dams, road-stream crossings), point source plumes, and burial/loss of UAEs can result in reduced longitudinal and lateral connectivity within urban watersheds (Kaushal and Belt 2012; Thornhill et al. 2018), preventing access to critical spawning and nursery habitats

 DOI: 10.4324/9781003016120-23

Figure 20.1 Examples of urban aquatic ecosystems (UAEs). UAEs have different potential to harbor urban biodiversity depending on hydrology, habitat, connectivity, and water quality. Photo credits: A.C. Engman.

and altering metapopulation dynamics in ponds. Conversely, infrastructure (e.g. stormwater pipes) and increased permanence can result in hyper-connectivity, further homogenizing urban waterbodies. Ultimately, the fishes and macroinvertebrates that can survive, reproduce, and disperse in these urban environments comprise the UAE biodiversity described in this chapter.

Knowledge of fish and aquatic macroinvertebrate diversity in cities has been primarily informed by field studies and experiments that use rural to urban land cover gradients to investigate the effects of urbanization (Paul and Meyer 2001; Brown et al. 2005; Walsh et al. 2005; Wenger et al. 2009). Urbanization typically results in biotic homogenization, whereby compositional similarity increases as endemic and sensitive biota are replaced with introduced and generally tolerant species (Petsch 2016; Toussaint et al. 2016). Species with traits that confer resistance and resilience to disturbance, such as fast colonization rates, short life cycles, omnivory, air breathing, and broad habitat preferences, typically proliferate in UAEs whereas habitat or trophic specialists and those with complex life history requirements disappear from UAEs (Walsh et al. 2005; Berger et al. 2018). UAE biota also include species that are intentionally (e.g. for recreational fishing, biotic control, or aesthetics) or unintentionally (e.g. pet disposal) introduced and organisms that proliferate in degraded habitats and water quality, and many urban ponds.

Fish state of the knowledge

Species

Urban fish assemblages generally have low diversity because urbanization extirpates sensitive taxa and increases dominance of tolerant, generalist taxa (Walsh et al. 2005). In the most extreme cases, urbanization results in extirpation of all native fishes. More commonly, some native and naturalized fishes persist, while others are lost, decreasing species richness. As some species are lost, other species increase in abundance, thus reducing evenness. For example, redbreast sunfish (*Lepomis auritus*) and guppy (*Poecilia reticulata*) exhibit disproportionately high relative abundance in urbanized waters of their native ranges (Kennen et al. 2005; Terra et al. 2013).

Persistence through urbanization is related to functional and life-history traits. Fishes with broad diets and those with non-specific habitat requirements for spawning, growth, or survival are more likely to persist in UAEs. For example, the relative abundance of trophic generalist fish species (i.e. omnivores) is correlated with urbanization in the United States and Brazil (Cunico et al. 2011). Trait plasticity also appears to confer persistence in urban systems. Flexibility in life history, diet, and habitat use were all linked to resilience to urbanization in *Awaous stamineus* (a freshwater goby) on the island of Oahu, Hawaii (Lisi et al. 2018).

High relative abundance of non-native species and biotic homogenization are common characteristics of urban fish assemblages (Marchetti et al. 2006). Fish biotic homogenization occurs at global and regional scales via non-native species introductions and regional species translocations (Toussaint et al. 2016). Fishes with economic value (i.e. those favored in aquaculture, aquaria, or fishing) are most likely to be introduced (Rahel 2007), and many traits associated with economic value (e.g. thermal and hypoxia tolerance, fast growth rate, and short generation time) may also confer propensity to establish and spread in urban environments.

Additionally, as centers of economic activity, cities are susceptible to unintentional introductions through import processes (Leprieur et al. 2008).

Although many urban fish assemblages are low-diversity and dominated by non-native species, UAEs can be important habitat for native fish conservation. Some UAEs harbor relatively diverse, native-dominated fish assemblages. The Río Piedras is the most urbanized river on the Caribbean island of Puerto Rico, and it contains all native fish taxa along with many introduced and potentially invasive species (Engman and Ramírez 2012). The predominant life history trait of Puerto Rico's native stream fishes – diadromy, or migration between salt and fresh water – may confer resistance and resilience to urbanization because of tolerance to high conductivity, common in UAEs, or decoupling of sensitive life stages from urban stressors. Alternatively, larvae exported from non-urbanized streams could subsidize urban fish populations (Ramírez et al. 2012). UAEs occasionally function as critical habitat for rare or endangered fish species. The only remaining native populations of Chapultapec Splitfin, *Girardinichthys viviparus*, an IUCN red list critically endangered species endemic to the valley of Mexico basin, occur almost entirely within Mexico City, a metropolitan area of ~20 M people. Although much of this species' native habitat has been lost to development, reproducing and apparently healthy populations still exist in several heavily impacted lakes within this megacity, including Lake Alameda Oriente, a treated wastewater effluent dominated system (Gómez Márquez et al. 2013).

Stressors

Flow regime alterations associated with urbanization affect fish survival and persistence at multiple life stages and may increase the strength of negative ecological interactions (Table 20.1). The high and variable water velocities of a flashy flow regime may cause mortality or displacement in susceptible early life history stages (e.g. eggs and larvae). For many fish species phenology – the timing of key life history events like spawning, hatching, or migrations – is cued by flows. Flow alterations that disrupt phenology can cause mismatches with resource availability, predator abundances, or temperature regimes, which reduce recruitment success or increase mortality (Lytle and Poff 2004). Reduced baseflows also decrease or eliminate habitat volume, which can result in increased competitive interactions (Stradmeyer et al. 2008) and susceptibility to avian predation (Allouche and Gaudin 2001) and increased morbidity and mortality.

Lateral and longitudinal fragmentation in urban systems interrupts critical life history processes and alters fish assemblage structure. Channelization and incision of urban streams severs lateral connectivity with peripheral habitats like floodplains, oxbows, and backwaters that are used as for spawning and as nurseries (Boët et al. 1999). Dams are the principal anthropogenic barriers to longitudinal connectivity globally and can affect urban fish assemblage even when the dam is not within a city. For example, a large downstream dam may exclude all native diadromous fishes from an urban watershed, including species that are tolerant to other urban stressors (Ramírez et al. 2012). The effects of road crossings on longitudinal connectivity vary with crossing design and fish traits. Culverts – especially those that create rapid water velocities; have become clogged with wood, sediments, and trash; or have high hanging-height – have the greatest impacts on fish movement. Species traits such as jumping, climbing, or high-velocity swimming ability have been used to mechanistically explain assemblage differences above and below barriers (Favaro et al. 2014).

Physical habitat structure, usually degraded in urban aquatic ecosystems, is directly related to fish assemblage structure. Habitat heterogeneity and fish diversity have been shown to covary in UAEs. Loss of variation in substrate size, water velocity, and depth yield species-poor stream fish assemblages (Engman and Ramírez 2012), and littoral habitat complexity is positively related to intolerant fish richness and negatively related to tolerant fish richness in developed lakes (Kaufmann et al. 2014). The loss of off-channel and backwater microhabitats can reduce individual survival and growth in species that use them for refugia and feeding (Uerling et al. 2019). Increased pool depth and the loss of cover, like coarse wood, augments fish mortality through predation (Allouche 2002). Siltation, a common physical-habitat stressor in urban systems, can negatively affect fish diet and reproduction. Siltation inhibits algal biofilm growth and may favor unpalatable algae, which could exclude grazing fishes (Lisi et al. 2018) and affects fish recruitment by filling interstitial spaces used for spawning and early development (Peoples et al. 2011).

Water quality stressors in urban systems are a complex mix of pollutants and physiochemical processes that interact to produce lethal and sub-lethal effects on fishes. Water temperature and nutrient loading interact to dictate dissolved oxygen concentration, one of the most critical water quality parameters for fish survival (Davis 1975). Wastewater, lawn fertilizer, and pet waste are major sources of excess nutrients in urban waters that increase biological oxygen demand (BOD). Impervious surfaces and canopy cover-loss increase water temperature, further increasing BOD, while decreasing water oxygen solubility but increasing metabolic demand for oxygen in fish. Collectively, these stressors cause mass mortality and high variability in fish assemblage structure in the River Seine in Paris, France, where hypoxia is linked to summertime pulses of municipal wastes via stormwater runoff (Boët et al. 1999). Similar results occur during stormflows in the rainy season in the Rio das Velhas in Belo Horizonte, Brazil (Alves and Pompeu 2005). There are many other water pollutants, including polychlorinated biphenyls (PCBs), polyaromatic hydrocarbons (PAHs), current-use and legacy pesticides, personal care products, and heavy metals that are linked to urbanization and have suspected or demonstrated effects on fishes (Monosson 1999; Louiz et al. 2009; Lozano et al. 2012). However, disentangling the negative effects of multiple water-quality stressors, in conjunction with hydrologic alteration and other stressors, is challenging (Bryant and Carlisle 2012). Commonly, the combined effects of all urban water quality stressors on fishes are documented using histopathology or other health endpoints (Camargo and Martinez 2007) or with multimetric indices to assess whole-assemblage effects (Yoder and Smith 1998).

Macroinvertebrate state of the knowledge

Species

Macroinvertebrate assemblages in UAEs may be more or less diverse than less disturbed counterparts, depending on the ecosystem (Gál et al. 2019). Streams in urban watersheds have lower richness and more dominant taxa that are tolerant of pollutants than streams in rural watersheds (Morley and Karr 2002; Morse et al. 2003; Roy et al. 2003; Cuffney et al. 2010). Taxa in the sensitive orders Ephemeroptera (mayflies), Plecoptera (stoneflies), and Trichoptera (caddisflies) are rare in urban streams, although some families of mayflies (e.g. Baetidae) and caddisflies (e.g. Hydropsychidae) may be common. While some studies of urban lakes and wetlands show lower diversity than undisturbed systems (Gledhill et al. 2008), urban ponds with low nutrients, high transparency, and high cover of submerged vegetation may have high richness (Vermonden et al. 2009; Hassall 2014). In the United Kingdom, urban ponds have unique and diverse macroinvertebrate fauna compared to non-urban ponds and contribute to high regional biodiversity (Hill et al. 2016, 2017).

Macroinvertebrates in UAEs often have traits that allow them to tolerate urban stressors or confer resilience to disturbance (Berger et al. 2018). For example, macroinvertebrates with adult aquatic life stages and those that can breathe air are tolerant to low oxygen conditions that are common in benthic areas of UAEs. Multivoltine or polyvoltine taxa (those with greater than two generations per year) and taxa with aquatic or aerial passive dispersal can colonize temporary waters and persist in UAEs despite variable hydrology. Likewise, taxa that survive drying and freezing, such as cladocerans with dormant propagules and parthenogenic reproduction, contribute to temporary pond biodiversity (Pinel-Alloul and Mimouni 2013). Feeding preferences for dead animals and piercers are consistently associated with tolerant taxa in UAEs (Berger et al. 2018), whereas shredders are rare in urban waterbodies (Violin et al. 2011). While less is known about invasive macroinvertebrates compared to fishes, introduced bivalves, gastropods, planarians, oligochaetes, and crustaceans can be abundant in UAEs, likely because of tolerance to stressors such as low oxygen and high nutrients (Vermonden et al. 2009).

Constructed aquatic ecosystems in urban environments can also serve as important habitat for native and rare aquatic wildlife (Kowarik 2011; Palta et al. 2017). Drainage ditches, stormwater retention basins, and impoundments all harbor indigenous biota and may provide important anthropogenic refuges for regional populations (Chester and Robson 2013). These UAEs also include rare and threatened macroinvertebrate species. For example, six rare species of leeches that are on the Polish Red List (Koperski 2010) and two species of rare bivalves (Kołodziejczyk and Lewandowski 2017) are found in lakes and ponds in Warsaw, Poland. Two species on the IUCN Red List were found in UAEs in the Netherlands: *Planaria torva* (Tricladida) and *Loptocerus tineiformis* (Trichoptera) (Vermonden et al. 2009).

Stressors

Hydrologic alteration is a primary stressor of macroinvertebrate assemblages in urban lotic and lentic waterbodies. Catchments with more connected imperviousness produce streams with high volume, flashy, polluted storm flows that displace macroinvertebrates, thereby lowering macroinvertebrate diversity (Walsh and Webb 2016). Some urban headwater streams may experience increased hydrologic permanence with urbanization (Roy et al. 2009), thereby promoting macroinvertebrate colonization; however, urbanization can also cause drying and more variable flows (Bhaskar et al. 2016). In wetlands and temporary ponds, hydroperiod determines species persistence. Shorter hydroperiods prevent colonization by some species and unpredictability in inundation frequency and intensity affects biota (Oertli and Parris 2019). Finally, water withdrawals for drinking water, hydropower, and lake management can alter littoral macroinvertebrate assemblages in urban lakes (Zohary and Ostrovsky 2011).

The loss of connectivity in UAEs also affects macroinvertebrates, although impacts are usually less dramatic than for fishes. Drift is an important mechanism for downstream insect dispersal in streams, and dams and associated impoundments can interrupt this process. Stream network fragmentation can also be a barrier to overland dispersal (Smith et al. 2009); many emergent aquatic insects travel along streams, but dams, culverts, and buried streams can prevent long-distance movement. Longitudinal connectivity is also critical for freshwater mussels that rely on fishes for long-distance dispersal and population persistence (Modesto et al. 2018). Other dimensions of connectivity are also important for macroinvertebrates; urban streams commonly have greatly diminished riparian wetlands and hyporheic areas that serve as important refuges and breeding habitat. Even in waterbodies that are not directly connected via overland flow paths, like lakes and wetlands, urban infrastructure can impede overland dispersal, and fewer ponds can negatively impact dispersal potential and metapopulation dynamics of aquatic macroinvertebrates (Sun et al. 2019).

Urbanization causes numerous impacts to the physical habitat of UAEs, with associated alterations of macroinvertebrate assemblages. Urbanization can cause both fine sediment deposition, which can reduce clingers (Morley and Karr 2002) and favor burrowers, and scouring, which may favor filterers and taxa that can tolerate high flows. Reduced habitat complexity is a typical symptom of UAEs, which limits sensitive taxa (Violin et al. 2011). The loss of forest cover associated with urbanization reduces allochthonous material, resulting in few shredders in urban streams (Violin et al. 2011). Developed shorelines in lakes have less wood than forested shorelines, which can reduce macroinvertebrate diversity and alter assemblage composition. In lentic waterbodies, the presence of aquatic vegetation is important for supporting macroinvertebrates, thus the effect of urbanization on plants, which varies by UAE, influences macroinvertebrate diversity (Chester and Robson 2013).

Altered water quality is an important mechanism for reduced macroinvertebrate diversity in UAEs. Macroinvertebrates are regularly used for assessing impairment of waterbodies, largely because of their variable sensitivity to pollutants among taxa (Feio et al. 2021). Improved water quality associated with wastewater treatment has led to increased richness and more pollution-sensitive taxa throughout the world (e.g. Vaughan and Ormerod 2012). However, macroinvertebrate assemblages remain impaired in urban areas, often as a result of high conductivity, nutrients, and toxicants from hydrologically hyperconnected stormwater drainages. In urbanized areas in cold climates, road salt increases conductivity, with associated losses of sensitive macroinvertebrate taxa such as mayflies (Wallace and Biastoch 2016). Although, Hassall and Anderson (2015) found similar macroinvertebrate diversity between stormwater and unmanaged urban ponds despite differences in water chemistry, such as road salt. Nutrient enrichment may lead to proliferation of herbivores and contributes to dominance by invasive species (Vermonden et al. 2009). Finally, toxic metals and

Table 20.1 Four stressor categories and examples of the mechanisms through which each stressor affects fishes and macroinvertebrates, leading to a loss of urban biodiversity

Stressors	*Fish responses*	*Macroinvertebrate responses*
Hydrology		
Increased water level fluctuations	Mortality and displacement due to high water velocity Loss or alteration of cues for reproductive and migration phenology	Mortality and displacement due to high water velocity
Reduced baseflows/ permanence	Increased competitive interactions and predation	Dominance by multivoltine and desiccation-resistant species
Connectivity		
Reduced longitudinal connectivity in streams	Loss of migratory pathways for diadromous species Loss of genetic diversity	Loss of aerial-dispersers that use stream corridors Loss of genetic diversity
Reduced lateral connectivity in streams	Reduced growth and survival of juveniles Spawning failure Lost access to food resources	Reduced abundances of predators and pool taxa
Reduced number of lentic waterbodies		Loss of genetic diversity with fewer metapopulations
Physical habitat		
Sedimentation	Altered food resources Spawning failure for benthic nest spawners	Reduced clingers Increased burrowers
Reduced wood and cover	Increased predation	Reduced abundances of predators
Reduced habitat heterogeneity	Species homogenization Increased competitive interactions	Species homogenization
Increased macrophytes		Increased plant-associated species
Water quality		
Increased nutrients	Increased herbivores due to algal proliferation	Increased herbivores due to algal proliferation
Increased temperature	Mortality of cold-water species	Mortality of thermally sensitive species
Reduced dissolved oxygen	Mortality and displacement in hypoxia	Dominance by air breathers
Increased conductivity	Mortality and morbidity related to osmotic regulation	Mortality from poor osmotic regulation
Contaminants	Mortality, morbidity, and reproductive effects of toxicity	Mortality from toxicity

other contaminants in urban environments can reduce macroinvertebrate diversity (Sun et al. 2019) and may facilitate proliferation of toxicants in food webs (Table 20.1). As with fish assemblages, it is often difficult to determine which stressor is most limiting macroinvertebrate diversity.

Knowledge gaps

While research on UAEs has proliferated in recent decades, many knowledge gaps in our understanding of urban aquatic biodiversity remain. Wenger et al. (2009) identified 26 research questions in urban stream ecology, including impacts to biota, and many of these remain understudied. Here, we identify ten topic areas requiring further research for understanding urban aquatic biodiversity (#1–3), characteristics of biota and biotic interactions in UAEs (#4–6), and issues affecting urban aquatic biodiversity into the future (#7–10).

1 *Multiple threats and stressors that filter biodiversity.* The multiple and interacting threats and stressors in UAEs present challenges to identifying primary constraints on urban biota (Craig et al. 2017). Study design and analytical approaches may help to tease apart causal mechanisms. Alternatively, identifying and implementing restoration and management actions that address multiple, watershed-scale threats may best increase urban biodiversity.
2 *Global/physiographic differences in urban biota.* Most urban aquatic research is from temperate regions of North America, Europe, and Australia, with relatively limited work in Asia, Africa, and South America. Given the global diversity of urban impacts in UAEs (Booth et al. 2016), research in understudied regions is needed for a holistic understanding of urban biodiversity.
3 *Impacts of legacy land use and stressors on present-day assemblages.* Urban development commonly occurs on formerly disturbed lands; these land use legacies can serve as a filter for current assemblages. Even where threats (e.g. dams, culverts, source inputs) have been removed, the continued influence of stressors associated with former disturbances (e.g. sediment and metal accumulation, species loss) can affect ecosystems and food webs for decades to centuries (Walter and Merritts 2008). Understanding these legacies is helpful for developing realistic endpoints for restoration (see #9).
4 *Constructed ecosystems as critical habitat for biota and influences on natural ecosystems.* Constructed, anthropogenically created UAEs can support urban biota and may provide critical habitat for sustaining metapopulations (Kowarik 2011; Chester and Robson 2013), and there may be potential for increased benefits (e.g. in garden ponds) if explicitly managed for aquatic biota (Davies et al. 2009). However, constructed UAEs may increase the prevalence and persistence of invasive species, with negative consequences for natural ecosystems. An understanding of factors determining whether constructed UAEs have net benefits or negative impacts to adjacent ecosystems is critical for informing restoration efforts.
5 *Species interactions and trophic effects in UAEs.* Most studies of biotic interactions focus in undisturbed areas, and ecological concepts derived from natural areas may not apply to UAEs. Further investigation is needed to understand how urban-induced alteration of species presence and abundances, including invasive species, mediates presence and abundance of other species due to changes in predation and competitive interactions.
6 *Behavioral, reproductive, physiological, and other sublethal effects of urbanization.* Where biota persist in UAEs, they may have phenotypes that are different than rural systems that allow them to survive and thrive in degraded habitats (Marques et al. 2019). An understanding of intraspecific differences in physiological tolerance, reproductive traits, and behaviors between disturbed and undisturbed areas can help to elucidate mechanisms of impacts and biotic persistence in UAEs (Maas-Hebner et al. 2014).
7 *Prevalence and impacts of diseases on urban aquatic biota.* Land use change, including urbanization, increases infectious disease transmission between wildlife and humans (Gottdenker et al. 2014). Urban biota are not only vectors of diseases but may also be increasingly susceptible to pathogens. Information on the prevalence, susceptibility, spread, and impacts of diseases on biota in UAEs is needed.
8 *Climate change mediation of urban impacts on biota.* UAEs are impacted by climate change to varying degrees, and the combined threats of climate change and urbanization may exacerbate

impacts to biota. Urban stormwater management may also offer climate adaptation co-benefits, including urban cooling and flood mitigation. Understanding of urban interactions with climate stressors is needed to develop biotic expectations and temporal trajectories of change with changing climates and restoration.

9 *Biotic expectations following management and restoration.* Given land use legacies, climate change, and ongoing urban threats, it is unrealistic for UAEs to harbor biodiversity matching minimally disturbed systems. Developing realistic expectations of restoration aligned with the scale of restoration (local, watershed) is critical for evaluating ecological values of different restoration activities. Societal endpoints should also be included and assessed for a more holistic understanding of the benefits of urban management and restoration (Smith et al. 2016).

10 *Services provided by urban biota.* Biota in UAEs, including constructed urban ecosystems (Palta et al. 2017), provide critical ecosystem services, including provisioning services (e.g. fish consumption), biogeochemical cycling, cultural benefits, and aesthetics. Studies of the services and disservices that humans derive from urban ecosystems, including angler preferences and risks of contamination that could create environmental health inequities, are needed to assess environmental racism and create sustainable, resilient urban ecosystems and communities.

Conservation and management

UAE biotic integrity (e.g. individual population health, diversity, assemblage structure, and/or function) can be conserved or augmented through remediation of urban physical habitat, water quality, and connectivity stressors, actions that are commonly referred to as UAE restoration. Although there are many challenges associated with restoring and conserving UAE biotic integrity, practitioners can maximize the likelihood of project success by carefully considering project extent and limitations.

Most UAE restoration efforts occur at the local (i.e. reach, site) scale. Local-extent restorations directly remediate physical habitat degradation (e.g. channel incision, bank erosion), improve water quality (e.g. pond aeration to increase dissolved oxygen), or add specific habitat features (e.g. coarse wood for fish cover; Hughes et al. 2014a). The effectiveness of local-extent restorations is not fully understood, largely because of the lack of rigorous pre- and post-restoration monitoring and evaluation (Bernhardt et al. 2005). Limited studies have shown no changes in macroinvertebrate assemblage structure (Violin et al. 2011), possibly because of the lack of source populations or persistent urban stressors (Hughes et al. 2014a). Restoration success is also limited by space constraints from surrounding infrastructure that prevent optimal floodplain and riparian area reconstruction.

UAE restoration may be best addressed at the catchment extent through stormwater and wastewater management. Improvements in urban sewage treatment is related to significant improvements in fish assemblage structure over decades (Yoder and Smith 1998; Yoder et al. 2005; Happel and Gallagher 2020). Low-impact development (LID), green infrastructure, and water-sensitive urban design are synonymous terms that describe environmentally sensitive stormwater management technologies and strategies used to approximate natural flow paths, restore water balance, and remove pollutants in urbanized catchments (Askarizadeh et al. 2015; Walsh et al. 2016). LID may have the greatest desired effect if it is implemented during the process of urbanization (Hughes et al. 2014b). LID can be retrofitted to urbanized catchments, although effectiveness is still limited (Roy et al. 2014; Fanelli et al. 2019). Additionally, constructed ecosystems such as stormwater retention ponds engineered with LID principles have been shown to increase aquatic biodiversity and may even contribute to the conservation of imperiled species (Schaeffer et al. 2012; Oertli and Parris 2019). However, the apparent benefits of these constructed ecosystems must be tempered with the risk that they could function as ecological traps that attract organisms and compromise individual fitness or population viability (Hale et al. 2019).

Even with effective local- and catchment-extent UAE restoration and management, some biotic condition goals will not be achieved without restored connectivity. Longitudinal connectivity is especially critical for migratory (e.g. diadromous, potamodromous) fish and unionid mussels with migratory fish hosts (Modesto et al. 2018). Restoration of longitudinal connectivity can be achieved through removing instream barriers like dams, weirs, and road crossings and facilitating barrier passage with construction of bottomless or baffled culverts, rock ramps, barrier notches, and bypass channels (Choy et al. 2018). For isolated waterbodies, such as ponds and some wetlands, and for species with overland dispersal, biodiversity can be maximized in a high-density network of waterbodies on urban landscapes that are connected by dispersal corridors such as green areas, wetlands, and waterways (Oertli and Parris 2019; Zamora-Marín et al. 2021).

Fisheries management is also a critical component of biotic conservation in UAEs. UAE recreational, subsistence, and commercial fisheries yield valuable ecosystem services to urban people. Therefore, consideration of distinct stakeholder demographics, user preferences, and human health concerns in urban fisheries is important for their effective management. Urban recreational fisheries include greater proportions of underrepresented and underserved groups of anglers than rural fisheries, but they receive disproportionately little management effort and research focus (Hickley et al. 2004; Ivasauskas et al. 2017). In the United States, 67% of recreational anglers live in urban areas (U.S. Department of the Interior 2016), and in some instances, urban anglers are more likely from minority groups than rural anglers (e.g. Hutt and Neal 2010). Urban angler preferences and needs are varied, but management actions such as habitat enhancement, access and amenities at fishing locations, stock enhancement, and education and outreach programs may improve the urban fishing experience (Hickley et al. 2004). Sourcing food fish and invertebrates from UAEs may contribute to the global sustainability of cities and fisheries. For example, urban aquaculture that employs municipal wastewater and industry by products (e.g. brewery wastes) for water and feed simultaneously reduces local contamination of UAEs, global nutrient pollution, pressure on wild fish stocks, and global production and transportation emissions (Hoan et al. 2005; Bunting and Little 2015). Commercial wastewater-fed aquaculture is a prominent activity and source of food for the cities of many populous countries such as China, Germany, India, Indonesia, and Vietnam (Hoan et al. 2005). Artisanal inland fisheries like the Lago Paranoá fishery in Brasilia, Brazil, can be a principal source of affordable protein and an important revenue stream for some low-income residents in large cities (Walter and Petrere 2007). Although there are social and environmental benefits, consuming animals from UAEs has the potential to harm human health because urban pollutants can accumulate in animal tissues to levels that exceed health-based guidelines for consumption (Buttermore et al. 2018). A vital role of urban fisheries managers is to measure contaminant levels in food fishes and invertebrates and effectively communicate risk to diverse user groups through consumption advisories (Gray et al. 2020).

Conclusions

Some of the most intense human land use on Earth occurs around UAEs, severely altering their hydrology, physical habitat, connectivity, and water quality and thereby reducing their biodiversity. However, UAEs can also be critical habitat for endemic and endangered fish and invertebrate species and these biota contribute valuable ecosystem services to a growing portion of the world's human population. If managed properly as part of a network of habitats, UAEs have the potential to support regional biodiversity conservation and contribute to the well-being of city-dwelling people, globally. Recognition of the value of natural and constructed UAEs can propel research and monitoring to understand ecological processes, engage student and public scientists (França et al. 2019), and lead to new strategies to maintain and enhance their biota, including in the context of global change.

Key companion papers

Chester, E.T., B.J. Robson. 2013. Anthropogenic refuges for freshwater biodiversity: Their ecological characteristics and management. *Biological Conservation* 166:64–75.

Gál, B., I. Szivák, J. Heino, D. Schmera. 2019. The effect of urbanization on freshwater macroinvertebrates – Knowledge gaps and future research directions. *Ecological Indicators* 104:357–364.

Hughes, R.M., et al. 2014. A review of urban water body challenges and approaches: (1) Rehabilitation and remediation. *Fisheries* 39:18–29.

Oertli, B., K.M. Parris. 2019. Review: Toward management of urban ponds for freshwater biodiversity. *Ecosphere* 10:e02810.

Toussaint, A., et al. 2016. Worldwide freshwater fish homogenization is driven by a few widespread non-native species. *Biological Invasions* 18:1295–1304.

Walsh, C.J., et al. 2005. The urban stream syndrome: Current knowledge and the search for a cure. *Journal of the North American Benthological Society* 24:706–723.

Acknowledgments

ACE and AHR contributed equally to this manuscript.

References

Allouche, S. 2002. Nature and functions of cover for riverine fish. *Bulletin Français de la Pêche et de la Pisciculture* 365–366:297–324.

Allouche, S., P. Gaudin. 2001. Effects of avian predation threat, water flow and cover on growth and habitat use by chub, *Leuciscus cephalus*, in an experimental stream. *Oikos* 94:481–492.

Alves, C.B.M., P.S. Pompeu. 2005. Historical changes in the Rio das Velhas fish fauna – Brazil. *American Fisheries Society Symposium* 2005(45):587–602.

Askarizadeh, A., M.A. Rippy, T.D. Fletcher, D.L. Feldman, J. Peng, P. Bowler, A.S. Mehring, B.K. Winfrey, J.A. Vrugt, A. AghaKouchak, S.C. Jiang, B.F. Sanders, L.A. Levin, S. Taylor, S.B. Grant. 2015. From rain tanks to catchments: Use of low-impact development to address hydrologic symptoms of the urban stream syndrome. *Environmental Science and Technology* 49(19):11264–11280.

Berger, E., P. Haase, R.B. Schäfer, A. Sundermann. 2018. Towards stressor-specific macroinvertebrate indices: Which traits and taxonomic groups are associated with vulnerable and tolerant taxa? *Science of the Total Environment* 619–620:144–154.

Bernhardt, E.S., M.A. Palmer, J.D. Allan, G. Alexander, K. Barnas, S. Brooks, J. Carr, S. Clayton, C. Dahm, J. Follstad-Shah, D. Galat, S. Gloss, P. Goodwin, D. Hart, B. Hassett, R. Jenkinson, S. Katz, G.M. Kondolf, P.S. Lake, R. Lave, J.L. Meyer, T.K. O'Donnell, L. Pagano, B. Powell, E. Sudduth. 2005. Synthesizing U.S. river restoration efforts. *Science* 308(5722):636–637.

Bhaskar, A.S., L. Beesley, M.J. Burns, T.D. Fletcher, P. Hamel, C.E. Oldham, A.H. Roy. 2016. Will it rise or will it fall? Managing the complex effects of urbanization on base flow. *Freshwater Science* 35(1):293–310.

Boët, P., J. Belliard, R. Berrebi-dit-Thomas, E. Tales. 1999. Multiple human impacts by the city of Paris on fish communities in the Seine river basin, France. *Hydrobiologia* 410:59–68.

Booth, D.B., A.H. Roy, B. Smith, K.A. Capps. 2016. Global perspectives on the urban stream syndrome. *Freshwater Science* 35:412–420.

Brown, L.R., R.H. Gray, R.M. Hughes, M.R. Meador. 2005. Effects of urbanization on stream ecosystems. *American Fisheries Society Symposium* 47:1–8.

Bryant, W.L., D.M. Carlisle. 2012. The relative importance of physicochemical factors to stream biological condition in urbanizing basins: Evidence from multimodel inference. *Freshwater Science* 31(1):154–166.

Bunting, S.W., D.C. Little. 2015. Urban aquaculture for resilient food systems. Pages 312–335 in H. de Zeeuw, P. Drechsel, editors. *Cities and Agriculture: Developing Resilient Urban Food Systems*. Routledge, Abingdon, Oxon, Oxfordshire, UK.
Buttermore, E.N., W.G. Cope, T.J. Kwak, P.B. Cooney, D. Shea, P.R. Lazaro. 2018. Contaminants in tropical island streams and their biota. *Environmental Research* 161:615–623.
Camargo, M.M.P., C.B.R. Martinez. 2007. Histopathology of gills, kidney and liver of a Neotropical fish caged in an urban stream. *Neotropical Ichthyology* 5(3):327–336.
Chester, E.T., B.J. Robson. 2013. Anthropogenic refuges for freshwater biodiversity: Their ecological characteristics and management. *Biological Conservation* 166:64–75.
Choy, M., D. Lawrie, C.B. Edge. 2018. Measuring 30 years of improvements to aquatic connectivity in the Greater Toronto Area. *Aquatic Ecosystem Health & Management* 21:342–351.
Craig, L.S., J.D. Olden, A.H. Arthington, S. Entrekin, C.P. Hawkins, J.J. Kelly, T.A. Kennedy, B.M. Maitland, E.J. Rosi, A.H. Roy, D.L. Strayer, J.L. Tank, A.O. West, M.S. Wooten. 2017. Meeting the challenge of interacting threats in freshwater ecosystems: A call to scientists and managers. *Elementa: Science of the Anthropocene* 5:1–15.
Cuffney, T.F., R.A. Brightbill, J.T. May, I.R. Waite. 2010. Responses of benthic macroinvertebrates to environmental changes associated with urbanization in nine metropolitan areas. *Ecological Applications* 20(5):1384–1401.
Cunico, A.M., J.D. Allan, A.A. Agostinho. 2011. Functional convergence of fish assemblages in urban streams of Brazil and the United States. *Ecological Indicators* 11(5):1354–1359.
Davies, Z.G., R.A. Fuller, A. Loram, K.N. Irvine, V. Simms, K.J. Gaston. 2009. A national scale inventory of resource provision for biodiversity within domestic gardens. *Biological Conservation* 142:761–771.
Davis, J.C. 1975. Minimal dissolved oxygen requirements of aquatic life with emphasis on Canadian species: A review. *Journal of the Fisheries Research Board of Canada* 32:2295–2332.
Engman, A.C., A. Ramírez. 2012. Fish assemblage structure in urban streams of Puerto Rico: The importance of reach- and catchment-scale abiotic factors. *Hydrobiologia* 693:141–155.
Fanelli, R.M., K.L. Prestegaard, M.A. Palmer. 2019. Urban legacies: Aquatic stressors and low aquatic biodiversity persist despite implementation of regenerative stormwater conveyance systems. *Freshwater Science* 38:818–833.
Favaro, C., J.W. Moore, J.D. Reynolds, M.P. Beakes. 2014. Potential loss and rehabilitation of stream longitudinal connectivity: Fish populations in urban streams with culverts. *Canadian Journal of Fisheries and Aquatic Sciences* 71:1805–1816.
Feio, M.J., R.M. Hughes, M. Callisto, S.J. Nichols, O.N. Odume, B.R. Quintella, M. Kuemmerlen, F.C. Aguiar, S.F.P. Almeida, P. Also-EguíaLis, F.O. Arimoro, F.J. Dyer, J.S. Harding, S. Jang, P.R. Kaufmann, S. Lee, J. Li, D.R. Macedo, A. Mendes, N. Mercado-Silva, W. Monk, K. Nakamura, G.G. Ndiritu, R. Ogden, M. Peat, T.B. Reynoldson, B. Rios-Touma, P. Segurado, A.G. Yates. 2021. The biological assessment and rehabilitation of the world's rivers: An overview. *Water* 13(3):371.
França, J.S., R. Solar, R.M. Hughes, M. Callisto. 2019. Student monitoring of the ecological quality of neotropical urban streams. *Ambio* 48:867–878.
Gál, B., I. Szivák, J. Heino, D. Schmera. 2019. The effect of urbanization on freshwater macroinvertebrates – Knowledge gaps and future research directions. *Ecological Indicators* 104:357–364.
Gledhill, D.G., P. James, D.H. Davies. 2008. Pond density as a determinant of aquatic species richness in an urban landscape. *Landscape Ecology* 23(10):1219–1230.
Gómez Márquez, J.L., B. Peña Mendoza, J.L. Guzmán Santiago. 2013. Occurrence of the fish *Girardinichthys viviparus* (Cyprinodontiformes: Goodeidae) in an urban lake at Mexico city. *UNED Research Journal/Cuadernos de Investigación UNED* 5(1):89–95.
Gottdenker, N.L., D.G. Streicker, C.L. Faust, C.R. Carroll. 2014. Anthropogenic land use change and infectious diseases: A review of the evidence. *EcoHealth* 11(4):619–632.
Gray, K.M., C.E. LePrevost, W.G. Cope. 2020. Anglers' views on using signs to communicate fish consumption advisories. *Fisheries* 45:307–316.
Happel, A., D. Gallagher. 2020. Chicago's fish assemblage over ~30 years – More fish and more native species. *Urban Ecosystems* 24:311–325.
Hale, R., S.E. Swearer, M. Sievers, R. Coleman. 2019. Balancing biodiversity outcomes and pollution management in urban stormwater treatment wetlands. *Journal of Environmental Management* 233:302–307.
Hassall, C. 2014. The ecology and biodiversity of urban ponds. *Wiley Interdisciplinary Reviews: Water* 1:187–206.

Hassall, C., S. Anderson. 2015. Stormwater ponds can contain comparable biodiversity to unmanaged wetlands in urban areas. *Hydrobiologia* 745:137–149.
Hickley, P., R. Arlinghaus, R. Tyner, M. Aprahamian, K. Parry, M. Carter. 2004. Rehabilitation of urban lake fisheries for angling by managing habitat: General overview and case studies from England and Wales. *International Journal of Ecohydrology and Hydrobiology* 4(4):365–378.
Hill, M.J., J. Biggs, I. Thornhill, R.A. Briers, D.G. Gledhill, J.C. White, P.J. Wood, C. Hassall. 2017. Urban ponds as an aquatic biodiversity resource in modified landscapes. *Global Change Biology* 23(3):986–999.
Hill, M.J., D.B. Ryves, J.C. White, P.J. Wood. 2016. Macroinvertebrate diversity in urban and rural ponds: Implications for freshwater biodiversity conservation. *Biological Conservation* 201:50–59.
Hoan, V., P. Edwards, B. Costa-Pierce, A. Desbonnet, D. Baker. 2005. Wastewater reuse through urban aquaculture in Hanoi, Vietnam: Status and prospects. Pages 103–118 in B.A. Costa-Pierce, A. Desbonnet, P. Edwards, D. Baker, editors. *Urban Aquaculture*. CAB International, Oxfordshire, Oxford.
Hopkins, K.G., N.B. Morse, D.J. Bain, N.D. Bettez, N. Grimm, J.L. Morse, M.M. Palta, W.D. Shuster, A.R. Bratt, A.H. Suchy. 2015. Assessment of regional variation in streamflow responses to urbanization and the persistence of physiography. *Environmental Science and Technology* 49(5):2724–2732.
Hughes, R.M., S. Dunham, K.G. Maas-Hebner, J.A. Yeakley, C.B. Schreck, M. Harte, N. Molina, C.C. Shock, V.W. Kaczynski. 2014b. A review of urban water body challenges and approaches: (2) Mitigation and research needs. *Fisheries* 39:30–40.
Hughes, R.M., S. Dunham, K.G. Maas-Hebner, J.A. Yeakley, C.B. Schreck, M. Harte, N. Molina, C.C. Shock, V.W. Kaczynski, J. Schaeffer. 2014a. A review of urban water body challenges and approaches: (1) Rehabilitation and remediation. *Fisheries* 39(1):18–29.
Hutt, C.P., J.W. Neal. 2010. Arkansas urban resident fishing site preferences, catch related attitudes, and satisfaction. *Human Dimensions of Wildlife* 15(2):90–105.
Ivasauskas, T.J., W.N. Xiong, A.C. Engman, J.R. Fischer, T.J. Kwak, K.R. Rundle. 2017. Relationships among catch, angler satisfaction, and fish assemblage characteristics of an urban small impoundment fishery. *Journal of the Southeastern Association of Fish and Wildlife Agencies* 4:31–38.
Kaufmann, P.R., R.M. Hughes, T.R. Whittier, S.A. Bryce, S.G. Paulsen. 2014. Relevance of lake physical habitat assessment indices to fish and riparian birds. *Lake & Reservoir Management* 30:177–191.
Kaushal, S.S., K.T. Belt. 2012. The urban watershed continuum: Evolving spatial and temporal dimensions. *Urban Ecosystems* 15:409–435.
Kaushal, S.S., W.H. McDowell, W.M. Wollheim, T.A. Newcomer Johnson, P.M. Mayer, K.T. Belt, M.J. Penninov. 2015. Urban evolution: The role of water. *Water* 7(8):4063–4087.
Kennen, J.G., M. Chang, B.H. Tracy. 2005. Effects of landscape change on fish assemblage structure in a rapidly growing metropolitan area in North Carolina, USA. *American Fisheries Society Symposium* 47:39–52.
Kołodziejczyk, A., K. Lewandowski. 2017. Water bodies of a city park as habitats of rare, protected and alien species of molluscs. *Folia Malacologica* 25:257–265.
Koperski, P. 2010. Urban environments as habitats for rare aquatic species: The case of leeches (*Euhirudinea, Clitellata*) in Warsaw freshwaters. *Limnologica – Ecology and Management of Inland Waters* 40:233–240.
Kowarik, I. 2011. Novel urban ecosystems, biodiversity, and conservation. *Environmental Pollution* 159:1974–1983.
Leprieur, F., O. Beauchard, S. Blanchet, T. Oberdorff, S. Brosse. 2008. Fish invasions in the world's river systems: When natural processes are blurred by human activities. *PLoS Biology* 6(2):e28.
Lisi, P.J., E.S. Childress, R.B. Gagne, E.F. Hain, B.A. Lamphere, R.P. Walter, J.D. Hogan, J.F. Gilliam, M.J. Blum, P.B. McIntyre. 2018. Overcoming urban stream syndrome: Trophic flexibility confers resilience in a Hawaiian stream fish. *Freshwater Biology* 63(5):492–502.
Louiz, I., M. Ben-Attia, O.K. Ben-Hassine. 2009. Gonadosomatic index and gonad histopathology of *Gobius niger* (Gobiidea, Teleost) from Bizerta lagoon (Tunisia): Evidence of reproduction disturbance. *Fisheries Research* 100:266–273.
Lozano, N., C.P. Rice, J. Pagano, L. Zintek, L.B. Barber, E.W. Murphy, T. Nettesheim, T. Minarik, H.L. Schoenfuss. 2012. Concentration of organic contaminants in fish and their biological effects in a wastewater-dominated urban stream. *Science of the Total Environment* 420:191–201.
Lytle, D.A., N. LeRoy Poff. 2004. Adaptation to natural flow regimes. *Trends in Ecology & Evolution* 19(2):94–100.
Maas-Hebner, K.G., R.M. Hughes, C.B. Schreck. 2014. Wild salmonids in the urban environment: Lethal and sublethal effects. Pages 169–182 in J.A. Yeakley, K.G. Maas-Hebner, R.M. Hughes, editors. *Wild Salmonids in the Urbanizing Pacific Northwest*. Springer, New York.

Marchetti, M.P., J.L. Lockwood, T. Light. 2006. Effects of urbanization on California's fish diversity: Differentiation, homogenization and the influence of spatial scale. *Biological Conservation* 127:310–318.
Marques, P.S., L.R. Manna, R. Mazzoni, R. El-Sabaawi. 2019. Intraspecific trait variation in urban stream ecosystems: Toward understanding the mechanisms shaping urban stream communities. *Freshwater Science* 38(1):1–11.
Modesto, V., M. Ilarri, A.T. Souza, M. Lopes-Lima, K. Douda, M. Clavero, R. Sousa. 2018. Fish and mussels: Importance of fish for freshwater mussel conservation. *Fish and Fisheries* 19(2):244–259.
Monosson, E. 1999. Reproductive and developmental effects of PCBs in fish: A synthesis of laboratory and field studies. *Reviews in Toxicology* 3:25–75.
Morley, S.A., J.R. Karr. 2002. Assessing and restoring the health of urban streams in the Puget Sound basin. *Conservation Biology* 16:1498–1509.
Morse, C.C., A.D. Huryn, C. Cronan. 2003. Impervious surface area as a predictor of the effects of urbanization on stream insect communities in Maine, U.S.A. *Environmental Monitoring and Assessment* 89:95–127.
Oertli, B., K.M. Parris. 2019. Review: Toward management of urban ponds for freshwater biodiversity. *Ecosphere* 10:e02810.
Palta, M.M., N.B. Grimm, P.M. Groffman. 2017. 'Accidental' urban wetlands: Ecosystem functions in unexpected places. *Frontiers in Ecology and the Environment* 15:248–256.
Paul, M.J., J.L. Meyer. 2001. Streams in the urban landscape. *Annual Review of Ecology and Systematics* 32:333–365.
Peoples, B.K., M.B. Tainer, E.A. Frimpong. 2011. Bluehead chub nesting activity: A potential mechanism of population persistence in degraded stream habitats. *Environmental Biology of Fishes* 90(4):379s391.
Petsch, D.K. 2016. Causes and consequences of biotic homogenization in freshwater ecosystems. *International Review of Hydrobiology* 101:113–122.
Pinel-Alloul, B., E.-A. Mimouni. 2013. Are cladoceran diversity and community structure linked to spatial heterogeneity in urban landscapes and pond environments? *Hydrobiologia* 715:195–212.
Rahel, F.J. 2007. Biogeographic barriers, connectivity and homogenization of freshwater faunas: It's a small world after all. *Freshwater Biology* 52:696–710.
Ramírez, A., A. Engman, K.G. Rosas, O. Perez-Reyes, D.M. Martinó-Cardona. 2012. Urban impacts on tropical island streams: Some key aspects influencing ecosystem response. *Urban Ecosystems* 15(2):315–325.
Roy, A.H., A.L. Dybas, K.M. Fritz, H.R. Lubbers. 2009. Urbanization affects the extent and hydrologic permanence of headwater streams in a midwestern US metropolitan area. *Journal of the North American Benthological Society* 28(4):911–928.
Roy, A.H., L.K. Rhea, A.L. Mayer, W.D. Shuster, J.J. Beaulieu, M.E. Hopton, M.A. Morrison, A. St Amand. 2014. How much is enough? Minimal responses of water quality and stream biota to partial retrofit stormwater management in a suburban neighborhood. *PLoS ONE* 9(1):e85011.
Roy, A.H., A.D. Rosemond, M.J. Paul, D.S. Leigh, J.B. Wallace. 2003. Stream macroinvertebrate response to catchment urbanisation (Georgia, U.S.A.). *Freshwater Biology* 48:329–346.
Schaeffer, J.S., J.K. Bland, J. Janssen. 2012. Use of a storm water retention system for conservation of regionally endangered fishes. *Fisheries* 37(2):66–75.
Smith, R.F., L.C. Alexander, W.O. Lamp. 2009. Dispersal by terrestrial stages of stream insects in urban watersheds: A synthesis of current knowledge. *Journal of the North American Benthological Society* 28(4):1022–1037.
Smith, R.F., R.J. Hawley, M.W. Neale, G.J. Vietz, E. Diaz-Pascacio, J. Herrmann, A.C. Lovell, C. Prescott, B. Rios-Touma, B. Smith, R.M. Utz. 2016. Urban stream renovation: Incorporating societal objectives to achieve ecological improvements. *Freshwater Science* 35(1):364–379.
Stradmeyer, L., J. Höjesjö, S.W. Griffiths, D.J. Gilvear, J.D. Armstrong. 2008. Competition between brown trout and Atlantic salmon parr over pool refuges during rapid dewatering. *Journal of Fish Biology* 72(4):848–860.
Sun, Z., E. Sokolova, J.E. Brittain, S.J. Saltveit, S. Rauch, S. Meland. 2019. Impact of environmental factors on aquatic biodiversity in roadside stormwater ponds. *Scientific Reports* 9(1):5994.
Terra, B.D.F., R.M. Hughes, M.R. Francelino, F.G. Araújo. 2013. Assessment of biotic condition of Atlantic Rain Forest streams: A fish-based multimetric approach. *Ecological Indicators* 34:136–148.
Thornhill, I., L. Batty, M. Hewitt, N.R. Friberg, M.E. Ledger. 2018. The application of graph theory and percolation analysis for assessing change in the spatial configuration of pond networks. *Urban Ecosystems* 21:213–225.
Toussaint, A., O. Beauchard, T. Oberdorff, S. Brosse, S. Villéger. 2016. Worldwide freshwater fish homogenization is driven by a few widespread non-native species. *Biological Invasions* 18:1295–1304.

Uerling, C.C., M.J. Hamel, M.A. Pegg. 2019. Fish community response to habitat variables in two restored side channels of the lower Platte River, Nebraska. *River Research and Applications* 35(2):178–187.

U.S. Department of the Interior, U.S. Fish and Wildlife Service, U.S. Department of Commerce, U.S. Census Bureau. 2016. *2016 National Survey of Fishing, Hunting, and Wildlife-Associated Recreation*. Washington, DC.

Vaughan, I.P., S.J. Ormerod. 2012. Large-scale, long-term trends in British river macroinvertebrates. *Global Change Biology* 18(7):2184–2194.

Vermonden, K., R.S.E.W. Leuven, G. van der Velde, M.M. van Katwijk, J.G.M. Roelofs, A. Jan Hendriks. 2009. Urban drainage systems: An undervalued habitat for aquatic macroinvertebrates. *Biological Conservation* 142:1105–1115.

Violin, C.R., P. Cada, E.B. Sudduth, B.A. Hassett, D.L. Penrose, E.S. Bernhardt. 2011. Effects of urbanization and urban stream restoration on the physical and biological structure of stream ecosystems. *Ecological Applications* 21(6):1932–1949.

Wallace, A.M., R.G. Biastoch. 2016. Detecting changes in the benthic invertebrate community in response to increasing chloride in streams in Toronto, Canada. *Freshwater Science* 35(1):353–363.

Walsh, C.J., D.B. Booth, M.J. Burns, T.D. Fletcher, R.L. Hale, L.N. Hoang, G. Livingston, M.A. Rippy, A.H. Roy, M. Scoggins, A. Wallace. 2016. Principles for urban stormwater management to protect stream ecosystems. *Freshwater Science* 35(1):398–411.

Walsh, C.J., A.H. Roy, J.W. Feminella, P.D. Cottingham, P.M. Groffman, R.P. Morgan II. 2005. The urban stream syndrome: Current knowledge and the search for a cure. *Journal of the North American Benthological Society* 24(3):706–723.

Walsh, C.J., J.A. Webb. 2016. Interactive effects of urban stormwater drainage, land clearance, and flow regime on stream macroinvertebrate assemblages across a large metropolitan region. *Freshwater Science* 35(1):324–339.

Walter, R.C., D.J. Merritts. 2008. Natural streams and the legacy of water-powered mills. *Science* 319(5861): 299–304.

Walter, T., M. Petrere. 2007. The small-scale urban reservoir fisheries of Lago Paranoá, Brasília, DF, Brazil. *Brazilian Journal of Biology* 67(1):9–21.

Wenger, S.J., A.H. Roy, C.R. Jackson, E.S. Bernhardt, T.L. Carter, S. Filoso, C.A. Gibson, W.C. Hession, S.S. Kaushal, E. Marti, J.L. Meyer, M.A. Palmer, M.J. Paul, A.H. Purcell, A. Ramrez, A.D. Rosemond, K.A. Schofield, E.B. Sudduth, C.J. Walsh. 2009. Twenty-six key research questions in urban stream ecology: An assessment of the state of the science. *Journal of the North American Benthological Society* 28(4):1080–1098.

Yoder, C.O., E.T. Rankin, M.A. Smith, B.C. Alsdorf, D.J. Altfater, C.E. Boucher, R.J. Miltner, D.E. Mishne, R.E. Sanders, R.F. Thoma. 2005. Changes in fish assemblage status in Ohio's nonwadeable rivers and streams over two decades. Pages 399–430 in J.N. Rinne, R.M. Hughes, B. Calamusso, editors. *Historical Changes in Large River Fish Assemblages of the Americas*. American Fisheries Society Symposium 45, Bethesda, MD.

Yoder, C.O., M.A. Smith. 1998. Using fish assemblages in a state biological assessment and criteria program: Essential concepts and considerations. Pages 17–56 in T.P. Simon, editor. *Assessing the Sustainability and Biological Integrity of Water Resources Using Fish Communities*. CRC Press, Boca Raton, FL.

Zamora-Marín, J.M., C. Ilg, E. Demierre, N. Bonnet, A. Wezel, J. Robin, D. Vallod, J.F. Calvo, F.J. Oliva-Paterna, B. Oertli. 2021. Contribution of artificial waterbodies to biodiversity: A glass half empty or half full? *Science of the Total Environment* 753:141987.

Zohary, T., I. Ostrovsky. 2011. Ecological impacts of excessive water level fluctuations in stratified freshwater lakes. *Inland Waters* 1(1):47–59.

PART 4

Urban biodiversity practice: management, planning, and design for healthy communities

21
URBAN GREEN INFRASTRUCTURE, BIODIVERSITY, POVERTY, AND EQUITY

Charlie M. Shackleton

Introduction

It is now well-established that urban greening and biodiversity provide a wide range of ecosystem goods and services that contribute to urban sustainability, livability, and individual and community wellbeing (Breuste et al. 2015; Sharifi 2021). The benefits that urban dwellers obtain from urban nature span the cultural, provisioning, and regulating functions, including, but not limited to, social cohesion, sense of place, recreation, spirituality, nutrition, energy, and physical and mental health. As such, approaches for the integration and management of biodiversity in urban contexts must be viewed as a cross-cutting challenge much larger than simply conservation concerns. Thus, individual and cross-sectoral policies and strategies to promote the various dimensions of human wellbeing (such as health, security, social relations, basic needs, etc.) and urban sustainability need to integrate biodiversity benefits as a core element. This is especially so in situations where the provision of biodiversity, and hence it's associated benefits, is not equally distributed within and between urban settings.

In many cities, one sector of urban society that commonly has reduced access to biodiversity benefits is the urban poor. Frequently, poor neighbourhoods have less urban green infrastructure (UGI; synonymous with urban green spaces in many contexts) than more affluent ones, reflected in measures such as less coverage of public green spaces, greater distances to parks, fewer street trees, and smaller or no home gardens (Rigolon et al. 2018a; Venter et al. 2020). Additionally, public UGI in poorer neighbourhoods typically is of lower quality due to less investment and maintenance (Rigolon et al. 2018b; Manyani et al. 2021). The reasons for such disparities in the provision of UGI in relation to neighbourhood socioeconomic status are complex and context specific, including historical legacies, higher housing densities in poorer areas reducing the space available for public and private UGI, inadequate planning and lack of investment (or purposeful disinvestment), through to lack of awareness about the situation by some city officials and planners.

The commonly lower provision and perhaps quality of UGI in poorer neighbourhoods (Nero 2017; Nesbitt et al. 2019; Venter et al. 2020) is paradoxical because poor communities and households frequently have a greater reliance on public UGI and biodiversity than do more affluent households. This is because of the lower availability of private UGI (due to the higher density of

 DOI: 10.4324/9781003016120-25

dwellings) and the prevalence of income poverty and livelihood vulnerability. It is this intersection between UGI and urban poverty that is the focus of this chapter.

Understanding poverty

Before exploring the inequities in the distribution of biodiversity benefits due to poverty, it is first necessary to briefly define poverty for the purposes of this chapter. Despite there being numerous definitions of poverty (Mabughi and Selim 2006), it is mostly widely described or defined in income terms (Barrett 2005) because income terms are widely understood across different disciplines, can be measured with relative ease, and can be translated across contexts and scales (Angelsen and Wunder 2003). Thus, individuals or households are considered poor if they earn less than a certain internationally or nationally stipulated income (which takes into account the costs of living for the region), which is the sum total of cash and non-cash receipts over a defined period. However, there can be other measures of poverty, such as asset poverty, energy poverty, food poverty, and the like. More holistic definitions of poverty are multidimensional and more akin to the concept of general wellbeing (Savadogo et al. 2015). For example, the World Bank (2001) defines poverty as "a pronounced deprivation of wellbeing related to lack of material income or consumption, low levels of education and health, vulnerability and exposure to risk, no opportunity to be heard, and powerlessness." Thus, in this chapter, poverty is considered as a bundle of human wellbeing facets relating to income, dignity, belonging, freedom to make decisions, voice, and participation (especially in community and higher scale activities and decisions). As such it is commonly associated with high levels of vulnerability, whereby poor individuals, households, and communities have greater exposure to shocks and stresses and reduced capacity and options to respond to them. Whilst it is possible to define and identify poor households and communities by different measures, it is necessary to acknowledge that the prevalence and intensity of poverty changes in time and place in response to household, local, and higher-level dynamics (Ward 2016).

The links between urban poverty and biodiversity

Shackleton (2021a) conceptualised the direct and indirect links between local, public UGI and poverty alleviation. It was presented as a unidirectional flow of benefits from UGI that could be used to mitigate or alleviate urban poverty. However, the relationship is more complex because, as previously mentioned, the poverty status of a neighbourhood in turn influences both (1) the extent, quality, and maintenance of UGI and the biodiversity found there and (2) the types of uses and extent of reliance on local UGI. These can then become reinforcing as depicted in Figure 21.1. The rest of this chapter considers each of the different components in Figure 21.1.

Poor households and communities express greater reliance on public UGI and biodiversity

Poverty can result in greater reliance on locally available natural resources as substitutes for purchased commodities (Petersen et al. 2012; Kaoma and Shackleton 2015; Shackleton 2021b). For example, the use of locally collected firewood rather than the purchase of electricity or gas, or the collection of wild foods (e.g. fruits, vegetables, insects, fish) as a means to reduce money spent on purchased foods. Collection of such locally available biodiversity resources is not restricted to just consumptive purposes but may also include items with decorative, cultural, or spiritual significance. It is assumed that the more diverse the local environment is, the greater might be the use of

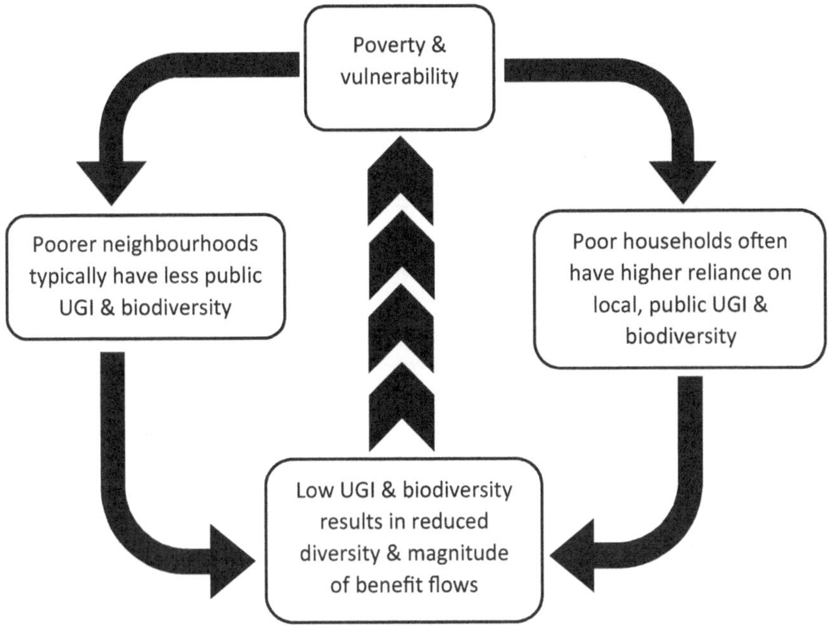

Figure 21.1 The dual and reinforcing links between poverty and vulnerability and significance of biodiversity resources from local urban green infrastructure (UGI).

biodiversity resources, in either quantity or diversity of species, but this remains to be empirically verified. Overall, the use of biodiversity resources collected from public and private UGI can be used in different ways to alleviate or avoid poverty (Figure 21.2).

Biodiversity resources for household consumption

The use of locally-sourced wild resources by urban populations is a relatively recent research topic, increasingly referred to as urban foraging (Shackleton et al. 2017b). Resources collected by urban communities include wild or traditional foods (e.g. fruits, seeds, nuts, leafy vegetables, mushrooms, honey, bushmeat), energy (e.g. firewood, charcoal, kindling), shelter/protection (e.g. wooden poles for housing and fences; grass, reeds, or palm fronds for roofing), medicines (e.g. bark, roots, seeds, leaves), fibres (e.g. lianas, palm fronds, rattans, grasses, bark), and wood for crafting or implements. Although a relatively new research field, it is clear that urban foraging is a widespread practice across Global North and Global South contexts and across communities and households of widely differing socioeconomic status (Shackleton 2021b). For example, a recent survey by Lovrić et al. (2020) of more than 17,000 people across 28 countries in Europe found that approximately one-quarter (26%) had collected one or more biodiversity resources in the previous 12 months (mostly wild fruits, nuts, and mushrooms, as well as medicinal and aromatic herbs and decorative materials), with a similar proportion (27%) reported by Short Gianotti and Hurley (2016) in central and eastern Massachusetts, USA. A higher participation rate was reported by Golos and Kaliszewski (2016) amongst urban dwellers in Poland, where 54% reported collecting wild berries, although the proportion harvested in urban and peri-urban landscapes was unclear. Considering Global South contexts, Vuola et al. (2018) revealed that 95% of urban respondents

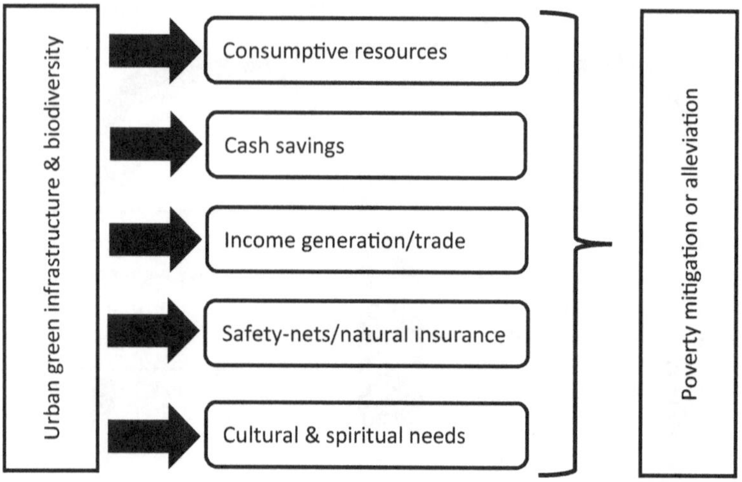

Figure 21.2 The contributions of urban green infrastructure and biodiversity to livelihoods and wellbeing of the urban poor.

in Belém, Brazil, used one or more of seven "edible" biodiversity resources (e.g. tree fruits, nuts, medicinal oils), whereas Schlesinger et al. (2015) reported that 50% of urban dwellers in six towns across five different sub-Saharan countries used one or more of nine different biodiversity resources (firewood, wild fruits, wild vegetables, wild mushrooms, honey, bushmeat, insect, fish, medicinal plants). In comparison, Joos-Vanderwalle et al. (2018) found that all (100%) urban and peri-urban households in northern Botswana reported using one or more of 19 different biodiversity resources (the same as Schlesinger et al. 2015, plus grasses, palms or reeds for weaving, roofing and brushes, along with wood for carving, household utensils, building, and fencing).

Urban biodiversity resources for cash saving

A further benefit from collecting and using local biodiversity resources from public and private UGI is that it represents a saving of the limited cash resources that define most poor households. Thus, the collection of foods, firewood, medicines, and the like for free means that poor households do not have to buy such goods and needs. For example, Kaoma and Shackleton (2015) calculated that the value of tree products (mostly fruits and firewood, but also traditional medicines and building timber) collected by poorer households in three towns in South Africa contributed a mean of 20% of household cash and non-cash income. This means that cash resources can be reserved for non-substitutable needs such as school fees or uniforms, transport to and from a place of employment, or to purchase stock to trade. Additionally, when the contributions of locally harvested tree products were incorporated into household income, the proportion of households living below the poverty line reduced by 5–10% per town (Kaoma and Shackleton 2015).

Urban biodiversity resources for income generation

Some poor urban households collect local biodiversity resources to sell in raw or processed form as a means of earning cash income. The amount of income earned depends upon a variety of factors, including the nature of the resource, season, the extent of processing or value addition, the quantities sold, the nature of the markets where it is traded, and the level of engagement in collecting and trading. Thus, it can range from relatively small amounts that provide supplementary income to a household to providing a significant proportion of the total household income (Mahonya et al. 2019). Irrespective, it is clear that many urban people around the world earn some cash income from trading in wild biodiversity products collected in urban and peri-urban spaces; see for example Imami et al. (2015) regarding medicinal and aromatic plants in Austria, or Grabbattin et al. (2011) on sweet grass basketry in South Carolina (USA), and Shackleton et al. (2006) for firewood in South Africa.

Urban biodiversity resources as safety nets during times of stress or shock

Locally available biodiversity resources can represent vital natural insurance for helping poor urban households cope during adverse times, such as retrenchment, illness or death of a breadwinner, or loss of assets due to disasters such as storms, floods, fires, theft, pandemics, or even armed conflicts. For example, Dalu and Shackleton (2018) investigated the coping responses of urban dwellers of informal housing zones in three coastal towns in South Africa to a major flooding event resulting from excessive rainfall. Locally collected timber for building and fencing, and thatch for roofing provided the equivalent of 57% of the costs of repairing or rebuilding the damaged structures, ranging between 46% to 70% in the three towns. In another example, a survey of 420 non-timber forest product traders (of which 65% were in urban settings) in five southern African countries, Weyer et al. (2018) found that almost one-third had started trading in wild biodiversity resources because of death or illness of the main income earner in the household. A particular telling example is the increased reliance on wild foods during armed conflicts, such as the siege of Sarajevo in Bosnia (Redžic 2010). During such adverse times, poor households might (1) increase their use of biodiversity resources as a means to save scarce cash, (2) substitute normally purchased commodities with biodiversity resources they can collect for free, or (3) commence selling one or more biodiversity resources as a means to earn cash, until such time they are able to adapt to or recover from the original shock.

Urban biodiversity resources for cultural and spiritual needs

Viewing poverty as a multidimensional concept (and experiential reality) means that some households or communities might be considered as poor if they experience limited access to biodiversity valued for cultural or spiritual needs and practices. For example, many indigenous worldviews or cosmologies revere nature, and individuals need access to nature generally or to specific species or sites, for one or both of their physical and spiritual or psychological wellbeing or rites of passage (Kepe et al. 2015; Cocks et al. 2016). At a species level, many urban people make use of a range of specific species for culturally mandated spiritual needs and beliefs, such as the warding off of perceived evil influences or bad luck or invoking intercedence of the ancestors (Cocks and Møller 2002; Fraser et al. 2016). Such culturally driven uses of biodiversity are not restricted to the urban poor, but it is the poor who most commonly experience constraints in accessing the necessary species or sites due to their income poverty and the generally lower extent of UGI in poor areas (Astell-Burt et al. 2014; Mayan Huerta 2022).

Poor neighbourhoods often have less UGI and biodiversity

The greater reliance of poor households and communities on local biodiversity resources can be compromised or constrained by the limited extent and quality of UGI in many poor neighbourhoods.

This is interpreted to be a consequence of one or two reasons. The first is the generally greater housing densities in poor communities than in more affluent ones, resulting in less space available for UGI (Gopal and Nagendra 2014; Gwedla 2020). The typically higher housing densities may be planned for or be a result of "vacant" spaces becoming occupied by informal housing or businesses (Douglas 2018). The second is some degree of neglect or lack of awareness by authorities regarding the needs of the poor for access to UGI in home yards and/or proximate, accessible public spaces. In terms of the first, there are multiple studies indicating that poor communities typically have less UGI per capita in both Global North (Astell-Burt et al. 2014; Kabisch and Haase 2014; Li et al. 2015) and Global South settings (Nero 2017; Venter et al. 2020). This includes private garden space, public parks or other green spaces, and street trees. In South Africa, several studies have shown smaller yards in poorer areas, fewer, smaller, and poorly maintained public parks and no or few street trees, and all being markedly less than found in the more affluent areas (McConnachie and Shackleton 2010; Kuruneri-Chitepo and Shackleton 2011; Shackleton and Blair 2013; Gwedla and Shackleton 2017). It is not unusual for poverty to intersect with the status of being an ethnic or racial minority in the country under consideration. For example, the current extent of tree cover in areas historically zoned for racial minorities in 37 cities across the USA is almost half that currently found in neighbourhoods promoted for white occupation (Locke et al. 2021). In two cities in Israel, Omer and Or (2005) found that the Arab minority had significantly lower access to public green space than the Jewish majority.

The disparities in UGI according to socioeconomic status are not restricted to just quantity of public UGI, but also to quality. For example, in South Africa, McConnachie et al. (2008) found that mean park size was smaller in poor towns and neighbourhoods, and most of the woody species were invasive aliens, which was not the case in the affluent areas. Residents in poor neighbourhoods are more likely to complain than those in more affluent areas about lack of maintenance of formal and informal green spaces, and lack of consultation about what sorts of green spaces, amenities, and species they would like and where (Shackleton and Blair 2013; Shackleton and Njwaxu 2021), illustrating the lack of being heard as a defining characteristic of poverty (Narayan 2000).

Despite the constraints of space and financial resources, some poor communities and households still seek ways to nurture and access biodiversity resources of value to them. In private spaces, households maintain species of consumptive, cultural, or aesthetic value to them in containers of various materials and sizes (Gopal and Nagendra 2014). These may be kept on windowsills, in doorways, and on roofs. Alternatively, valuable trees, shrubs, and herbs may be planted and tended in minute yards if space is available. For example, 73% of households in informal housing in two towns in northern South Africa had planted a tree in the space adjacent to their home (Shackleton et al. 2015). In the slums of Bangalore, India, Gopal and Nagendra (2014) recorded 46 different tree species and 95 shrub and herb species. Close to half of the tree species had one or more medicinal uses, and a third of the tree species provided edible fruits.

Food and nutritional security can be a significant concern amongst income-poor, urban populations (e.g. Chakona and Shackleton 2017; Bowden et al. 2018; Boonyabancha et al. 2019). Some contribution can be made from biodiversity resources by either cultivating food through engagement in urban agriculture or regularly collecting from urban spaces through urban foraging (e.g. Gockowski et al. 2003; Garekae and Shackleton 2020). Urban agriculture can be a particularly important strategy for some poor households (Zezza and Tasciotti 2010; Orsini et al. 2013). In home spaces, approaches range from growing food in containers and on rooftops to small beds in the yard. When there is insufficient private space, some households may cultivate ostensibly unused spaces in the neighbourhood or wider urban matrix, such as along riparian fringes or sidewalks, servitudes, institutional grounds (e.g. schools, clinics), and undeveloped lands. It might be done individually, or as part of a formal or informal collective. Urban agriculture adds to managed UGI in cities, but is frequently dominated by non-native species, particularly if focused on food production. The magnitude of the contribution of urban agriculture to food security and cash saving amongst poor households is debated (Zezza and Tasciotti 2010; Badami and Ramankutty 2015) and ultimately is dependent on a host of local factors (Shackleton 2021a), such as how much land and labour are available, whether rainfed or irrigated, the cost of inputs (e.g. seeds, fertilisers, pesticides), the local cost of equivalent food, and security of access and usufruct of the land cultivated. For example, Mupeta et al. (2020) calculated that urban agriculture contributed between 13% and 19% of household income (cash and non-cash) in Zambia. Nevertheless, even in situations where the regular contributions from urban agriculture might be limited, they can be ramped up as a temporary coping strategy or safety-net during times of household or community stress and can be a crucial cash saving strategy, as illustrated by Masauvre (2016) in Harare, Zimbabwe, after economic structural adjustment programmes devalued the local currency and employment prospects.

Regular collection of wild foods such as mushrooms, fruits, nuts, honey, tubers, and leafy vegetables add to dietary and nutritional diversity, especially with respect to micronutrients and vitamins. However, no estimates have been made of the proportion that foraged foods contribute to the total food intake of urban households that engage in such practices. The same applies with respect to quantification of the nutritional benefits of foraged foods. However, some urban citizens forage for food several times a week (Garekae and Shackleton 2020), which is likely to indicate a degree of reliance on foraged foods as a substantive part of their diet.

Livestock husbandry is a particular form of urban agriculture that is most common amongst the urban poor in particular regions of the world. For example, 34% of plots surveyed in Belém, Brazil are used for urban livestock (Madaleno 2000) and 20% of households in Nakuru, Kenya keep large stock (Foeken and Owuor 2008). In a survey of six cities in five Sub-Saharan countries, Shackleton et al. (2020) reported that on average 25% of urban and 38% of peri-urban households kept large stock and/or poultry. Livestock husbandry in towns and cities of the Global North generally involves smaller species, such as rabbits and poultry (e.g. McClintock et al. 2014). Despite being widespread, the relationships between animal husbandry and UGI or biodiversity have hardly been explored. Urban livestock require food resources, which may be obtained via free-ranging, or the fodder might be harvested by the owner or herder and transported to the animals kept in some enclosure. But there has been no comprehensive or systematic examination of the effects of urban livestock on urban biodiversity, such as what forage species are favoured by the animals or the herders and what the impacts of domestic animals are on urban biodiversity from the direct grazing or browsing, as well as through trampling and compaction. For example, the depredations of urban livestock can be a significant factor in reducing the survival of trees planted in parks or along streets (Richardson and Shackleton 2014; Shackleton et al. 2017a), and deterring some citizens away from particular urban green spaces (Shackleton and Njwaxu 2021), and yet the dung from livestock might be used as a fuel for cooking and space heating or added as manure to food plots (e.g. Paudel et al. 2018).

Despite the many innovative ways in which the urban poor access, nurture, and use specific spaces, species, and products for their consumptive and non-consumptive purposes, many households experience shortages. They must then employ a range of coping strategies, shaped by the assets at their disposal and the nature of the resource in short supply. For example, a shortage of firewood stocks means that poor households in Mozambique either purchased it (or substitutes) or adopted one or more coping strategies such as cooking fewer meals in the day or the cooking of lower quality foods that could be cooked faster (Brouwer et al. 1989). All these may cascade through to impoverished diets in quantity or quality. Or in New Zealand, Maori elders lamented the implications for local culture by the constrained supply of what they regarded as crucial species for dying, weaving, and carving of cultural artifacts (Wehi and Wehi 2010).

Poor neighbourhoods are generally more vulnerable to the effects of urban ecosystem disservices

The preceding sections have focused on the benefits that the urban poor may obtain from urban green spaces and the biodiversity therein and that the inequitable distribution of UGI means that, in many situations, the urban poor do not benefit optimally or equitably from urban biodiversity. These inequities may be further magnified when urban ecosystem disservices (UED) are taken into account. Ecosystem disservices are viewed as "ecosystem generated functions, processes and attributes that result in perceived or actual negative impacts on human wellbeing" (Shackleton et al. 2016). In urban settings, they span a wide range of ecosystem functions and products such as allergens from urban trees, waterborne or insect-vector diseases, damage to infrastructure from falling branches, or roots cracking paving (Lyytimäki et al. 2008; Von Döhren and Haase 2015). However, the effects of UED vary markedly depending on the vulnerability of the affected individual, household, or community (Davoren and Shackleton 2021). Vulnerability is a complex concept and needs to be disaggregated into the constituent components of risk, sensitivity, and coping capacity (Adger 2006).

Risk considers the likelihood of a particular UED occurring; for example, drought risk, fire risk, or disease risk. At a community level, poor communities and those in informal housing are more likely to live in risk-prone situations such as on steep slopes or floodplains (O'Hare and Rivas 2005; Dalu et al. 2018). And as already observed, poor neighbourhoods also typically have lower UGI (Venter et al. 2020) to mitigate against some UED, such as high ambient air temperatures or mudslides.

Sensitivity to a specific UED is dependent on the nature, timing, and magnitude of the UED, as well as the susceptibility of the affected household or community. For example, many poor households suffer from lower general health than affluent ones (e.g. Reijneveld et al. 2000; Habib et al. 2019) due to a range of potential causes, and thus when an infectious or water-borne disease emerges, the higher impacts on poor households are partially due to their already poorer general health.

The capacity to respond to UED is also a function of the nature, timing, and magnitude of the UED and also the assets and capabilities of the affected person, household, or community. But for a UED of equivalent nature and magnitude, poor households generally have fewer options on how to respond and fewer assets to mobilise as a coping response than do better-off households or communities. Because of this, they may well also experience a higher frequency of UED.

Conclusions: recognising the complex links between urban biodiversity and poverty

This chapter has demonstrated the many and intimate uses and links that poor households have with UGI and the biodiversity found there. Similar links undoubtedly also exist for more affluent households, but the degree or intensity of reliance is likely to be different. This is because (1) poor households lack the resources to purchase substitutes when the supply of a particular biodiversity resource is constrained, and (2) they have less access to private UGI. However, despite potentially higher reliance on local public UGI by poor households, the availability and quality of public UGI is frequently lower in poor neighbourhoods than in less poor ones.

The higher reliance on, but lower supply of, some biodiversity resources can in some contexts lead to degradation of UGI and excessive harvesting of specific species in urban and peri-urban spaces, for example, the decline in firewood stocks around Gaborone in Botswana (Nkambwe and Sekhwela 2006), or the higher rate of debarking and cutting of *Warburgia ugandensis* and *Elaeodendron buchananii* for medicinal purposes near low-income settlements in Nairobi, Kenya (Furukawa et al. 2016). The risk of overharvesting may be exacerbated by land use change associated with development of formal or informal infrastructure on UGI. Such changes require that users and harvesters constantly reassess the local biodiversity resources and may need to renegotiate or reconfigure their access as land use, access rights, or ownership change (Grabbattin et al. 2011; Garekae and Shackleton 2021). Thus, poverty can shape particular needs and uses of urban biodiversity, but in a reciprocal fashion, the constrained availability or lack of particular species or biodiversity resources can exacerbate poverty.

The most meaningful intervention that urban planners and officials can undertake is to address the commonly occurring inequities in the distribution, abundance, and quality of UGI that can exacerbate urban poverty in all its dimensions. Precisely how much UGI, what types, and what species are preferred need to be determined through listening to the voices of the poor. Secondly, green space managers need to incorporate local needs and preferences into management plans and actions. This should include planting of important food, medicinal, or cultural species in parks, along streets or road verges, and medians, such as done with sweet grass (*Muhlenbergia sericea*) in South Carolina (USA), or urban food forests in several cities in Brazil (Vannozzi Brito and Borelli 2020). The growing concept of "nature-based solutions" has relevance here (i.e. working with nature to reduce or eliminate poverty and inequity in all its manifestations).

Key companion papers

Kaoma, H., C.M. Shackleton. 2015. The direct-use value of urban tree non-timber forest products to household income in poorer suburbs in South African towns. *Forest Policy and Economics* 61:104–112.

Rigolon, A., H.E. Browning, K., Lee, S. Shin. 2018. Access to urban green space in cities of the global south: A systematic literature review. *Urban Science* 2(3):67.

Shackleton, C.M. 2021a. Provisioning services in Global South cities. Pages 203–226 in C.M. Shackleton, S.S. Cilliers, E. Davoren, M. du Toit, editors. *Urban Ecology in the Global South*. Springer, Cham, Switzerland.

Shackleton, C.M. 2021b. Urban green infrastructure for poverty alleviation: Evidence synthesis and conceptual considerations. *Frontiers in Sustainable Cities* 3:710549.

Shackleton, C.M., A. Drescher, J. Schlesinger. 2020. Urbanisation reshapes gendered engagement in land-based livelihood activities in mid-sized African towns. *World Development* 103:104946.

Shackleton, C.M., P.T. Hurley, A. Dahlberg, M.R. Emery, H. Nagendra. 2017. Urban foraging: A ubiquitous human practice but overlooked by urban planners, policy and research. *Sustainability* 9(10):1884.

Acknowledgements

I was supported to write this chapter by the South African Research Chairs Initiative of the Dept of Science and Technology and the National Research Foundation of South Africa (grant no. 84379). Any opinion, finding, conclusion or recommendation expressed in this material is that of the author.

References

Adger, N. 2006. Vulnerability. *Global Environmental Change* 16:268–281.

Angelsen, A., S. Wunder. 2003. *Exploring the Forest – Poverty Link: Key Concepts, Issues and Research Implications*. CIFOR Occasional Paper 40, Center for International Forestry Research (CIFOR), Jakarta, Indonesia.

Astell-Burt, T., X. Feng, S. Mavoa, H.M. Badland, B. Giles-Corti. 2014. Do low-income neighbourhoods have the least green space? A cross-sectional study of Australia's most populous cities. *BMC Public Health* 14:292.

Badami, N.G., A. Ramankutty. 2015. Urban agriculture and food security: A critique based on an assessment of urban land constraints. *Global Food Security* 4:8–15.

Barrett, C.B. 2005. Rural poverty dynamics: Development policy implications. *Agricultural Economics* 32(s1):45–60.

Boonyabancha, S., T. Kerr, L. Joshi, C. Tacoli. 2019. How the urban poor define and measure food security in Cambodia and Nepal. *Environmental and Urbanization* 31(2):517–532.

Bowden, R., E. Even-Zahav, C. Kelly. 2018. Innovative food procurement strategies of women living in Khayelitsha, Cape Town. *Urban Forum* 29(3):315–332.

Breuste, J., M. Artmann, J. Li, M. Xie. 2015. Special issue on green infrastructure for urban sustainability. *Journal of Urban Planning and Development* 141(3):A2015001.

Brouwer, I.D., L.M. Nederveen, A.P. den Hartog, A.H. Vlasveld. 1989. Nutritional impacts of an increasing fuelwood shortage in rural households in developing countries. *Progress in Food and Nutrition Science* 13(3–4):349–361.

Chakona, G., C.M. Shackleton. 2017. Voices of the hungry: A qualitative measure of household food access and food insecurity in South Africa. *Agriculture and Food Security* 6(1):66.

Cocks, M., J. Alexander, L. Mogano, S. Vetter. 2016. Ways of belonging: Meanings of 'nature' among Xhosa-speaking township residents in South Africa. *Journal of Ethnobiology* 36(4):820–841.

Cocks, M., C. Møller. 2002. Use of indigenous and indigenised medicines to enhance personal well-being: A South African case study. *Social Science and Medicine* 54(4):387–397.

Dalu, M.T., C.M. Shackleton. 2018. The potential use of natural resources in urban informal settlements as substitutes for financial capital during flooding emergencies. *Physics and Chemistry of the Earth, Parts A/B/C* 104:18–27.

Dalu, M.T., C.M. Shackleton, T. Dalu. 2018. Influence of land cover, proximity to streams and household topographical location on flooding impact in the Eastern Cape, South Africa. *International Journal of Disaster Risk Reduction* 28:481–490.

Davoren, E., C.M. Shackleton. 2021. Ecosystem disservices in Global South cities. Pages 265–292 in C.M. Shackleton, S.S. Cilliers, E. Davoren, M. du Toit, editors. *Urban Ecology in the Global South*. Springer, Cham, Switzerland.

Douglas, I. 2018. The challenge of urban poverty for the use of green infrastructure on floodplains and wetlands to reduce flood impacts in intertropical Africa. *Landscape and Urban Planning* 180:262–272.

Foeken, D.W., S.O. Owuor. 2008. Farming as a livelihood source for the urban poor of Nakuru, Kenya. *Geo-Forum* 39:1978–1990.

Fraser, J.A., M. Diabaté, W. Narmah, P. Beavogui, K. Guilavogui, H. De Foresta, A.B. Junqueira. 2016. Cultural valuation and biodiversity conservation in the Upper Guinea forest, West Africa. *Ecology and Society* 21(3):36.

Furukawa, T., S.K. Kiboi, P.B. Mutiso, K. Fujiwara. 2016. Multiple use patterns of medicinal trees in an urban forest in Nairobi, Kenya. *Urban Forestry & Urban Greening* 18:34–40.

Garekae, H., C.M. Shackleton. 2020. Urban foraging of wild plants in two medium-sized South African towns: People, perceptions and practices. *Urban Forestry & Urban Greening* 49:126581.

Garekae, H., C.M. Shackleton. 2021. Knowledge of formal and informal regulations affecting wild plant foraging practices in urban spaces in South Africa. *Society and Natural Resources* 34:1546–1565.

Gockowski, J., J. Mabazo'o, G. Mbah, T.F. Toulende. 2003. African traditional leafy vegetables and the urban and peri-urban poor. *Food Policy* 28(3):221–235.

Golos, P., A. Kaliszewski. 2016. Economic importance of selected non-wood forest products in Poland. *Sylwan* 160:336–343.

Gopal, D., H. Nagendra. 2014. Vegetation in Bangalore's slums: Boosting livelihoods, wellbeing and social capital. *Sustainability* 6:2459–2473.

Grabbattin, B., P.T. Hurley, A. Halfacre. 2011. 'I still have the old tradition': The co-production of sweetgrass basketry and coastal development. *Geoforum* 42(6):638–649.

Gwedla, N. 2020. *Barriers to, and Enablers of Tree Planting in Low-Cost Housing Areas: Lessons from Participatory Learning Processes in South Africa*. PhD Thesis, Rhodes University, Makhanda, South Africa.

Gwedla, N., C.M. Shackleton. 2017. Population size and development history determine street tree distribution and composition within and between Eastern Cape towns, South Africa. *Urban Forestry & Urban Greening* 25:11–18.

Habib, R.R., S. Hojeij, K. Elzein, J. Chaaban. 2019. Self-rated health and relative socioeconomic deprivation in the Palestinian refugee communities of Lebanon. *Journal of Immigrant and Minority Health* 21(6):1257–1265.

Imami, D., A. Ibraliu, N. Fasillia, N. Gruda, E. Skreli. 2015. Analysis of the medicinal and aromatic plants value chain in Albania. *Gesunde Pflanzen* 67(4):155–164.

Joos-Vanderwalle, S., R. Wynberg, K.A. Alexander. 2018. Dependencies on natural resources in transitioning urban centres of northern Botswana. *Ecosystem Services* 30(C):342–349.

Kabisch, N., D. Haase. 2014. Green justice or just green? Provision of urban green spaces in Berlin, Germany. *Landscape and Urban Planning* 122:129–139.

Kaoma, H., C.M. Shackleton. 2015. The direct-use value of urban tree non-timber forest products to household income in poorer suburbs in South African towns. *Forest Policy and Economics* 61:104–112.

Kepe, T., G. McGregor, P. Irvine. 2015. Rights of 'passage' and contested land use: Gendered conflict over urban space during ritual performance in South Africa. *Applied Geography* 57:91–99.

Kuruneri-Chitepo, C., C.M. Shackleton. 2011. The distribution, abundance and composition of street trees in selected towns of the Eastern Cape, South Africa. *Urban Forestry & Urban Greening* 10(3):247–254.

Li, X., C. Zhang, W. Li, Y.A. Kuzovkina, D. Weiner. 2015. Who lives in greener neighbourhoods? The distribution of street greenery and its association with residents' socioeconomic conditions in Hartford, Connecticut, USA. *Urban Forestry & Urban Greening* 14:751–759.

Locke, D.H., B. Hall, J.M. Grove, S.T.A. Pickett, L.A. Ogden, C. Aoki, C.G. Boone, J.P.M. O'Neil-Dunne. 2021. Residential housing segregation and urban tree canopy in 37 US cities. *NPJ Urban Sustainability* 1:15.

Lovrić, M., R. Da Re, E. Vidale, I. Prokofieva, J. Wong, D. Pettenella, P.J. Verkerk, R. Mavsar. 2020. Non-wood forest products in Europe: A quantitative overview. *Forest Policy and Economics* 116:102175.

Lyytimäki, J., L.K. Petersen, B. Normander, P. Bezák. 2008. Nature as a nuisance? Ecosystem services and disservices to urban lifestyle. *Environmental Science* 5(3):161–172.

Mabughi, N., T. Selim. 2006. Poverty as social deprivation: A survey. *Review of Social Economy* 62(2):181–204.

Madaleno, I.M. 2000. Urban agriculture in Belém, Brazil. *Cities* 17(1):73–77.

Mahonya, S., C.M. Shackleton, K. Schreckenberg. 2019. Non-timber forest product use and market chains along a deforestation gradient in southwest Malawi. *Frontiers in Forests and Global Change* 2:71.

Manyani, A., C.M. Shackleton, M.L. Cocks. 2021. Attitudes and preferences towards elements of formal and informal public green spaces in two South African towns. *Landscape and Urban Planning* 214:104147.

Masauvre, S. 2016. Coping with food poverty in cities: The case of urban agriculture in Glen Norah township in Harare. *Renewable Agriculture and Food Systems* 31:202–213.

Mayan Huerta, C. 2022. Rethinking the distribution of urban green spaces in Mexico City: Lessons from the COVID-19 outbreak. *Urban Forestry & Urban Greening* 70:127525.

McClintock, N., E. Pallana, H. Wooton. 2014. Urban livestock ownership, management and regulation in the United States: An exploratory survey and research agenda. *Land Use Policy* 38:426–440.

McConnachie, M.M., C.M. Shackleton. 2010. Public green space inequality in small towns in South Africa. *Habitat International* 34(2):244–248.

McConnachie, M.M., C.M. Shackleton, G.K. McGregor. 2008. Extent of public green space and alien species in ten small towns in the thicket biome, South Africa. *Urban Forestry & Urban Greening* 7(1):1–13.

Mupeta, M., E. Kuntashula, T. Kalinda. 2020. Impact of urban agriculture on household income in Zambia: An economic analysis. *Asian Journal of Agriculture and Rural Development* 10(2):550–562.

Narayan, D. 2000. Poverty is powerlessness and voicelessness. *Finance and Development* 37(4):18–21.

Nero, B.F. 2017. Urban green space dynamics and socio-environmental inequity: Multi-resolution and spatiotemporal data analysis of Kumasi, Ghana. *International Journal of Remote Sensing* 38(23):6993–7020.

Nesbitt, L., M.J. Meitner, C. Girling, S.R.J. Sheppard, Y. Lu. 2019. Who has access to urban vegetation? A spatial analysis of distributional green equity in 10 US cities. *Landscape and Urban Planning* 181:51–79.

Nkambwe, M., M.B.M. Sekhwela. 2006. Utilization characteristics and importance of woody biomass resources on the rural-urban fringe in Botswana. *Environmental Management* 37(2):281–296.

O'Hare, G., S. Rivas. 2005. The landslide hazard and human vulnerability in La Paz, Bolivia. *Geographical Journal* 171(3):239–258.

Omer, I., U. Or. 2005. Distributive environmental justice in the city: Differential access in two mixed Israeli cities. *Tijdschrift voor Economische en Sociale Geografie* 96(4):433–443.

Orsini, F., R. Kahane, R. Nono-Womdin, G. Gianquinto. 2013. Urban agriculture in the developing world: A review. *Agronomy for Sustainable Development* 33(4):695–720.

Paudel, U., U. Khatri, K.P. Pant. 2018. Understanding the determinants of household cooking fuel choice in Afghanistan: A multinomial logit estimation. *Energy* 156:55–62.

Petersen, L.M., E.J. Moll, R. Collins, M.T. Hockings. 2012. Development of a compendium of local, wild-harvested species used in the informal economy trade, Cape Town, South Africa. *Ecology and Society* 17(2):26.

Redžic, S. 2010. Use of wild and semi-wild edible plants in nutrition and survival of people in 1430 days of siege of Sarajevo during the war in Bosnia and Herzegovina (1992–1995). *Collegium Anthropology* 34:551–570.

Reijneveld, S., R. Verheij, D.H. de Bakker. 2000. The impact of area deprivation on differences in health: Does the choice of the geographical classification matter? *Journal of Epidemiology and Community Health* 54(4):306–313.

Richardson, E., C.M. Shackleton. 2014. The extent, causes and local perceptions of street tree damage in small towns in the Eastern Cape, South Africa. *Urban Forestry & Urban Greening* 13:425–432.

Rigolon, A., M.H.E. Browning, K. Lee, S. Shin. 2018a. Access to urban green space in cities of the global south: A systematic literature review. *Urban Science* 2.

Rigolon, A., M.H.E., Browning, V. Jennings. 2018b. Inequities in the quality of urban park systems: An environmental justice investigation of cities in the United States. *Landscape and Urban Planning* 178:156–169.

Savadogo, G., A. Souarès, A. Sié, D. Parmar, G. Bibeau, R. Sauerborn. 2015. Using a community-based definition of poverty for targeting poor households for premium subsidies in the context of a community health insurance in Burkina Faso. *BMC Public Health* 15:84.

Schlesinger, J., A. Drescher, C.M. Shackleton. 2015. Socio-spatial dynamics in the use of wild natural resources: Evidence from six rapidly growing medium-sized cities in Africa. *Applied Geography* 56:107–115.

Shackleton, C.M. 2021a. Urban green infrastructure for poverty alleviation: Evidence synthesis and conceptual considerations. *Frontiers in Sustainable Cities* 3:710549.

Shackleton, C.M. 2021b. Provisioning services in Global South cities. Pages 203–226 in C.M. Shackleton, S.S. Cilliers, E. Davoren, M. du Toit, editors. *Urban Ecology in the Global South*. Springer, Cham, Switzerland.

Shackleton, C.M., A. Blair. 2013. Perceptions and use of public green space is influenced by its relative abundance in two small towns in South Africa. *Landscape and Urban Planning* 113:104–112.

Shackleton, C.M., A. Drescher, J. Schlesinger. 2020. Urbanisation reshapes gendered engagement in land-based livelihood activities in mid-sized African towns. *World Development* 103:104946.

Shackleton, C.M., J. Guild, B. Bromham, S. Impey, M. Jarrett, M. Ngubane, K. Steijl. 2017a. How compatible are urban livestock and urban green spaces and trees? An assessment in a medium-sized South African town. *International Journal of Sustainable Urban Development* 9:243–252.

Shackleton, C.M., P.T. Hurley, A. Dahlberg, M.R. Emery, H. Nagendra. 2017b. Urban foraging: A ubiquitous human practice but overlooked by urban planners, policy and research. *Sustainability* 9(10):1884.

Shackleton, C.M., M. McConnachie, M.I. Chauke, J. Mentz, F. Sutherland, J. Gambiza, R. Jones. 2006. Urban fuelwood demand and markets in a small town in South Africa: Livelihood vulnerability and alien plant control. *International Journal of Sustainable Development and World Ecology* 13:481–491.

Shackleton, C.M., A. Njwaxu. 2021. Does the absence of community involvement underpin the demise of urban neighbourhood parks in the Eastern Cape, South Africa? *Landscape and Urban Planning* 207:104006.

Shackleton, C.M., S. Ruwanza, G.K. Sinasson Sanni, S. Bennett, P. De Lacy, R. Modipa, N. Mtati, M. Sachikonye, G. Thondhlana. 2016. Unpacking Pandora's box: Understanding and categorising ecosystem disservices for environmental management and human wellbeing. *Ecosystems* 19(4):587–600.

Shackleton, S., A. Chinyimba, P. Hebinck, C.M. Shackleton, H. Kaoma. 2015. Multiple benefits and values of trees in urban landscapes in two towns in northern South Africa. *Landscape and Urban Planning* 136:76–86.

Sharifi, A. 2021. Urban sustainability assessment: An overview and bibliometric analysis. *Ecological Indicators* 121:107102.

Short Gianotti, A.G., P.T. Hurley. 2016. Gathering plants and fungi along the urban-rural gradient: Uncovering differences in the attitudes and practices among urban, suburban and rural landowners. *Land Use Policy* 57:555–563.

Vannozzi Brito, V., S. Borelli. 2020. Urban food forestry and its role to increase food security: A Brazilian overview and its potentialities. *Urban Forestry & Urban Greening* 56:126835.

Venter, Z.S., C.M. Shackleton, F. van Staden, O. Selomane, V.A. Masterson 2020. Green Apartheid: Urban green space remains unequally distributed across income and race geographies in South Africa. *Landscape and Urban Planning* 203:103889.
Von Döhren, P., D. Haase. 2015. Ecosystem disservices research: A review of the state of the art with a focus on cities. *Ecological Indicators* 52:490–497.
Vuola, M., S.C. Bauch, E.O. Sills 2018. The regional market for non-timber forest products. *Desenvolvimento e Meio Ambiente* 48:498–511.
Ward, P.S. 2016. Transient poverty, poverty dynamics and vulnerability to poverty: An empirical analysis using a balanced panel from rural China. *World Development* 78:541–553.
Wehi, P.M., W.L. Wehi. 2010. Traditional plant harvesting in contemporary fragmented and urban landscapes. *Conservation Biology* 24(2):594–604.
Weyer, D., C.M. Shackleton, Y.O. Adam. 2018. HIV/AIDS and other household shocks as catalysts of local commercialisation of non-timber forest products. *Development Policy Review* 36(S1):O285–O301.
World Bank. 2001. *Poverty and Inequality Analysis.* World Bank, New York.
Zezza, A., L. Tasciotti. 2010. Urban agriculture, poverty, and food security: Empirical evidence from a sample of developing countries. *Food Policy* 35:265–273.

22

MANAGING VEGETATION FOR URBAN BIODIVERSITY

Jun Yang, Sonja Knapp, and Wayne Zipperer

Introduction

Urban vegetation harbors abundant plant species and urban wildlife. This chapter explores the ways to manage urban vegetation to preserve and enhance urban biodiversity. Management actions can be taken to conserve plant biodiversity in natural and semi-natural vegetation and cultivated vegetation guided by knowledge of forces driving the vegetation dynamics in urban areas. Urban vegetation can also be managed to support urban wildlife biodiversity. A key lesson from existing projects is that long-term planning and management consistency are necessary to achieve desired results. Furthermore, regular monitoring and assessment should be in place to ensure that vegetation management works for urban biodiversity.

The shaping forces of biodiversity in urban vegetation

Urban vegetation is the total assemblage of plants within and on the perimeter of cities and towns (Carne 1994). Urban vegetation serves multiple functions, including critical habitats for threatened and endangered species (Ives et al. 2016), food and shelter for wildlife species (Hostetler et al. 2011), and ecological structure for producing ecosystem functions (Bolund and Hunhammar 1999). Active management and monitoring are required to maintain these functions. However, urban vegetation management is not a simple task since myriad natural and anthropogenic forces influence urban vegetation (Knapp and Zipperer, this volume). These forces include regional climate and biogeography, human-mediated biotic interchange, urban form and development history, socioeconomic and cultural influences, local human facilitation, and species interactions (Aronson et al. 2014; Williams et al. 2015; Aronson et al. 2016). The interaction effects of these forces shape the composition, abundance, and distribution of plants that form urban vegetation and wildlife that depend on urban vegetation. Management actions based on a sound understanding of those forces can preserve, restore, and increase urban vegetation biodiversity and the wildlife that relies on it.

DOI: 10.4324/9781003016120-26

Managing urban vegetation for biodiversity

The overall goal and strategies

Urban vegetation management should improve the biodiversity of plants and animals as a system rather than focusing on a particular species. The concept of "Biodiversity quality" (Jalkanen et al. 2020), which evaluates biodiversity comprehensively from the aspects of species richness, biomass, population density, evenness, rarity, support for specialist species, and regional representativeness of the species assemblages in an area, provides an anchoring point for setting up the overall goal. Management strategies include five principal components: (1) implement citywide planning; (2) preserve and restore existing natural and semi-natural vegetation; (3) create an ecological network of corridors; (4) reduce degradation of existing vegetation; and (5) be creative in finding solutions.

First, citywide plans need to incorporate all types of urban nature, including cultivated, and novel habitats that might not be in the focus of classic nature conservation approaches. This plan includes biodiversity-friendly and cost-effective vegetation management in cultivated green spaces such as yards or parks (Friedman 2021; Aronson et al. 2017), for example, by reducing mowing intensity. Beyond, citywide plans should involve various stakeholders and decision-makers to include public and privately owned lands across the metropolitan area. A comprehensive approach to citywide management reduces risks of piecemeal, uncoordinated, and ineffective approaches, particularly for biophysical systems that may occur with local-scale planning (Bush et al. 2020). Tools such as the Institutional Analysis and Development Framework (IAD) can be used to understand the roles of stakeholders in inhibiting or supporting biodiversity conservation in urban vegetation to make better plans (MacKenzie and Gibbons 2019). Nevertheless, it should be noted that having a citywide biodiversity conservation plan itself is meaningless if no savvy political and governance measures exist to enact funding and policies that support the implementation of the plan.

Second, preserving natural or semi-natural vegetation needs to be made a top priority (Friedman 2021). Natural or semi-natural vegetation has a higher biodiversity value than cultivated vegetation (Barrico et al. 2018). They also support a higher diversity of insects, birds, and other wildlife (Müller et al. 2018).

Third, areas occupied by urban vegetation needs to be increased in cities. As a first step, cities should know the amount of existing urban vegetation. Data on vegetation types and extant at fine resolution (<10 m) can be collected through field surveys or interpretation of remotely sensed data. Furthermore, urban vegetation should be connected through a network of corridors that spread across urban-rural landscapes. This network can help decrease the detrimental effects of the small size of many inner-urban vegetation patches on species. It has been found that the impact of patch size and corridors on biodiversity is more important than vegetation structure and local management. Large areas of habitats (>50 ha) are essential for preventing rapid loss of area-sensitive species (Beninde et al. 2015).

Fourth, policies should be created to guide the design and construction of developments to reduce the impact on nearby natural vegetation and enhance the biodiversity of vegetation in developed lands (Hostetler et al. 2011). For example, biodiversity-friendly designing practices such as preserving vegetation on the development sites or using native species in planting should be encouraged.

Finally, the complex socio-ecological processes involved in urban vegetation management demand creative solutions. New ideas should be encouraged, tested, and duplicated if proved effective. Temporary conservation, which aims to maintain a spatiotemporal mosaic of different successional stages ranging from pioneer to pre-forest communities, may help cope with the highly dynamic urban landscapes (Kattwinkel et al. 2011). Another new idea is passive rewilding, that is, to allow for spontaneous ecological processes in urban landscapes, which can potentially reduce management intensity and achieve higher biodiversity in urban vegetation (Müller et al. 2018).

Managing natural and semi-natural vegetation

Natural and semi-natural vegetation include patches of remnant forests and other original vegetation types unaltered or moderately modified by human activities. They harbor higher shares of plant taxa with high conservation and ecological values than vegetation more strongly affected by humans (Barrico et al. 2018; Kowarik and von der Lippe 2018). Spontaneous species are the main components of natural and semi-natural vegetation. These areas also support a higher diversity of invertebrate and bird species than cultivated vegetation (Müller et al. 2018). Consequently, maintaining the structural and functional integrity of natural and semi-natural vegetation should be prioritized in urban vegetation management.

Intensive anthropogenic activities in urban landscapes frequently degrade the structure and function of natural and semi-natural habitats. Restoration is often required to revive the ecological integrity of native habitats (Johnson and Handel 2016). Restoration activities represent a continuum from passive to active management depending on site conditions and management objectives (Mostert et al. 2018). Passive restoration activities emphasize the need to facilitate the spontaneous natural dynamics of vegetation and are generally faster and less costly than active restoration measures (Capotorti et al. 2017). Active restoration requires extensive human interventions. Examples of active restoration methods include 'close-to-nature' silviculture techniques, such as creating small gaps to simulate gap-dynamics (Teobaldelli et al. 2020), restoration of seed sources, ways to enhance functional connectivity for biogeographically representative and ecologically coherent trees (Capotorti et al. 2017), and targeted seeding of native grassland species into urban wasteland sites (Fischer et al. 2013). In their restoration framework, Mostert et al. (2018) consider the extent of invasive species coverage, indigenous vegetation, and other biotic and abiotic factors to prioritize sites for different types of restoration.

No matter which restoration approach is taken, a reference condition or state is often the goal of restoring a habitat. The reference condition is normally the pristine condition of the habitat, a late-successional sere, or some historical condition. However, such conditions are often challenging to reach in urban landscapes because of the altered biophysical environment, landscape configuration, and disturbance regimes (Johnson and Handel 2019). Also, species and communities consistently recover, reassemble, adapt, and evolve in urban environments due to the complex interactions among different influencing factors (Kowarik 2011; McDonald et al. 2016). The urban landscape requires vegetation restoration to take a more dynamic, multidimensional approach than conventional restoration and realize that restored vegetation may not necessarily recover its former state (Johnson and Handel 2019). Recognizing the need to have a reference condition, McNellie et al. (2020) proposed a contemporary reference state framework to set goals and benchmarks for restoration activities in human-altered landscapes. The framework identifies the contemporary "Best on-Offer" reference state based on a standardized assessment of the social and ecological context, providing a way for policymakers and practitioners to select a reference state that maximizes conservation and restoration outcomes in contemporary ecosystems (McNellie et al. 2020). Moreover, when actively restoring the urban landscape, consistent management is necessary to achieve desired long-term outcomes such as maintenance of biodiversity and ensuring the structural and functional integrity of resident plant communities. It is unrealistic to expect an urban habitat fragment to quickly reach the original ecological status that took centuries to form (Johnson and Handel 2019).

The managing agency should also pay attention to preventing new damage to natural and semi-natural vegetation in cities. Urban residents frequently use natural and semi-natural vegetation sites for recreation, dumping yard waste and garbage, and excavation (Sharpe et al. 1986). Residents' unregulated use of informal trails in natural and semi-natural vegetation can degrade the vegetation structure (Ballantyne and Pickering 2015). Those activities should be regulated. Residents can be organized to serve as stewards of natural and semi-natural vegetation.

Management of cultivated vegetation

Cultivated vegetation in gardens, lawns, parks, streetscapes, and similar green spaces composes a major component of urban vegetation. Besides generating ecosystem services, which benefit people, cultivated vegetation can be managed to enhance biodiversity in cities. Aronson et al. (2017) point out that managing cultivated vegetation can significantly affect biodiversity in urban landscapes both negatively and positively. For instance, a homogeneous structure of vegetation dominated by lawns was created by homeowners, which at the same time are similar in cities across the globe (Ignatieva et al. 2020). Similarly, many cities focus on managing individual units rather than managing lands collectively for publicly owned green spaces. A high degree of habitat fragmentation and increased spatial heterogeneity potentially create multiple sinks for meta-population dynamics (Aronson et al. 2017). In addition, multiple owners often introduce a high number of non-native species for ornamental and horticultural uses. Nonetheless, private and public land managers can recognize the benefits of coordinating management efforts. For example, coordinated actions have been taken to convert lawns into native gardens (Ignatieva and Hedblom 2018) and enhance corridors between green spaces (Pirnat and Hladnik 2016). Not only do these activities increase native plant species richness in our cities, but they also enhance faunal species such as butterflies and bees (Goddard et al. 2010) and human well-being (Frumkin et al. 2017).

Lawns represent a central component of cultivated vegetation (Ignatieva et al. 2020), and their management can significantly affect biodiversity. For instance, lawns with low use of biocides and limited public access were found to have higher species richness and rarity than intensively managed lawns (Politi Bertoncini et al. 2012). Similarly, reduced mowing frequency could enhance biodiversity, aesthetics, and pollination services (O'Sullivan et al. 2017) but often conflict with local ordinances. Still, cities are starting to recognize the combined benefits of reduced mowing intensity for biodiversity and management costs (e.g. cities cooperating in a national network of biodiversity-oriented municipalities "Kommunen für biologische Vielfalt" in Germany). At the community and city level, with smart planning, large areas of lawns can be set aside for less frequent or no mowing while making the rest of the lawns accessible to the public for recreational use (Ignatieva and Hedblom 2018). Another approach to increase lawn biodiversity value is to create gaps in existing lawns and planting the gaps with seeds or seedlings of native meadow species as applied in Sweden (Mårtensson 2017).

Private gardens also provide considerable opportunities to support urban biodiversity. Private gardens often are planted for their aesthetic and utilitarian values (e.g. edible or medicinal plants), and many of them are exotic species (Clarke et al. 2014; Klepacki and Kujawska 2018). Planting native species should be encouraged by providing the owners with knowledge of biodiversity and sources for buying native seeds and plants of regional provenance. Doody et al. (2010) showed that residential gardens in Christchurch, New Zealand, can support the regeneration of native woody species from the nearby forest if the owners were given the species' information and control over the planting locations. Similarly, Goddard et al. (2010) report the need for a landscape approach to enhancing biodiversity with green spaces across the metropolitan area.

Urban forestry provides a framework to unify all management activities on cultivated trees and shrubs in urban areas. Management options focusing on increasing biodiversity can be incorporated into all aspects of the urban forest (Alvey 2006). These options include but are not restricted to survey the baseline situation of biodiversity of urban forests, increase the use of native species in tree planting programs, retention of existing trees, create artificial gaps in the old-growth urban forest, and preserve the interconnections between urban and suburban forests (Alvey 2006; Kowarik and von der Lippe 2018; Massad et al. 2019). Cities worldwide are expanding their urban forests, for example, the Million Trees program in New York City had a goal of adding one million new trees in a decade (Onaindia and Fisher 2020). Incorporating a biodiversity dimension into urban forest planning and management at the very beginning is essential.

Wastelands and vacant lots exist in many cities and provide an opportunity to increase biodiversity (Kowarik 2011; Fischer et al. 2013). Both active and passive management options can be employed to maintain or enhance biodiversity objectives on these parcels. For instance, passive management practices that keep extant soils and resident vegetation and allow the sites to develop naturally often lead to novel communities composed of native and non-native species (Kowarik 2011; Zipperer 2002). Over time, the composition of such novel communities can become similar to natural remnants (Kowarik and von der Lippe 2018). Active management may include directly seeding and planting native species. For instance, Kövendi-Jakó et al. (2019) introduced propagules of native grassland into degraded sites in Nyírség, Hungary, to grow communities similar to those of reference grasslands. They observed that within three years, the restored sites were similar to reference sites. Equally important are the social considerations of restoring wastelands and vacant lots. Managers must balance the need for biodiversity and people's perceptions. A vacant lot with a mixture of spontaneous vegetation may be viewed as unmaintained and distracting by residents of one neighborhood, but valued for its educational and ecological benefits by residents in another neighborhood (Ignatieva and Hedblom 2018). An acceptable compromise is not to leave spontaneous vegetation in wastelands and vacant lots unattended but to actively manage it for biodiversity and ecological benefits.

Invasive species management

Urban landscapes are a source and sink for invasive species, and managing these species requires different approaches than used in rural areas (Gaertner et al. 2017). Urban residents may not support the management measures due to the species' utilitarian values (Novoa et al. 2017) or do not perceive invasive plant management as a high priority relative to other environmental problems (Potgieter et al. 2018). A vital element of invasive species management is educating the public on the negative impacts of invasive species on natural communities and the types of management used to control or eradicate targeted species on public and private lands (Novoa et al. 2017). Dearborn and Kark (2010) emphasize that managers must convey the importance of native species and the management of invasive species to create habitats (e.g. using wastelands and vacant lots) for local biodiversity and human well-being. Successful management plans focus not only on the ecological but also the human benefits.

These concerns suggest that invasive species management in urban areas should be based on biological factors, stakeholder views, and the social consequences of management actions (Dearborn and Kark 2010). Based on the social and ecological context, Gaertner et al. (2017) proposed three management approaches: tolerance, active engagement, and control priority. Non-native species that have a low potential for negative ecological impact but can benefit people may be tolerated. In contrast, active control measures should be applied to species with a relatively high potential for negative impact and deliver relatively low benefits. However, rapid actions need to be taken to control an invasive species if there is no adequate information on its impacts and benefits, preferably at the early invasion stage.

Invasive plant species can be removed from urban vegetation using biological control, mechanical removal, and chemical control. The effect of biological control tends to vary among regions. For example, biological control agents, including insects and fungus, have been applied to control common ragweed (*Ambrosia artemisiifolia* L.) in Australia, Russia, China, and some European countries with varying success (Igrc et al. 1995; Gerber et al. 2011; Kovalev et al. 2015). The biological control agents' effectiveness is determined by many factors such as climate, biology of the biological control agents, and interactions between the agents and the host plant species.

The effectiveness of mechanical removal and chemical treatment is more predictable than that of biological control. For instance, Bierzychudek (2020) found that treating *Hedera* spp. on the ground with biocides and cutting the plants' woody stems increased the abundance and the diversity of native herbs and woody seedlings substantially in natural areas in Portland, Oregon. However, adverse effects on biodiversity across taxa are apparent with repeated application of mixtures of biocides (Brittain et al. 2010; Schäfer et al. 2019; Sattler et al. 2020). Weeding was found to control the population of *Impatiens glandulifera* successfully and increased the native species richness in a riparian zone in London (Cockel et al. 2014). In New York, the combined use of mechanical removal, chemical treatment, and native tree planting has resulted in significantly lower invasive species abundance, more complex forest structure, and greater native tree recruitment in native forests in urban parks (Johnson and Handel 2016). Johnson and Handel (2016) found that active planting of native species after removing invasive species enhanced native species and vegetation structure. Consistent management over extended time frames is needed to shape the structure and composition of the plant communities. Adverse effects of treatments, especially the harmful effects of biocides on biodiversity and ecosystems, need to be weighed against the threats posed by invasive species. Those management approaches that have the smallest (or best no) adverse effects on biodiversity should be favored.

Managing urban vegetation for wildlife

Generally, more urban vegetation cover means more habitats for wildlife. Studies have shown a positive association between urban vegetation cover and species richness and abundance of wildlife such as arthropods and birds (Turrini and Knop 2015; Pena et al. 2017). Increasing the overall amount of urban vegetation cover is a viable management objective for biodiversity conservation. However, not all vegetation benefits urban biodiversity to the same extent. Urban vegetation's value to wildlife is affected by its size, origin, structure, species composition, and surrounding environments. These observations should form the basis for designing management actions.

Maintaining natural or semi-natural vegetation should be the priority, and cultivated vegetation only used as a second option to increase urban vegetation (Chong et al. 2014). Large areas of intact connected urban vegetation need to be conserved. Intensive development should be avoided in adjacent areas to reduce light and noise pollution, or a buffer should be created around the habitat to minimize human activities and intrusions (McDonald et al. 2014).

Creating multiple strata and adding a greater diversity of native plants will increase habitats for a broader range of flora and fauna species in cultivated vegetation (Burgin 2016). For instance, lawns can be mowed less frequently, creating vertical strata, and native meadow species can be added to enhance species richness. Studies have shown that these practices increased diversity for plants, invertebrates, and soil microbes (Norton et al. 2019; Mata et al. 2017). Similarly, in private gardens and vacant lots, non-native ornamental plants that attract pollinators and sustenance-oriented food crops can be planted to support pollinator populations (Lowenstein et al. 2019; O'Connell et al. 2020). Non-native species can serve as essential food sources for pollinators (e.g. white clover, *Trifolium repens*) in private gardens and vacant lots when native species are unavailable. As long as these species are not invasive, they can be crucial for pollinator conservation (Turo et al. 2020; Lowenstein et al. 2019).

Urban forest structure can be improved by increasing species richness through planting native tree species, preserving large-diameter trees, and increasing the complexity of understory vegetation structures (Johnson and Handel 2019). Also, adding features such as standing water and watercourses increase spatial complexity of the urban forest. These practices have been shown to positively impact birds, bats, and invertebrates (Ferenc et al. 2014; Ikin et al. 2015; Archibald et al. 2017; Pena et al. 2017; Threlfall et al. 2017).

Although threatened and endangered species may require special management attention, management, in general, should focus on the overall biodiversity of the urban landscape rather than on a small range of species. The focus on a single species may have unintended consequences, such as the decline in amenities for humans and threatened endemic species (Burgin 2016). For instance, *Grevillea* and *Callistemon* spp. have been widely planted in Australian cities to respond to Birds Australia's call for providing food plants for birds. Although the intended outcome was to enhance bird diversity by providing food, the actions led to establishing honeyeaters in Australian cities, creating an ongoing pest management problem for managers. Smaller-bodied birds were absent when the more aggressive honeyeaters were present (Burgin 2016). A solution to minimizing unintended consequences is to construct ecological species profiles, which enable land managers to manage a wide range of species according to their habitat requirements and disturbance responses simultaneously (Opdam et al. 2001; Garden et al. 2007).

Monitoring and assessment

Survey biodiversity of urban vegetation

Continuous monitoring is essential for managing urban vegetation for biodiversity. While all methods developed for monitoring biodiversity in natural environments can be adapted for use in urban areas, methods considering social and biophysical aspects of urban areas should be preferred. One of these methods is urban biotope mapping, which classifies urban areas into homogeneous habitats based on land use types and biotic and abiotic conditions (Sukopp and Weiler 1988). Urban biotope mapping is conceptually well-structured to meet the demand for maintaining biodiversity in urban areas (Qiu et al. 2010). A similar method, urban habitats biodiversity assessment (UrHBA), provides a standardized way to incorporate habitat information into a spatial data framework. The method includes three steps: Interpreting drawings of the main habitat patches, recording the characteristics of those patches, and constructing maps to link with all recorded data

(Farinha-Marques et al. 2017). Detailed information on urban biodiversity can be obtained using the cited approaches. However, applications of those approaches are time-consuming and require considerable resources.

Rapid biodiversity assessment tools such as the one developed by Tzoulas and James (2010) can be deployed to get a quick overview of biodiversity in urban vegetation. These tools are simple to use, and the results are easy to interpret. They are suitable for aiding managers to develop management strategies and to reach out to the public. Nevertheless, they are not suitable for generating data that cover biodiversity in the studied area completely.

Some emerging methods can be adopted to supplement monitoring works based on conventional field surveys. The use of unmanned aerial vehicles (UAV) in urban vegetation monitoring is especially promising. UAV can capture high-resolution optical images of urban vegetation using cameras. The images can be used alone or together with point data captured using airborne LIDAR boarding UAV to extract information on the size, shape, distribution, species composition, and structure of urban vegetation in urban areas (Feng et al. 2015; Egerer et al. 2020; Schiefer et al. 2020). Besides the images captured using UAV, freely available satellite images can also be used in urban vegetation monitoring to increase data availability and efficiency. For example, the Weed Aerial Surveillance Program framework developed in Australia takes a structured approach to integrate multiple remote sensing technologies into urban and peri-urban weed management efforts (Sheffield and Dugdale 2020).

Using apps based on smartphones and tablets to enlist citizen scientists to assist in biodiversity monitoring work is another promising method. Citizen science projects based on apps have been successfully implemented in many cities to monitoring invasive plant species, urban tree populations, and wildlife living in urban vegetation patches (Hawthorne et al. 2015; Roman et al. 2017; Prudic et al. 2018). These projects are a cost-effective way to supplement the monitoring works and an excellent way to engage residents in biodiversity conservation works.

Assess the effectiveness of management measures

Continuous monitoring of an urban management plan provides the data needed for assessing the effectiveness of managing urban vegetation for biodiversity. An assessment program typically contains four major steps: (1) determine the purpose and scale of the assessment, (2) develop an indicator system, (3) conduct the assessment, (4) communicate the result to policymakers and the public. Defining the setting and scales at which urban vegetation is being assessed is the first step. Due to a wide range of social and ecological factors that affect urban vegetation, it is crucial to assess urban management at multiple scales and include the urban residents' perception of biodiversity associated with urban vegetation (Nilon 2011).

A set of suitable indicators are the core of an assessment. Biophysical indicators and indicators of the social-economic systems, such as measures of attitudes and ecosystem benefit indicators, should be included in that indicator system (Bastian et al. 2020). Frameworks are available to guide the cities to select indicators suitable to local conditions. An example is the City biodiversity index (CBI), which was released by the Convention on Biological Diversity in 2010. The 23 indicators of the CBI measure three aspects of urban biodiversity: (1) native biodiversity in the city (e.g. % of natural areas in the city), (2) ecosystem services provided by biodiversity in the city (e.g. regulation of the quantity of water), and (3) governance and management of biodiversity in the city (e.g. budget allocated to biodiversity). CBI serves as a basis for local governments to build their indicator systems and has been successfully applied in cities worldwide (Kohsaka et al. 2013; Uchiyama and Kohsaka this volume). Two indicators worth being included in every set of indicators are connectivity among vegetation patches (Pirnat and Hladnik 2016) and naturalness (Bastian et al. 2020). They offer critical information on habitat network integrity and the difference between actual vegetation and potential natural vegetation.

After completing the assessment, reporting findings to policymakers and the public is essential. Five principles of reporting should be followed. First, to tailor the results to the target audience. Second, the result must be presented in a clear, transparent, and easy-to-understand manner. Third, confidence in indicator assessments must be clearly communicated. Fourth, to state the progress towards predetermined targets and goals. Finally, co-develop indicators with policymakers can ensure the usefulness of the assessment products (McQuatters-Gollop et al. 2019). Through the assessment and feedback from both the public and decision-makers, managers can adapt their management objectives to address concerns and improve strategies to meet those concerns and potentially new objectives.

Key companion papers

Aronson, M.F.J., et al. 2017. Biodiversity in the city: Key challenges for urban green space management. *Frontiers in Ecology and the Environment* 15:189–196.

Beninde, J., M. Veith, A. Hochkirch. 2015. Biodiversity in cities needs space: A meta-analysis of factors determining intra-urban biodiversity variation. *Ecology Letters* 18(6):581–592.

Chong, K.Y., et al. 2014. Not all green is as good: Different effects of the natural and cultivated components of urban vegetation on bird and butterfly diversity. *Biological Conservation* 171:299–309.

Goddard, M.A., A.J. Dougill, T.G. Benton. 2010. Scaling up from gardens: Biodiversity conservation in urban environments. *Trends in Ecology & Evolution* 25(2):90–98.

Ignatieva, M., M. Hedblom. 2018. An alternative urban green carpet. *Science* 362(6411):148–149.

References

Alvey, A.A. 2006. Promoting and preserving biodiversity in the urban forest. *Urban Forestry & Urban Greening* 5(4):195–201.

Archibald, C.L., M. McKinney, K. Mustin, D.F. Shanahan, H.P. Possingham. 2017. Assessing the impact of revegetation and weed control on urban sensitive bird species. *Ecology and Evolution* 7(12):4200–4208.

Aronson, M.F.J., F.A. La Sorte, C.H. Nilon, M. Katti, M.A. Goddard, C.A. Lepczyk, P.S. Warren, N.S.G. Williams, S. Cilliers, B. Clarkson, C. Dobbs, R. Dolan, M. Hedblom, S. Klotz, J.L. Kooijmans, I. Kühn, I. MacGregor-Fors, M. McDonnell, U. Mörtberg, P. Pyšek, S. Siebert, J. Sushinsky, P. Werner, M. Winter. 2014. A global analysis of the impacts of urbanization on bird and plant diversity reveals key anthropogenic drivers. *Proceedings of the Royal Society B: Biological Sciences* 281(1780):20133330.

Aronson, M.F., C.H. Nilon, C.A. Lepczyk, T.S. Parker, P.S. Warren, S.S. Cilliers, M.A. Goddard, A.K. Hahs, C. Herzog, M. Katti. 2016. Hierarchical filters determine community assembly of urban species pools. *Ecology* 97:2952–2963.

Aronson, M.F.J., C.A. Lepczyk, K.L. Evans, M.A. Goddard, S.B. Lerman, J.S. MacIvor, C.H. Nilon, T. Vargo. 2017. Biodiversity in the city: Key challenges for urban green space management. *Frontiers in Ecology and the Environment* 15(4):189–196.

Ballantyne, M., C.M. Pickering. 2015. Differences in the impacts of formal and informal recreational trails on urban forest loss and tree structure. *Journal of Environmental Management* 159:94–105.

Barrico, L., H. Castro, A.P. Coutinho, M.T. Gonçalves, H. Freitas, P. Castro. 2018. Plant and microbial biodiversity in urban forests and public gardens: Insights for cities' sustainable development. *Urban Forestry & Urban Greening* 29:19–27.

Bastian, O., P. Cudlín, V. Pechanec, P. Brzoska, L. Štěrbová, R. Včeláková, J. Purkyt, K. Grunewald. 2020. Assessments of biodiversity and habitat services in cities – Exemplified by Dresden (Germany) and Liberec (Czech Republic). *Ekologia(Bratislava)* 39(2):174–189.

Beninde, J., M. Veith, A. Hochkirch. 2015. Biodiversity in cities needs space: A meta-analysis of factors determining intra-urban biodiversity variation. *Ecology Letters* 18(6):581–592.

Bierzychudek, P. 2020. Passive recovery of an urban forest in the Pacific Northwest after removal of invasive plants. *Urban Ecosystems* 23(5):1023–1038.
Bolund, P., S. Hunhammar. 1999. Ecosystem services in urban areas. *Ecological Economics* 29(2):293–301.
Brittain, C.A., M. Vighi, R. Bommarco, J. Settele, S.G. Potts. 2010. Impacts of a pesticide on pollinator species richness at different spatial scales. *Basic and Applied Ecology* 11(2):106–115.
Burgin, S. 2016. What about biodiversity? Redefining urban sustainable management to incorporate endemic fauna with particular reference to Australia. *Urban Ecosystems* 19:669–678.
Bush, J., B. Coffey, S. Fastenrath. 2020. Governing urban greening at a metropolitan scale: An analysis of the Living Melbourne strategy. *Australian Planner* 56(2):95–102.
Capotorti, G., E.D. Vico, I. Anzellotti, L. Celesti-Grapow. 2017. Combining the conservation of biodiversity with the provision of ecosystem services in urban green infrastructure planning: Critical features arising from a case study in the metropolitan area of Rome. *Sustainability* 9(1):10.
Carne, R.J. 1994. Urban vegetation: Ecological and social value. Pages 211–25 in M.A. Scheltema, editor. *National Greening Australia Conference*. Greening Australia Ltd, Canberra, Australia.
Chong, K.Y., S. Teo, B. Kurukulasuriya, Y.F. Chung, S. Rajathurai, H.T.W. Tan. 2014. Not all green is as good: Different effects of the natural and cultivated components of urban vegetation on bird and butterfly diversity. *Biological Conservation* 171:299–309.
Clarke, L.W., L. Li, G.D. Jenerette, Z. Yu. 2014. Drivers of plant biodiversity and ecosystem service production in home gardens across the Beijing Municipality of China. *Urban Ecosystems* 17:741–760.
Cockel, C.P., A.M. Gurnell, J. Gurnell. 2014. Consequences of the physical management of an invasive alien plant for riparian plant species richness and diversity. *River Research and Applications* 30(2):217–229.
Dearborn, D.C., S. Kark. 2010. Motivations for conserving urban biodiversity. *Conservation Biology* 24(2):432–440.
Doody, B.J., J.J. Sullivan, C.D. Meurk, G.H. Stewart, H.C. Perkins. 2010. Urban realities: The contribution of residential gardens to the conservation of urban forest remnants. *Biodiversity and Conservation* 19:1385–400.
Egerer, M.H., B. Wagner, B.B. Lin, D. Kendal, K. Zhu. 2020. New methods of spatial analysis in urban gardens inform future vegetation surveying. *Landscape Ecology* 35:761–778.
Farinha-Marques, P., C. Fernandes, F. Guilherme, J.M. Lameiras, P. Alves, R.G.H. Bunce. 2017. Urban Habitats Biodiversity Assessment (UrHBA): A standardized procedure for recording biodiversity and its spatial distribution in urban environments. *Landscape Ecology* 32:1753–1770.
Feng, Q., J. Liu, J. Gong. 2015. UAV Remote sensing for urban vegetation mapping using random forest and texture analysis. *Remote Sensing* 7(1):1074–1094.
Ferenc, M., O. Sedláček, R. Fuchs. 2014. How to improve urban greenspace for woodland birds: Site and local-scale determinants of bird species richness. *Urban Ecosystems* 17(2):625–640.
Fischer, L.K., M. von der Lippe, M.C. Rillig, I. Kowarik. 2013. Creating novel urban grasslands by reintroducing native species in wasteland vegetation. *Biological Conservation* 159:119–126.
Friedman, A. 2021. Urban design for biodiversity. Pages 245–255 in A. Friedman, editor. *Fundamentals of Sustainable Urban Design*. Springer, Cham, Switzerland.
Frumkin, H., G.N. Bratman, S.J. Breslow, B. Cochran, P.H. Kahn Jr., J.J. Lawler, P.S. Levin, P.S. Tandon, U. Varanasi, K.L. Wolf, S.A. Wood. 2017. Nature contact and human health: A research agenda. *Environmental Health Perspectives* 125:075001.
Gaertner, M., J.R.U. Wilson, M.W. Cadotte, J.S. MacIvor, R.D. Zenni, D.M. Richardson. 2017. Non-native species in urban environments: Patterns, processes, impacts and challenges. *Biological Invasions* 19:3461–3469.
Garden, J.G., C.A. McAlpine, H.P. Possingham, D.N. Jones. 2007. Habitat structure is more important than vegetation composition for local-level management of native terrestrial reptile and small mammal species living in urban remnants: A case study from Brisbane, Australia. *Austral Ecology* 32(6):669–685.
Gerber, E., U. Schaffner, A. Gassmann, H.L. Hinz, M. Seier, H. Müller-Schärer. 2011. Prospects for biological control of *Ambrosia artemisiifolia* in Europe: Learning from the past. *Weed Research* 51(6):559 573.
Goddard, M.A., A.J. Dougill, T.G. Benton. 2010. Scaling up from gardens: Biodiversity conservation in urban environments. *Trends in Ecology & Evolution* 25(2):90–98.
Hawthorne, T.L., V. Elmore, A. Strong, P. Bennett-Martin, J. Finnie, J. Parkman, T. Harris, J. Singh, L. Edwards, J. Reed. 2015. Mapping non-native invasive species and accessibility in an urban forest: A case study of participatory mapping and citizen science in Atlanta, Georgia. *Applied Geography* 56:187–198.
Hostetler, M., W. Allen, C. Meurk. 2011. Conserving urban biodiversity? Creating green infrastructure is only the first step. *Landscape and Urban Planning* 100(4):369–371.

Ignatieva, M., F. Eriksson, T. Eriksson, T. Kätterer, P. Tidåker, J. Wissman, K. Ahrné, J. Bengtsson, M. Hedblom. 2020. Pros and cons of transdisciplinary research: A case study of Swedish lawns and their sustainable alternatives. *Urban Forestry & Urban Greening* 56:126799.
Ignatieva, M., M. Hedblom. 2018. An alternative urban green carpet. *Science* 362(6411):148–149.
Igrc, J., C.J. Deloach, V. Zlof. 1995. Release and establishment of *Zygogramma suturalis* F. (coleoptera: Chrysomelidae) in Croatia for control of common ragweed (*Ambrosia artemisiifolia* L). *Biological Control* 5:203–208.
Ikin, K., D.S. Le Roux, L. Rayner, N.R. Villaseñor, K. Eyles, P. Gibbons, A.D. Manning, D.B. Lindenmayer. 2015. Key lessons for achieving biodiversity-sensitive cities and towns. *Ecological Management & Restoration* 16(3):206–214.
Ives, C.D., P.E. Lentini, C.G. Threlfall, K. Ikin, D.F. Shanahan, G.E. Garrard, S.A. Bekessy, R.A. Fuller, L. Mumaw, L. Rayner, R. Rowe, L.E. Valentine, D. Kendal. 2016. Cities are hotspots for threatened species. *Global Ecology and Biogeography* 25(1):117–126.
Jalkanen, J., K. Vierikko, A. Moilanen. 2020. Spatial prioritization for urban biodiversity quality using biotope maps and expert opinion. *Urban Forestry & Urban Greening* 49:126586.
Johnson, L.R., S.N. Handel. 2016. Restoration treatments in urban park forests drive long-term changes in vegetation trajectories. *Ecological Applications* 26(3):940–956.
Johnson, L.R., S.N. Handel. 2019. Management intensity steers the long-term fate of ecological restoration in urban woodlands. *Urban Forestry & Urban Greening* 41:85–92.
Kattwinkel, M., R. Biedermann, M. Kleyer. 2011. Temporary conservation for urban biodiversity. *Biological Conservation* 144(9):2335–2343.
Klepacki, P., M. Kujawska. 2018. Urban allotment gardens in Poland: Implications for botanical and landscape diversity. *Journal of Ethnobiology* 38(1):123–137.
Kohsaka, R., H.M. Pereira, T. Elmqvist, L. Chan, R. Moreno-Peñaranda, Y. Morimoto, T. Inoue, M. Iwata, M. Nishi, M. Da Luz Mathias, C. Souto Cruz, M. Cabral, M. Brunfeldt, A. Parkkinen, J. Niemelä, Y. Kulkarni-Kawli, G. Pearsell. 2013. Indicators for management of urban biodiversity and ecosystem services: City Biodiversity Index. Pages 699–718 in T. Elmqvist, M. Fragkias, J. Goodness, B. Guneralp, P.J. Marcotullio, R.I. McDonald, S. Parnell, M. Schewenius, M. Sendstad, K.C. Seto, C. Wilkinson, editors. *Urbanization, Biodiversity and Ecosystem Services: Challenges and Opportunities: A Global Assessment.* Springer, Dordrecht, Netherlands.
Kovalev, O.V., Y.V. Tyutyunov, O.E. Arkhipova, N.A. Kachalina, L.P. Iljina, L.I. Titova. 2015. On assessment of the large-scale effect of introduction of the ragweed leaf beetle *Zygogramma suturalis* F. (Coleoptera, Chrysomelidae) on the phytocenoses of South Russia. *Entomological Review* 95:1–14.
Kövendi-Jakó, A., M. Halassy, A. Csecserits, K. Hülber, K. Szitár, T. Wrbka, K. Török. 2019. Three years of vegetation development worth 30 years of secondary succession in urban-industrial grassland restoration. *Applied Vegetation Science* 22(1):138–149.
Kowarik, I. 2011. Novel urban ecosystems, biodiversity, and conservation. *Environmental Pollution* 159(8–9):1974–1983.
Kowarik, I., M. von der Lippe. 2018. Plant population success across urban ecosystems: A framework to inform biodiversity conservation in cities. *Journal of Applied Ecology* 55(5):2354–2361.
Lowenstein, D.M., K.C. Matteson, E.S. Minor. 2019. Evaluating the dependence of urban pollinators on ornamental, non-native, and 'weedy' floral resources. *Urban Ecosystems* 22:293–302.
MacKenzie, A., P. Gibbons. 2019. Enhancing biodiversity in urban green space: An exploration of the IAD framework applied to ecologically mature trees. *Urban Science* 3(4):103.
Mårtensson, L.M. 2017. Methods of establishing species-rich meadow biotopes in urban areas. *Ecological Engineering* 103(A):134–140.
Massad, T.J., G.L. Williams, M. Wilson, C.E. Hulsey, E. Deery, L.E. Bridges. 2019. Regeneration dynamics in old-growth urban forest gaps. *Urban Forestry & Urban Greening* 43:126364.
Mata, L., C.G. Threlfall, N.S.G. Williams, A.K. Hahs, M. Malipatil, N.E. Stork, S.J. Livesley. 2017. Conserving herbivorous and predatory insects in urban green spaces. *Scientific Reports* 7:40790.
McDonald, R., B. Guneralp, W. Zipperer, P. Marcotullio. 2014, November–December. The future of global urbanization and the environment. *Solutions*:60–69.
McDonald, T., G.D. Gann, J. Jonson, K.W. Dixon. 2016. *International Standards for the Practice of Ecological Restoration-Including Principles and Key Concepts*. Society for Ecological Restoration, Washington, DC.

McNellie, M.J., I. Oliver, J. Dorrough, S. Ferrier, G. Newell, P. Gibbons. 2020. Reference state and benchmark concepts for better biodiversity conservation in contemporary ecosystems. *Global Change Biology* 26(12):6702–6714.

McQuatters-Gollop, A., I. Mitchell, C. Vina-Herbon, J. Bedford, P.F.E. Addison, C.P. Lynam, P.N. Geetha, E.A. Vermeulan, K. Smit, D.T.I. Bayley, E. Morris-Webb, H.J. Niner, S.A. Otto. 2019. From science to evidence – How biodiversity indicators can be used for effective marine conservation policy and management. *Frontiers in Marine Science* 6:109.

Mostert, E., M. Gaertner, P.M. Holmes, P.J. O'Farrell, D.M. Richardson. 2018. A multi-criterion approach for prioritizing areas in urban ecosystems for active restoration following invasive plant control. *Environmental Management* 62(6):1150–1167.

Müller, A., P.K. Bøcher, C. Fischer, J.-C. Svenning. 2018. 'Wild' in the city context: Do relative wild areas offer opportunities for urban biodiversity? *Landscape and Urban Planning* 170:256–265.

Nilon, C.H. 2011. Urban biodiversity and the importance of management and conservation. *Landscape and Ecological Engineering* 7:45–52.

Norton, B.A., G.D. Bending, R. Clark, R. Corstanje, N. Dunnett, K.L. Evans, D.R. Grafius, E. Gravestock, S.M. Grice, J.A. Harris, S. Hilton, H. Hoyle, E. Lim, T.G. Mercer, M. Pawlett, O.L. Pescott, J.P. Richards, G.E. Southon, P.H. Warren. 2019. Urban meadows as an alternative to short mown grassland: Effects of composition and height on biodiversity. *Ecological Applications* 29(6):e01946.

Novoa, A., K. Dehnen-Schmutz, J. Fried, G. Vimercati. 2017. Does public awareness increase support for invasive species management? Promising evidence across taxa and landscape types. *Biological Invasions* 19(12):3691–3705.

O'Connell, M., Z. Jordan, E. McGilvray, H. Cohen, H. Liere, B.B. Lin, S.M. Philpott, S. Jha. 2020. Reap what you sow: Local plant composition mediates bumblebee foraging patterns within urban garden landscapes. *Urban Ecosystems* 24(2):391–404.

Onaindia, M., J. Fisher. 2020. Urban growth and biodiversity conservation. Pages 1069–1079 in W.L. Filho, A.M. Azul, L. Brandli, A. Lange Salvia, T. Wall, editors. *Life on Land: Encyclopedia of the UN Sustainable Development Goals*. Springer, Cham, Switzerland.

Opdam, P., R. Foppen, C. Vos. 2001. Bridging the gap between ecology and spatial planning in landscape ecology. *Landscape Ecology* 16:767–779.

O'Sullivan, O.S., A.R. Holt, P.H. Warren, K.L. Evans. 2017. Optimising UK urban road verge contributions to biodiversity and ecosystem services with cost-effective management. *Journal of Environmental Management* 191:162–171.

Pena, J.C.D.C., F. Martello, M.C. Ribeiro, R.A. Armitage, R.J. Young, M. Rodrigues. 2017. Street trees reduce the negative effects of urbanization on birds. *PLoS ONE* 12(3):e0174484.

Pirnat, J., D. Hladnik. 2016. Connectivity as a tool in the prioritization and protection of sub-urban forest patches in landscape conservation planning. *Landscape and Urban Planning* 153:129–139.

Politi Bertoncini, A., N. Machon, S. Pavoine, A. Muratet. 2012. Local gardening practices shape urban lawn floristic communities. *Landscape and Urban Planning* 105(1–2):53–61.

Potgieter, L.J., M. Gaertner, U.M. Irlich, P.J. O'Farrell, L. Stafford, H. Vogt, D.M. Richardson. 2018. Managing urban plant invasions: A multi-criteria prioritization approach. *Environmental Management* 62(6):1168–1185.

Prudic, K.L., J.C. Oliver, B.V. Brown, E.C. Long. 2018. Comparisons of citizen science data-gathering approaches to evaluate urban butterfly diversity. *Insects* 9(4):186.

Qiu, L., T. Gao, A. Gunnarsson, M. Hammer, R. von Bothmer. 2010. A methodological study of biotope mapping in nature conservation. *Urban Forestry & Urban Greening* 9(2):161–166.

Roman, L.A., B.C. Scharenbroch, J.P.A. Östberg, L.S. Mueller, J.G. Henning, A.K. Koeser, J.R. Sanders, D.R. Betz, R.C. Jordan. 2017. Data quality in citizen science urban tree inventories. *Urban Forestry & Urban Greening* 22:124–135.

Sattler, C., A.T. Gianuca, O. Schweiger, M. Franzen, J. Settele. 2020. Pesticides and land cover heterogeneity affect functional group and taxonomic diversity of arthropods in rice agroecosystems. *Agriculture, Ecosystems & Environment* 297:106927.

Schäfer, R.B., M. Liess, R. Altenburger, J. Filser, H. Hollert, M. Roß-Nickoll, A. Schäfer, M. Scheringer. 2019. Future pesticide risk assessment: Narrowing the gap between intention and reality. *Environmental Sciences Europe* 31:21.

Schiefer, F., T. Kattenborn, A. Frick, J. Frey, P. Schall, B. Koch, S. Schmidtlein. 2020. Mapping forest tree species in high resolution UAV-based RGB-imagery by means of convolutional neural networks. *ISPRS Journal of Photogrammetry and Remote Sensing* 170:205–215.

Sharpe, D.M., F. Stearns., L.A. Leitner, J.R. Dorney. 1986. Fate of natural vegetation during urban development of rural landscapes in southeastern Wisconsin. *Urban Ecology* 9(3–4):267–287.

Sheffield, K., T. Dugdale. 2020. Supporting urban weed biosecurity programs with remote sensing. *Remote Sensing* 12:2007.

Sukopp, H., S. Weiler. 1988. Biotope mapping and nature conservation strategies in urban areas of the Federal Republic of Germany. *Landscape and Urban Planning* 15(1–2):39–58.

Teobaldelli, M., F. Cona, A. Stinca, L. Saulino, E. Anzano, D. Giordano, A. Migliozzi, G. Bonanomi, G. D'Urso, S. Mazzoleni, A. Saracino. 2020. Improving resilience of an old-growth urban forest in Southern Italy: Lesson(s) from a stand-replacing windstorm. *Urban Forestry & Urban Greening* 47:126521.

Threlfall, C.G., L. Mata, J.A. Mackie, A.K. Hahs, N.E. Stork, N.S.G. Williams, S.J. Livesley. 2017. Increasing biodiversity in urban green spaces through simple vegetation interventions. *Journal of Applied Ecology* 54(6):1874–1883.

Turo, K.J., M.R. Spring, F.S. Sivakoff, Y.A. Delgado de la flor, M.M. Gardiner. 2020. Conservation in post-industrial cities: How does vacant land management and landscape configuration influence urban bees? *Journal of Applied Ecology* 58(1):58–69.

Turrini, T., E. Knop. 2015. A landscape ecology approach identifies important drivers of urban biodiversity. *Global Change Biology* 21(4):1652–1667.

Tzoulas, K., P. James. 2010. Making biodiversity measures accessible to non-specialists: An innovative method for rapid assessment of urban biodiversity. *Urban Ecosystems* 13(1):113–127.

Williams, N.S.G., A.K. Hahs, P.A. Vesk. 2015. Urbanisation, plant traits and the composition of urban floras. *Perspectives in Plant Ecology Evolution and Systematics* 17:78–86.

Zipperer, W.C. 2002. Species composition and structure of regenerated and remnant forest patches within an urban landscape. *Urban Ecosystems* 6:271–290.

23

THE ROLE OF ENGINEERED GREEN INFRASTRUCTURE IN SUPPORTING URBAN BIODIVERSITY

J. Scott MacIvor, Nicholas S.G. Williams, and Jeremy Lundholm

Introduction 1

Green infrastructure (GI) typically refers to any vegetated area that contributes ecosystem services. The biodiversity associated with green infrastructure is partly controlled by designers (e.g. which plant species are planted) but also includes wild flora, fauna, and fungi that spontaneously colonise and use these systems for food, nesting, or other resources, further contributing to ecosystem services. The biodiversity of familiar green infrastructure such as parks, gardens and remnant vegetation is well studied, at least in some regions, but because engineered GI is relatively new, reviews of the biodiversity utilising it are needed.

Engineered GI such as green roofs, living walls, stormwater ponds, and green streetscapes are embedded into (or onto) built infrastructure (i.e. buildings, paved areas) to ameliorate urban environmental problems resulting from a loss of habitat, such as runoff from impermeable surfaces, excess urban heat, and habitat loss. By adding 'new' habitat into urban areas, engineered GI are often considered to contribute to biodiversity.

This chapter synthesises the current state of knowledge on the role of engineered GI elements in supporting biodiversity in cities. We consider four types of engineered GI: (1) green roofs; (2) living walls and façades; (3) bioretention systems, including stormwater ponds, rain gardens, and bioswales; and (4) green streetscapes (Figure 23.1). We then account for a series of nine hierarchical environmental filters that act on regional species pools and create local ecological communities to interpret design considerations, challenges, and opportunities for engineered GI.

Green roofs

Green roofs consist of vegetation in growing media on top of buildings (Figure 23.1). These provide many benefits: reduced building energy consumption in both hot and cold seasons, amelioration of the urban heat island effect, stormwater capture, and benefits to human wellbeing and biodiversity (Shafique et al. 2018; Williams et al. 2021). 'Extensive' green roofs are those with shallow substrate depths and drought-tolerant low-growing plants (e.g. succulents, herbs), and 'intensive' green roofs have deeper growing media that support a greater variety of plant types such as lawn, shrubs, and even small trees, but require irrigation and maintenance. However, these categories fail to capture the full range of functions and biodiversity provided by green roofs (Kotze et al. 2020).

 DOI: 10.4324/9781003016120-27

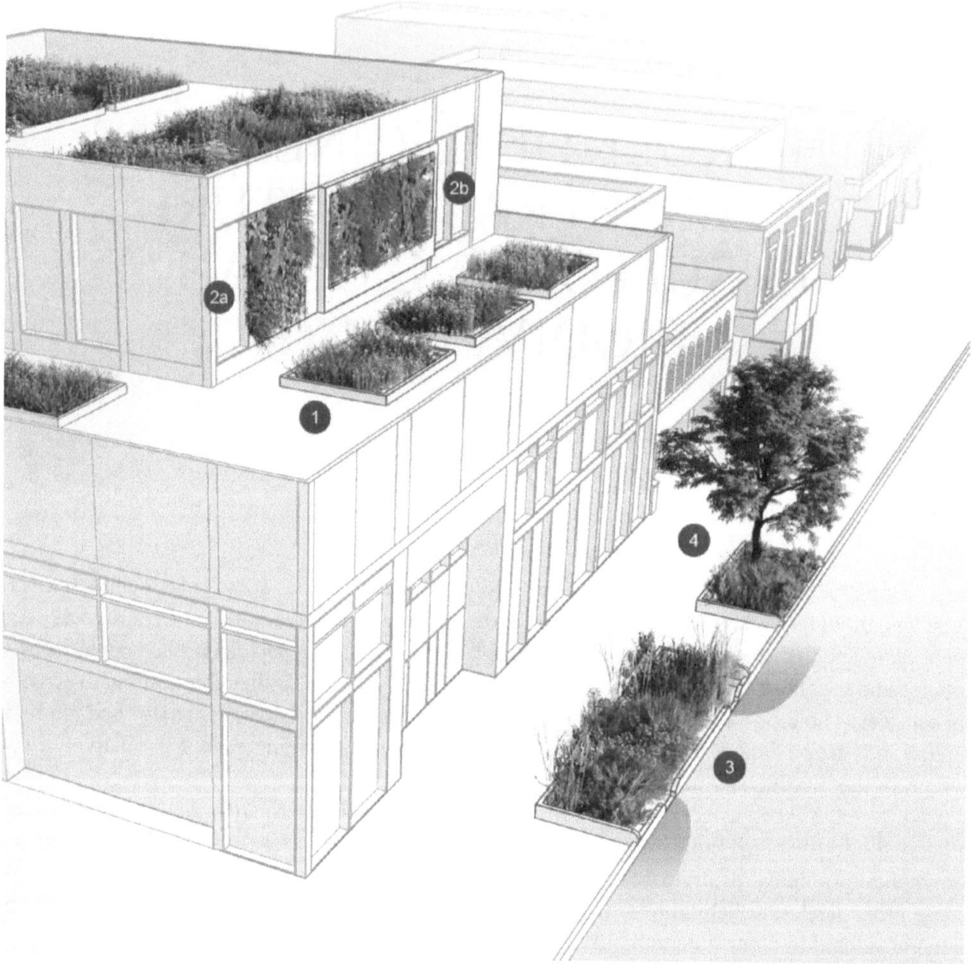

Figure 23.1 An illustration depicting where engineered green infrastructure is implemented: (1) green roofs, (2a) green façade, (2b) living walls, (3) a bioswale for water retention, and (4) green streetscape with street trees (illustration credit: Vicky Nguyen).

Green roofs are visited or colonised by a diverse range of fauna and have potential to enhance urban biodiversity (Williams et al. 2014). Green roofs support greater bird (Partridge and Clark 2018; Belcher et al. 2019) and arthropod (Berthon et al. 2015) species richness than conventional roofs as they provide foraging and breeding opportunities otherwise absent (Williams et al. 2014). However, the species richness of green roof avian (Washburn et al. 2016), bee (Colla et al. 2009; Braaker et al. 2017; Kratschmer et al. 2018), and other arthropod communities (MacIvor and Lundholm 2011; Braaker et al. 2017) is generally lower than in similar nearby ground-level habitats. Nevertheless, in some cases green roofs support regionally rare species (Wang et al. 2017).

Invertebrates are widespread in cities, occupying a variety of niches and contributing to ecosystem services; as a result, these are the best studied animal group on green roofs (MacIvor and Ksiazek 2015). Bee communities are particularly well studied and usually represent a subset of generalist species found at ground-level (Passaseo et al. 2020). Colonisation has been shown to decline with roof height (MacIvor 2016), which may relate to body size (Braaker et al. 2017). Ground nesting bees appear to be more common than species with other nesting requirements (Brenneisen 2005; Braaker et al. 2017; Passaseo et al. 2020), indicating that green roof growing media could be made attractive to this group, particularly if they are deep and finely textured (Kratschmer et al. 2018).

Butterflies and moths are highly mobile insects regularly caught as part of broader arthropod surveys on green roofs, although they generally have low abundances (Berthon et al. 2015; Partridge and Clark 2018; Dromgold et al. 2020). Finally, beetles (Coleoptera) were shown to have higher diversity on green roofs with higher plant diversity compared to stormwater roofs (i.e. lower plant diversity) and unmanaged ground-level sites (Gonsalves et al. 2021). Beetle abundance and richness are positively affected by vegetation cover, particularly forbs and grasses, and negatively affected by roof height and age (Kyrö et al. 2018; Mills and Rott 2020). While green roofs are mostly colonised by mobile species adapted to open dry habitats (Kyrö et al. 2018), some can also support rare beetle species (Pétremand et al. 2018; Mills and Rott 2020).

Despite their abundance in urban areas, there have been very few records and no studies of how small mammals such as rats, mice, and possums use green roofs, which is a significant research gap. Pest control and building management companies in North America warn of the dangers of rats and mice on green roofs, suggesting these ubiquitous urban species frequent them. Microbats are the one mammal taxa that has been studied, and activity was higher over green roofs than conventional roofs. In New York City, three of five species detected had higher activity over green roofs (Partridge et al. 2020) with the activity of the most commonly recorded species, the Eastern red bat (*Lasiurus borealis*), positively correlated with green roof moth abundance (Parkins and Clark 2015). However, in London, UK, only 'biodiverse' green roofs planted with or colonised by a variety of wildflowers, herbs, *Sedums*, mosses, and grasses had higher total and feeding bat activity. There was no difference between conventional (i.e. bare) roofs and green roofs dominated by *Sedum* species (Pearce and Walters 2012).

Green roofs are used by both resident and migratory birds (Deng and Jim 2017; Partridge and Clark 2018). Most bird species recorded using green roofs are either insectivores or omnivores with a smaller proportion of granivores (Brenneisen 2003; Deng and Jim 2017; Partridge and Clark 2018). Successful nesting has been documented on both tropical and temperate green roofs (Baumann 2006; Wang et al. 2017) but may require additional habitat engineering (e.g. by providing a water source on the green roof).

Although the relatively hot and dry environment of green roofs is potentially suitable for reptiles, there are few records (e.g. Davies et al. 2010) and no scientific studies of them using the habitat. However, this could be a function of the only relatively recent uptake of green roofs by cities in warmer climates, where reptile diversity and abundance are higher. Although many reptiles have climbing ability, they are unlikely to spontaneously colonise green roofs that are not connected to suitable ground-level habitat. The lack of permanent water and the vertical isolation from ground means that most green roofs are unlikely to provide suitable habitat for amphibians unless they are deliberately designed to include ponds and introduced populations.

Living walls and façades

Living walls and façades refer to vertical vegetation covering the walls of buildings (Figure 23.1). Living walls contribute to urban temperature amelioration, building cooling (Hunter et al. 2014), trapping of pollutants (Weerakkody et al. 2017), and improving the visual amenity in cities (Collins et al. 2017). There is variation in the methods used to attach vegetation to the building wall. Living wall systems have growing media for plants contained in planters or pockets of different lightweight materials, such as felt hydroponic blankets attached to the wall, while living façades use climbing plants rooted at the base of the wall or growing in planter boxes at different levels (Manso et al. 2021). The plants used on green walls are mainly herbaceous species, but self-clinging or trellis-supported vines are common on green façades (Manso and Castro-Gomes 2015). Irrigation is a key concern for living walls in hot or dry climates and can be a considerable cost for this type of engineered GI. Green façades with mature vegetation planted in-ground will require less irrigation over time.

Although living walls and façades have biodiversity benefits compared to conventional 'bare' walls of brick, concrete, or steel (Filazzola et al. 2019), few biological surveys have been conducted. The vertical orientation of living walls and green façades limits access for many animal species and affects plant type, litter accumulation, substrate volume, and water availability, further reducing the potential type and number of colonising species. Madre et al. (2015) found that aerially dispersing spiders and beetles preferred living walls, whereas green façades of mature trellised flowering vines supported more pollinators. For example, common ivy (*Hedera helix*) is a vine used in green façades for building cooling (Sternberg et al. 2011) and in the UK supports wild bees, including the specialist *Colletes hederae* (Hennessy et al. 2021). Two other vines commonly used on green façades, Virginia creeper (*Parthenocissus* spp.) and hops (*Humulus lupulus*), provide habitat for many species and families of butterflies and moths (Robinson et al. 2010). Despite many pollinators being highly mobile and able to forage effectively on flowering living wall plants, wind velocities increase with height, and this may hamper flying ability especially for smaller species (Freeman 1945).

The leafy vegetation of living walls can provide habitat for nesting, perching, and foraging birds. For example, sparrows in many cities will nest among the vegetation in green façades (Bolhuis Casas 2016). Chiquet et al. (2013) surveyed 27 green walls in the UK and identified nine bird species associated with them, but no species were present on bare walls of brick and tile. Two species of conservation concern in the study region, the house sparrow (*Passer domesticus*) and European starling (*Sturnus vulgaris*), were primary users of living walls. However, these species are considered invasive in other parts of the world, so increasing living wall adoption could support their populations.

Biofiltration and retention systems

Biofiltration and retention systems include stormwater ponds, bioswales, and rain gardens. They are usually a ground level basin or container with a growing medium and plants to intercept stormwater running off the surface of roads or other impermeable surfaces (Figure 23.1). Water flows into the system and may infiltrate into the underlying substrate or re-enter the drainage infrastructure depending on the design. These systems tend to use plants such as wetland grasses, sedges, or shrubs that can tolerate both wet conditions and drier periods between rain events. Stormwater ponds, bioswales, and rain gardens collectively provide urban habitat for invertebrates, birds, and amphibians. However, due to their different designs, function, hydroperiod, and scale, each category may contribute differently, and comparisons need to be made carefully.

Constructed stormwater ponds are artificial wetlands that typically hold water permanently. They can support comparable invertebrate abundance (Hassall and Anderson 2015; Stephansen et al. 2016), dragonfly diversity (Perron and Pick 2020), and amphibian assemblages (Holtmann et al. 2017) to natural water bodies. However, this depends strongly on design, which can vary considerably. For example, amphibian communities are particularly influenced by the amount of aquatic vegetation present (Scheffers and Paszkowski 2013). Further, bioaccumulation of pollutants such as heavy metals, chemicals, and microplastics is higher in urban stormwater ponds compared to natural ponds, which can potentially impact biodiversity (Olesen et al. 2019). Mackintosh et al. (2015) found that arthropod communities in stormwater ponds in Melbourne, Australia, were relatively homogenous, consisting of dominant species (e.g. non-biting midges) that can tolerate anoxic and polluted conditions. In regions with cold winters, salts used to de-ice roads also accumulate in stormwater ponds (Marsalek 2003) and cause changes from less tolerant to more tolerant species in invertebrates (Wallace and Biastoch 2016) and amphibians (Gallagher et al. 2014). Degraded stormwater ponds may also become a source of invasive plant species (Sinclair et al. 2020).

Bioswales and rain gardens are typically smaller than stormwater ponds and may cycle between being inundated with stormwater to being completely dry, normally within 36 hours to prevent mosquito populations (Ishimatsu et al. 2017). This extreme variation in hydroperiod necessitates well-draining substrates and a plant community that can thrive in these conditions. Fewer studies have characterised the fauna in bioswales and rain gardens than stormwater ponds. Soil fauna have been found to be particularly rich and are important contributors to nutrient turnover (Mehring et al. 2016). Small bioretention basins and bioswales located on suburban streets in Melbourne, Australia, show higher invertebrate diversity with greater plant diversity and leaf litter depth and higher diversity than traditional street verges due to higher vegetation cover and greater abundance of flowering plants (Kazemi et al. 2009, 2011).

Research has concentrated on the biodiversity contribution of biofiltration and retention systems alongside roads. The utility of residential rain gardens or those near buildings, and in parks, is understudied and necessitates assessment for urban biodiversity taking into account differing hydroperiods and design (Hale et al. 2015).

Green streetscapes

Green streetscapes refer to the roadside and median strips planted with street trees in most cities (Figure 23.1). They commonly have an understory of mown turf but can contain more complex vegetation such as resident verge gardens (Marshall et al. 2020), municipal planting beds, and spontaneous vegetation (Bonthoux et al. 2019b). Although narrow and fragmented, the cumulative green space provided by streetscapes can be substantial (Marshall et al. 2019). However, there can be great inequality between and within cities with lower socioeconomic areas often having significantly less vegetated streetscapes (Wood and Esaian 2020). Streetscapes provide amenity and many ecosystem services, including microclimatic cooling, stormwater absorption, and air pollution mitigation (Säumel et al. 2016). They can also provide important biodiversity habitat and provide connectivity across urban landscapes.

Street tree plantings can support different avian communities. In Australia, those planted with native trees typically have higher species richness and abundance of native birds than streets with exotic trees or newly developed streets with no trees. Differences being largely due to the progressive loss of insectivorous and nectar-feeding species and increase of exotic species (Sewell and Catterall 1998; Ikin et al. 2013). Similar results have been reported from Brazil (De Castro Pena et al. 2017) and California (Wood and Esaian 2020), where streets with more native trees and large trees had higher bird abundance and richness.

Increasing vegetation structural complexity and the proportion of native vegetation on streets is likely to increase the abundance and richness of multiple taxa (Threlfall et al. 2017). This could be spontaneous vegetation growing in pavements and at the base of street trees which significantly contributes to streetscape plant biodiversity (Wittig and Becker 2010; Bonthoux et al. 2019b) and can be positively perceived by residents (Weber et al. 2014; Bonthoux et al. 2019a). Designed vegetation that includes native urban wildflower meadows can be more highly preferred than mown grass (Southon et al. 2017) and has been consistently shown to support higher levels of insect and pollinator abundance and diversity (Norton et al. 2019; Mody et al. 2020). However, with the exception of Mody et al. (2020), few of these studies have exclusively investigated urban streetscapes where harsh conditions and impacts such as street lighting and pollution (Leonard and Hochuli 2017) can negatively affect plant growth and habitat for insect biodiversity. Consequently, there is a need for ecologists to work with city governments to investigate the longer-term impact of interventions designed to increase streetscape biodiversity (Tan et al. 2022).

Ecological concepts

Diversity of the spontaneous taxa colonising and using GI is determined by a hierarchical series of environmental filters (Concept 1; all Concepts reviewed in Table 23.1) that act on regional species pools and create local ecological communities. Filters include the connectivity of the designed and broader green infrastructure network (Concept 2), the size of the GI elements (Concept 3) (Beninde et al. 2015), as well as their interactions with the adjacent grey infrastructure in the urban matrix (Concept 4) (Aronson et al. 2016), and human influences such as plant choice and vegetation management (Aronson et al. 2017; Threlfall et al. 2017). Due to their small size, and highly fragmented nature, the communities that use engineered GI often comprise species that are generalists with traits that facilitate dispersal (Concept 5) (Fournier et al. 2020).

Engineered GI embedded in urban hardscapes are limited in size due to competing land uses, the size of buildings (e.g. for green roofs and walls), and cost. This restricts their ability to support biodiversity as island biogeography theory suggests that smaller patches can support fewer species (Crowe 1979). Small populations in isolated GI elements can go extinct due to random mortality more easily than larger populations (Dornier and Cheptou 2012), suggesting species turnover in these systems could be high, but requires further study.

Increasing habitat complexity by integrating a diversity of native plant species can enhance the number and functional diversity of species that engineered GI can support (Concept 6). Organisms differ in their resource requirements, so a spatially heterogeneous patch that contains variation in resources and other features is likely to support more species (Concept 7) (Stein et al. 2014). While many construction guidelines recommend incorporating habitat complexity, there are few studies that demonstrate that these techniques increase biodiversity or support species of conservation concern in urban settings (but see Threlfall et al. 2017).

The small size of engineered GI types may limit the range of habitats possible for wildlife. Landscape complexity may also contribute to the problem of small patch sizes, as urban factors such as construction, building configuration, or building height may reduce biodiversity access to these engineered GIs. However, at larger spatial scales, the mosaic of these could have significant benefits for urban wildlife, but more research is needed to investigate this.

Long-term monitoring is needed to determine whether certain GI might act as ecological traps for some species (Concept 8), such as polluted stormwater wetlands (Hale et al. 2015). An ecological trap occurs when habitat conditions are perceived as supportive by the individual, but reproductive fitness is impacted by an unseen threat. GI examples include killdeers repeatedly nesting on green roofs in Switzerland that had very low fecundity due to harsh conditions and predators (Baumann 2006) and frogs breeding in polluted urban stormwater ponds having lower tadpole survival (Sievers et al. 2018).

Finally, a variety of factors might lead to non-native species establishment and proliferation in urban GI (Gaertner et al. 2017), and these will vary by GI type and degree of invasion (Concept 9). Invasive species establishment and spread may result even when good intentions drive creation

Table 23.1 Nine concepts from the ecological sciences having design implications for engineered green infrastructure and its ability to support urban biodiversity

Concept	*Design application*	*Examples*	*Limitations*
1. Environmental filtering restricts the number and kinds of species (Kraft et al. 2015)	• Resource supplementation ↑ survival • Plant and maintain native species unable to establish naturally (Cadotte et al. 2021) to support dependent animal species	• Designed reservoir in green roof supports rare coastal species (Nagase and Tashiro-Ishii 2018)	• Anthropogenic factors outside GI design (e.g. proportion of urban land cover or population density) influence species assembly
2. Landscape connectivity can ↑ biodiversity by allowing species to move from patch to patch (Fahrig 2003)	• Location of GI close to natural and semi-natural ecosystems ↑ ability for species to colonise GI • Plan for clusters of GI type and number (e.g. multiple green roofs and ground level greening in same area)	• Mobile beetle species can use green roofs as stepping stone habitats. For less mobile species, connectivity with ground-level habitats is important (Braaker et al. 2017) • Proximity to forest, ↑ value of stormwater ponds for amphibians (Clevenot et al. 2018)	• May be impossible in urban core • Scale of planning (e.g. building-scale) does not match relevant scale for biodiversity • Same processes drive invasive species proliferation
3. Small patch size can ↓ biodiversity as small populations are more dynamic (Hanski 1994)	• Many small patches can collectively support ↑ biodiversity (↑ beta diversity despite ↑ turnover in individual patches)	• Small GI patches in close proximity (<200 m apart) could ↑ wildflower diversity in green streetscapes (Vega and Küffer 2021) • Mobile species (e.g. bees) benefit from dispersed floral resources but limited by nesting opportunities	• Small GI size, ↓ core habitat, ↑ edge-interior ratio and opportunity to escape poor conditions • Small populations less likely to persist vs large populations
4. Between-patch heterogeneity can ↑ biodiversity (Tews et al. 2004)	• Proximity of a GI to different GI can support species with complex resource requirements (e.g. nesting and foraging) (Colding 2007) • Variations in GI design can support more species at the landscape scale	• Different kinds of living walls with different moisture conditions support different spider and beetle assemblages (Madre et al. 2015)	• Scale of planning (e.g. building-scale) does not match relevant scale for biodiversity • Communication between different property owners, GI managers difficult to coordinate

(*Continued*)

Table 23.1 (Continued)

Concept	*Design application*	*Examples*	*Limitations*
5. Urban species are mostly generalists, not species of conservation concern (Gray 1989)	• GI design can promote ↑ abundance of habitat specialists or urban avoiders by including their required resources	• Green roofs support ↑ urban avoiding bird species than conventional roofs (Partridge and Clark 2018)	• Most engineered GI have limited value for specialists because they have already disappeared from local urban landscape
6. Biodiversity begets biodiversity (Haddad et al. 2011)	• ↑ plant species richness correlated with ↑ arthropod species richness (Siemann et al. 1998) • Attracting species that disperse seeds, invertebrates to GI enhance diversity	• Different plant species offer resources to different bird species (Deng and Jim 2017) • ↑ plant diversity correlates with ↑ invertebrate diversity in bioretention basins (Kazemi et al. 2009)	• This can also be a negative as birds may transport undesirable plant species into GI
7. Within-patch complexity help support biodiversity (Lundholm and Richardson 2010)	• Adding habitat heterogeneity, such as different substrate types on green roof projects could encourage biodiversity (Lundholm and Richardson 2010)	• Often recommended to ↑ biodiversity in green roofs and stormwater management GI (Monberg et al. 2018)	• Small patch size limits amount of complexity that can be built in GI • Limited evidence that habitat complexity ↑ diversity in small GI
8. Ecological traps attract species but lead to negative consequences for populations (Battin 2004)	• Higher green roof heights result in ↓ reproductive success for urban bees (MacIvor 2016) • Polluted stormwater retention ponds can be ecological traps for frogs and fish (Hale et al. 2019)	• Artificial structures (e.g. growing media for ground-nesting bees) or resource supplementation (e.g. water source on green roofs) could result in positive outcomes • Take landscape context into account	• Factors influencing net benefit to any population are complex; may be difficult to predict (Clevenot et al. 2018) • Species-specific habitat may support some, not other species
9. Urban species pools are overrepresented by non-native biodiversity (McKinney 2004)	• Include non-natives that provide valuable functions if they are not invasive • Plant invasion-resistant GI	• Non-native *Sedum* on extensive green roofs are visited by urban pollinators (MacIvor et al. 2015) • Green roof design standards specify high plant cover, reducing colonisation by weeds	• Spontaneous colonisation of GI can be dominated by undesirable non-natives resulting in maintenance costs or lowered functionality

of urban green spaces. In cities, propagule pressure is high, and there is a need to monitor GI and manage for invasive species to mitigate the environmental impacts (Sinclair et al. 2020) and economic costs (Diagne et al. 2021).

Conclusion

Engineered GI can support biodiversity, but this depends on the type, size, and design. Due to the limitations for plant selection and lack of substrate, when considering biodiversity habitat in urban design, engineered GI should be considered only as 'complementary' habitat, used in addition to other GI types and existing green space conservation, rather than as replacement for green space lost due to development. It is possible that further developments in design, creativity, and ecological knowledge will lead to even more ways to modify and integrate GI into many urban conditions. As this mosaic develops, a goal for engineered GI should be to mitigate urbanisation impacts on biodiversity to the benefit of people and nature.

Key companion papers

Filazzola, A., N. Shrestha, J.S. MacIvor. 2019. The contribution of constructed green infrastructure to urban biodiversity: A synthesis and meta-analysis. *Journal of Applied Ecology* 56(9):2131–2143.

Hale, R., R. Coleman, V. Pettigrove, S.E. Swearer. 2015. Identifying, preventing and mitigating ecological traps to improve the management of urban aquatic ecosystems. *Journal of Applied Ecology* 52(4):928–939.

Kotze, D., et al. 2020. A revised terminology for vegetated rooftops based on function and vegetation. *Urban Forestry & Urban Greening* 49:126644.

Lundholm, J.T., P.J. Richardson. 2010. Habitat analogues for reconciliation ecology in urban and industrial environments. *Journal of Applied Ecology* 47(5):966–975.

Madre, F., P. Clergeau, N. Machon, A. Vergnes. 2015. Building biodiversity: Vegetated façades as habitats for spider and beetle assemblages. *Global Ecology and Conservation* 3(3):222–233.

Williams, N.S., et al. 2009. A conceptual framework for predicting the effects of urban environments on floras. *Journal of Ecology* 97:4–9.

References

Aronson, M.F.J., C.A. Lepczyk, K.L. Evans, M.A. Goddard, S.B. Lerman, J.S. MacIvor, C.H. Nilon, T. Vargo. 2017. Biodiversity in the city: Key challenges for urban green space management. *Frontiers in Ecology and the Environment* 15(4):189–196.

Aronson, M.F.J., C.H. Nilon, C.A. Lepczyk, T.S. Parker, P.S. Warren, S.S. Cilliers, M.A. Goddard, A.K. Hahs, C. Herzog, M. Katti, F.A. La Sorte, N.S.G. Williams, W. Zipperer. 2016. Hierarchical filters determine community assembly of urban species pools. *Ecology* 97(11):2952–2963.

Battin, J. 2004. When good animals love bad habitats: Ecological traps and the conservation of animal populations. *Conservation Biology* 18(6):1482–1491.

Baumann, N. 2006. Ground-nesting birds on green roofs in Switzerland: Preliminary observations. *Urban Habitats* 4(1):37–44.

Belcher, R.N., K.R. Sadanandan, E.R. Goh, J.Y. Chan, S. Menz, T. Schroepfer. 2019. Vegetation on and around large-scale buildings positively influences native tropical bird abundance and bird species richness. *Urban Ecosystems* 22:213–225.

Beninde, J., M. Veith, A. Hochkirch. 2015. Biodiversity in cities needs space: A meta-analysis of factors determining intra-urban biodiversity variation. *Ecology Letters* 18(6):581–592.

Berthon, K., D. Nipperess, P. Davies, M. Bulbert. 2015. *Confirmed at Last: Green Roofs Add Invertebrate Diversity*. State of Australian Cities Conference 2015. https://apo.org.au/sites/default/files/resource-files/2015-12/apo-nid63205.pdf

Bolhuis Casas, E.I. 2016. *Bird Use of Living Walls in the City of Bogotá, Colombia*. Thesis, Universidad de Ciencias Aplicadas y Ambientales, Bogotá, Colombia.

Bonthoux, S., S. Chollet, I. Balat, N. Legay, L. Voisin. 2019a. Improving nature experience in cities: What are people's preferences for vegetated streets? *Journal of Environmental Management* 230:335–344.
Bonthoux, S., L. Voisin, S. Bouché-Pillon, S. Chollet. 2019b. More than weeds: Spontaneous vegetation in streets as a neglected element of urban biodiversity. *Landscape and Urban Planning* 185:163–172.
Braaker, S., M.K. Obrist, J. Ghazoul, M. Moretti. 2017. Habitat connectivity and local conditions shape taxonomic and functional diversity of arthropods on green roofs. *Journal of Animal Ecology* 86:521–531.
Brenneisen, S. 2003. *The Benefits of Biodiversity from Green Roofs – Key Design Consequences. 1st Annual Greening Rooftops for Sustainable Communities Conference, Awards and Trade Show*. The Cardinal Group, Toronto, Chicago.
Brenneisen, S. 2005. *The Natural Roof (NADA): Research Project Report on the Use of Extensive Green Roofs by Wild Bees*. University of Wädenswil, Zürich, Switzerland.
Cadotte, M.W., L.J. Potgieter, C.J. Wang, J.S. MacIvor. 2021. Invasion theory as a management tool for increasing native biodiversity in urban ecosystems. *Journal of Applied Ecology* 58(11):2394–2403.
Chiquet, C., J.W. Dover, P. Mitchell. 2013. Birds and the urban environment: The value of green walls. *Urban Ecosystems* 16:453–462.
Clevenot, L., C. Carré, P. Pech. 2018. A review of the factors that determine whether stormwater ponds are ecological traps and/or high-quality breeding sites for amphibians. *Frontiers in Ecology and Evolution* 6:40.
Colding, J. 2007. 'Ecological land-use complementation' for building resilience in urban ecosystems. *Landscape and Urban Planning* 81(1):46–55.
Colla, S.R., E. Willis, L. Packer. 2009. Can green roofs provide habitat for urban bees (Hymenoptera: Apidae)? *Cities and the Environment (CATE)* 2:4.
Collins, R., M. Schaafsma, M.D. Hudson. 2017. The value of green walls to urban biodiversity. *Land Use Policy* 64:114–123.
Crowe, T. 1979. Lots of weeds: Insular phytogeography of vacant urban lots. *Journal of Biogeography* 6(2):169–181.
Davies, R., R. Simcock, G. Ussher, R. Toft, M. Boult, C. deGroot. 2010. *Elevated Enclaves – Living Roof Biodiversity Enhancement Through Prosthetic Habitats*. CitiesAlive: 8th Annual Green Roof and Wall Conference. Green Roofs for Healthy Cities. Vancouver, British Columbia, Canada.
De Castro Pena, J.C., F. Martello, M.C. Ribeiro, R.A. Armitage, R.J. Young, M. Rodrigues. 2017. Street trees reduce the negative effects of urbanization on birds. *PLoS ONE* 12(3):e0174484.
Deng, H., C.Y. Jim. 2017. Spontaneous plant colonization and bird visits of tropical extensive green roof. *Urban Ecosystems* 20:337–352.
Diagne, C., B. Leroy, A.-C. Vaissière, R.E. Gozlan, D.A. Roiz, I. Jarić, J.-M. Salles, C.J.A. Bradshaw, F. Courchamp. 2021. High and rising economic costs of biological invasions worldwide. *Nature* 592(7855):571–576.
Dornier, A., P.-O. Cheptou. 2012. Determinants of extinction in fragmented plant populations: *Crepis sancta* (Asteraceae) in urban environments. *Oecologia* 169:703–712.
Dromgold, J.R., C.G. Threlfall, B.A. Norton, N.S.G. Williams. 2020. Green roof and ground-level invertebrate communities are similar and are driven by building height and landscape context. *Journal of Urban Ecology* 6:juz024.
Fahrig, L. 2003. Effects of habitat fragmentation on biodiversity. *Annual Review of Ecology, Evolution, and Systematics* 34:487–515.
Filazzola, A., N. Shrestha, J.S. MacIvor. 2019. The contribution of constructed green infrastructure to urban biodiversity: A synthesis and meta-analysis. *Journal of Applied Ecology* 56(9):2131–2143.
Fournier, B., D. Frey, M. Moretti. 2020. The origin of urban communities: From the regional species pool to community assemblages in city. *Journal of Biogeography* 47(3):615–629.
Freeman, J.A. 1945. Studies in the distribution of insects by aerial currents. *Journal of Animal Ecology* 14:128–154.
Gaertner, M., J.R.U. Wilson, M.W. Cadotte, J.S. MacIvor, R.D. Zenni, D.M. Richardson. 2017. Non-native species in urban environments: Patterns, processes, impacts and challenges. *Biological Invasions* 19:3461–3469.
Gallagher, M.T., J.W. Snodgrass, A.B. Brand, R.E. Casey, S.M. Lev, R.J. Van Meter. 2014. The role of pollutant accumulation in determining the use of stormwater ponds by amphibians. *Wetlands Ecology and Management* 22(5):551–564.

Gonsalves, S., O. Starry, A. Szallies, S. Brenneisen. 2021. The effect of urban green roof design on beetle biodiversity. *Urban Ecosystems* 25:205–219.
Gray, J.S. 1989. Effects of environmental stress on species rich assemblages. *Biological Journal of the Linnean Society* 37(1–2):19–32.
Haddad, N.M., G.M. Crutsinger, K. Gross, J. Haarstad, D. Tilman. 2011. Plant diversity and the stability of foodwebs. *Ecology Letters* 14:42–46.
Hale, R., R. Coleman, V. Pettigrove, S.E. Swearer. 2015. Identifying, preventing and mitigating ecological traps to improve the management of urban aquatic ecosystems. *Journal of Applied Ecology* 52(4):928–939.
Hale, R., S.E. Swearer, M. Sievers, R. Coleman. 2019. Balancing biodiversity outcomes and pollution management in urban stormwater treatment wetlands. *Journal of Environmental Management* 233:302–307.
Hanski, I. 1994. A practical model of metapopulation dynamics. *Journal of Animal Ecology* 63(1):151–162.
Hassall, C., S. Anderson. 2015. Stormwater ponds can contain comparable biodiversity to unmanaged wetlands in urban areas. *Hydrobiologia* 745:137–149.
Hennessy, G., C. Uthoff, S. Abbas, S.C. Quaradeghini, E. Stokes, D. Goulson, F.L.W. Ratnieks. 2021. Phenology of the specialist bee Colletes hederae and its dependence on Hedera helix L. in comparison to a generalist, Apis mellifera. *Arthropod-Plant Interactions* 15(2):183–195.
Holtmann, L., K. Philipp, C. Becke, T. Fartmann. 2017. Effects of habitat and landscape quality on amphibian assemblages of urban stormwater ponds. *Urban Ecosystems* 20:1249–1259.
Hunter, A.M., N.S.G. Williams, J.P. Rayner, L. Aye, D. Hes, S.J. Livesley. 2014. Quantifying the thermal performance of green façades: A critical review. *Ecological Engineering* 63:102–113.
Ikin, K., E. Knight, D.B. Lindenmayer, J. Fischer, A.D. Manning. 2013. The influence of native versus exotic streetscape vegetation on the spatial distribution of birds in suburbs and reserves. *Diversity and Distributions* 19(3):294–306.
Ishimatsu, K., K. Ito, Y. Mitani, Y. Tanaka, T. Sugahara, Y. Naka. 2017. Use of rain gardens for stormwater management in urban design and planning. *Landscape and Ecological Engineering* 13:205–212.
Kazemi, F., S. Beecham, J. Gibbs. 2011. Streetscape biodiversity and the role of bioretention swales in an Australian urban environment. *Landscape and Urban Planning* 101(2):139–148.
Kazemi, F., S. Beecham, J. Gibbs, R. Clay. 2009. Factors affecting terrestrial invertebrate diversity in bioretention basins in an Australian urban environment. *Landscape and Urban Planning* 92(3–4):304–313.
Kotze, D.J., K. Kuoppamäki, J. Niemikapee, M. Mesimäki, V. Vaurola, S. Lehvävirta. 2020. A revised terminology for vegetated rooftops based on function and vegetation. *Urban Forestry & Urban Greening* 49:126644.
Kraft, N.J.B., P.B. Adler, O. Godoy, E.C. James, S. Fuller, J.M. Levine. 2015. Community assembly, coexistence and the environmental filtering metaphor. *Functional Ecology* 29(5):592–599.
Kratschmer, S., M. Kriechbaum, B. Pachinger. 2018. Buzzing on top: Linking wild bee diversity, abundance and traits with green roof qualities. *Urban Ecosystems* 21:429–446.
Kyrö, K., S. Brenneisen, D.J. Kotze, A. Szallies, M. Gerner, S. Lehvävirta. 2018. Local habitat characteristics have a stronger effect than the surrounding urban landscape on beetle communities on green roofs. *Urban Forestry & Urban Greening* 29:122–130.
Leonard, R.J., D.F. Hochuli. 2017. Exhausting all avenues: Why impacts of air pollution should be part of road ecology. *Frontiers in Ecology and the Environment* 15(8):443–449.
Lundholm, J.T., P.J. Richardson. 2010. Habitat analogues for reconciliation ecology in urban and industrial environments. *Journal of Applied Ecology* 47(5):966–975.
MacIvor, J.S. 2016. Building height matters: Nesting activity of bees and wasps on vegetated roofs. *Israel Journal of Ecology and Evolution* 62(1–2):88–96.
MacIvor, J.S., K. Ksiazek. 2015. Invertebrates on green roofs. Pages 333–355 in R.K. Sutton, editor. *Green Roof Ecosystems*. Springer, Cham, Switzerland.
MacIvor, J.S., J. Lundholm. 2011. Insect species composition and diversity on intensive green roofs and adjacent level-ground habitats. *Urban Ecosystems* 14(2):225–241.
MacIvor, J.S., A. Ruttan, B. Salehi. 2015. Exotics on exotics: Pollen analysis of urban bees visiting Sedum on a green roof. *Urban Ecosystems* 18(2):419–430.
Mackintosh, T.J., J.A. Davis, R.M. Thompson. 2015. The influence of urbanisation on macroinvertebrate biodiversity in constructed stormwater wetlands. *Science of the Total Environment* 536:527–537.
Madre, F., P. Clergeau, N. Machon, A. Vergnes. 2015. Building biodiversity: Vegetated façades as habitats for spider and beetle assemblages. *Global Ecology and Conservation* 3:222–233.

Manso, M., J. Castro-Gomes. 2015. Green wall systems: A review of their characteristics. *Renewable and Sustainable Energy Reviews* 41(C):863–871.
Manso, M., I. Teotónio, C.M. Silva, C.O. Cruz. 2021. Green roof and green wall benefits and costs: A review of the quantitative evidence. *Renewable and Sustainable Energy Reviews* 135:110111.
Marsalek, J. 2003. Road salts in urban stormwater: An emerging issue in stormwater management in cold climates. *Water Science and Technology* 48(9):61–70.
Marshall, A.J., M.J. Grose, N.S.G. Williams. 2019. From little things: More than a third of public green space is road verge. *Urban Forestry & Urban Greening* 44:126423.
Marshall, A.J., M.J. Grose, N.S.G. Williams. 2020. Of mowers and growers: Perceived social norms strongly influence verge gardening, a distinctive civic greening practice. *Landscape and Urban Planning* 198:103795.
McKinney, M.L. 2004. Measuring floristic homogenization by non-native plants in North America. *Global Ecology and Biogeography* 13(1):47–53.
Mehring, A.S., B.E. Hatt, D. Kraikittikun, B.D. Orelo, M.A. Rippy, S.B. Grant, J.P. Gonzalez, S.C. Jiang, R.F. Ambrose, L.A. Levin. 2016. Soil invertebrates in Australian rain gardens and their potential roles in storage and processing of nitrogen. *Ecological Engineering* 97:138–143.
Mills, W.P., A. Rott. 2020. Vertical life: Impact of roof height on beetle diversity and abundance on wildflower green roofs. *Journal of Urban Ecology* 6:1–8.
Mody, K., D. Lerch, A.K. Müller, N.K. Simons, N. Blüthgen, M. Harnisch. 2020. Flower power in the city: Replacing roadside shrubs by wildflower meadows increases insect numbers and reduces maintenance costs. *PLoS ONE* 15(6):e0234327.
Monberg, R.J., A.G. Howe, H.P. Ravn, M.B. Jensen. 2018. Exploring structural habitat heterogeneity in sustainable urban drainage systems (SUDS) for urban biodiversity support. *Urban Ecosystems* 21(6):1159–1170.
Nagase, A., Y. Tashiro-Ishii. 2018. Habitat template approach for green roofs using a native rocky sea coast plant community in Japan. *Journal of Environmental Management* 206:255–265.
Norton, B.A., G.D. Bending, R. Clark, R. Corstanje, N. Dunnett, K.L. Evans, D.R. Grafius, E. Gravestock, S.M. Grice, J.A. Harris, S. Hilton, H. Hoyle, E. Lim, T.G. Mercer, M. Pawlett, O.L. Pescott, J.P. Richards, G.E. Southon, P.H. Warren. 2019. Urban meadows as an alternative to short mown grassland: Effects of composition and height on biodiversity. *Ecological Applications* 29(6):e01946.
Olesen, K.B., D.A. Stephansen, N. van Alst, J. Vollertsen. 2019. Microplastics in a stormwater pond. *Water* 11:1466.
Parkins, K.L., J.A. Clark. 2015. Green roofs provide habitat for urban bats. *Global Ecology and Conservation* 4:349–357.
Partridge, D.R., J.A. Clark. 2018. Urban green roofs provide habitat for migrating and breeding birds and their arthropod prey. *PLoS ONE* 13(8):e02022Z98.
Partridge, D.R., K.L. Parkins, S.B. Elbin, J. Alan Clark. 2020. Bat activity correlates with moth abundance on an urban green roof. *Northeastern Naturalist* 27:77–89.
Passaseo, A., G. Pétremand, S. Rochefort, E. Castella. 2020. Pollinator emerging from extensive green roofs: Wild bees (Hymenoptera, Antophila) and hoverflies (Diptera, Syrphidae) in Geneva (Switzerland). *Urban Ecosystems* 23:1079–1086.
Pearce, H., C.L. Walters. 2012. Do green roofs provide habitat for bats in urban areas? *Acta Chiropterologica* 14:469–478.
Perron, M.A.C., F.R. Pick. 2020. Stormwater ponds as habitat for Odonata in urban areas: The importance of obligate wetland plant species. *Biodiversity and Conservation* 29:913–931.
Pétremand, G., Y. Chittaro, S. Braaker, S. Brenneisen, M. Gerner, M.K. Obrist, S. Rochefort, A. Szallies, M. Moretti. 2018. Ground beetle (Coleoptera: Carabidae) communities on green roofs in Switzerland: Synthesis and perspectives. *Urban Ecosystems* 21(1):119–132.
Robinson, G.S., P.R. Ackery, I.J. Kitching, G.W. Beccaloni, L.M. Hernández. 2010. *HOSTS-a Database of the World's Lepidopteran Hostplants*. Natural History Museum, London. www.nhm.ac.uk/our-science/data/hostplants/
Säumel, I., F. Weber, I. Kowarik. 2016. Toward livable and healthy urban streets: Roadside vegetation provides ecosystem services where people live and move. *Environmental Science & Policy* 62:24–33.
Scheffers, B.R., C.A. Paszkowski. 2013. Amphibian use of urban stormwater wetlands: The role of natural habitat features. *Landscape and Urban Planning* 113:139–149.
Sewell, S.V., C.P. Catterall. 1998. Bushland modification and styles of urban development: Their effects on birds in south-east Queensland. *Wildlife Research* 25(1):41–63.

Shafique, M., R. Kim, M. Rafiq. 2018. Green roof benefits, opportunities and challenges – A review. *Renewable and Sustainable Energy Reviews* 90:757–773.

Siemann, E., D. Tilman, J. Haarstad, M. Ritchie. 1998. Experimental tests of the dependence of arthropod diversity on plant diversity. *The American Naturalist* 152(5):738–750.

Sievers, M., K.M. Parris, S.E. Swearer, R. Hale. 2018. Stormwater wetlands can function as ecological traps for urban frogs. *Ecological Applications* 28(4):1106–1115.

Sinclair, J.S., C.R. Adams, A.J. Reisinger, E. Bean, L.S. Reisinger, A.L. Holmes, B.V. Iannone III. 2020. High similarity and management-driven differences in the traits of a diverse pool of invasive stormwater pond plants. *Landscape and Urban Planning* 201:103839.

Southon, G.E., A. Jorgensen, N. Dunnett, H. Hoyle, K.L. Evans. 2017. Biodiverse perennial meadows have aesthetic value and increase residents' perceptions of site quality in urban green-space. *Landscape and Urban Planning* 158:105–118.

Stein, A., K. Gerstner, H. Kreft. 2014. Environmental heterogeneity as a universal driver of species richness across taxa, biomes and spatial scales. *Ecology Letters* 17:866–880.

Stephansen, D.A., A.H. Nielsen, T. Hvitved-Jacobsen, M.L. Pedersen, J. Vollertsen. 2016. Invertebrates in stormwater wet detention ponds – Sediment accumulation and bioaccumulation of heavy metals have no effect on biodiversity and community structure. *Science of the Total Environment* 566:1579–1587.

Sternberg, T., H. Viles, A. Cathersides. 2011. Evaluating the role of ivy (Hedera helix) in moderating wall surface microclimates and contributing to the bioprotection of historic buildings. *Building and Environment* 46:293–297.

Tan, H., L. Harrison, J. Nelson, M. Lokic, J.P. Rayner, C.G. Threlfall, J. Baumann, A. Marshall, M. Callow, J. Peeler, R. Korossy-Horwood, C. Nicholson, N.S.G. Williams. 2022. Designing and managing biodiverse streetscapes: Key lessons from the city of Melbourne. *Urban Ecosystems* 25:733–740.

Tews, J., U. Brose, V. Grimm, K. Tielbörger, M.C. Wichmann, M. Schwager, F. Jeltsch. 2004. Animal species diversity driven by habitat heterogeneity/diversity: The importance of keystone structures. *Journal of Biogeography* 31(1):79–92.

Threlfall, C.G., L. Mata, J.A. Mackie, A.K. Hahs, N.E. Stork, N.S.G. Williams, S.J. Livesley. 2017. Increasing biodiversity in urban green spaces through simple vegetation interventions. *Journal of Applied Ecology* 54(6):1874–1883.

Vega, K.A., C. Küffer. 2021. Promoting wildflower biodiversity in dense and green cities: The important role of small vegetation patches. *Urban Forestry & Urban Greening* 62:127165.

Wallace, A.M., R.G. Biastoch. 2016. Detecting changes in the benthic invertebrate community in response to increasing chloride in streams in Toronto, Canada. *Freshwater Science* 35(1):353–363.

Wang, J.W., C.H. Poh, C.Y.T. Tan, V.N. Lee, A. Jain, E.L. Webb. 2017. Building biodiversity: Drivers of bird and butterfly diversity on tropical urban roof gardens. *Ecosphere* 8:e01905.

Washburn, B.E., R.M. Swearingin, C.K. Pullins, M.E. Rice. 2016. Composition and diversity of avian communities using a new urban habitat: Green roofs. *Environmental Management* 57(6):1230–1239.

Weber, F., I. Kowarik, I. Säumel. 2014. A walk on the wild side: Perceptions of roadside vegetation beyond trees. *Urban Forestry & Urban Greening* 13:205–212.

Weerakkody, U., J.W. Dover, P. Mitchell, K. Reiling. 2017. Particulate matter pollution capture by leaves of seventeen living wall species with special reference to rail-traffic at a metropolitan station. *Urban Forestry & Urban Greening* 27:173–186.

Williams, N.S.G., R.S. Bathgate, C. Farrell, K.E. Lee, C. Szota, J. Bush, K.A. Johnson, R.E. Miller, A. Pianella, L.D. Sargent, J. Schiller, K.J.H. Williams, J.P. Rayner. 2021. Ten years of greening a wide brown land: A synthesis of Australian green roof research and roadmap forward. *Urban Forestry & Urban Greening* 62:127179.

Williams, N.S.G., J. Lundholm, J.S. MacIvor. 2014. Do green roofs help urban biodiversity conservation? *Journal of Applied Ecology* 51(6):1643–1649.

Wittig, R., U. Becker. 2010. The spontaneous flora around street trees in cities: A striking example for the worldwide homogenization of the flora of urban habitats. *Flora: Morphology, Distribution, Functional Ecology of Plants* 205(10):704–709.

Wood, E.M., S. Esaian. 2020. The importance of street trees to urban avifauna. *Ecological Applications* 30(7):e02149.

24

GLOBAL BIODIVERSITY REQUIRES INTEGRATING SOCIAL AND ECOLOGICAL GOALS FOR URBAN BIODIVERSITY

Insights from ecological restoration

Lea R. Johnson

Introduction

People living in cities will decide how much of the diversity of life on Earth today will travel with us into the future. Because our experiences of nature shape perceptions and values that underpin the decisions we make about land, species, and natural resources – and because billions of people now experience only urban biodiversity – conservation, restoration, and enhancement of biodiversity in cities are needed to help humanity make better decisions about the future of the planet.

This need is now being recognized, and efforts in behalf of urban biodiversity are increasing in number across the globe. An array of approaches is being tried, from conservation of remnant habitats and ecological restoration projects aspiring to re-create native ecosystems to designed interventions enhancing the potential of built structures to support biodiversity. All require creativity, flexibility, and community engagement for success.

The first and most essential condition for any effort on behalf of urban biodiversity is that there be room for it. Large, high-quality, connected habitats are key to supporting the maximum number of species in cities (Beninde et al. 2015), making planning for preservation of large natural areas a primary concern for urban biodiversity. Preserving large natural areas is also important for people. The services urban ecosystems provide to people are recognized (Bolund and Hunhammar 1999; Elmqvist et al. 2015), and large wild places within cities provide particular benefits. For example, to develop connections with biodiversity, children's interactions with nature need to be broad and deep, including engagement with natural areas big enough to get lost in whenever possible (Kahn and Weiss 2017). Immersion in nature is now prescribed to treat stress and anxiety in people of all ages (Crnic and Kondo 2019; Koselka et al. 2019), and "forest bathing" has become an increasingly popular approach to decreasing stress and improving feelings of well-being (Song et al. 2016; Miyazaki 2018; Antonelli et al. 2019).

Maximizing opportunities for urban biodiversity also means looking beyond the boundaries of large parks and preserves to include privately held properties in conservation planning (Goddard et al. 2010) and considering the broad range of diversity in pattern and form in urban landscapes

DOI: 10.4324/9781003016120-28

(Cadenasso et al. 2007; Irwin and Bockstael 2007). Residential areas are one such key to urban biodiversity (Cook et al. 2012; Hostetler 2015; Padullés Cubino et al. 2018). Trees in front and back yards, medians, and along streets comprise an essential component of urban tree canopy (Nowak and Dwyer 2001; Ossola et al. 2019). Yards and gardens can increase habitat connectivity and provide resources for both resident and migratory species (Rudd et al. 2002; Gaston et al. 2005; Burghardt et al. 2009) – if homeowner association rules and other restrictive covenants can be designed to invite biodiverse plantings that support local wildlife (Lerman et al. 2012). In both residential and commercial zones, designing and retrofitting built structures to provide habitats on green roofs, walls, and terraces can enhance support for generalist species while providing important nature experiences to nearby people (Turner et al. 2004; Oberndorfer et al. 2007), while community gardens and urban farms provide a wide variety of benefits to people, from food security and nutrition to physical activity and social cohesion, while supporting cultural and biological diversity (Campbell and Wiesen 2011; Egli et al. 2016; Baldock et al. 2019).

Maximizing urban biodiversity also requires noticing, encouraging, and expanding opportunities for biodiversity in overlooked places. In "unintentional" landscapes (Gandy 2016) like vacant lots, train corridors, and roadsides, relict and regenerating patches of spontaneous vegetation dot the urban fabric and weave through neighborhoods. Small and fragmented patches of spontaneous vegetation are surprisingly valuable (Guldin et al. 1990; Godefroid and Koedam 2003; Bodin et al. 2006; Fahrig 2017). They make up a large proportion of the land available to support biodiversity in cities (e.g. Avins 2013) and can support regionally unique habitats and species (Kühn and Klotz 2006). Because they are interwoven with the places people live and work, they are key spaces in which to focus efforts to preserve and improve the ecological health of nearby nature – and of the urban communities surrounding them (Douglas 2012; J. Aronson et al. 2016; van den Bosch and Ode Sang 2017).

Wicked problems

Work on behalf of biodiversity in cities requires solving wicked ecological problems in an intensely populated social context with long and complex legacies of land use. Common challenges have been encountered as these efforts have become more widespread. Invasive species, for example, are a common challenge to biodiversity in urban natural areas (Pregitzer et al. 2019). Cities are hotspots of invasive species introductions (Von der Lippe et al. 2005), and changes to ecosystem processes, hydrology, soils, and local climate act as strong filters on the survival and reproduction of many native species (M. Aronson et al. 2016).

Shifts in wildlife communities caused by urbanization also influence plant communities, especially where herbivores become superabundant and predators, seed dispersers, and pollinators are rare or missing (Beauchamp et al. 2013; DiCicco 2014; Gharehaghaji et al. 2019; Johnson et al. 2020). Once extirpated by urbanization, reintroduction of keystone species that once regulated ecosystems, like bison or wolves, is likely to be neither successful nor desirable to nearby residents (Gobster 2010).

Similarly, important ecological disturbances that structure local ecosystems, such as fire and flooding, are difficult or impossible to reintroduce in cities due to the configuration, small size, and fixed boundaries of natural areas (Gobster 2007, 2012), while disturbances associated with urbanization like land clearing, soil movement, construction, demolition, and rerouting of water flows are frequent and intense, with long-lasting effects.

The necessity of land management across jurisdictional and ownership boundaries by groups with different goals and resources adds additional complexity (Belaire et al. 2011; McWilliam et al.

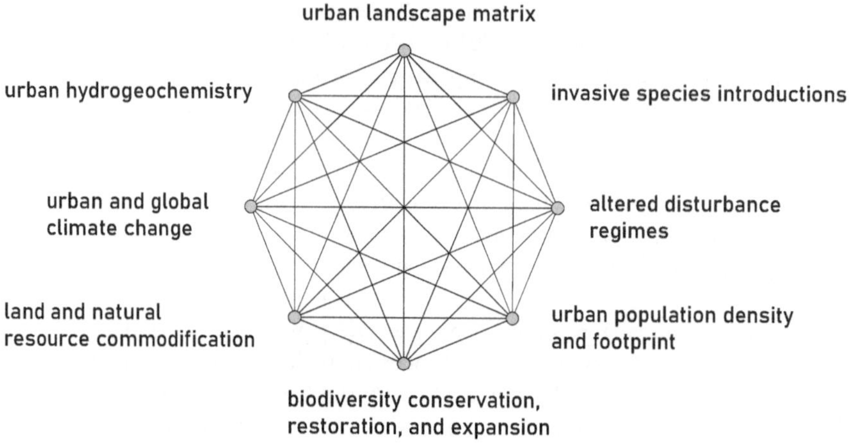

Figure 24.1 Axes of change in urban ecosystems, adapted from Falk (2017). Impacts of human activity on ecosystems – axes of change – interact across space and time, complicating efforts to mitigate these impacts and to restore ecological functions (Falk 2017). In urban ecosystems, all axes are present and intensified. Conservation, restoration, and expansion of biodiversity at the scale of cities and urban regions where most people live have the potential to influence perceptions and values that affect biodiversity at broader national to global scales.

2012). Organizations engaged in biodiversity-focused land management in cities are, like their rural counterparts, often short on both personnel and funding in relation to the land area to be managed, with the added responsibility of managing access to and uses by surrounding communities. Personnel limitations lead to reliance on volunteers – with impacts on accountability, survival of plantings, and timing (Gobster 2007; Pregitzer et al. 2019) that can reduce effectiveness of these interventions (Smith et al. 2016).

Expectations must also be managed. Ecological interventions may not go entirely as planned because understanding of both urban ecosystem dynamics and how to manipulate them on behalf of biodiversity are still young (Sukopp 2008; McDonnell 2011). Novel biophysical conditions and species assemblages limit our ability to predict the outcomes of urban ecological interventions, compared to places where people have observed typical patterns of development over longer periods of time. Models of ecosystem processes developed in more pristine sites may or may not apply (Ehrenfeld 2000), and new and complex interactions can result in surprises (Schilthuizen 2018).

Insights from ecological restoration

Because of urbanization's drastic effects on ecosystems, management of urban land to conserve and restore biodiversity often intersects with the practice of ecological restoration – defined as "the process of assisting the recovery of an ecosystem that has been degraded, damaged, or destroyed" (Gann et al. 2019). Ecological restoration aims to guide a trajectory of recovery within the context of local and global changes, supporting the persistence and evolution of an ecosystem's component species and making it possible for ecosystem processes to continue or begin again.

People engaged in work on behalf of biodiversity in cities must grapple with both the fundamental questions of ecological restoration – *What should the goals be? What methods should be used? How should success be evaluated?* – and with constraints and opportunities particular to urban environments. The practice of ecological restoration and the complementary science of restoration ecology have yielded insights, techniques, and approaches that are relevant to these questions.

Early pioneers of ecological restoration in the USA, Edith Roberts and Aldo Leopold, began experimenting with restoration of plant diversity in urban environments – at Vassar College in 1924 and at University of Wisconsin-Madison in 1934, respectively – but it was not until the 1980s that large-scale and citywide projects focused on restoring urban biodiversity began to increase in number and scope (Gobster 2010; Bounds et al. 2015). Early urban ecological restoration projects began in natural areas that had been protected over the course of the previous century by people motivated by different interests and concerns, ranging from open space preservation and opportunity for urban workers to access recreation and exercise to estates with picturesque landscape designs (Platt et al. 1994). In these spaces, what goals for ecological restoration are appropriate and attainable?

A key feature of ecological restoration is the use of reference ecosystems for assessment of success (Gann et al. 2019). Methods for selecting reference ecosystems have been a source of active discussion for decades (Moore et al. 1999), and how best to use reference sites to advance work on behalf of urban biodiversity remains an open question. While more intact ecosystems of the surrounding region may at first appear appropriate to target conditions, urban constraints can affect outcomes (Geist and Galatowitsch 1999), leading to a growing awareness that references for urban ecological restoration cannot be based on past or intact conditions alone (Gobster 2007; Hychka and Druschke 2017).

This realignment of perspective is congruent with increasing emphasis in ecological restoration thinking to focus on restoring ecosystem processes and trajectories in addition to recovery of historic, native species assemblages (SERI 2004; McDonald et al. 2016; Gann et al. 2019). This refocusing has inspired new approaches, including those that incorporate urban locations experiencing similar conditions and restrictions, but containing desired species or high levels of desired ecosystem functions and services (Klaus and Kiehl 2021). Current standards of practice for ecological restoration incorporate cultural ecosystems and traditional uses, new approaches to selecting reference systems, and a "restorative continuum" that defines a spectrum of nature-based solutions, recognizing co-benefits such as flood reduction, quality of life, and public health improvements (Smith et al. 2016; Hychka and Druschke 2017; Gann et al. 2019). This menu of options better aligns with the spectrum of restoration opportunities (Gobster 2007) that urban natural areas present. It also recognizes the importance of aligning ecological and social goals for biodiversity. When these goals are in conflict, resistance may arise. For example, people have protested restrictions on off-leash dog access to protect wildlife and rallied to prevent removal of charismatic invasive trees (Gobster 2007). Causes of project failure – such as budget constraints, limited land availability, and insufficient time and labor – also have underlying social causes (Geist and Galatowitsch 1999).

Integrating social and ecological goals

Cities are complex social-ecological systems (Alfsen et al. 2011) where social and ecological forces combine to co-produce the species composition and physical structure of urban biodiversity (Johnson et al. 2021). Success of efforts to conserve, restore, and enhance urban biodiversity depend upon community involvement and on understanding and consideration of community needs (Smith et al. 2016; Hychka and Druschke 2017). The work must be grounded in both sociocultural and biophysical context, recognizing that these are interdependent (Yocom 2014) and that consent

Figure 24.2 Conflicting social and ecological goals for urban biodiversity. Introduced Australian blue gum eucalyptus (*Eucalyptus globulus*) trees have become established with intentional afforestation and fire suppression in the San Francisco Bay area, USA. Removal of mature trees to reduce fire danger and re-establish native coastal oak scrub and dune ecosystems has been hotly contested by residents concerned about aesthetic values, uncertainty regarding fire dynamics, and use of herbicides (Gobster 2011; St. George 2016). By Brocken Inaglory – Own work, CC BY-SA 3.0, https://commons.wikimedia.org/w/index.php?curid=7927608

and active support of local communities and stakeholders is essential. As with reference ecosystems in ecological restoration, measures of project success developed in non-urban contexts may also not be adequate. For example, the size of an area benefiting from a project is a less useful yardstick of success where space available is limited by an urban landscape matrix (Hychka and Druschke 2017). At the same time, a small patch of urban habitat might provide important nature experience to a larger number of people than in a less populated place.

Several approaches have been developed to integrate biological and social goals for biodiversity, and these can be applied in cities. They include prioritization indices with social and ecological criteria (Bonilla-Rodríguez et al. 2021), ecosystem function and services evaluation (Elmqvist et al. 2015; Klaus and Kiehl 2021), and cost-benefit analysis (Martínez and López-Barrera 2008).

This integration has become a central focus of international efforts to restore global ecosystems. For example, guidance for reforestation undertaken as part of the United Nations Decade on Ecosystem Restoration (2021–2030). To make it more evergreen explicitly integrated biodiversity recovery with a focus on livelihoods and planetary-scale ecosystem services, emphasizing conservation of existing habitats, integration of multiple social and ecological goals, community participation, appropriate site, species, and plant material selection, and the importance of both planning ahead and learning from results (Di Sacco et al. 2021). Other approaches to global-scale biodiversity restoration extend further into the terrain of social benefit, advocating a more "people-centered" approach (Elias et al. 2021).

However, application of these general approaches can vary widely. For example, biodiversity plans are being developed for cities in many countries, but these plans often lack specific targets (Nilon et al. 2017). These plans may express their aspirations in a variety of ways, ranging from a general desire for better ecosystem health to highly specific targets for herbaceous plant cover. Clearly articulated targets are key to measuring progress and learning from outcomes, whether the aim is preserving or restoring habitats and species, designing built environments for enhanced biodiversity, providing ecosystem services, or fulfilling ethical responsibilities such as addressing environmental justice (Dearborn and Kark 2010).

Local community engagement is important in all phases of projects focused on biodiversity in cities and developing a sense of place and community during the early stages can build group cohesiveness that supports projects in the longer term (Geist and Galatowitsch 1999). Successful projects include measures that increase public participation (Bernhardt et al. 2007), such as transparent prioritization – particularly for publicly funded projects (Yocom 2014) – and material benefits like food and fiber, while job and livelihood creation are especially important where poverty is a driver of ecological degradation (Palomar 2010; Terada et al. 2010). Training that builds technical and organizational capacity at the local level can be important to providing these benefits (Hychka and Druschke 2017).

Given the developing nature of both science and practice, expectations for work on behalf of urban biodiversity should incorporate time for learning from projects, including retrospective or cross-scale analyses to promote adaptive management. More flexible timeframes for spending funds can also increase success (Folke et al. 2005; Hychka and Druschke 2017), and demonstration projects are one way to build support for larger projects by leveraging visible short-term outcomes to achieve larger and long-term objectives (Hychka and Druschke 2017). In the long term, adaptive management, education, and outreach are important to maintaining public engagement (Smith et al. 2016).

Inspiring care and stewardship

In a time of broad concern about the future of the planet, inspiring care can be a way to engage people in stewardship of biodiversity by connecting their responses to species they notice in everyday life to their effect on larger environmental systems (Nassauer 2011). Engaging in care for urban biodiversity can take many forms, from local tree planting to scientific data collection to coordinated advocacy, and effects of these activities reach beyond the impact of a planted tree. People are most likely to be interested in learning about and conserving nature when they experience it personally – especially as children – but most people now experience only species found in cities. This has been called the "pigeon paradox" (Dunn et al. 2006). While it is essential to biodiversity at a global scale that people in cities experience biodiversity in everyday life (Nilon 2011), unequal access means that many people rarely venture into areas of high biodiversity within cities

Figure 24.3 Community-led biodiversity protection: forest patches in Baltimore City, Maryland, USA. A wide array of organizations and communities participate in stewardship of natural areas in cities. Founded by community gardeners to protect a treasured garden from development, Baltimore Greenspace (https://baltimoregreenspace.org/) is a land trust that helps preserve community gardens, pocket parks, and patches of urban forest. They assist residents with skills and resources in response to community requests and engage with scientists and forest stewards to produce original research on the value of urban green spaces. (Photo of Baltimore oriole with nest by Andrew Weitzel – www.flickr.com/photos/21531749@N06/51169273100/, CC BY-SA 2.0, https://commons.wikimedia.org/w/index.php?curid=105254878)

(Livesley et al. 2016), and these patterns of inequality can be deeply rooted in historical injustice (Gerrish and Watkins 2018; Watkins and Gerrish 2018; Grove et al. 2018; Locke et al. 2021).

Growing interest has fortunately opened new avenues for engagement with biodiversity. Home and community gardening, which offer a rich array of nature experiences and many social benefits (Campbell and Wiesen 2011; Egli et al. 2016), have grown in popularity and experienced a major surge with the onset of the COVID-19 pandemic (Bulgari et al. 2021; Mullins et al. 2021; Theodorou et al. 2021). Urban environmental stewardship provides opportunities for people of all ages to engage in community-building action that enhances human well-being and civic engagement (Svendsen 2009; Fisher et al. 2015; Sorensen et al. 2018). People across the globe are engaged in stewardship networks ranging in scope from neighborhoods and municipalities to regions, countries, and the planet (Clemants et al. 2006; Aronson et al. 2020; INTERLACE 2021), inspired by an array of motivations (Svendsen and Campbell 2008; Belaire et al. 2011; Romolini 2012; Fisher et al. 2012; Asah and Blahna 2013; Connolly et al. 2014).

Urban nature does not always look beautiful, however, and not all places with potential for biodiversity may seem inviting. Frequent, intense ecological disturbances associated with building, demolition, and land use conversion are typical of cities. "Informal" green spaces (Rupprecht and Byrne 2014) created by these disturbances can provide regionally scarce refuge for early successional plants (Kattwinkel et al. 2009, 2011) and support pollinators (Buchholz et al. 2020), but urban grasslands (Fischer et al. 2013) and other recently disturbed plant communities have the disadvantage of looking “weedy” to many people and can include invasive species and weeds of lawns and gardens in addition to native plants (Johnson et al. 2018). One gap that must be joined to connect more people with urban biodiversity is an apparent difference between the typical vegetation of the region surrounding a city and the scruffy vegetation of these ecologically young

Figure 24.4 Urban environmental stewardship: forest restoration in Seattle, Washington, USA. Municipal governments in many cities have initiated volunteer programs that invite residents to participate in environmental stewardship. The Green Seattle Partnership (www.seattle.gov/parks/volunteer/urban-forest-restoration) is a partnership between the City of Seattle, community groups, nonprofit organizations, schools, and volunteers that maintains the city’s forested parklands. By Seattle Municipal Archives from Seattle, WA – Seattle Municipal Archives, CC BY 2.0, https://commons.wikimedia.org/w/index.php?curid=96265838

and awkward places. They can be approached using interventions that give "cues to care", such as mown strips and along walkways and roads and crisply delineated bands of vegetation that can increase acceptance of wildness by placing "messy ecosystems" in an "orderly frame" (Nassauer 1995). They can also be modified with organizing and beautifying interventions that make them more attractive as an alternative form of ornamental planting (Kühn 2006).

Eventually, abandoned urban land may undergo the process of ecological succession, resulting in the spontaneous growth of urban forest patches in temperate and tropical cities where ample rainfall and long growing seasons invite trees to grow. These forests contribute to ecological functioning and sustainability in cities (Kowarik 2011) and to human well-being (Tyrväinen et al. 2005; Van den Berg et al. 2007; Korpela et al. 2010; Hauru et al. 2012). Increasing awareness of their benefits has led to the implementation of forest restoration and afforestation projects in cities around the world (City of Seattle Parks and Recreation 2011; Bounds et al. 2015; Rink and Arndt 2016; Sanesi et al. 2017; Pregitzer et al. 2019; Wallace and Clarkson 2019), and restoration efforts can shift long-term trajectories of vegetation development toward desired native forest community structure (Johnson and Handel 2016).

Much of a city, however, is not so spontaneous as a forest in a vacant lot. How can we enhance biodiversity in built and designed landscapes? New solutions and techniques continually arise. Reconciliation ecology, for example, calls attention to the rapid transformation of the planet's terrestrial surface to human uses, and advocates for redesigning human habitats so that their use by people is compatible with use by a broad array of other species (Rosenzweig 2003). Examples of urban interventions that would fit this model abound, from green roofs to green streets (Oberndorfer et al. 2007; Andersson et al. 2014; Williams et al. 2014; van der Jagt et al. 2018).

These and many other complementary approaches are needed to "bring nature home" (Tallamy and Darke 2009) to the city and shorten the distance between people and experiences of biodiversity. Urban ecosystems can provide benefits to people while contributing to biodiversity conservation. They can and should contribute more. To amplify their potential, they must be recognized for both the benefits they already provide and what is possible. We need reference ecosystems to inform what is possible, what is missing, and what may be added back. We need to find wonder in weeds and inspire curiosity and care in the places where future generations will build their understanding of biodiversity.

Key companion papers

Geist, C., S.M. Galatowitsch. 1999. Reciprocal model for meeting ecological and human needs in restoration projects. *Conservation Biology* 13:970–979.

Gobster, P.H. 2010. Urban ecological restoration. *Nature and Culture* 5(3):227–230.

Hychka, K., C.G. Druschke. 2017. Adaptive management of urban ecosystem restoration: Learning from restoration managers in Rhode Island, USA. *Society & Natural Resources* 30:1358–1373.

Johnson, L.R., S.N. Handel. 1996. Restoration treatments in urban park forests drive long-term changes in vegetation trajectories. *Ecological Applications* 26:940–956.

Lerman, S.B., V.K. Turner, C. Bang. 2012. Homeowner associations as a vehicle for promoting native urban biodiversity. *Ecology and Society* 17.

Nassauer, J.I. 2011. Care and stewardship: From home to planet. *Landscape and Urban Planning* 100:321–323.

Acknowledgments

I would like to thank James Aronson for insightful comments on the draft of this chapter, and Ginny Johnson for being my first and most reliable reader.

References

Alfsen, C., A. Duval, T. Elmqvist. 2011. The urban landscape as a social-ecological system for governance of ecosystem services. Pages 213–218 in J. Niemelä, J.H. Breuste, G.R. Guntenspergen, N.E. McIntyre, T. Elmqvist, P. James, editors. *Urban Ecology: Patterns, Processes, and Applications*. Oxford University Press, New York.

Andersson, E., S. Barthel, S. Borgström, J. Colding, T. Elmqvist, C. Folke, Å. Gren. 2014. Reconnecting cities to the biosphere: Stewardship of green infrastructure and urban ecosystem services. *Ambio* 50:1636–1638.

Antonelli, M., G. Barbieri, D. Donelli. 2019. Effects of forest bathing (shinrin-yoku) on levels of cortisol as a stress biomarker: A systematic review and meta-analysis. *International Journal of Biometeorology* 63(8):1117–1134.

Aronson, J.C., C.M. Blatt, T.B. Aronson. 2016. Restoring ecosystem health to improve human health and well-being: Physicians and restoration ecologists unite in a common cause. *Ecology and Society* 21(4):39.

Aronson, J.C., N. Goodwin, L. Orlando, C. Eisenberg, A.T. Cross. 2020. A world of possibilities: Six restoration strategies to support the United Nation's Decade on Ecosystem Restoration. *Restoration Ecology* 28(4):730–736.

Aronson, M.F.J., C.H. Nilon, C.A. Lepczyk, T.S. Parker, P.S. Warren, S.S. Cilliers, M.A. Goddard, A.K. Hahs, C. Herzog, M. Katti, F.A. La Sorte, N.S.G. Williams, W. Zipperer. 2016. Hierarchical filters determine community assembly of urban species pools. *Ecology* 97(11):2952–2963.

Asah, S.T., D.J. Blahna. 2013. Practical implications of understanding the influence of motivations on commitment to voluntary urban conservation stewardship. *Conservation Biology* 27(4):866–875.

Avins, M. 2013. *Baltimore's Forest Patches: Emerald Assets for Ecosystem Services*. Baltimore Green Space, Baltimore, MD. https://baltimoregreenspace.org/wp-content/uploads/2022/02/ForestPatchesWeb.pdf

Baldock, K.C.R., M.A. Goddard, D.M. Hicks, W.E. Kunin, N. Mitschunas, H. Morse, L.M. Osgathorpe, S.G. Potts, K.M. Robertson, A.V. Scott, P.P.A. Staniczenko, G.N. Stone, I.P. Vaughan, J. Memmott. 2019. A systems approach reveals urban pollinator hotspots and conservation opportunities. *Nature Ecology & Evolution* 3(3):363–373.

Beauchamp, V.B., N. Ghuznavi, S.M. Koontz, R.P. Roberts. 2013. Edges, exotics and deer: The seed bank of a suburban secondary successional temperate deciduous forest. *Applied Vegetation Science* 16:571–584.

Belaire, J.A., A.K. Dribin, D.P. Johnston, D.J. Lynch, E.S. Minor. 2011. Mapping stewardship networks in urban ecosystems. *Conservation Letters* 4:464–473.

Beninde, J., M. Veith, A. Hochkirch. 2015. Biodiversity in cities needs space: A meta-analysis of factors determining intra-urban biodiversity variation. *Ecology Letters* 18(6):581–592.

Bernhardt, E.S., E.B. Sudduth, M.A. Palmer, J.D. Allan, J.L. Meyer, G. Alexander, J. Follastad-Shah, B. Hassett, R. Jenkinson, R. Lave, J. Rumps, L. Pagano. 2007. Restoring rivers one reach at a time: Results from a survey of U.S. river restoration practitioners. *Restoration Ecology* 15(3):482–493.

Bodin, Ö., M. Tengö, A. Norman, J. Lundberg, T. Elmqvist. 2006. The value of small size: Loss of forest patches and ecological thresholds in southern Madagascar. *Ecological Applications* 16(2):440–451.

Bolund, P., S. Hunhammar. 1999. Ecosystem services in urban areas. *Ecological Economics* 29:293–301.

Bonilla-Rodríguez, M., D. Arroyo-Lambaer, A. Castillo, L. Zambrano, Z. Cano-Santana. 2021. Urban ecological restoration: Setting priorities for restoring native vegetation in lava field remnants in Mexico City. *Frontiers in Sustainable Cities* 3:709357.

Bounds, K., M.J. Feller, J. Greenfeld, M. Heaviland, C. Pregitzer, T. Wenskus, B. Gunther, L.R. Johnson, K. King, J. Lu, M. Naidich, E. Pehek, B. Simmons, S. Stanley, E. Toth. 2015. *Guidelines for Urban Forest Restoration*. New York City Department of Parks and Recreation, Natural Resources Group, New York.

Buchholz, S., A.K. Gathof, A.J. Grossmann, I. Kowarik, L.K. Fischer. 2020. Wild bees in urban grasslands: Urbanisation, functional diversity and species traits. *Landscape and Urban Planning* 196:103731.

Bulgari, R., A. Petrini, G. Cocetta, C. Nicoletto, A. Ertani, P. Sambo, A. Ferrante, S. Nicola. 2021. The impact of COVID-19 on horticulture: Critical issues and opportunities derived from an unexpected occurrence. *Horticulturae* 7:124.

Burghardt, K.T., D.W. Tallamy, W. Gregory Shriver. 2009. Impact of native plants on bird and butterfly biodiversity in suburban landscapes. *Conservation Biology* 23(1):219–224.

Cadenasso, M.L., S.T.A. Pickett, K. Schwarz. 2007. Spatial heterogeneity in urban ecosystems: Reconceptualizing land cover and a framework for classification. *Frontiers in Ecology and the Environment* 5(2):80–88.

Campbell, L., A. Wiesen, editors. 2011. *Restorative Commons: Creating Health and Well-Being Through Urban Landscapes*. U.S. Department of Agriculture, Forest Service, Northern Research Station USDA Forest Service General Technical Report NRS-P-39. https://doi.org/10.2737/NRS-GTR-P-39

City of Seattle Parks and Recreation. 2011. *Urban Forest Restoration Program*. www.seattle.gov/parks/horticulture/forestrestoration.htm

Clemants, S.E., S.N. Handel, L.N. Sievert, C.A. De Sousa, A. Light. 2006. Restoring urban nature: Projects and process. Page 61 in R.H. Platt, editor. *The Humane Metropolis: People and Nature in the 21st-Century City*. Lincoln Institute of Land Policy, Cambridge, MA.

Connolly, J.J.T., E.S. Svendsen, D.R. Fisher, L.K. Campbell. 2014. Networked governance and the management of ecosystem services: The case of urban environmental stewardship in New York City. *Ecosystem Services* 10:187–194.

Cook, E.M., S.J. Hall, K.L. Larson. 2012. Residential landscapes as social-ecological systems: A synthesis of multi-scalar interactions between people and their home environment. *Urban Ecosystems* 15(1):19–52.

Crnic, M., M.C. Kondo. 2019. Nature Rx: Reemergence of pediatric nature-based therapeutic programs from the late 19th and early 20th centuries. *American Journal of Public Health* 109(10):1371–1378.

Dearborn, D.C., S. Kark. 2010. Motivations for conserving urban biodiversity. *Conservation Biology* 24(2):432–440.

DiCicco, J.M. 2014. Long-term urban park ecological restoration: A case study of Prospect Park, Brooklyn, New York. *Ecological Restoration* 32(3):314–326.

Di Sacco, A., K.A. Hardwick, D. Blakesley, P.H.S. Brancalion, E. Breman, L. Cecilio Rebola, S. Chomba, K. Dixon, S. Elliott, G. Ruyonga, K. Shaw, P. Smith, R.J. Smith, A. Antonelli. 2021. Ten golden rules for reforestation to optimize carbon sequestration, biodiversity recovery and livelihood benefits. *Global Change Biology* 27(7):1328–1348.

Douglas, I. 2012. Urban ecology and urban ecosystems: Understanding the links to human health and well-being. *Current Opinion in Environmental Sustainability* 4(4):385–392.

Dunn, R.R., M.C. Gavin, M.C. Sanchez, J.N. Solomon. 2006. The Pigeon Paradox: Dependence of global conservation on urban nature. *Conservation Biology* 20(6):1814–1816.

Egli, V., M. Oliver, E.-S. Tautolo. 2016. The development of a model of community garden benefits to wellbeing. *Preventive Medicine Reports* 3:348–352.

Ehrenfeld, J.G. 2000. Evaluating wetlands within an urban context. *Ecological Engineering* 15:253–265.

Elias, M., M. Kandel, S. Mansourian, R. Meinzen-Dick, M. Crossland, D. Joshi, J. Kariuki, L.C. Lee, P. McElwee, A. Sen, E. Sigman, R. Singh, E.M. Adamczyk, T. Addoah, G. Agaba, R.S. Alare, W. Anderson, I. Arulingam, S. Ḵung, V. Bellis, R. Birner, S. De Silva, M. Dubois, M. Duraisami, M. Featherstone, B. Gallant, A. Hakhu, R. Irvine, E. Kiura, C. Magaju, C. McDougall, G.D. McNeill, H. Nagendra, T.H. Nghi, D.K. Okamoto, A.M.P. Valencia, T. Pagella, O. Pontier, M. Post, G.W. Saunders, K. Schreckenberg, K. Shelar, F. Sinclair, R.S. Gautam, N.B. Spindel, H. Unnikrishnan, G.N. Wilson, L. Winowiecki. 2021. Ten people-centered rules for socially sustainable ecosystem restoration. *Restoration Ecology* 30(4):e13574.

Elmqvist, T., H. Setälä, S.N. Handel, S. van der Ploeg, J. Aronson, J.N. Blignaut, E. Gómez-Baggethun, D.J. Nowak, J. Kronenberg, R. de Groot. 2015. Benefits of restoring ecosystem services in urban areas. *Current Opinion in Environmental Sustainability* 14:101–108.

Fahrig, L. 2017. Ecological responses to habitat fragmentation per se. *Annual Review of Ecology, Evolution, and Systematics* 48:1–23.

Falk, D.A. 2017. Restoration ecology, resilience, and the axes of change. *Annals of the Missouri Botanical Garden* 102(2):201–216.

Fischer, L.K., L.M. von der Lippe, I. Kowarik. 2013. Urban land use types contribute to grassland conservation: The example of Berlin. *Urban Forestry & Urban Greening* 12:263–272.

Fisher, D.R., L.K. Campbell, E.S. Svendsen. 2012. The organisational structure of urban environmental stewardship. *Environmental Politics* 21:26–48.

Fisher, D.R., E.S. Svendsen, J. Connolly, E.S. Svendsen, J. Connolly. 2015. *Urban Environmental Stewardship and Civic Engagement: How Planting Trees Strengthens the Roots of Democracy*. Routledge, New York.

Folke, C., T. Hahn, P. Olsson, J. Norberg. 2005. Adaptive governance of social-ecological systems. *Annual Review of Environment and Resources* 30:441–473.

Gandy, M. 2016. Unintentional landscapes. *Landscape Research* 41(4):433–440.

Gann, G.D., T. McDonald, B. Walder, J. Aronson, C.R. Nelson, J. Jonson, J.G. Hallett, C. Eisenberg, M.R. Guariguata, J. Liu, F. Hua, C. Echeverría, E. Gonzales, N. Shaw, K. Decleer, K.W. Dixon. 2019. International principles and standards for the practice of ecological restoration, second edition. *Restoration Ecology* 27(S1):S1–S46.

Gaston, K.J., R.M. Smith, K. Thompson, P.H. Warren. 2005. Urban domestic gardens (II): Experimental tests of methods for increasing biodiversity. *Biodiversity & Conservation* 14:395.

Geist, C., S.M. Galatowitsch. 1999. Reciprocal model for meeting ecological and human needs in restoration projects. *Conservation Biology* 13(5):970–979.

Gerrish, E., S.L. Watkins. 2018. The relationship between urban forests and income: A meta-analysis. *Landscape and Urban Planning* 170:293–308.

Gharehaghaji, M., S. Kobal, R. Reklau, E.S. Minor. 2019. Management slows down invasion by non-native plants but does not prevent community change over 35 years in urban forests of the Midwestern USA. *Forest Ecology and Management* 448:424–431.

Gobster, P.H. 2007. Models for urban forest restoration: Human and environmental values. Pages 10–13 in J. Stanturf, editor. *Proceedings of the IUFRO Conference on Forest Landscape Restoration; 2007 May 14–19; Seoul, Korea.* International Union of Forestry Research Organizations and Korea Forest Research Institute. www.nrs.fs.usda.gov/pubs/jrnl/2007/nrs_2007_gobster_002.pdf

Gobster, P.H. 2010. Urban ecological restoration. *Nature and Culture* 5(3):227–230.

Gobster, P.H. 2011. Factors affecting people's responses to invasive species management. Pages 249–263 in I. Rotherham, R. Lambert, editors. *Invasive and Introduced Plants and Animals: Human Perceptions, Attitudes and Approaches to Management.* Earthscan, London.

Gobster, P.H. 2012. Alternative approaches to urban natural areas restoration: Integrating social and ecological goals. Pages 155–176 in J. Stanturf, D. Lamb, P. Madsen, editors. *Forest Landscape Restoration.* Springer Netherlands, Dordrecht, Netherlands.

Goddard, M.A., A.J. Dougill, T.G. Benton. 2010. Scaling up from gardens: Biodiversity conservation in urban environments. *Trends in Ecology & Evolution* 25:90–98.

Godefroid, S., N. Koedam. 2003. Distribution pattern of the flora in a peri-urban forest: An effect of the city – Forest ecotone. *Landscape and Urban Planning* 65:169–185.

Grove, M., L. Ogden, S. Pickett, C. Boone, G. Buckley, D.H. Locke, C. Lord, B. Hall. 2018. The legacy effect: Understanding how segregation and environmental injustice unfold over time in Baltimore. *Annals of the American Association of Geographers* 108(2):524–537.

Guldin, J.M., J.R. Smith, L. Thompson. 1990. Stand structure of an old-growth upland hardwood forest in Overton Park, Memphis, Tennessee. Pages 61–66 in R.S. Mitchell, C.J. Sheviak, D.J. Leopold, editors. *Ecosystem Management: Rare Species and Significant Habitats.* New York State Museum, Albany, NY.

Hauru, K., S. Lehvävirta, K. Korpela, D.J. Kotze. 2012. Closure of view to the urban matrix has positive effects on perceived restorativeness in urban forests in Helsinki, Finland. *Landscape and Urban Planning* 107(4):361–369.

Hostetler, M. 2015. *How Can We Engage Residents to Conserve Urban Biodiversity? Talk to Them.* www.thenatureofcities.com/2015/06/14/how-can-we-engage-residents-to-conserve-urban-biodiversity-talk-to-them/

Hychka, K., C.G. Druschke. 2017. Adaptive management of urban ecosystem restoration: Learning from restoration managers in Rhode Island, USA. *Society & Natural Resources* 30(11):1358–1373.

INTERLACE. 2021. *Urban Ecological Restoration – A Global Challenge.* https://interlace-project.eu/node/29

Irwin, E.G., N.E. Bockstael. 2007. The evolution of urban sprawl: Evidence of spatial heterogeneity and increasing land fragmentation. *Proceedings of the National Academy of Sciences* 104:20672–20677.

Johnson, A.L., D. Borowy, C.M. Swan. 2018. Land use history and seed dispersal drive divergent plant community assembly patterns in urban vacant lots. *Journal of Applied Ecology* 55:451–460.

Johnson, L.R., S.N. Handel. 2016. Restoration treatments in urban park forests drive long-term changes in vegetation trajectories. *Ecological Applications* 26(3):940–956.

Johnson, L.R., M.L. Johnson, M.F.J. Aronson, L.K. Campbell, M.E. Carr, M. Clarke, V. D'Amico, L. Darling, T. Erker, R.T. Fahey, K.L. King, K. Lautar, D.H. Locke, A.T. Morzillo, S. Pincetl, L. Rhodes, J.P. Schmit, L. Scott, N.F. Sonti. 2021. Conceptualizing social-ecological drivers of change in urban forest patches. *Urban Ecosystems* 24:633–648.

Johnson, L.R., T.L.E. Trammell, T.J. Bishop, J. Barth, S. Drzyzga, C. Jantz. 2020. Squeezed from all sides: Urbanization, invasive species, and climate change threaten riparian forest buffers. *Sustainability* 12(4):1448.

Kahn, P.H. Jr., T. Weiss. 2017. The importance of children interacting with big nature. *Children, Youth and Environments* 27(2):7–24.

Kattwinkel, M., R. Biedermann, M. Kleyer. 2011. Temporary conservation for urban biodiversity. *Biological Conservation* 144(9):2335–2343.

Kattwinkel, M., B. Strauss, R. Biedermann, M. Kleyer. 2009. Modelling multi-species response to landscape dynamics: Mosaic cycles support urban biodiversity. *Landscape Ecology* 24(7):929–941.
Klaus, V.H., K. Kiehl. 2021. A conceptual framework for urban ecological restoration and rehabilitation. *Basic and Applied Ecology* 52:82–94.
Korpela, K.M., M. Ylén, L. Tyrväinen, H. Silvennoinen. 2010. Favorite green, waterside and urban environments, restorative experiences and perceived health in Finland. *Health Promotion International* 25:200–209.
Koselka, E.P.D., L.C. Weidner, A. Minasov, M.G. Berman, W.R. Leonard, M.V. Santoso, J.N. de Brito, Z.C. Pope, M.A. Pereira, T.H. Horton. 2019. Walking green: Developing an evidence base for nature prescriptions. *International Journal of Environmental Research and Public Health* 16:4338.
Kowarik, I. 2011. Novel urban ecosystems, biodiversity, and conservation. *Environmental Pollution* 159(8–9):1974–1983.
Kühn, I., S. Klotz. 2006. Urbanization and homogenization – Comparing the floras of urban and rural areas in Germany. *Biological Conservation* 127(3):292–300.
Kühn, N. 2006. Intentions for the unintentional. *Journal of Landscape Architecture* 1(2):46–53.
Lerman, S.B., V.K. Turner, C. Bang. 2012. Homeowner associations as a vehicle for promoting native urban biodiversity. *Ecology and Society* 17(4):45.
Livesley, S.J., F.J. Escobedo, J. Morgenroth. 2016. The biodiversity of urban and peri-urban forests and the diverse ecosystem services they provide as socio-ecological systems. *Forests* 7:291.
Locke, D.H., B. Hall, J.M. Grove, S.T.A. Pickett, L.A. Ogden, C. Aoki, C.G. Boone, J.P.M. O'Neil-Dunne. 2021. Residential housing segregation and urban tree canopy in 37 US Cities. *NPJ Urban Sustainability* 1:1–9.
Martínez, M.L., F. López-Barrera. 2008. Special issue: Restoring and designing ecosystems for a crowded planet. *Écoscience* 15(1):1–5.
McDonald, T., G.D. Gann, J. Jonson, K.W. Dixon, editors. 2016. *International Standards for the Practice of Ecological Restoration – Including Principles and Key Concepts*, 1st edition. Society for Ecological Restoration, Washington, DC.
McDonnell, M.J. 2011. The history of urban ecology. Pages 5–13 in J. Niemelä, J.H. Breuste, T. Elmqvist, G.R. Guntenspergen, P. James, N.E. McIntyre, editors. *Urban Ecology: Patterns, Processes and Applications*. Oxford University Press, Oxford.
McWilliam, W., P. Eagles, M. Seasons, R. Brown. 2012. Evaluation of planning and management approaches for limiting residential encroachment impacts within forest edges: A Southern Ontario case study. *Urban Ecosystems* 15(3):753–772.
Miyazaki, Y. 2018. *Shinrin Yoku: The Japanese Art of Forest Bathing*. Timber Press, Portland, OR.
Moore, M.M., W. Wallace Covington, P.Z. Fulé. 1999. Reference conditions and ecological restoration: A southwestern ponderosa pine perspective. *Ecological Applications* 9(4):1266–1277.
Mullins, L., S. Charlebois, E. Finch, J. Music. 2021. Home food gardening in Canada in response to the COVID-19 pandemic. *Sustainability* 13(6):3056.
Nassauer, J.I. 1995. Messy ecosystems, orderly frames. *Landscape Journal* 14(2):161–170.
Nassauer, J.I. 2011. Care and stewardship: From home to planet. *Landscape and Urban Planning* 100(4):321–323.
Nilon, C.H. 2011. Urban biodiversity and the importance of management and conservation. *Landscape and Ecological Engineering* 7(1):45–52.
Nilon, C.H., M.F.J. Aronson, S.S. Cilliers, C. Dobbs, L.J. Frazee, M.A. Goddard, K.M. O'Neill, D. Roberts, E.K. Stander, P. Werner, M. Winter, K.P. Yocom. 2017. Planning for the future of urban biodiversity: A global review of city-scale initiatives. *BioScience* 67(4):332–342.
Nowak, D.J., J.F. Dwyer. 2001. Urban forest structure, benefits, and value. Pages 24–26 in C. Kollin, editor. *National Urban Forest Conference Proceedings*. American Forests, Washington, DC.
Oberndorfer, E., J. Lundholm, B. Bass, R.R. Coffman, H. Doshi, N. Dunnett, S. Gaffin, M. Köhler, K.K.Y. Liu, B. Rowe. 2007. Green roofs as urban ecosystems: Ecological structures, functions, and services. *BioScience* 57(10):823–833.
Ossola, A., D. Locke, B. Lin, E. Minor. 2019. Yards increase forest connectivity in urban landscapes. *Landscape Ecology* 34:2935–2948.
Padullés Cubino, J., J. Cavender-Bares, S.E. Hobbie, D.E. Pataki, M.L. Avolio, L.E. Darling, K.L. Larson, S.J. Hall, P.M. Groffman, T.L.E. Trammell, M.K. Steele, J.M. Grove, C. Neill. 2018. Drivers of plant

species richness and phylogenetic composition in urban yards at the continental scale. *Landscape Ecology* 34:63–77.

Palomar, C. 2010. From the ground up: Why urban ecological restoration needs environmental justice. *Nature and Culture* 5:277–298.

Platt, R.H., R.A. Rowntree, P.C. Muick. 1994. *The Ecological City: Preserving and Restoring Urban Biodiversity*. University of Massachusetts Press, Amherst, MA.

Pregitzer, C.C., S. Charlop-Powers, C. McCabe, A. Hiple, B. Gunther, M.A. Bradford, editors. 2019. *Untapped Common Ground: The Care of Forested Natural Areas in American Cities*. Natural Areas Conservancy, New York.

Rink, D., T. Arndt. 2016. Investigating perception of green structure configuration for afforestation in urban brownfield development by visual methods – A case study in Leipzig, Germany. *Urban Forestry & Urban Greening* 15:65–74.

Romolini, M. 2012. *What is Urban Environmental Stewardship? Constructing a Practitioner-Derived Framework*. Note PNW-RN-566. Department of Agriculture, Forest Service, Pacific Northwest Research Station, Portland, OR.

Rosenzweig, M.L. 2003. Reconciliation ecology and the future of species diversity. *Oryx* 37(2):194–205.

Rudd, H., J. Vala, V. Schaefer. 2002. Importance of backyard habitat in a comprehensive biodiversity conservation strategy: A connectivity analysis of Urban green spaces. *Restoration Ecology* 10(2):368–375.

Rupprecht, C.D.D., J.A. Byrne. 2014. Informal urban greenspace: A typology and trilingual systematic review of its role for urban residents and trends in the literature. *Urban Forestry & Urban Greening* 13:597–611.

Sanesi, G., G. Colangelo, R. Lafortezza, E. Calvo, C. Davies. 2017. Urban green infrastructure and urban forests: A case study of the Metropolitan Area of Milan. *Landscape Research* 42:164–175.

Schilthuizen, M. 2018. *Darwin Comes to Town: How the Urban Jungle Drives Evolution*. Picador, New York.

SERI. 2004. *The SER International Primer on Ecological Restoration*. The Society for Ecological Restoration International, Tucson, AZ. www.ser.org

Smith, R.F., R.J. Hawley, M.W. Neale, G.J. Vietz, E. Diaz-Pascacio, J. Herrmann, A.C. Lovell, C. Prescott, B. Rios-Touma, B. Smith, R.M. Utz. 2016. Urban stream renovation: Incorporating societal objectives to achieve ecological improvements. *Freshwater Science* 35:364–379.

Song, C., H. Ikei, Y. Miyazaki. 2016. Physiological effects of nature therapy: A review of the research in Japan. *International Journal of Environmental Research and Public Health* 13(8):781.

Sorensen, A.E., R.C. Jordan, G. Blaise, J. Brown, L.K. Campbell, M.F.J. Aronson, M.L. Johnson. 2018. Drivers of public participation in urban restoration stewardship programs: Linkages between environmental identity and knowledge, and motivations. *Arboriculture & Urban Forestry* 44(6):266–282.

St. George, Z. 2016, October 18. *How Flammable is a Eucalyptus Tree*? https://baynature.org/article/burning-question-east-bay-hills-eucalyptus-flammable-compared/

Sukopp, H. 2008. The city as a subject for ecological research. Pages 281–298 in J.M. Marzluff, E. Schulenberger, W. Endlicher, M. Alberti, G. Bradley, C. Ryan, U. Simon, C. ZumBrennen, editors. *Urban Ecology: An International Perspective on the Interaction Between Humans and Nature*. Springer, Cham, Switzerland.

Svendsen, E.S. 2009. Cultivating resilience: Urban stewardship as a means to improving health and well-being. Pages 58–87 in L. Campbell, A. Wiesen, editors. *Restorative Commons: Creating Health and Well-Being Through Urban Landscapes*. U.S. Department of Agriculture, Forest Service, Northern Research Station Gen. Tech Rep. NRS-P-39. https://doi.org/10.2737/NRS-GTR-P-39

Svendsen, E.S., L. Campbell. 2008. Urban ecological stewardship: Understanding the structure, function and network of community-based urban land management. *Cities and the Environment (CATE)* 1(1).

Tallamy, D.W., R. Darke. 2009. *Bringing Nature Home: How You Can Sustain Wildlife with Native Plants*. Timber Press, Portland, OR.

Terada, T., M. Yokohari, J. Bolthouse, N. Tanaka. 2010. 'Refueling' Satoyama woodland restoration in Japan: Enhancing restoration practice and experiences through woodfuel utilization. *Nature and Culture* 5:251–276.

Theodorou, A., A. Panno, G. Carrus, G.A. Carbone, C. Massullo, C. Imperatori. 2021. Stay home, stay safe, stay green: The role of gardening activities on mental health during the Covid-19 home confinement. *Urban Forestry & Urban Greening* 61:127091.

Turner, W.R., T. Nakamura, M. Dinetti. 2004. Global urbanization and the separation of humans from nature. *BioScience* 54:585–590.

Tyrväinen, L., S. Pauleit, K. Seeland, S. de Vries. 2005. Benefits and uses of urban forests and trees. Pages 81–114 in C. Konijnendijk, K. Nilsson, T. Randrup, J. Schipperijn, editors. *Urban Forests and Trees: A Reference Book*. Springer Berlin Heidelberg, Berlin, Heidelberg, Germany.

Van den Berg, A.E., T. Hartig, H. Staats. 2007. Preference for nature in urbanized societies: Stress, restoration, and the pursuit of sustainability. *Journal of Social Issues* 63(1):79–96.

van den Bosch, M., Å. Ode Sang. 2017. Urban natural environments as nature-based solutions for improved public health – A systematic review of reviews. *Environmental Research* 158:373–384.

van der Jagt, A.P.N., M. Smith, B. Ambrose-Oji, C.C. Konijnendijk, V. Giannico, D. Haase, R. Lafortezza, M. Nastran, M. Pintar, Š. Železnikar, R. Cvejić. 2018. Co-creating urban green infrastructure connecting people and nature: A guiding framework and approach. *Journal of Environmental Management* 233:757–767.

Von der Lippe, M., I. Saumel, I. Kowarik. 2005. Cities as drivers for biological invasions – The role of urban climate and traffic. *Die Erde: Zeitschrift der Gesellshaft für Erdkunde zu Berlin* 136(2):123.

Wallace, K.J., B.D. Clarkson. 2019. Urban forest restoration ecology: A review from Hamilton, New Zealand. *Journal of the Royal Society of New Zealand* 49(3):347–369.

Watkins, S.L., E. Gerrish. 2018. The relationship between urban forests and race: A meta-analysis. *Journal of Environmental Management* 209:152–168.

Williams, N.S.G., J. Lundholm, J.S. MacIvor. 2014. FORUM: Do green roofs help urban biodiversity conservation? *Journal of Applied Ecology* 51(6):1643–1649.

Yocom, K. 2014. Building watershed narratives: An approach for broadening the scope of success in urban stream restoration. *Landscape Research* 39(6):698–714.

25

CITIES ARE PLANNING FOR BIODIVERSITY

A global survey of city plans

Karen M. O'Neill, Myla F.J. Aronson, Charles H. Nilon, Sarel S. Cilliers, Cynnamon Dobbs, Lauren J. Frazee, Mark A. Goddard, Debra Roberts, Emilie K. Stander, Peter Werner, and Ken P. Yocom

Introduction

The importance of managing biodiversity and ecosystem services in cities and other built environments has been increasingly recognized (Bolund and Hunhammar 1999; Shwartz et al. 2014; McDonald et al. 2018). Ecological research shows cities in all bioregions support considerable biodiversity (Aronson et al. 2014), including threatened species (Ives et al. 2016). A variety of plant and animal species can be found in city parks, in remaining indigenous landscapes, and in wetland margins (Hostetler et al. 2011; Aronson et al. 2017), but also in highly managed or artificial habitats such as gardens (e.g. Goddard et al. 2010), roadsides (Cilliers and Bredenkamp 2000), and constructed wetlands (e.g. Holzer 2014). The loss of urban biodiversity and ecosystem services are examples of "local-cumulative" problems, which may be noticed first at the local level but can accumulate globally (Conca 2006). This also means cities that monitor and plan to protect ecological conditions could contribute to global-scale biodiversity and ecosystem services goals and to broader sustainability initiatives, such as the sustainable development goals (SDGs). Studies of the ways city plans discuss biodiversity and ecosystem services have included case studies of single cities and comparisons of cities within a region or country (Elander et al. 2005; Miller et al. 2009; Stokes et al. 2010; McPhearson et al. 2014; Kabisch 2015; Ahmed and Puppim de Oliveira 2016; but see Nilon et al. 2017). This chapter presents a broader comparison by analyzing planning documents for large cities around the world in five topical categories: comprehensive master plans, climate action plans, ecosystem services and sustainability plans, green infrastructure and open space plans, and wildlife conservation and biodiversity action plans.

The general research question here is whether cities that discuss biodiversity and ecosystem services in their planning documents include similar combinations of goals (statements of intent) and measurable targets (benchmarks for goals). Including a goal or target in a plan may not ultimately improve environmental conditions, but understanding how cities are discussing these topics is a step toward learning whether, and how, they are trying to do so. Analyses included comparing the contents of planning documents for cities in the Global North and the Global South and comparing the contents of the five categories of planning documents. This exploratory study assessed whether there is an emerging set of similar concerns, even in cities and in topical

 DOI: 10.4324/9781003016120-29

categories of plans where biodiversity and ecosystem services are not explicitly stated as top concerns. The next section describes how people working on ecological research and on conservation efforts in individual cities influenced recent international recommendations for urban planning. This is followed by a discussion of the concept of policy mobility from planning theory, which explores similarities in planning ideas and practices across cities. The results assess whether plan content is converging and indicate the extent to which international scientific guidelines for biodiversity and ecosystem services are included in the sampled planning documents.

Local and international promotion of urban biodiversity and ecosystem services

Cities manage a growing proportion of the world's people and land, making them crucial jurisdictions for sustaining biodiversity and ecosystem services (Dearborn and Kark 2010; Seto et al. 2012; Güneralp et al. 2015; McDonald et al. 2018). Even amidst this process of urbanization, there has been a long history of locally initiated urban biodiversity planning. Planners and scientists involved in these projects created environmental networks to share ideas and built local institutions responding to local environmental conditions (Bolund and Hunhammar 1999; Alario 2000; Sukopp 2002; Adams 2005; Shwartz et al. 2014). Advocates also used these local experiences and scientific research to urge international institutions to promote scientifically based standards for urban biodiversity planning (Puppim de Oliveira et al. 2011; Wilkinson et al. 2013). Urban biodiversity was eventually included in major international policy regimes. The Curitiba Declaration of 2007 and the Conference of Parties 10 (COP) in 2010 encouraged member countries in the Convention on Biological Diversity (CBD) to involve cities when they write national biodiversity plans (Puppim de Oliveira et al. 2011). During the 2012 UN Conference on Sustainable Development in Rio de Janeiro, Brazil, cities were recognized as capable of promoting economic, social, and environmental sustainability, especially through urban green spaces and improvements in water and air quality (Secretariat of the Convention on Biological Diversity 2012). The Intergovernmental Panel on Climate Change (IPCC 2014) also highlighted city-scale actions on ecosystems and biodiversity as one option for making cities more resilient to climate change. The first joint report of the Intergovernmental Science-Policy Platform on Biodiversity and Ecosystem Services (IPBES) and the IPCC identifies cities and other urbanized areas as critical for sustaining biodiversity and advises planning to integrate climate and biodiversity policies for these sites (Pörtner et al. 2021).

Ecosystem services have been discussed internationally as being related to biodiversity and to sustainable economic development (World Commission on Environment and Development 1987). Ecosystem services are defined in several ways by economists, biologists, and others. The Millennium Ecosystem Assessment (2003) conceptualized ecosystem services as supporting human needs for clean water and air, energy, and well-being (Elmqvist et al. 2013), and the international initiative on the Economics of Ecosystems and Biodiversity (TEEB 2011) created guidelines for assessing ecological services and valuing them economically. The IPBES uses measures for biodiversity and ecosystem services in its ongoing assessments of urban conditions (Montanarella et al. 2019). The United Nation's Sustainable Development Agenda for 2030 (2015) affirms ecosystem services by linking human rights with environmental concerns and development and by promoting sustainable development goals for cities (in Goal 11).

Cities are not formal parties to international agreements but are involved in international discussions about the environment because they receive aid, suggestions for model practices (Schewenius et al. 2014), and technical assistance from international institutions and scientific organizations (Müller and Werner 2010; Puppim de Oliveira et al. 2011). International organizations contributed to guidelines for writing Local Biodiversity Action Plans (ICLEI 2013) and to the City Biodiversity Index (Chan et al. 2021), helping cities to measure their efforts over time and to compare themselves to other cities. Some guidelines use ecosystem services concepts in their biodiversity planning documents (TEEB 2011), an approach cities in Europe have also widely adopted (Kabisch 2015). The mutual influence of players at different scales is reflected in initiatives like the Urban Biodiversity Hub (2019), which involves volunteers around the world advising cities how to create and assess biodiversity strategies and action plans related to the Convention on Biodiversity.

Policy mobility: understanding similarities in plans

It would be impractical and likely unhelpful to attempt to track whether one or another of the initiatives discussed previously influenced plan developers around the world to include similar ideas in their planning documents. The international initiatives overlapped, and their recommendations and outreach efforts changed over time. In addition, people who developed plans may have learned separately about conservation projects in other cities or about findings from ecological science. Acknowledging these processes, urban planning theory uses the concept of policy mobility to characterize how similarities emerge in planning. Theorists created this concept to analyze why some ideas for city development get widely copied (e.g. waterfront entertainment zones). But they also use policy mobility to investigate how planners can be inspired by ideas and tailor them to meet local needs (Sanyal 2005; Hoyt 2006; Harris and Moore 2013).

Research on policy mobility counters what Friedmann (1971) notes was a longtime presumption in planning education, that practices are devised by top planners in the Global North and adopted directly by less innovative planners in the Global South. Instead, empirical research finds that while promoters of model practices (e.g. consultants, international foundations, nongovernmental organizations) may spread some ideas, policy mobility also emerges from more grounded practices such as planners' day-to-day decisions as they select from the many ideas available to them locally at a particular moment (Hoyt 2006; Robinson 2015). Stylized stories about best practices therefore do emerge and spread, often being characterized by enterprising consultants as universally valuable. But local officials take varied lessons from these stories, adjusting them to suit their city and often to fit decisions they have already made (González 2011). Even when an idea from one locale is directly used in another locale, it is necessarily changed by local planning practices, histories, resource constraints, and institutions (Freestone 2000; McCann and Ward 2010; Peck and Theodore 2010; Peck 2011).

Cities in the Global South have distinctive forms, social processes, and ideas for planning that affect policy mobility (Robinson 2006). Indeed, practices and policies from developing countries that recognize cultural practices, social justice, subsistence uses of nature, and distinctive features of the local environment are increasingly being promoted and used by cities around the world (González 2011; Söderström and Geertman 2013; Robinson 2015). The policy mobility concept therefore takes a locally specific, multi-actor perspective to understanding planning.

Research approach and questions

Although a case study might trace the origin and pathway of a planning idea (e.g. González 2011), in the comparative analyses presented here, the aim was instead to consider whether local approaches to planning are converging, no matter what their origin. A supporting aim was to consider whether city plans include ideals affirmed by scientific and international institutions. One broad principle supported by the international institutions and agreements discussed previously is to encourage cities in all countries to do active and participatory planning, regardless of their development status (Cilliers et al. 2021). Another recommendation is to include biodiversity and ecosystem services in planning documents on other topics so these aims are "mainstreamed," rather than being pushed to the side while the city prioritizes other goals (ICLEI 2013; Cilliers et al. 2021). A third recommendation is that cities use concepts of ecosystem services to promote biodiversity (TEEB 2011). A fourth recommendation by planners (American Planning Association 2006), ecological scientists, and international initiatives like the City Biodiversity Index (Chan et al. 2021) is that city plans set measurable targets for environmental improvements, using quantitative measures such as the number of species in a conserved patch to assess whether conditions are improving or declining. These four ideals promote scientific principles, planning professionalism, and accountability to elite institutions.

Content analysis is a common method for assessing the qualities of planning documents (Stevens et al. 2014). This method determines whether preselected criteria are present in sampled plans (Lyles and Stevens 2014). Planning documents are important to study because they remain at the heart of urban planning practice (Friedmann 2005). Plans state intent and are important expressions of concern, even if they are not matched by a city's current actions (Berke and Godschalk 2009). The specific questions for this content analysis we derived from planning research and the concept of policy mobility are as follows:

1. *Are there distinctive patterns in the goals and measurable targets in planning documents for cities in the Global South versus cities in the Global North?* This addresses whether internationally circulating ideas have produced similar content in the plans of cities in the Global South and the Global North. We used principal components analysis (PCA) to analyze cities' selection of attributes (goals and targets) across all their plans. If cities in the Global South and Global North take divergent approaches to planning for biodiversity and ecosystem services by selecting different sets of goals and targets, the PCA results would show separate clusters of cities divided by development status.
2. *Are there distinctive patterns in the goals and measurable targets included in five topical categories of planning documents?* Using PCA, we assessed whether plan developers have kept biodiversity and ecosystem services as boutique topics limited to specialty planning documents on biodiversity and ecosystem services or have "mainstreamed" these concerns across plans on different topics. We analyzed five categories of plans on topics most likely to reflect efforts at mainstreaming (Table 25.1). We tabulated and analyzed the presence of goals and targets on broad topics (e.g. climate change adaptation) rather than analyzing locally specific goals and targets (e.g. planting drought-resistant species). If concern for biodiversity and ecosystem services remained limited to specialty plans on those topics, the PCA graph should show separate clusters of attributes (goals and targets) related to the plans' topics.

   2a. *Do the frequencies of types of* goals *vary across the five categories of planning documents?*
   2b. *Do the frequencies of the types of* measurable targets *vary across the categories of planning documents?*

Complementing the PCA done to answer question 2, frequencies provided detail about the goals and targets included in each of the five categories of planning documents. Frequencies also helped evaluate whether urban plans follow scientific recommendations to use concepts about ecosystem services and to set measurable targets.

Table 25.1 Five categories of urban planning documents that address biodiversity and/or ecosystem services conservation

Theme	*Main goals*	*Historical context, support, motivations, and/or examples*
Wildlife and biodiversity (n = 35)	To conserve and protect biological species and their associated habitats	– Stated motivations range from primarily benefitting nature (i.e. protecting an endangered yet economically unimportant frog species) to primarily benefitting humankind (i.e. protecting fish species that are important for recreation; Dearborn and Kark 2010). – Often produced at the national and sub-national scale because of the Convention on Biological Diversity (Puppim de Oliveira et al. 2011; Wilkinson et al. 2013). – ICLEI (Local Governments for Sustainability) provides support for city governments to produce Local Biodiversity Strategies and Action Plans (LBSAPs; Avlonitis et al. 2013). For example, the United Kingdom outlines specific processes necessary for their municipalities to produce LBSAPs (UKLIAG 1997).
Green infrastructure and open space plans (n = 31)	To increase and improve the form and structure of green space and blue corridors for recreational and ecological benefits	– Protecting green areas and corridors in the city are typical in regional and urban conservation planning strategies (Bolund and Hunhammar 1999; Hostetler et al. 2011). – Spatial planning for green networks in many European cities deals with elements from studies of landscape ecology (e.g. patch size and connectivity; Baycan-Levent and Nijkamp 2009). For example, greening became popular in the US between 2004 and 2008 as nine large cities and municipalities proposed to plant a total of 11 million new trees during that time (Young 2011).
Sustainability and ecosystem services (ESS) (n = 29)	To manage the provisioning of ecosystem services through functional green space conservation and restoration	– Urban land managers and decision-makers have become increasingly aware of the services that nature provides to human beings (Kabisch 2015). – The TEEB Manual for Cities gives cities practical guidelines and information for incorporating ecosystem services conservation into planning and management practices (TEEB 2011). Often highlights nature-based solutions to problems such as stormwater control. Plans in this category may extend to ecosystem services attributes that we did not code and count for this study because they do not directly support biodiversity, such as city purchasing practices and waste reduction (Bolund and Hunhammar 1999; Plieninger et al. 2013; Kabisch 2015). For example, an ESS framework is the backbone of recent urban planning documents in Berlin (Kabisch 2015).

(*Continued*)

Table 25.1 (Continued)

Theme	*Main goals*	*Historical context, support, motivations, and/or examples*
Comprehensive and master development (n = 18)	To outline and delegate sustainable management goals and duties for all economic, social, political, historical, and environmental city resources	– Provide the overall vision and/or action plans for land development, community development, and a range of other concerns in a city (Friedmann 1971). – Describe long-term development objectives, informs and instructs specialist planners in their objectives, and support inter-agency collaborations to best further the common good (Haar 1955; Altschuler 1965). – Can be pivotal in directing land use and development at local scales (Norton 2008). – Includes provisioning land use patterns, employing maintenance measures, and enacting the fiscal and administrative programming necessary to achieve local needs (Friedmann 1971).
Climate action, adaptation, and energy (n = 22)	To mitigate and adapt to the impacts of a changing climate and the urban heat island effect	– ICLEI proposes that climate change planning involve inventorying emissions, setting emissions targets, developing new policies, implementing policy changes, and establishing a monitoring regime, yet evidence of the final two steps is scarce in "first generation" climate planning documents (Wheeler 2008). – Often aims to reduce a city's production of greenhouse gases and to improve a city's capacity to adapt to climate change effects that are already unavoidable and to recommend policy changes in sectors such as transportation, land use, and food production (Wheeler 2008).

Methods

We initiated a four-stage process: plan selection, coding for content analysis, plan characterization, and analysis of plans' goals and targets.

Plan selection and coding

Because only some cities are planning for biodiversity and ecosystem services, we took a purposive sample of cities that have at least one plan at the city or metropolitan scale that includes a goal related to biodiversity. A random sample of cities to determine the prevalence of these issues in planning would be productive only if the practices become more universal. Data come from large cities (populations between 200,000 to 20,000,000). Ecological scientists are especially interested in learning about the plans of cities that have biological baseline data, and so we began the dataset with cities that plan for biodiversity and have data included in biodiversity databases, as described by Aronson and colleagues (Aronson et al. 2014) and by Dobbs and colleagues (Dobbs et al. 2014). We added cities from bioregions underrepresented in these ecological databases, asking ecological scientists and planners about cities likely to have suitable plans. The 40 cities selected are broadly distributed geographically and are in countries with varied levels of economic development (details about city characteristics are given in Nilon et al. 2017).

The aim of this study is to consider the general circulation of ideas, no matter who wrote them. Authorship of the plans depends on domestic political arrangements. In some countries, cities have autonomy over planning for biodiversity. In other countries, national priorities may drive all or most of a city's planning for the environment, often with the aim of making urban plans consistent with national laws or with the Convention on Biodiversity or other international agreements (Elander et al. 2005). For instance, the United Kingdom, China, Vietnam, and other countries require or encourage their cities to write local biodiversity action plans, and South African cities align city- and provincial-level bioregional planning (UK Local Issues Advisory Group (UKLIAG) 1997; Pisupati 2007; Puppim de Oliveira et al. 2011). These differences in plan writing illustrate the complex pathways of ideas that led scholars to develop the policy mobility concept.

From January to December 2014, the team searched for all publicly available master plans and plans on the specialty topics of conservation, green infrastructure, ecosystem services, and climate, yielding 135 documents (see Nilon et al. 2017). Plans were accessed through online searches or provided by city employees or consultants. For most cities, we were able to contact a local official to confirm our collection of current and relevant plans. We were unable to identify city- or metropolitan-level plans that included biodiversity conservation or ecosystem services for cities in the Middle East and North Africa, although such plans may exist. Most plans were written in English, but we also selected and coded plans in Portuguese, Spanish, Dutch, German, French, Chinese, and Italian. Plans in languages other than English were assessed by team members with knowledge of those languages or were translated by others fluent in the language and then assessed by research team members.

Content analysis

Each plan was scored for the presence of 34 goals and measurable targets for biodiversity and ecosystem services that we identified in the course of coding (Table 25.2). Plan analysts are experts in biodiversity science or urban biodiversity conservation. Analysts were provided with keywords and concepts to use as identifiers and were trained in coding protocols. Multiple researchers evaluated each plan, with final discussions completed by the entire team in person as needed, to assure reliability in coding (Stevens et al. 2014). Biodiversity goals and targets included topics such as controlling invasive species, collecting baseline data on species, and increasing connectivity along corridors between land parcels (attributes mentioned in reports such as ICLEI 2013). Ecosystem services goals and targets were based on the Millennium Ecosystem Assessment classifications of regulating services (air and water quality, carbon sequestration, and urban heat island effect), provisioning services (urban agriculture), and cultural services (TEEB 2011).

Plan characterization

We selected and analyzed all plans available from each city likely to be in the five categories of planning documents and placed each plan into the category fitting its overarching goal (Table 25.1). Most plans were easily categorized. For several plans that could have been assigned to multiple categories, all research team members were consulted to select the appropriate category (e.g. ecosystem service plans often overlap with climate action plans or green infrastructure plans).

Countries' development status

Concepts based on economic metrics continue to affect international discourse and are therefore useful, if imperfect and politically inflected, indicators of ideas about the Global North and Global

Table 25.2 Biodiversity and ecosystem attributes in 135 planning documents from 40 cities globally (stewardship and monitoring, spatial, and species; see Nilon et al. 2017) and three categories of ecosystem services goals (regulating, provisioning, and cultural services)

Attribute	*Category*	*Attribute description*
Biodiversity Goals		Does the plan have specific and/or general biodiversity goals?
Connectivity	Spatial	Corridors and/or increasing connectivity for ecological purposes or to create a green network.
Species	Species	Conservation of native/indigenous species.
Habitats	Spatial	Conserve or restore/maintain/manage natural and semi-natural habitats.
Education	Stewardship/ Monitoring	Formal and informal education and/or outreach related to biodiversity conservation.
Stewardship	Stewardship/ Monitoring	Encourage volunteer groups/NGOs/community engagement/citizen science related to biodiversity conservation.
Monitoring	Stewardship/ Monitoring	Species and habitat monitoring, ecological research, adaptive management.
Invasive species	Species	Management of invasive alien species.
Constructed habitats	Spatial	Constructed habitats such as bioswales, green roofs, green streets, rain gardens, and gardens/yards.
Ecosystem Service Goals		Does the plan have specific and/or general ecosystem service goals?
Water quality	Regulating	Water quality amelioration and flood retention, including stormwater, freshwater wetlands, lakes, salt marshes, floodplains, and riparian areas.
Air quality	Regulating	Increasing vegetation/tree cover for air pollution amelioration.
Carbon sequestration	Regulating	Tree planting efforts, conservation of forests for carbon storage or sequestration purposes.
Urban heat island	Regulating	Tree planting efforts, conservation of forests for urban heat island amelioration.
Urban agriculture	Provisioning	Food production, urban gardens.
Cultural	Cultural	Nature conservation for sense of place, education, stewardship, recreation.
Biodiversity Targets		
Species		Targets for increasing populations of species identified by the plan for conservation.
Habitat		Targets for increasing habitat area identified by the plan for conservation.
Taxa		Targets for particular taxa identified for conservation.
Invasive		Targets for decreasing invasive/alien/non-native species.
Critical habitat		Targets for increasing critical biodiversity habitats.
Constructed		Quantitative for increasing the number or area of constructed habitats.
Other		Other targets related to biodiversity conservation.
Ecosystem Service Targets		
Water quality		Targets for reduction in water pollutants or increase in wetland habitat.
Air quality		Targets for reduction of air pollutants by increasing vegetation or tree cover.

Attribute	*Category*	*Attribute description*
Carbon sequestration		Targets for increasing the number of trees/biomass for carbon storage and sequestration purposes?
Urban heat island		Targets for tree planting, conservation of forests, mentioned for climate amelioration.
Urban agriculture		Targets for food production and/or urban gardens.
Cultural		Ecosystem service conservation for sense of place, education, stewardship, recreation.
Miscellaneous		
City Data		Does the plan use baseline data collected from within the city?
Habitats		Baseline data for specific habitats.
Species		Baseline data for specific species.
Commitment to implementation		Dedicated funds or actions for implementation of the biodiversity or ecosystem services goals in the plan.
Regulatory elements		Mandated elements of the plan.

South. The United Nations identifies 47 Least Developed Countries (United Nations Committee for Development Policy 2018) but has no universally designated lists for other categories. We used an unofficial list of developed and developing countries from the UN to categorize cities in the Global North versus the Global South (United Nations Conference on Trade and Development 2018). This yielded 13 of 40 cities in our sample that are in developing countries (i.e. the Global South), including Monrovia, Liberia, the only city in our sample that is also on the Least Developed Country list. To test this approach to categorizing development status, we also ran analyses using other approaches, as discussed in the results section later.

Data analysis

We answered questions 1 and 2 using principal components analysis (PCA) to visually identify possible clusters of goals and targets across the five categories of plans, using PCORD 6.08 (MjM Software, Gleneden Beach, OR). PCA is similar to factor analysis and is often a step in identifying typologies within a dataset that might reflect the actions of distinct social or environmental phenomena. Distinctive groupings, either across the five planning document categories or across cities in the Global South or Global North, could indicate cities are taking distinctive approaches to planning rather than sharing similar approaches through policy mobility. The PCA illustrated how planning documents grouped according to the 34 attributes (goals and targets) and identified which factors (i.e. attributes) drove the differences between the attribute composition of plans in multivariate space. Frequencies were calculated for questions 2a and 2b to allow several comparisons. For simpler visual display (Table 25.2), each goal was grouped into one of six categories: spatial, species, stewardship/monitoring, regulating services, provisioning services, and cultural services. Because we found no other studies with these analyses, we had no specific quantitative guidelines for determining whether plans are convergent or divergent. We used general guidelines for assessing clusters in the PCA runs and compared results across our analyses.

Results

PCA analysis for question 1 found that while cities differed in their selection of goals and targets, these differences did not result in distinct clusters based on the development status of each city's country (Figure 25.1). In the PCA, 67% of the variance was explained by the attributes included in the analysis (10 vectors with eigenvalues >1). Loading factors for the attributes in each component did not reach eigenvalues greater than 0.3, showing there was no dominance of any attribute in defining differences between cities in the Global South and cities in the Global North, as informally categorized by the UN (United Nations Conference on Trade and Development 2018). We also found no distinct clusters when we modified the informal UN list by switching cities in recently booming countries, like Brazil and South Africa, into the category of developed countries (figures for that analysis are not presented here).

For the second set of questions, results showed that plans in different topical categories also did not have markedly different combinations of goals and measurable targets (Figure 25.2). The five plan categories differed in their proportions of goals and targets for biodiversity and ecosystem services, but they did not divide into clear clusters of plans emphasizing either biodiversity or ecosystem services. The main components resulting from the PCA for Question 2 were selected using eigenvalues > 1. Variability in the types of attributes included in planning documents was wide. In the PCA, 67% of the variance was explained by the attributes included in the analysis (10 vectors with eigenvalues >1). Loading factors for the attributes in each component did not reach

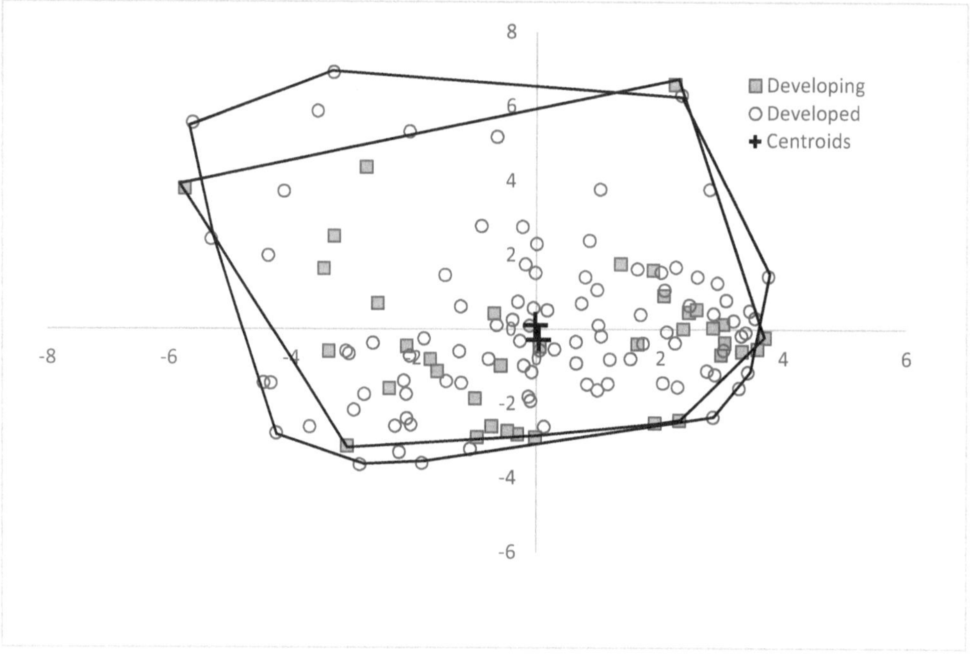

Figure 25.1 PCA ordination of attributes addressed in 135 plans from 40 cities. Convex hull polygons encircle points. The polygons overlap, indicating that there is no difference in the goals and targets that plans from the Global South (developing) and Global North (developed) address.

eigenvalues greater than 0.3, showing there was no dominance of any attribute in defining the type of planning document. The plans did not separate into distinct clusters defined by the plans' topics (Figure 25.2).

For question 2a, frequencies showed that some groups of goals were "mainstreamed" across all five planning document categories, with goals for spatially based biodiversity conservation (50–94%) and goals for regulating ecosystem services (45–96%) being most included (Figure 25.3). As would be expected, topical plans included goals specific to their topics more frequently. Significant differences in the frequencies between goals for biodiversity versus ecosystem services were therefore found among planning document categories (Pearson's Chi-squared test: $X^2 = 70.21$, df = 20, $p < 0.0001$). Biodiversity goals related to species conservation were significantly more frequent in biodiversity plans than would be expected by chance alone ($p < 0.05$). Ecosystem service goals related to regulating services were significantly more frequent in climate plans and significantly less frequent in biodiversity plans than would be expected by chance ($p < 0.05$).

Plans had fewer measurable targets than goals (question 2b). Overall, 52% of the planning documents included a target for biodiversity (Figure 25.4). Habitat conservation was the most common biodiversity target, appearing in 27–50% of the plans in each category. Significant differences in the frequencies of biodiversity targets were found among categories of planning documents (Pearson's Chi-squared test: $X^2 = 92.87$, df = 28, $p < 0.0001$). Targets related to species

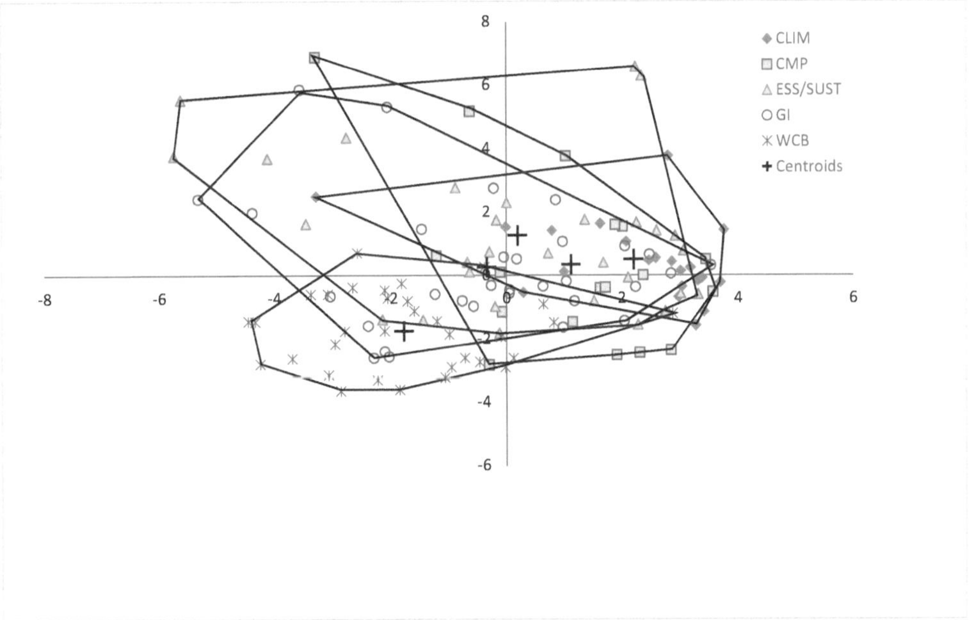

Figure 25.2 PCA ordination of attributes addressed in 135 plans from 40 cities. Convex hull polygons encircle points. The polygons overlap, indicating that there is no difference in the goals and targets that different types of plans address. CLIM: climate adaptation plans, CMP: comprehensive master plans, ESS/SUST: ecosystem services and sustainability, GI: green infrastructure, WCB: wildlife conservation and biodiversity plans.

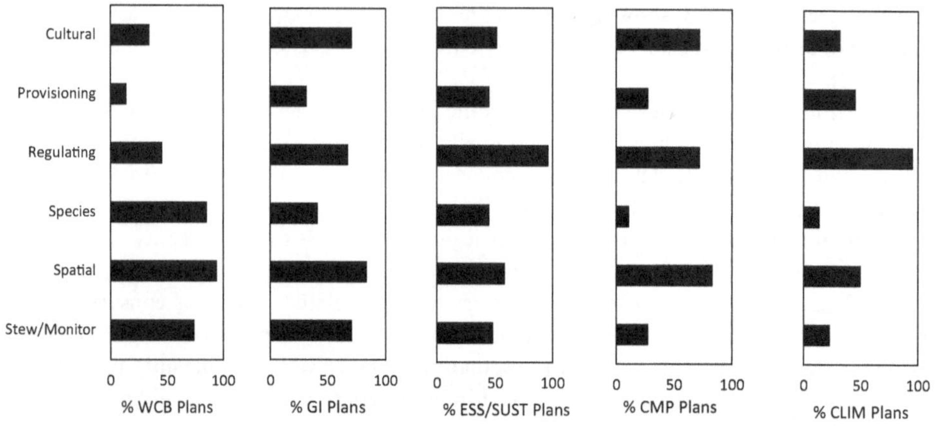

Figure 25.3 Frequency of goals presence by category of planning document. Significant differences in the frequency of goals were found among categories (Pearson's Chi-squared test: $X^2 = 70.21$, df = 20, $p < 0.0001$). CLIM: climate adaptation plans, CMP: comprehensive master plans, ESS/SUST: ecosystem services and sustainability, GI: green infrastructure, WCB: wildlife conservation and biodiversity plans. See Table 25.2 for goal descriptions.

conservation, invasive species management, and critical habitat conservation were found significantly more frequently in biodiversity plans than would be expected by chance ($p < 0.05$).

Approximately 37% of the planning documents included a target for ecosystem services that support biodiversity (Figure 25.4). Air and water quality were the most common of the ecosystem services targets, found in 3–41% and 3–45%, respectively, across the five plan types studied. Ecosystem services/sustainability plans had the most targets for ecosystem services. Significant differences in the frequencies of ecosystem service targets were found among categories of planning documents (Pearson's Chi-squared test: $X^2 = 91.60$, df = 20, $p < 0.0001$).

Discussion

Similarities in the attributes included in the five categories of plans and across cities in the Global North and Global South would not be expected if planning processes were shaped by distinctive local processes alone. From the PCAs, there were no clearly patterned associations or clusters of attributes that would indicate distinct typologies or approaches to planning for biodiversity and ecosystem services for the Global South versus the Global North or for the five categories of planning documents. There were also similarities in the composition of biodiversity and ecosystem services goals and targets across the five categories of planning documents, despite statistically significant differences in frequencies of some of the attributes. Most differences in the frequencies of goals and targets across plan categories reflected the plans' topical focus. For instance, biodiversity plans had a higher frequency of species conservation goals and targets than did plans on other topics. Cities' plans included a mix of ecosystem services attributes and biodiversity attributes. And while measurable targets were not as widely adopted as other elements, they appeared in many planning documents. We did not investigate the motivation for cities to plan for biodiversity and ecosystem services, nor did we investigate why certain goals and targets were included.

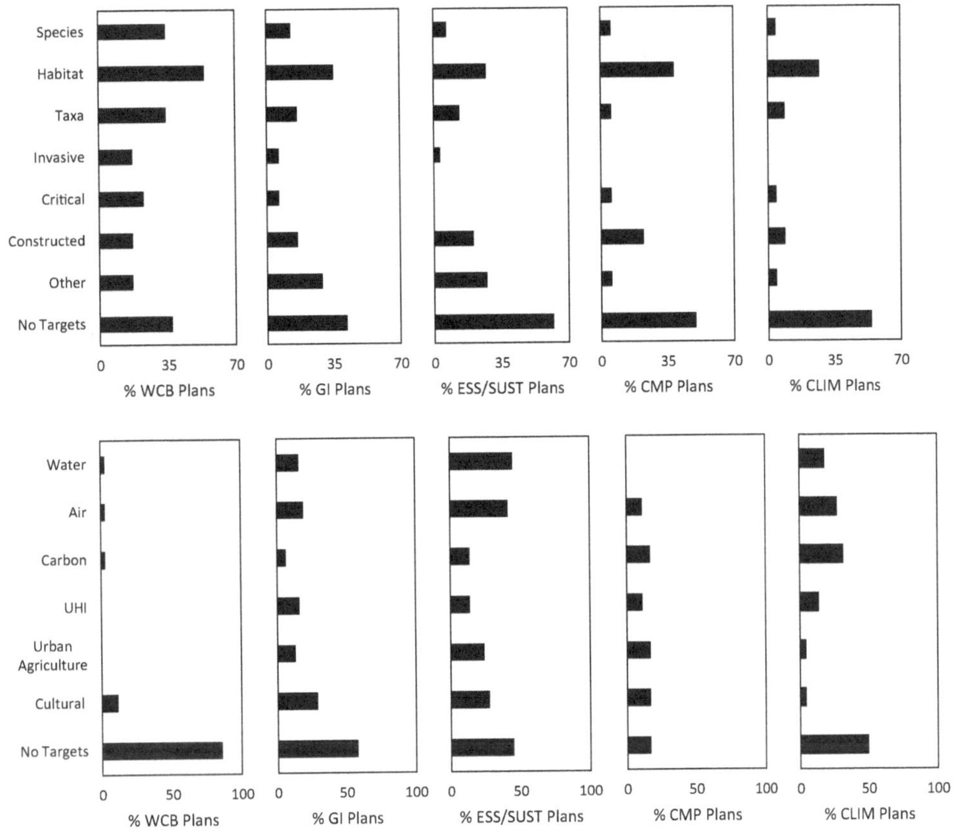

Figure 25.4 (Top) Cumulative frequency of biodiversity target type by planning document category. Significant differences in the frequencies of biodiversity targets were found among categories (Pearson's Chi-squared test: $X^2 = 92.87$, df = 28, $p < 0.0001$). (Bottom) Cumulative frequency of ecosystem service target type by planning document category. Plans with no targets are also shown. Significant differences in the frequencies of ecosystem service targets were found among document planning categories (Pearson's Chi-squared test: $X^2 = 91.60$, df = 20, $p < 0.0001$). CLIM: climate adaptation plans, CMP: comprehensive master plans, ESS/SUST: ecosystem services and sustainability, GI: green infrastructure, WCB: wildlife conservation and biodiversity plans. See Table 25.2 for target type descriptions.

Recommendations from international scientific institutions appeared in many of the planning documents: encouraging similar standards for biodiversity planning in all countries, regardless of development status, "mainstreaming" biodiversity and ecosystem services into plans on other topics, using ecosystem services concepts to promote biodiversity conservation, and to a lesser extent, setting measurable targets for improvement.

This study presents two indicators of policy mobility, first, that biodiversity and ecosystem services are mobile across planning documents on different topics, and second, that policies are mobile across cities in the Global North and Global South. A third indicator of policy mobility is whether biodiversity and ecosystem services are treated similarly across cities, which we addressed in a companion paper (Nilon et al. 2017). In that article, we reported broad similarities in the attributes chosen across our sample of cities, finding that only a few cities focused on ecosystem services only or on biodiversity attributes only. On the other hand, the differences in planning that we observed across these two studies suggest that planning remains locally grounded even as it incorporates issues of global concern. This is consistent with urban planning research about policy mobility that while ideas may be widely shared within and across cities, ideas are selected based on local contexts and are changed considerably as they are developed locally. One example of context-based planning in the Global South according to Cilliers and colleagues (2021) is the focus on systematic conservation planning in South African cities like Durban (McLean et al. 2016) and Cape Town (Holmes et al. 2012), which are situated in global biodiversity hot spots. Systematic conservation plans of these cities are often included in spatial development frameworks, which are legally binding in South Africa (Cilliers et al. 2021).

Beyond the specific research questions, another pattern emerged from the data concerning goals and targets most and least commonly included in plans. Biodiversity and ecosystem services attributes most frequently included were consistent with urban planning's existing structures and practices. For instance, planners and ecologists share an interest in promoting functional spatial patterns and landscape connectivity that would promote biodiversity and support human values (Opdam et al. 2006; Baycan-Levent and Nijkamp 2009; Hansen and Pauleit 2014; Beninde et al. 2015; Artmann et al. 2017). By comparison, other attributes that are common in natural resources plans for rural sites – gathering of baseline data, long-term monitoring, and setting measurable targets – appeared in few of the city-level plans analyzed here. This may reflect that scientific and management practices in rural planning have been motivated primarily for economic use, such as sustained yield from forests, or for conservation use, such as species restoration projects in nature reserves. We did not set out to investigate differences between urban and rural management, but these results suggest city planning presents different sets of concerns from rural planning that could be further researched.

Conclusions

These analyses of planning documents for cities and metropolitan areas indicated planning remains locally grounded but has incorporated internationally circulating scientific concerns. As planning advances, random sample studies could determine the prevalence of biodiversity and ecosystem services goals and targets across cities and across categories of planning documents, and trends in the mobility of key ideas could be tracked. There may be regional differences in planning cultures (Friedmann 2005) affecting how biodiversity and ecosystem services goals and targets are taken up. Other planning instruments that shape land use decisions more directly, such as building codes and zoning, could be compared to analyze how biodiversity concerns are integrated into daily decisions. Such analyses would be part of a broader effort of weighing whether urban planning tools reflect new scientific understandings of urban biodiversity that may differ from approaches created for rural areas (Ahmed and Puppim de Oliveira 2016). The ultimate importance of urban biodiversity and ecosystem services planning may be that they reflect and are motivating changes in conceptions about the social and ecological gradient, from urban to rural areas.

Key companion papers

Cilliers, S.S., C. Breed, E.J. Cilliers, L.G. Lategan. 2021. Urban ecological planning and design in the Global South. Pages 365–401 in C.M. Shackleton, S.S. Cilliers, E. Davoren, M.J. du Toit, editors. *Urban Ecology in the Global South*. Springer, Cham, Switzerland.

Kabisch, N. 2015. Ecosystem service implementation and governance challenges in urban green space planning – The case of Berlin, Germany. *Land Use Policy* 42:557–567.
Nilon, C.H., et al. 2017. Planning for the future of urban biodiversity: A global review of city-scale initiatives. *BioScience* 67(4):332–342.
Puppim de Oliveira, J.A., et al. 2011. Cities and biodiversity: Perspectives and governance challenges for implementing the convention on biological diversity (CBD) at the city level. *Biological Conservation* 144(5):1302–1313.
TEEB. 2011. *TEEB Manual for Cities: Ecosystem Services in Urban Management.* The Economics of Ecosystems and Biodiversity (TEEB), Geneva, Switzerland. www.teebweb.org

References

Adams, L.W. 2005. Urban wildlife ecology and conservation: A brief history of the discipline. *Urban Ecosystems* 8:139–156.
Ahmed, A., J.A. Puppim de Oliveira. 2016. Integration of biodiversity in urban planning instruments in developing countries: The case of Kumasi Metropolitan Assembly, Ghana. *Journal of Environmental Planning and Management* 60(10):1–24.
Alario, M. 2000. Urban and ecological planning in Chicago: Science, policy and dissent. *Journal of Environmental Planning and Management* 43(4):489–504.
Altschuler, A. 1965. The goals of comprehensive planning. *Journal of the American Institute of Planners* 31(3):186–195.
American Planning Association. 2006. *Planning and Urban Design Standards.* Wiley, Hoboken, NJ.
Aronson, M.F.J., F.A. La Sorte, C.H. Nilon, M. Katti, M.A. Goddard, C.A. Lepczyk, P.S. Warren, N.G.S. Williams, S. Cilliers, B. Clarkson, C. Dobbs, R. Dolan, M. Hedblom, S. Klotz, J.L. Kooijmans, I. Kühn, I. MacGregor-Fors, M. McDonnell, U. Mörtberg, P. Pyšek, S. Siebert, J. Sushinsky, P. Werner, M. Winter. 2014. A global analysis of the impacts of urbanization on bird and plant diversity reveals key anthropogenic drivers. *Proceedings of the Royal Society of London B: Biological Sciences* 281(1780).
Aronson, M.F.J., M.V. Patel, K.M. O'Neill, J.G. Ehrenfeld. 2017. Urban riparian systems function as corridors for both native and invasive plant species. *Biological Invasions* 19(12):3645–3657.
Artmann, M., O. Bastian, K. Grunewald. 2017. Using the concepts of green infrastructure and ecosystem services to specify leitbilder for compact and green cities: The example of the landscape plan of Dresden (Germany). *Sustainability* 9(2):198.
Avlonitis, G., C. Doll, R. Galt, A. Mader, R. Moreno-Penarada, S. Patrickson, J. Puppim de Oliveira, W. Shih. 2013. *Local Biodiversity Strategy and Action Plan Guidelines: An Aid to Municipal Planning and Biodiversity Conservation.* ICLEI – Local Governments for Sustainability, United Nations University Institute of Advanced Studies, Secretariat of the Convention on Biological Diversity, Cape Town, South Africa.
Baycan-Levent, T., P. Nijkamp. 2009. Planning and management of urban green spaces in Europe: Comparative analysis. *Journal of Urban Planning and Development* 135(1):1–12.
Beninde, J., M. Veith, A. Hochkirch. 2015. Biodiversity in cities needs space: A meta-analysis of factors determining intra-urban biodiversity variation. *Ecology Letters* 18(6):581–592.
Berke, P., D. Godschalk. 2009. Searching for the good plan: A meta-analysis of plan quality studies. *Journal of Planning Literature* 23(3):227–240.
Bolund, P., S. Hunhammar. 1999. Ecosystem services in urban areas. *Ecological Economics* 29(2):293–301.
Chan, L., O. Hillel, P. Werner, N. Holman, I. Coetzee, R. Galt, T. Elmqvist. 2021. *Handbook on the Singapore Index on Cities' Biodiversity (also Known as the City Biodiversity Index).* Montreal: Montreal: Secretariat of the Convention on Biological Diversity and Singapore: National Parks Board, Singapore.
Cilliers, S.S., G.J. Bredenkamp. 2000. Vegetation of road verges on an urbanisation gradient in Potchefstroom, South Africa. *Landscape and Urban Planning* 46(4):217–239.
Cilliers, S.S., C. Breed, E.J. Cilliers, L.G. Lategan. 2021. Urban ecological planning and design in the Global South. Pages 365–401 in C.M. Shackleton, S.S. Cilliers, E. Davoren, M.J. du Toit, editors. *Urban Ecology in the Global South.* Springer, Cham, Switzerland.
Conca, K. 2006. *Governing Water: Contentious Transnational Politics and Global Institution Building.* MIT Press, Cambridge, MA.
Dearborn, D.C., S. Kark. 2010. Motivations for conserving urban biodiversity. *Conservation Biology* 24:432–440.

Dobbs, C., C.R. Nitschke, D. Kendal. 2014. Global drivers and tradeoffs of three urban vegetation ecosystem services. *PLoS ONE* 9(11):e113000.
Elander, I., E.L. Alm, B. Malbert, U.G. Sandström. 2005. Biodiversity in urban governance and planning: Examples from Swedish cities. *Planning Theory & Practice* 6(3):283–301.
Elmqvist, T., M. Fragkias, J. Goodness, B. Güneralp, P.J. Marcotullio, R.I. McDonald, S. Parnell, M. Schewenius, M. Sendstad, K.C. Seto, C. Wilkinson, editors. 2013. *Urbanization, Biodiversity and Ecosystem Services: Challenges and Opportunities, A Global Assessment*. Springer Open, Dordrecht, The Netherlands.
Freestone, R. 2000. Learning from planning's histories. Pages 1–19 in R. Freestone, editor. *Urban Planning in a Changing World: The Twentieth Century Experience*. Routledge, Abingdon.
Friedmann, J. 1971. The future of comprehensive urban planning: A critique. *Public Administration Review* 31:315–326.
Friedmann, J. 2005. Planning cultures in transition. Pages 29–44 in B. Sanyal, editor. *Comparative Planning Cultures*. Taylor & Francis Group, New York.
Goddard, M.A., A.J. Dougill, T.G. Benton. 2010. Scaling up from gardens: Biodiversity conservation in urban environments. *Trends in Ecology & Evolution* 25(2):90–98.
González, S. 2011. Bilbao and Barcelona 'in motion.' How urban regeneration 'models' travel and mutate in the global flows of policy tourism. *Urban Studies* 48(7):1397–1418.
Güneralp, B., A.S. Perlstein, K.C. Seto. 2015. Balancing urban growth and ecological conservation: A challenge for planning and governance in China. *Ambio* 44(6):532–543.
Haar, C.M. 1955. In accordance with a comprehensive plan. *Harvard Law Review* 68(7):1154–1175.
Hansen, R., S. Pauleit. 2014. From multifunctionality to multiple ecosystem services? A conceptual framework for multifunctionality in green infrastructure planning for urban areas. *Ambio* 43(4):516–529.
Harris, A., S. Moore. 2013. Planning histories and practices of circulating urban knowledge. *International Journal of Urban and Regional Research* 37(5):1499–1509.
Holmes, P.M., A.G. Rebelo, C. Dorse, J. Wood. 2012. Can Cape Town's unique biodiversity be saved? Balancing conservation imperatives and development needs. *Ecology and Society* 17(2):28.
Holzer, K.A. 2014. Amphibian use of constructed and remnant wetlands in an urban landscape. *Urban Ecosystems* 17(4):955–968.
Hostetler, M., W. Allen, C. Meurk. 2011. Conserving urban biodiversity? Creating green infrastructure is only the first step. *Landscape and Urban Planning* 100(4):369–371.
Hoyt, L. 2006. Importing ideas: The transnational transfer of urban revitalization policy. *International Journal of Public Administration* 29(1–3):221–243.
ICLEI. 2013. *Local Biodiversity Strategy and Action Plan Guidelines: An Aid to Municipal Planning and Biodiversity Conservation*. ICLEI-Local Governments for Sustainability, United Nations University Institute of Advanced Studies, Secretariat of the Convention on Biological Diversity, Century City, South Africa.
Intergovernmental Panel on Climate Change. 2014. Urban areas. In climate change 2014 – Impacts, adaptation and vulnerability: Part A: Global and sectoral aspects. Pages 535–612 in *Working Group II Contribution to the IPCC Fifth Assessment Report*. Cambridge University Press, Cambridge, MA.
Ives, C.D., P.E. Lentini, C.G. Threlfall, K. Ikin, D.F. Shanahan, G.E. Garrard, S.A. Bekessy, R.A. Fuller, L. Mumaw, L. Rayner, R. Rowe, L.E. Valentine, D. Kendal. 2016. Cities are hotspots for threatened species. *Global Ecology and Biogeography* 25(1):117–126.
Kabisch, N. 2015. Ecosystem service implementation and governance challenges in urban green space planning – The case of Berlin, Germany. *Land Use Policy* 42:557–567.
Lyles, W., M. Stevens. 2014. Plan quality evaluation 1994–2012: Growth and contributions, limitations, and new directions. *Journal of Planning Education and Research* 34(4):433–450.
McCann, E., K. Ward. 2010. Relationality/territoriality: Toward a conceptualization of cities in the world. *Geoforum* 41(2):175–184.
McDonald, R.I., M.L. Colbert, M. Hamann, R. Simkin, B. Walsh. 2018. *Nature in the Urban Century: A Global Assessment of Important Areas for Safeguarding Biodiversity and Human Well-Being*. The Nature Conservancy, Arlington, VA. www.nature.org/urban100
McLean, C.T., L.E. Ground, R.G. Boon, D.C. Roberts, N. Govender, A. McInnes. 2016. *Durban's Systematic Conservation Assessment*. Durban, EThekwini Municipality: EThekwini Municipality, Environmental Planning and Climate Protection Department, Durban, South Africa.
McPhearson, T., Z.A. Hamstead, P. Kremer. 2014. Urban ecosystem services for resilience planning and management in New York City. *Ambio* 43(4):502–515.

Millennium Ecosystem Assessment CFWG. 2003. *Ecosystems and Human Well-Being: A Framework for Assessment*. Island Press, Washington, DC.

Miller, J.R., M. Groom, G.R. Hess, T. Steelman, D.L. Stokes, J. Thompson, T. Bowman, L. Fricke, B. King, R. Marquardt. 2009. Biodiversity conservation in local planning. *Conservation Biology* 23(1):53–63.

Montanarella, L., R. Scholes, A. Brainich, editors. 2019. *The IPBES Assessment Report on Land Degradation and Restoration*. Secretariat of the Intergovernmental Science-Policy Platform on Biodiversity and Ecosystem Services (IPBES), Bonn.

Müller, N., P. Werner. 2010. Urban biodiversity and the case for implementing the Convention on Biological Diversity in towns and cities. Pages 3–34 in N. Müller, P. Werner, J.G. Kelcey, editors. *Urban Biodiversity and Design*. Wiley-Blackwell, Oxford.

Nilon, C.H., M.F.J. Aronson, S.S. Cilliers, C. Dobbs, L.J. Frazee, M.A. Goddard, K.M. O'Neill, D. Roberts, E.K. Stander, P. Werner, M. Winter, K.P. Yocom. 2017. Planning for the future of urban biodiversity: A global review of city-scale initiatives. *BioScience* 67(4):332–342.

Norton, R.K. 2008. Using content analysis to evaluate local master plans and zoning codes. *Land Use Policy* 25(3):432–454.

Opdam, P.F.M., E.G. Steingröver, S.A.M. van Rooij. 2006. Ecological networks: A spatial concept for multi-actor planning of sustainable landscapes. *Landscape and Urban Planning* 75(3):322–332.

Peck, J. 2011. Geographies of policy: From transfer-diffusion to mobility-mutation. *Progress in Human Geography* 35(6):773–797.

Peck, J., N. Theodore. 2010. Mobilizing policy: Models, methods, and mutations. *Geoforum* 41(2):169–174.

Pisupati, B. 2007. *Effective Implementation of NBSAPs: Using a Decentralized Approach-Guidelines for Developing Sub-National Biodiversity Action Plans*. United Nations University, UNU-IAS Institute of Advanced Studies, Yokohama, Japan.

Plieninger, T., S. Dijks, E. Oteros-Rozas, C. Bieling. 2013. Assessing, mapping, and quantifying cultural ecosystem services at community level. *Land Use Policy* 33:118–129.

Pörtner, H.O., R.J. Scholes, J. Agard, E. Archer, A. Arneth, X. Bai, D. Barnes, M. Burrows, L. Chan, W.L. Cheung, S. Diamond, C. Donatti, C. Duarte, N. Eisenhauer, W. Foden, M.A. Gasalla, C. Handa, T. Hickler, O. Hoegh-Guldberg, K. Ichii, U. Jacob, G. Insarov, W. Kiessling, P. Leadley, R. Leemans, L. Levin, M. Lim, S. Maharaj, S. Managi, P.A. Marquet, P. McElwee, G. Midgley, T. Oberdorff, D. Obura, E. Osman, R. Pandit, U. Pascual, A.P.F. Pires, A. Popp, V. ReyesGarcía, M. Sankaran, J. Settele, Y.J. Shin, D.W. Sintayehu, P. Smith, N. Steiner, B. Strassburg, R. Sukumar, C. Trisos, A.L. Val, J. Wu, E. Aldrian, C. Parmesan, R. Pichs-Madruga, D.C. Roberts, A.D. Rogers, S. Díaz, M. Fischer, S. Hashimoto, S. Lavorel, N. Wu, H.T. Ngo. 2021. *Scientific Outcome of the IPBES-IPCC Co-Sponsored Workshop on Biodiversity and Climate Change*. IPBES Secretariat, Bonn, Germany.

Puppim de Oliveira, J.A., O. Balaban, C.N.H. Doll, R. Moreno-Peñaranda, A. Gasparatos, D. Iossifova, A. Suwa. 2011. Cities and biodiversity: Perspectives and governance challenges for implementing the convention on biological diversity (CBD) at the city level. *Biological Conservation* 144(5):1302–1313.

Robinson, J. 2006. *Ordinary Cities: Between Modernity and Development Abingdon Oxon*. Routledge, London.

Robinson, J. 2015. 'Arriving At' urban policies: The topological spaces of urban policy mobility. *International Journal of Urban and Regional Research* 39(4):831–834.

Sanyal, B., editor. 2005. *Comparative Planning Cultures*. Routledge, New York.

Schewenius, M., T. McPhearson, T. Elmqvist. 2014. Opportunities for increasing resilience and sustainability of urban social – Ecological systems: Insights from the URBES and the Cities and Biodiversity Outlook Projects. *Ambio* 43(4):434–444.

Secretariat of the Convention on Biological Diversity. 2012. *Cities and Biodiversity Outlook*. Montreal, Canada.

Seto, K.C., B. Güneralp, L.R. Hutyra. 2012. Global forecasts of urban expansion to 2030 and direct impacts on biodiversity and carbon pools. *Proceedings of National Academies of Sciences* 109:16083–16088.

Shwartz, A, A. Turbé, R. Julliard, L. Simon, A.-C. Prévot. 2014. Outstanding challenges for urban conservation research and action. *Global Environmental Change* 28:39–49.

Söderström, O., S. Geertman. 2013. Loose threads: The translocal making of public space policy in Hanoi. *Singapore Journal of Tropical Geography* 34(2):244–260.

Stevens, M.R., W. Lyles, P.R. Berke. 2014. Measuring and reporting intercoder reliability in plan quality evaluation research. *Journal of Planning Education and Research* 34(1):77–93.

Stokes, D.L., M.F. Hanson, D.D. Oaks, J.E. Straub, A.V. Ponio. 2010. Local land-use planning to conserve biodiversity: Planners' perspectives on what works. *Conservation Biology* 24(2):450–460.

Sukopp, H. 2002. On the early history of urban ecology in Europe. *Presilia* 74(373–393).

TEEB. 2011. *TEEB Manual for Cities: Ecosystem Services in Urban Management.* The Economics of Ecosystems and Biodiversity (TEEB), Geneva, Switzerland. www.teebweb.org

UK Local Issues Advisory Group (UKLIAG). 1997. *Guidance for Local Biodiversity Action Plans.* UKLIAG, London.

United Nations. 2015. *Transforming Our World: The 2030 Agenda for Sustainable Development A/RES/70/1.* UN, New York.

United Nations Committee for Development Policy. 2018. *List of Least Developed Countries (as of March 2018).* United Nations Committee for Development Policy; Development Policy and Analysis Division; Department of Economic and Social Affairs, Geneva, Switzerland. https://unctad.org/topic/least-developed-countries/list

United Nations Conference on Trade and Development. 2018. *Development Status Groups and Composition.* United Nations Conference on Trade and Development, UNCTADSTAT. United Nations Publications, New York.

Urban Biodiversity Hub. 2019. *About Us.* www.ubhub.org/aboutUs

Wheeler, S.M. 2008. State and municipal climate change plans: The first generation. *Journal of the American Planning Association* 74(4):481–496.

Wilkinson, C., M. Sendstad, S. Parnell, M. Schewenius. 2013. Urban governance of biodiversity and ecosystem services. Pages 539–587 in T. Elmqvist, M. Fragkias, J. Goodness, B. Güneralp, P.J. Marcotullio, R.I. McDonald, S. Parnell, M. Schewenius, M. Sendstad, K.C. Seto, editors. *Urbanization, Biodiversity and Ecosystem Services: Challenges and Opportunities: A Global Assessment.* Springer, Dordrecht, Netherlands.

World Commission on Environment and Development. 1987. *Our Common Future.* Oxford University Press, New York.

Young, R.F. 2011. Planting the living city. *Journal of the American Planning Association* 77(4):368–381.

26

DEVELOPMENT AND EVOLUTION OF URBAN BIODIVERSITY INDICATORS AND ASSESSMENT TOOLS

Yuta Uchiyama and Ryo Kohsaka

Introduction

International process of urban biodiversity conservation

Historically, efforts to conserve biodiversity in the international arena have focused on the policymaking process as in: Biodiversity Liaison Group (Convention on Biological Diversity [CBD], Convention on International Trade in Endangered Species of Wild Fauna and Flora [CITES], Convention on Migratory Species [CMS], Ramsar, and World Heritage) as well as the Global Biodiversity Information Facility (GBIF), Future Earth, or the Intergovernmental Science-Policy Platform on Biodiversity and Ecosystem Services (IPBES) (also Mechanism of Scientific Expertise on Biodiversity [IMoSEB] in the history) in scientific communities. Amongst these, the initiatives most pertinent to urban biodiversity were a range of decisions since 2018 under the auspices of the CBD, led by the Singapore government. In 2008, the first decision related to urban biodiversity (or "cities and biodiversity" in CBD contexts), Decision IX/28, was adopted by the Conference of the Parties to the Convention on Biological Diversity (hereafter COP) at its 9th meeting. The decision promoted engagement of cities and local authorities in seeking to reverse the decline of biodiversity at local level. The government of Singapore led the discussion as a nation and a city by organizing regular expert meetings. Since 2010 when the 10th meeting of COP adopted a revised and updated Strategic Plan for Biodiversity, including Aichi Biodiversity Targets for the 2011–2020 (Decision X/2), biodiversity conservation has further been discussed in forums related to sustainability, green infrastructure, and disaster risk reduction, where urban farming played a role. The Cities and Biodiversity Outlook, the first global white paper on the topic, was published in 2013 and has facilitated research focusing on biodiversity in urban areas. Until such decisions and publications, urban biodiversity was not necessarily the mainstream in the dialogue, the main fields being protected areas, suburban rural areas, and natural areas. Urban biodiversity conservation became recently highlighted in those conventional fields since the late 2000s.

More recently, the sustainable development goals (SDGs) were adopted by all United Nations Member States in 2015. Goal 11 promotes sustainable cities and settlements, while Goals 14 and 15 are focused on terrestrial (including freshwater) and marine biodiversity. In total, there are 17

DOI: 10.4324/9781003016120-30

major goals that are directly or indirectly related to both urban environmental management and biodiversity conservation. Relevant themes to various types of goals include green infrastructure for disaster risk reduction in urban areas.

International forums on infrastructure and development have considered biodiversity through the lens of disaster prevention. Symbolically, the third United Nations World Conference on Disaster Risk Reduction held in Sendai in 2015, four years after the Great East Japan earthquake, was a milestone event on urban disaster prevention and urban settlement and shed light on the role of green infrastructure. In the conference, the Sendai Framework for Disaster Risk Reduction 2015–2030 was adopted. The following year, at the 3rd United Nations Conference on Human Settlements – Habitat 3 in 2016, the "New Urban Agenda" was proposed and adopted. In both conferences, biodiversity was highlighted as a key element for quantitative and qualitative improvements in multiple aspects of cities related to the environment, economy, and society, such as disaster risk reduction and quality of life improvement and biodiversity in cities. Thus, it is clear that biodiversity conservation was being mainstreamed beyond the environmental arena in the mid-2010s with key terms such as green infrastructures, nature-based solutions (NbS), and ecosystem-based disaster risk reduction (EcoDRR).

The roles of urban biodiversity in terms of quality of life, sustainability, and disaster prevention and mitigation need to be considered in the conservation policies and actions, and there is a movement to merge biodiversity conservation with the broader discussion of global sustainability. Discussions on the shrinking phase of cities have also begun (Haase 2013). Biodiversity itself is affected by multiple driving forces and resulting dynamics including urban shrinkages. The need for cross-departmental efforts as well as cooperation among government departments responsible for environmental conservation and management are one of the lessons learned. The creation of specific indicators for different aspects of urban systems itself has potentials to facilitate cross-departmental dialogue (Uchiyama and Kohsaka 2019). Furthermore, dialogue between scientists and policymakers and participatory management that involve citizens have been treated as international priorities. For example, science and policy dialogue is discussed in IPBES. One of us (RK) participated as a member of the government delegates and as a coordinating lead author (CLA) of the report for the Asia Pacific region in the past, contributing to the aspects including policy frameworks and ecosystem service evaluations for urban biodiversity. Kohsaka was also elected as an external review panel member at its review of the platform at the end of its first work program or so-called "IPBES External Review Panel" (Stevance et al. 2020), which is a milestone review for the science-policy platform that can be an essential part of implementation of urban biodiversity policies. Furthermore, both authors (Kohsaka and Uchiyama) have been engaged in developing and revising the City Biodiversity Index in Singapore (SCBD 2010).

Indicators for assessing city biodiversity

As described in the previous section, the history of urban biodiversity indicators is a relatively short one and started with COP9 in 2008 and the proposal of Singapore indicators (Chan and Djoghlaf 2009; Chan et al. 2010; Kohsaka 2010). The original indicators are categorized into three groups: (1) the status of native biodiversity, (2) ecosystem services, and (3) the governance and management of biodiversity conservation in cities. In total it had 23 indicators. In 2021, the revised version of the indicators was published, and it now has 28 indicators, including the ones related to urban agriculture and climate change mitigation and adaptation (Chan et al. 2021). Given the rich debate related to other urban aspects such as disasters and well-being, the debate over biodiversity indicators is somewhat younger than other global debates, such as on population growth and urbanization. Due to its unique characteristics as a "city-state", Singapore spearheaded the discussion of city and biodiversity and its indicators. As clearly outlined in the key message of the CBO (City Biodiversity Outlook), "urbanization is both a challenge and an opportunity to manage ecosystem services globally". Cities are part of solutions and are no longer regarded merely as entities to provide negative impacts on biodiversity and ecosystems. Cities can contribute to global and local biodiversity conservation, and methods to evaluate and manage both aspects are being discussed. For example, cities are regarded as key players that can contribute to global biodiversity conservation by improving production and consumption patterns (Kohsaka et al. 2013).

After the Singapore Index was proposed, science and policy platforms and international contributions to urban biodiversity conservation have been discussed intensively (Kohsaka et al. 2013). Based on the Singapore Index, cities in different regions of the world have attempted to develop their own City Biodiversity Index, and Japan is also conducting practical assessments on the application of the index. However, challenges were identified, ranging from the definition of terms to methodological questions related to the quantification of ecosystem services (Kohsaka and Okumura 2014; Kohsaka and Uchiyama 2017). A larger scale assessment based on CBI was conducted encompassing 665 municipalities across Japan in 2016. The evaluation was based on the "simplified" biodiversity index published by the Ministry of Land, Infrastructure, Transport and Tourism as an example of a nationwide evaluation of city-level indicators for ecosystems and biodiversity. The assessment itself was conducted by an independent body set by the Ministry. The project was pioneering in nature and its scale and practical methodology.

Trends in urban biodiversity indicators

There are further cases where Singapore Indexes have been applied beyond Asian contexts. The results of their applications have been disclosed and disseminated mainly in European and North American cities. However, cities in developing countries and emerging economies, where economic growth is a top priority, have been sluggish in their applications.

Sixty-two cities have applied the indicators as of November 2020 (the previously mentioned nationwide evaluation of Japan with 665 municipalities are not counted in the website showing the status of application of the indicators published by Singapore, because the applied indicators were simplified and revised version of Singapore Index, considering the practicality of the application of the indicators) (Figure 26.1). There are variations in the institutions applying the indicators, with 36 cities assessed by city governments, 14 by academic institutions, and 12 in the process of applying the indicators.

The number of cities in each country where the indicator was applied by urban government initiatives, India (8 cities) and Thailand (4 cities) were relatively active countries, followed by the United Kingdom, Portugal, New Zealand, Canada, and Mexico with two cases. Although there are many cities around the world where the indicators are used, there are few examples in China and African countries, where there are a large number of cities, and the spread of the indicators is still in its infancy. India published several city-level reports at COP11 of the Convention on Biological Diversity, and the number of cities which applied the indicators is increasing.

Out of 14 Japanese cities, 12 are evaluated by Canadian academic institutions. This example of international collaboration with research institutions in the application of indicators points to future opportunities for increased collaboration to increase the application of self-evaluation tools. Further, cities that were assessed by academic institutions tended to be relatively proactive in formulating policies and strategies for biodiversity conservation. There are exceptions however with Chiba City, for example, lacking a regional biodiversity strategy, yet their indicators were evaluated by academic institutions. The city of Sagamihara, in contrast, applied its own indicators.

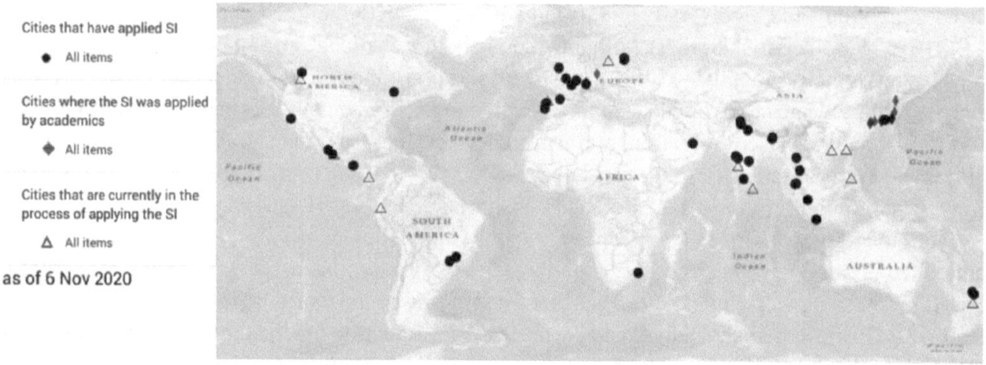

Figure 26.1 Status of application of Singapore indicators by city.

The countries that applied the Singapore Index listed in the previous section are not limited to developed countries nor countries with serious urbanization challenges. Therefore, the indicators were used based on the city government's awareness of the need for assessment rather than the influence of national-level policies. Significantly, the Canadian representative at an expert workshop stated "GIS has given us an overview of the city and increased our awareness" as positive changes within the organizational culture as a result of the applications.

To date, there has been little discussion related to disaster risk reduction during the development of the indicators. However, recently, increased recognition of ecosystem-based disaster response, or EcoDRR, has been identified in the New Urban Agenda. In addition, the ecosystem approach based on biodiversity is beginning to attract attention as a complementary approach to infrastructure development in situations where the vulnerability of modern and uniform man-made infrastructure is being exposed. In this context, the ecosystem approach based on biodiversity is gaining salience as a complementary approach to infrastructure development under the keywords green infrastructures (Spanò et al. 2017) and EcoDRR (UNDRR 2020).

Academic research and biodiversity assessment in urban areas

Trends in academic research on urban biodiversity indicators

Besides the scientific validity of the indicators, practicality has been emphasized, and discussions and evaluations of natural areas have focused on the percentage extent of natural areas, including forests and grasslands. However, the need to properly evaluate connectivity of natural areas and other qualitative aspects in addition to the percentage extent of the area has recently been emphasized with indicators proposed to evaluate green corridors in terms of their connectivity and other characteristics (Deslauriers et al. 2017; Sahani and Raghavaswamy 2018).

Research projects to disseminate practical indicators include, for example, the European Space Agency's supported research project "Earth Observation in Support of The City Biodiversity Index" (2015–2017). The project identified that it is challenging for city governments to collect data and undertake assessments on biodiversity and the natural areas with limited human resources, expertise, and budget. Therefore, the project aims to overcome both problems in cooperation with city governments. Specifically, the city governments are in charge of collecting existing data on biodiversity and the natural environment, while the project side calculates indicator values by combining the collected data with satellite images and other data. The method of calculating the indicator values is also shared with the city governments to support sustainable monitoring and evaluation. Major cities in Europe and the US, including Barcelona, are collaborating in this project, and Japanese cities have also received requests to collaborate.

However, Japanese cities have been unwilling to participate in this international project. The language barrier and the complexity of the decision-making process appear to be the main reasons. To ensure the City Biodiversity Index is utilized as a platform for international knowledge sharing, both technical aspects of data collection and dissemination, calculation methods, and cultural aspects of overcoming knowledge and administrative barriers need to be addressed. By overcoming these challenges, it is possible to establish appropriate guidelines for handling data on biodiversity and the natural areas.

Biodiversity assessment at the urban street scale

Cities can be evaluated at the street scale as well the city scale (such as with the Singapore Index–based approaches), and for planning and implementing projects, this is a necessity to facilitate conservation actions and policies in multiple scales (Yilmaz et al. 2020). There is also a need to incorporate broader views for comprehensive evaluation of various aspects, including the environment, economy, and society. While biodiversity is being evaluated as a part of the environmental indicator system, it can also be applied to other dimensions. Real estate and city districts are increasingly attracting attention as long-term investment targets from the perspective of environmental conservation, and environmental indicators are becoming important indicators in the evaluation of the various values of city districts, regardless of whether they are in developing countries where urbanization is advancing or developed countries where shrinkages are the trends (Koutra et al. 2018).

Specific evaluation indicators include the Comprehensive Assessment System for Built Environment Efficiency (CASBEE) from Japan and the Leadership in Energy and Environmental Design (LEED) from the United States. Both systems were developed as evaluation and certification systems for the built environment (CASBEE and LEED were transitioned in 2002 and 1998, respectively, and their evaluation systems are being developed and updated), and they also deal with international evaluation indicators at the district and city levels.

Biodiversity is included in the indicators of CASBEE-urban areas, and the detailed items include items to evaluate the conservation status and items to evaluate the restoration and creation. The improvement of the quality of the urban environment, including biodiversity, spills over to the surrounding urban areas and requires management over a wide area. From this perspective, it is the intention of the operator of CASBEE (the Institute for Building Environment and Energy Conservation) that the operation and certification of CASBEE-urban areas be used in projects led by public entities such as local governments. CASBEE itself is being used by 24 municipalities to evaluate the built environment. However, since its introduction in 2006, only four cases of CASBEE-urban areas have been certified.

Biodiversity assessments are explicitly stated in CASBEE–Urban Area, but not necessarily in LEED-ND (Neighborhood Development). This is probably due to the technical difficulties in assessing biodiversity itself, and also due to regional differences between Europe, the US, and Asia. Specifically, in the urban areas of Europe and the United States, the central areas are dominated by artificial built environments. Even though there are large and small parks and street trees, there is no diversity of vegetation and ecosystems as in the tropics, and there is little agricultural land such as rice paddies and fields. Alternatively, Asian cities, especially in monsoon Asia, are established by multi-layered suburban-rural and urban environments to such an extent that it is difficult to clearly define the urban territory in the first place (McGee 1991). Therefore, there is probably a lot of room for biodiversity to grow in urban areas. In some studies of biodiversity from urban to suburban areas in Europe and the United States, it is reported that the diversity in urban areas is not necessarily low. However, in Asian cities where the weight of biodiversity within cities is relatively high, it is meaningful to evaluate the diversity in urban districts, including the city center. LEED itself is being used in many parts of the world other than the US, and the challenge is to develop a flexible indicator system that takes into account regional differences, a challenge that is also shared by CASBEE.

In Europe and the US, many evaluation programs have been introduced to assess biodiversity at the level of buildings and city blocks. For example, "BiodiverCity", an evaluation program for cities in Europe (Maes et al. 2021), is trying to evaluate biodiversity from a social perspective, besides environmental ones, particularly from the viewpoint of promoting biophilia, which is the philosophy of living in harmony with nature as the basis of life in both urban and suburban areas. Biophilia is not just about increasing greenery and organisms in our living environment but is also connected to biomimetics, a technology that has recently been applied from the molecular level to the urban scale, to mimic the structure and function of organisms and ecosystems. It is a philosophy that supports the pursuit of ecological cities that are not dependent on fossil fuels.

Future challenges

In promoting biodiversity conservation in cities at a range of scales, there is a need to develop a cross-scale evaluation system in the future. In addition to the evaluation of a city, city blocks within the city, and the built environment, the "city-region" evaluation is largely overlooked to date. Specifically, wide-area assessments are required to include the neighborhoods that are directly affected by cities, as well as the relationships with remote areas that are indirectly but significantly affected through consumption activities in cities. For example, the Ecological Footprint has been used as an evaluation indicator which encompasses consumption activities and tele-coupling (environmental impact and burden of one area to areas in remote areas) to evaluate city units. However, the indicator cannot assess the relationships amongst cities nor the regions. If we consider that a city cannot exist without its region, we should consider the issue of urban biodiversity conservation as a problem of the system that supports the city and which is impacted by the city. The development of indicators to evaluate the relationship between cities and regions is an urgent issue.

One of the main technical challenges of evaluation indicators is the setting of weighting on individual indicators. In particular, when conducting a comprehensive evaluation of multiple indicators, it is frequently conducted by standardizing and weighting on individual indicators to conduct a comprehensive evaluation. Weighting may have a scientific basis or may be depending on the perception and preference of stakeholders. For example, in the case of the conservation of a type of vegetation, values of evaluation indicator of the vegetation, which is obviously above or under a certain level, might not mean a scientific problem. However, in practice, the interpretations of the indicator values can be depending on the stakeholders who are managing the vegetation and using the indicators. In order to ensure scientific evaluation, appropriate interpretation of the indicators is required. Although the perceptions of stakeholders need to be considered in the conservation actions, scientifically accurate interpretation of the indicators and their values is essential for sustainable management of urban biodiversity.

As international efforts are made to assess urban biodiversity at various scales, the knowledge and experience sharing in assessment methods and policy decisions are desirable to create a promising cycle that will lead to improvements in assessment indicators and measures. In doing so, it will be useful to understand the characteristics of the basic environment, such as land use, of each municipality, and to promote knowledge sharing among municipalities with similarities (Uchiyama et al. 2015). There are existing frameworks for inter-municipal cooperation, and there is further need to establish a cooperation system that reflects the diversity and similarity of local authorities. The establishment of integrated policy frameworks for actions, indicator-based assessment, and outputs is also needed (Pierce et al. 2020; Portman 2013; Hou et al. 2014).

As a future development, we proposed collaboration to comprehend the diversity and similarity of local governments, but when rankings of indicator values are announced, there is a tendency to focus only on the rankings. As an initial step, the characteristics of urban ecosystems and similarities need to be considered (Uchiyama et al. 2015; Uchiyama and Kohsaka 2019). The main purpose of evaluating with indicators is to understand the timing, sequence and success of policy implementation, and to inform future policy development and implementation. Therefore, indicator assessment tools can be viewed as a means to an end and a form of adaptive management to achieve the desired outcome of protecting biodiversity. The focus of self-assessment results can be best focused on improved performance. When globally applicable urban biodiversity indicators are developed, it will be necessary to evaluate the indicator system itself from the perspective of whether effective measures have been prescribed using indicators based on scientific findings and whether biodiversity conservation can be promoted widely in new fields, industries, and contexts. In addition, concepts such as green infrastructure and shrinking cities have not yet been actively discussed to elaborate practical indicators.

From our perspective, the concept of biophilia holds the greatest promise conceptually and in encouraging transformative change. As mentioned earlier, Asia, including Japan, has spatial characteristics that make it difficult to clearly separate urban areas from suburban and rural areas, and Japan has a culture that does not separate human life from the natural environment in their perceptions. Therefore, the concept of biophilia can be accepted as compatible with the local context. As mentioned in the beginning of this section, the interdependence of cities and regions needs to be properly assessed in terms of biodiversity conservation, and the biophilia concept can be a tool to foster perceptions of integrated assessment (Barau et al. 2013; Fang et al. 2016) and management of cities and regions.

Key companion papers

Chan, L., et al. 2021. *Handbook on the Singapore Index on Cities' Biodiversity (also Known as the City Biodiversity Index)*. CBD Technical Series No. 98. Secretariat of the Convention on Biological Diversity, Montreal, Canada, and National Parks Board, Singapore.

Kohsaka, R. 2010. Developing biodiversity indicators for cities: Applying the DPSIR model to Nagoya and integrating social and ecological aspects. *Ecological Research* 25:925–936.

Kohsaka, R., et al. 2013. Indicators for management of urban biodiversity and ecosystem services: City Biodiversity Index. Pages 699–718 in T. Elmqvist et al., editors. *Urbanization, Biodiversity and Ecosystem Services: Challenges and Opportunities*. Springer, Dordrecht, Netherlands.

Nilon, C.H., et al. 2017. Planning for the future of urban biodiversity: A global review of city-scale initiatives. *BioScience* 67(4):332–342.

Uchiyama, Y., Kohsaka, R. 2019. Application of the city Biodiversity Index to populated cities in Japan: Influence of the social and ecological characteristics on indicator-based management. *Ecological Indicators* 106:105420.

Acknowledgments

This study is funded by the JSPS KAKENHI Grant Numbers JP23H03605; JP22H03852; JP21K18456; JP20K12398; JP16KK0053; JP17K02105; JST RISTEX Grant Number JPM-JRX20B3; Kurita Water and Environment Foundation [20C002].

References

Barau, A.S., A.N.M. Ludin, I. Said. 2013. Socio-ecological systems and biodiversity conservation in African city: Insights from Kano Emir's Palace gardens. *Urban Ecosystems* 16(4):783–800.

Chan, L., E. Calcaterra, T. Elmqvist, O. Hillel, N. Holman, M. Mader, P. Werner. 2010. *User's Manual for the City Biodiversity Index. ver. 27 September 2010.* Second Expert Workshop on the Development of the City Biodiversity Index, Singapore.

Chan, L., A. Djoghlaf. 2009. Invitation to help compile an index of biodiversity in cities. *Nature* 460:33.

Chan, L., O. Hillel, P. Werner, N. Holman, I. Coetzee, R. Galt, T. Elmqvist. 2021. *Handbook on the Singapore Index on Cities' Biodiversity (also Known as the City Biodiversity Index).* Technical Series No. 98. Secretariat of the Convention on Biological Diversity, Montreal, Canada, and National Parks Board, Singapore.

Deslauriers, M.R., A. Asgary, N. Nazarnia, J.A.G. Jaeger. 2017. Implementing the connectivity of natural areas in cities as an indicator in the City Biodiversity Index (CBI). *Ecological Indicators* 94(2):91–113.

Fang, C., H. Liu, G. Li. 2016. International progress and evaluation on interactive coupling effects between urbanization and the eco-environment. *Journal of Geographical Sciences* 26(8):1081–1116.

Haase, D. 2013. Shrinking cities, biodiversity and ecosystem services. Pages 253–274 in T. Elmqvist, M. Fragkias, J. Goodness, B. Güneralp, P.J. Marcotullio, R.I. McDonald, S. Parnell, M. Schewenius, M. Sendstad, K.C. Seto, C. Wilkinson, editors. *Urbanization, Biodiversity and Ecosystem Services: Challenges and Opportunities.* Springer, Dordrecht.

Hou, Y., S. Zhou, B. Burkhard, F. Müller. 2014. Socioeconomic influences on biodiversity, ecosystem services and human well-being: A quantitative application of the DPSIR model in Jiangsu, China. *Science of the Total Environment* 490:1012–1028.

Kohsaka, R. 2010. Developing biodiversity indicators for cities: Applying the DPSIR model to Nagoya and integrating social and ecological aspects. *Ecological Research* 25(5):925–936.

Kohsaka, R., S. Okumura. 2014. Greening the cities with biodiversity indicators: Experience and challenges from Japanese cities with CBI. Pages 409–424 in S. Nakano, T. Yahara, T. Nakashizuka, editors. *Integrative Observations and Assessments.* Springer Japan, Tokyo, Japan.

Kohsaka, R., H.M. Pereira, T. Elmqvist, L. Chan, R. Moreno-Peñaranda, Y. Morimoto, T. Inoue, M. Itawa, M. Nishi, M. Mathias, C. Souto Cruz, M. Cabral, M. Brundfeldt, A. Parkkinen, J. Niemelä, Y. Kulkarni-Kawli, G. Pearsall. 2013. Indicators for management of urban biodiversity and ecosystem services: City biodiversity index. Pages 699–718 in T. Elmqvist, M. Fragkias, J. Goodness, B. Güneralp, P.J. Marcotullio, R.I. McDonald, S. Parnell, M. Schewenius, M. Sendstad, K.C. Seto, C. Wilkinson, editors. *Urbanization, Biodiversity and Ecosystem Services: Challenges and Opportunities: A Global Assessment.* Springer, Dordrecht.

Kohsaka, R., Y. Uchiyama. 2017. Motivation, strategy and challenges of conserving urban biodiversity in local contexts: Cases of 12 municipalities in Ishikawa, Japan. *Procedia Engineering* 198:212–218.

Koutra, S., V. Becue, M.-A. Gallas, C.S. Ioakimidis. 2018. Towards the development of a net-zero energy district evaluation approach: A review of sustainable approaches and assessment tools. *Sustainable Cities and Society* 39:784–800.

Maes, J., A. Quaglia, A. Martinho Guimarães Pereira, M. Tokarski, G. Zulian, F. Marando, S. Schade. 2021. *BiodiverCities: A Roadmap to Enhance the Biodiversity and Green Infrastructure of European Cities by 2030.* Publications Office of the European Union, Luxembourg.

McGee, T.G. 1991. The emergence of desakota regions in Asia: Expanding a hypothesis. Pages 3–25 in N. Ginsburg, B. Koppel, T.G. McGee, editors. *The Extended Metropolis: Settlement Transition in Asia.* University of Hawaii Press, Honolulu, Hawaii.

Pierce, J.R., M.A. Barton, M.M.J. Tan, G. Oertel, M.D. Halder, P.A. Lopez-Guijosa, R. Nuttall. 2020. Actions, indicators, and outputs in urban biodiversity plans: A multinational analysis of city practice. *PLoS ONE* 15(7):e0235773.

Portman, M.E. 2013. Ecosystem services in practice: Challenges to real world implementation of ecosystem services across multiple landscapes – A critical review. *Applied Geography* 45:185–192.

Sahani, S., V. Raghavaswamy. 2018. Analyzing urban landscape with City Biodiversity Index for sustainable urban growth. *Environmental Monitoring and Assessment* 190(8):1–18.

SCBD: Secretariat of the Convention on Biological Diversity. 2010. *User's Manual for the City Biodiversity Index (CBI), SCBD.* www.cbd.int/subnational/partners-and-initiatives/city-biodiversity-index

Spanò, M., F. Gentile, C. Davies, R. Lafortezza. 2017. The DPSIR framework in support of green infrastructure planning: A case study in Southern Italy. *Land Use Policy* 61:242–250.

Stevance, A.-S., P. Bridgewater, S. Louafi, N. King, T.D. Beard, A.S. Van Jaarsveld, Z. Ofir, R. Kohsaka, K. Jenderedijan, M.R. Benites, K.J. Mulongoy, K.L. Chaudhari, J. Painter, A. Meter. 2020. The 2019 review of IPBES and future priorities: Reaching beyond assessment to enhance policy impact. *Ecosystems and People* 16(1):70–77.

Uchiyama, Y., K. Hayashi, R. Kohsaka. 2015. Typology of cities based on city biodiversity index: Exploring biodiversity potentials and possible collaborations among Japanese cities. *Sustainability* 7 (10):14371–14384.

Uchiyama, Y., R. Kohsaka. 2019. Application of the City Biodiversity Index to populated cities in Japan: Influence of the social and ecological characteristics on indicator-based management. *Ecological Indicators* 106:105420.

UNDRR. 2020. *Ecosystem-Based Disaster Risk Reduction: Implementing Nature-Based Solutions for Resilience*. United Nations Office for Disaster Risk Reduction – Regional Office for Asia and the Pacific, Bangkok.

Yilmaz, F.C., M. Zengin, C.T. Cure. 2020. Determination of ecologically sensitive areas in Denizli province using geographic information systems (GIS) and analytical hierarchy process (AHP). *Environmental Monitoring and Assessment* 192(9):1–13.

27

URBAN BIODIVERSITY

Opportunities and challenges in the Paris Region

Marc Barra

Introduction

The Paris Region, also called *Île-de-France*, is the smallest province but the most populated in France, representing 18% of the nation's population (12 million inhabitants). It includes Paris City and 1,288 other cities and towns. Urban areas account for 23% of the land area of the region and are characterized by a population density of 1,022 inhabitants/km², reinforcing pressure on biodiversity, climate change adaptation, and health of populations. Over the last decades, the abundance of butterflies and birds has fallen by more than 20% in urban spaces compared to the 2000s (Muratet 2016). Common urban bird species such as the house martin (*Delichon urbicum*) and house sparrow (*Passer domesticus*) have decreased by 64% and 28% respectively over the same period. The abundance of urban specialist birds have fallen by 30% in 11 years (Muratet 2016). Pollinator insects such as beetles, diptera, hymenoptera, and lepidoptera also gradually decreased in dense urban areas (Deguines et al. 2016). Specialist species are gradually being replaced by generalist species as homogenization of urban biodiversity is occurring in urban areas.

Over this same period, the Paris Region is facing potentially dramatic changes in climate, and the impacts are already being experienced by plant and animal species and residents. According to projected climate scenarios, the occurrence of heat waves will increase temperatures from 2–4°C by the end of the century in Paris City and its suburbs. During heat waves, strong territorial contrasts are observed between the inner suburbs and peripheral areas, especially during summer nights. For example, there was a 7–8°C difference during an extreme heatwave in Paris in 2003. Further, several flood events occurred in the past years (2016 and 2018) exacerbated by runoff from significant areas, impervious land cover, and modified agricultural locations surrounding Paris. Increases in the frequency and intensity of flood events as well as shortages in water supply are expected. Though little change in total annual precipitation is expected, projections from the French National Meteorological Service reveal greater seasonal contrasts.

Biodiversity loss and climate change are interlinked challenges of a similar magnitude and urgency. These rapid changes lead political authorities to implement reforms to reduce urbanization and more effectively protect urban and peri-urban green spaces. The concept of nature-based solutions (NBS) is gaining support as a response to these challenges, whether to curb biodiversity decline, to adapt to climate change, or to improve the well-being of the inhabitants. To be effective,

DOI: 10.4324/9781003016120-31

this broad concept should go beyond "greenery" and integrate more knowledge and principles of urban ecology. Data from local initiatives and experiments provide feedback and give us the opportunity to assess the effectiveness of NBS and integrate them in the framework of the Paris Region Master Plan.

Towards a "zero land take" objective

In France from 2009 to 2019, an average of 27,000 ha/year are consumed annually by urbanization (Fosse et al. 2019). In 2021, the French Government established a "Zero Net Land Take" policy as part of the Climate and Resilience Law. Land take must now be gradually reduced to 50% by 2031, while the Zero Net Land Take objective should be achieved by 2050. This objective must be reflected in planning documents at a regional level, before being then applied at inter-municipal and municipal levels. It refers to (1) reducing urban sprawl by encouraging urban renewal and density and (2) rewilding the equivalent amount of land consumed by urbanization. The Net Zero Land Take goal sees rewilding as a strategy to restore disturbed land to its initial natural state; however, as rewilding would cost billions of euros, land take should be avoided or strongly reduced in priority.

In the Paris Region, despite a slowdown between 2012 and 2021, land consumption still accounts for an average of 880 ha/year, mainly on agricultural land according to the Paris Region Institute. Considering the national objectives, land take in the Paris Region should be reduced by 50% by 2031. Approved 25 years ago, the Paris Region Master Plan is currently being updated, and the Zero Net Land Take is a unique opportunity to build a new paradigm and provide more green space in the future.

Urban densification policies could threaten biodiversity

The implementation of a Zero Net Land Take objective would greatly contribute to reducing the pressure on existing natural land, but the associated densification policies could threaten intra-urban green space. In the Paris Region, urban wastelands are often considered the main opportunity for reducing land take. However, several scientists have shown that these abandoned sites with spontaneous vegetation have a real potential to contribute to biodiversity conservation in urban areas (Bonthoux et al. 2014). In the Paris Region, the diversity of plants, birds, and butterflies in wastelands is higher than in any other public-managed green space (Muratet 2006; Baude et al. 2011). Other research has shown the importance of connectivity between brownfields and their role in hosting urban-avoider species (Shwartz et al. 2014). Apart from ecological services, spontaneous woodlands could support a range of social, cultural, and health services (Trentanovi et al. 2021). Even if the total area of wasteland has decreased over the last decade, the Paris Region Institute has recently identified more than 2,700 brownfields representing a potential of 4,200 hectares, raising hope among stakeholders for fostering urban development while reducing the land take (Delaville et al. 2021). This potential includes a great variety of wastelands in size and soil type, from former factories to brownfields covered with dense vegetation. Urban planners should reconsider the role of brownfields and protect those which support biodiversity and contribute to ecological connectivity in cities.

Densification policies could also lead to a drastic reduction in size coupled with increasing fragmentation of urban green spaces. The density of Paris and its immediate suburbs is significantly higher than in other European metropolises and has tended to increase since the 1990s, according to the French National Institute of Statistics and Economic Studies. The total surface area of parks and gardens in the metropolitan core has grown more slowly than the population over recent years. Today, the Paris metropolitan area, including Paris City and its suburbs, has one of the lowest amounts of green space/inhabitant in Europe and below the World Health Organization's recommendation of 10 m²/inhabitant. In the Paris Region, 935 municipalities are considered green space deprived, while an additional 57 (6%) are very deprived (Gueymard and Laruelle 2017). In 2017, the Paris Region authority launched a Green Plan to create more green spaces and allow everyone access to a green space within a 15-minute walking radius. This Green Plan consists of creating 500 hectares of new green and wooded spaces by 2021.

While this policy only targets population health and well-being, it should also address the importance of green spaces for biodiversity as research has shown that biodiversity decreases in abundance and richness as urban density increases (Clergeau et al. 1998). Green space patch size remains one of the most important factors for structuring the biodiversity in the city (Spotswood et al. 2019). At the city level, small contiguous habitat patches with an average size of 4.4 hectares can minimize the loss of urban-adapter species. (Beninde et al. 2015). By contrast, an average size of 53.3 hectares or more is necessary to conserve threatened or urban-avoider species (Beninde et al. 2015). Smaller patches within a city also have the potential to connect populations of highly mobile and small animals (e.g. butterflies) with source habitat located in peri-urban areas (Rudd et al. 2002). For land sharing strategies supporters, large non-urbanized spaces can help to conserve significant biodiversity (Soga et al. 2014). On the other hand, land sharing supporters have shown that residential areas with low-density can support several groups of species (Lososová et al. 2011). The Paris Region could take advantage of this situation as the urban matrix is a combination of high-density areas at the heart of the metropolitan and low-density residential areas in the suburbs. The intra-urban scale and the regional scale are therefore connected and complementary. While this approach creates legal and jurisdictional challenges, it could be understood and planned in the Paris Region by linking green space deprivation with the need to increase habitat size for biodiversity throughout the Paris metropolitan area.

Apart from habitat size, a growing body of evidence has shown that landscape connectivity enhances biodiversity in fragmented urban habitats (e.g. Shanahan et al. 2011). Wildlife corridor strategies are now widely used by municipalities in planning and are useful tools for enhancing biodiversity in urban areas. In the Paris Region, the first Regional Ecological Connectivity Plan drawn up in 2013 establishes objectives for the preservation or restoration of "ecological continuities" (reservoirs of biodiversity interconnected by corridors). It offers, on a fairly detailed scale, an overall understanding of the regional green-blue grid (Laruelle and Barra 2020). However, this framework has been implemented at the regional level, and more work remains to be done to identify habitat patches and corridors in dense urban landscapes. This is the ambition of the High Park (Parc des Hauteurs) project in eastern Paris, which aims to reconnect several green spaces made up of forests, public parks and gardens, private woods, orchards, and collective gardens that are now fragmented. This project of ecological, urban, and cultural transformation of a large territory is unique in Île-de-France by its scale. The objective is to reach 320 hectares of easily accessible and interconnected nature areas by 2030. Other cities outside the Paris Region can provide inspiration for more accurate models of green-blue networks in dense urban environments. In 2014, the city of Strasbourg in eastern France initiated a project to identify and reconnect urban green spaces. The connectivity of the landscape was determined based on the optimal habitat of the red squirrel (*Sciurus vulgaris*), its travel preferences and its mobility in the city. This modeling made it possible to highlight the importance of small elements of nature to be preserved or restored within the city (Schwoertzig et al. 2015).

A recent review provides more information about the influence of urban characteristics such as building heights and morphology on biodiversity (Flégeau et al. 2021). At the intra-urban scale, this review highlights the importance of building heterogeneity and intermediate urban forms presenting high levels of vegetation for biodiversity. In a study of Parisian birds, Pellissier et al. (2012) showed that the abundance of omnivorous, ground, and tree nester species are sensitive to urban form and individual building characteristics. Species abundance was greater in areas with a high density of medium-height buildings (Haussmannian buildings). Areas with a high proportion of shrubs, and located near park areas, benefits grain-eating species. This research suggests that urban planning could be bird-friendly in city centers by varying the heights of buildings in urban renewal projects rather than clustering buildings of similar form and height.

Increases in biodiversity can be identified in every type of urban density type – high, intermediate, or low – as soon as unbuilt spaces within neighborhoods and connectivity between patches are planned or restored. Research has shown that a low vegetation cover below can cause rapid declines in species richness (Radford et al. 2005). Increasing vegetation cover could be achieved by developing and applying a green index to urban planning, design, and construction requirements. Similar urban green indexes have been developed and implemented in major metropolitan areas around the world, such as the Berlin Biotope Area Factor, the Malmo Green Factor, the Seattle Green Factor, the Singapore Index's Green Plot Ratio, and the Poland Ratio of Biologically Vital Area. These green indexes are defined as a ratio of the area covered by greenery compared to the total site area (Szulczewska et al. 2014). The Paris Region could also experiment with a green factor to encourage controlled densification while protecting more green cover in planning and design of neighborhoods. However, indexes are not equivalent regarding the type of vegetation cover they promote. In France, many municipalities have applied the Berlin Biotope Area Factor, which allows substituting ground-level vegetation by slab or roof-level vegetation. Even if there is a lack of research regarding ground-level versus building-level biodiversity, cities should prioritize the protection of soils and ground-level vegetation and avoid massive densification. A review of several planning documents in the Paris Region showed that municipalities are struggling to define ground-level green spaces versus slab or roof-level vegetation and that lots of work remains to be done to integrate soil protection in planning (Cornet and Coquière 2020).

From nature-based solutions to ecology-based solutions

Nature-based solutions (NBS) have gained popularity as an integrated approach that can address climate change and biodiversity loss, as well as taking into account the benefits for the people (Seddon et al. 2021). In France, the second National Climate Change Adaptation Plan 2018–2022 and the Biodiversity Plan (2018) provide a favorable framework to encourage nature-based solutions. In the Paris Region, NBS are reflected in the Regional Biodiversity Strategy approved in 2019. The Paris Region is also involved in the European H2020 project, REGREEN, that aims to foster nature-based solutions in urban environments and is also one of the partners of the European integrated LIFE project ARTISAN (Achieving Resiliency by Triggering Implementation of Nature-based Solutions for Climate Adaptation at a National Scale), coordinated by the French Office for Biodiversity. While REGREEN focuses on urban environments, ARTISAN also covers agricultural and forestry environments. These projects are a unique opportunity to mainstream nature-based solutions in planning and convince policymakers and constituents that these solutions could cheaply fulfill some of the functions that more traditional "gray" infrastructure strategies can only perform at considerable cost.

NBS refer to a large range of actions from protecting, restoring, or even creating new ecosystems. For local stakeholders, this broad concept can be vague and lead to very different understanding and applications. In urban areas, NBS are often confused with traditional landscape design and the assumption that NBS will automatically support biodiversity conservation goals by virtue of being green (Connop 2018). Many so-called NBS projects refer to urban "greener" and focus on a few aspects (e.g. aesthetics, stormwater management) rather than an integrated approach based on science in ecology. While more explanation is needed to clarify the links between biodiversity and NBS design, NBS approaches have the potential to move towards biodiversity-led approaches by considering opportunities for increasing biodiversity and ecosystem functioning. Some researchers have informed policymakers and practitioners that NBS should be explicitly designed to provide measurable benefits for biodiversity (Seddon et al. 2021). More co-construction between urban ecologists and landscape planners could help avoid what Stuart Connop and colleagues call *blandscaping*, defined as landscaping that uses the same designs, and often the same species, across different urban regions nationally and globally (Connop 2018). Many "green" solutions have been standardized in cities (e.g. beehives) to the detriment of habitats for wild pollinators, ready-to-use vegetation systems for green roofs or façades, bioswales designed for water infiltration and not as habitats for wildlife, urban micro-forests with attractive marketing instead of wild urban forests protection. More effort should be made to design NBS with diverse habitat types and varied vegetation strata and structures that support native biodiversity, use of local resources, and aims to connect habitat patches. In the future, local guidelines designed by urban ecologists and landscape architects can help cities to integrate these urban ecology principles in NBS design and management.

NBS also provides a framework for how we manage urban green areas. The intensity of management by mowing, pruning of trees and shrubs, removal of dead wood, and mulching can lead to simplification of habitats and reduction of biodiversity (Aronson et al. 2017). In France, even though the heritage of horticultural intensive management is still widespread, management practices have greatly improved towards biodiversity conservation. In 2017, France approved a ban on pesticide use in public and private green spaces, and most cities have recognized the importance of reducing management intensity. In the Paris Region, the richness of plant diversity in urban interstices has increased by over 90% in just seven years (Muratet 2016). This underlines the positive effects of the pesticide ban on biodiversity and a greater acceptance of weed species in urban spaces. Most urban gardeners have been educated on the ecological management of green spaces and have reduced mowing or increased the use of native vegetation in planting. A French initiative called "Vegetal Local" (native vegetation) allows municipalities to source native plants and seeds for green space design based on sub-climate divisions. In the Paris Region, more than 307 green spaces are now certified EcoJardin, a recognition of a global approach to ecological management of green spaces.

One of the remaining challenges is to spatialize NBS into a priority framework for areas in the Master Plan of the Paris Region. A planned strategy for NBS would greatly contribute to make NBS more efficient and make local opportunities more coherent and systematic. For example, flooding data from the Paris Region Institute could be used to identify potential sectors in which wetlands should be protected or restored. In many cities, urban rivers could be daylighted as well as riverbanks and riverine forests restored. The Paris Region Institute has provided data on urban heat island effect over the entire territory of the Greater Paris Metropolis and part of its fringes (Cordeau 2017). These identified sectors could largely benefit from NBS targeting air cooling, such as peri-urban forests or enhanced tree cover in dense urban areas, even if greening is probably insufficient to limit health impacts of extreme heat (Pascal et al. 2021). NBS could also target the recovery of wildlife corridors, which will be important for the adaptation of species to climate change. This data could usefully be used for the implementation of NBS into the planning process, whether it is in the zoning system or in guidelines associated with planning schemes, such as green and blue frameworks or regulations for biodiverse design and management.

A host of initiatives in urban areas

A growing number of NBS initiatives are taking place throughout France and the Paris Region. In 2017, the French Office for Biodiversity created the "Regions committed to nature" initiative. This initiative recognizes and assists municipalities in developing and implementing a biodiversity action plan over a three-year period, with the support of regional biodiversity agencies. Since 2019, 124 cities have been recognized as "committed to nature" in the Paris Region. This initiative is connected to the "French Capital of Biodiversity competition", which has been open to every French municipality since 2010. The competition identifies, promotes, and disseminates the best biodiversity-related practices supported by French municipalities. As part of the RE-GREEN European program, several of nature-based solutions have been identified and mapped as "best practices" in the Paris Region. Most of these projects are related to the use of sustainable urban drainage systems (SUDS), such as bioswales, ditches, basins, green roofs or street trees. In France, SUDS are increasingly used in urban planning, mostly because of Water Agency incentives. In the Paris Region, several municipalities are gradually replacing gray infrastructure (water concrete tanks, artificial basins) with SUDS and urban green spaces. The Parc des Guillaumes, in the city of Noisy-le-Sec, was designed to catch and store urban runoff over an area of 5.8 hectares of heterogeneous vegetation (dry areas, meadow, orchard, edge forest, wetland) and extensive management. This urban park meets the definition of NBS by offering a space for recreation and relaxation while being an open-air stormwater runoff storage site, through a retention basin and a lagoon area. The Parc des Guillaumes is also part of the continuity of a green corridor at the inter-municipal scale. Research has shown that SUDs have the potential to provide habitat for biodiversity in urban green areas (Monberg et al. 2018). However, researchers also point to the need for improvement in designing and managing SUDs. If well designed and managed, urban ponds have the potential to support a much greater biodiversity than current levels (Oertli and Parris 2019). Wetland management practices can unintentionally impact non-target species such as mosquitoes, and potentially public health (Hanford et al. 2020). Public policy and guidelines should also encourage removal of barriers between SUDS and connectivity with other wetlands or green spaces to improve species movements (Ahn and Schmidt 2019). A better association between landscape designers and urban ecologists could help to adjust design and management practices for more effective biodiversity conservation (Barra 2020). While the ecologist formulates recommendations with regard to the biodiversity issues identified, the landscape architect implements them, provides guarantees on the technical feasibility, and inserts these recommendations into a landscape and aesthetic approach.

In France, many urban rivers have been heavily modified by being buried, channelized, or through streambank modification. These rivers were often incorporated into the sewerage networks under the road network, now causing regular flooding problems in many cities. In recent years, reforms have made it possible to restore many urban rivers by reopening, realigning, and rehabilitating habitat for aquatic species. The benefits of these projects are multiple although the urban environment presents many constraints. In the city of Sarcelles, the daylighting of the Petit Rosne is a great example of urban river restoration, despite limited space available and dense urbanization. In 1992, the city center of Sarcelle was submerged under 1.50 m of water during a flood, and the Petit Rosne could no longer accommodate runoff. After several years of studies, the Syndicat Intercommunal d'Aménagement Hydraulique (SIAH) has reopened and restored the river over a stretch of 165 m. The restoration of the Petit Rosne is a combination of techniques from ecological engineering applied to a hydraulic context and technologies to monitor the watercourse in real time. The restoration objective could only be met with a co-construction process with local people and strong communication about the public acceptance of a more NBS approach to the design of their city.

Rivers and streams in the Paris Region are also heavily armored with many kilometers of artificial riverbanks (e.g. riprap, concrete dykes, sheet piles), especially along the River Seine. Armoring has resulted in the loss of habitat for wildlife. In urban areas, an association called Espace is responsible for the management of nearly 12 km of banks of the Seine since 1999. Espace uses ecological engineering techniques such as willow fascines and helophyte benches to limit erosion and restore habitats. In some areas, they have worked on the creation of reed beds or vegetated rafts. In the Île Saint Germain sector alone, 575 m of shoreline and 300 m of embankment have been restored using techniques derived from NBS and plant engineering. These actions help to improve the quality of water and can contribute to an increase in the diversity of wildlife species such as fish in the Seine (four species of fish in the 1970s compared to approximately 30 currently) and endangered birds such as the European kingfisher (Belliard et al. 2009).

In relation to flood management, the Vignois flood expansion zone is one of the best examples of NBS explicitly designed to provide measurable benefits for biodiversity. In the city of Gonesse, the SIAH established a flood expansion zone to protect citizens from flooding caused by river overflow and stormwater runoff in 2019. Although the main objective remains the management of the river's hydrology, this wetland offers a diversity of habitats for biodiversity on 12 hectares and can accommodate up to 55,000 m^3 of water through a connected system of wetlands and open water ponds. Collaboration between ecologists, landscape architects, and planners made it possible to design different regimes of hydrology and heterogeneous habitats such as meadows, reed beds, willows, and groves. The place is not lit at night in order to maintain a "darkness corridor" for birds and bats. Understory vegetation was planted, but existing trees were protected, and the spontaneous vegetation was accepted. Since 2020, biodiversity monitoring is carried out on several taxa (moths, dragonflies, orthoptera, reptiles, amphibians, birds, bats). This assessment is based on standardized protocols from participatory science programs, in order to have a comparison over time. After three years of implementation, the initial results show increasing trends in biodiversity across taxa and confirm that restoration of wetlands have significant impact on biodiversity and climate change adaptation (Rivallin 2021).

In dense urban areas, like in the center of Paris, several actions have been implemented to protect biodiversity despite strong competition for land use. One of the most emblematic sites in Paris is the Petite Ceinture, a former railway line of 34 km following the inner contour of the capital, totaling more than 40 hectares. Over time, the tracks have become an ecological corridor, in which vegetation and wildlife have been able to inhabit away from the noise and disturbance of the city. This sanctuary of urban nature is a real asset for biodiversity and exhibits a great diversity of varied natural habitats, such as groves, meadows, forest edges, and pioneer vegetation of ballasts and walls. In 2006, the City of Paris and the national railway company committed to gradually open to the public several portions of this railway while ensuring that some places remain closed to public access. The Petite Ceinture of Paris is a complex location, very different from a landscaped park or managed green spaces in Paris. The Petite Ceinture is a privileged ground for questioning the relationship of inhabitants to biodiversity and the idea of cohabitation with a wilder environment (Scapino 2016). An inventory carried out by the city of Paris in 2013 identified 590 species, including 254 plants, 36 birds, and 5 species of bats.

Several initiatives have also emerged in recent years to design buildings that are more permeable and hospitable for biodiversity. Architects have joined forces with urban ecologists and are experimenting with wildlife solutions through architecture design. In Boulogne-Billancourt, the School of Sciences and Biodiversity is one such building, the result of a partnership between Chartier-Dalix Architects and urban ecologists. The rooftop vegetation has been designed with a variation in substrate depth ranging from 30 cm to 100 cm, allowing the creation of meadows and urban "micro-forests" ecosystems. This space is accessible to children and educational teams for workshops. The facade of the building has been designed to evolve over time and accommodate spontaneous vegetation. The variable geometry of the walls provides nesting sites for different species of birds and will eventually allow for bats. Regular inventories of flora and fauna have been set up to follow the evolution of biodiversity. The roof is currently being renovated with seeds collected in wild meadows in the outskirts of Paris to change the floristic composition towards a larger quantity of perennial, flowering, local species adapted to the conditions of the site. Another example, in the city of Rosny-sous-Bois, the Boutours school group (nursery and elementary school) is one of the most successful examples of straw-bale construction addressing biodiversity and adaptation to climate change while taking care of the health of the children. It was designed by a team of architects and engineers from the city, through a participatory and educational approach. The landscape design considered the protection of existing trees as much as possible, and the roof terraces are planted with perennial berries and flowery meadows. These examples illustrate a new dialogue between architects and urban ecologists and a new process of architecture design based on applied ecology.

In Epinay-sur-Seine, in the Northern part of Paris Region, the municipality has turned a former wasteland threatened by a development project into a green space to maintain biodiversity and create an accessible green space for citizens. The 1.5 hectare site, La Réserve Écologique, has been divided into several zones with different levels of accessibility: places open to all, allotments and shared gardens, more restricted access, and finally, areas left for wildlife including woods in the center of the park. The park is managed ecologically, and the city uses this place to organize nature events and to educate the inhabitants and schoolchildren on the importance of urban biodiversity. Many remarkable species have been observed in the park, including relatively rare dragonflies and grasshoppers. This project emphasizes the need for more consideration of urban wastelands and spontaneous wild areas. In France, several projects of constructed urban tiny forests, or Miyawaki forests, inspired by the Japanese botanist Akira Miyawaki, are gaining interest among cities. His method consists of dense plantations of various tree species, generally on an area of less than one hectare. However, urban open space policies must also evolve towards better consideration of spontaneous urban forests and the acceptance of spontaneous flora in urban spaces (Kowarik 2005).

Sealed areas such as oversized parking lots, school or building yards, concrete banks, and public paved space represent a huge potential for restoring soils and enhancing biodiversity in urban dense areas. As part of the European H2020 REGREEN project, the Paris Region Institute studied this potential in the city of Aulnay-Sous-Bois, in the northeast part of the Paris Region. In order to identify priority sectors for renaturation, the most vulnerable areas of the city were mapped using three criteria: biodiversity, adaptation to climate change, and health and well-being of populations. Once identified, paved areas were analyzed using a predefined typology. The first results show a great potential of paved areas for ecological restoration (hundreds of hectares at the city level). Although more fieldwork is required, the model could help cities to identify a large number of potentially paved areas and plan for NBS in high-priority locations.

Assessing biodiversity and ecosystem services provided by nature-based solutions

While many municipalities are experimenting with NBS, there is a growing need for evaluation and assessment of these newly established environments to assess their potential regarding their efficiency to address climate change and biodiversity recovery. The Agency for Biodiversity in the

Paris Region has conducted several research projects in order to assess biodiversity changes and the ecosystem services provided by NBS in order to compare different design and management practices. This work is aimed at providing local managers with guidelines on designing and planning for biodiversity.

From 2017 to 2019, a study called Grooves (for Green Roofs Verified Ecosystem Services) evaluated the benefits of green roofs in Paris and its suburbs (Barra and Johan 2021). Varied green roof typologies (extensive, semi-intensive, intensive, and wildroof) were compared based on plants and invertebrate (including pollinators) monitoring, substrate samples to study soil pollution, microorganisms, and water retention capacity. A total of 400 species of plants were monitored on the 36 roofs studied, and nearly 70% of the plants settled on these roofs spontaneously, carried by the wind or by wildlife. More than 600 species of invertebrates, mainly beetles, hymenoptera (bees, wasps, ants), and hemiptera (grasshoppers) were monitored. These results confirm that green roofs can be suitable habitats for a wide range of urban species. The rooftops' diversity of pollinators was higher than many ground-level green spaces. This diversity varied according to the roof types: extensive roofs, with essentially thin mineral substrates, had a lower richness than semi-intensive and intensive roofs which benefit from a deeper substrate. Floristic richness increases with substrate thickness up to 25 cm thick, while pollinator diversity continued to increase beyond this threshold. Regarding ecosystem services, only five of the 26 roofs studied were able to regulate an average ten-year rainfall of 48 mm in four hours. Regarding the cooling potential, only six roofs out of the 14 provide sufficient evapotranspiration to cool the roof surface, but not necessarily beyond, which limits the contribution of green roofs to urban cooling. These studies confirm that design and management principles are likely to influence biodiversity and ecosystem services provided by green roofs.

Since the 2000s, urban agriculture has grown exponentially in France. In the Paris Region, an observatory of urban agriculture carried out in 2018 produced a first map of the diversity of practices (urban harvesting, collective and shared gardens, urban farms, greenhouses, educational gardens). The regional biodiversity agency has studied 20 urban agriculture sites ranging from 200 m^2 to 40,000 m^2 to investigate the links between agricultural farming and biodiversity in cities (unpublished data, Paris Region Institute 2021). Inventories were done using participatory science protocols to compare the diversity of these spaces with that of other urban environments. The initial results reveal the mean diversity of plants as higher than in other urban ecosystems (urban parks, green roofs), including 315 spontaneous plants and 158 cultivated species. Nearly 150 taxa of insects and 535 species of arthropods were also monitored. At this stage, this study confirms that urban agriculture can benefit biodiversity in cities. Urban agricultural sites with a diversity of micro-habitats such as bare soils for solitary bees, piles of stones for lizards, and wooden boxes covered with leaves for hedgehogs were more attractive for biodiversity. Regarding agricultural techniques, the diversification of cultivated species and varieties of fruits, vegetables, or herbs, as well as keeping uncultivated areas and reducing mowing enables insects to carry out their entire life cycle. Planting shrub species can help to feed other species (birds, mammals, etc.). These recommendations can help managers and gardeners to improve design principles of urban farming sites in the future.

A similar evaluation was recently conducted on cemeteries, known as quiet places, where human activities are limited. In France, old cemeteries are traditionally graveled, but an evolution of management practices and the new design approaches for cemeteries give more space for increasing biodiversity. The Paris Region Biodiversity Agency and its partners are actively studying 45 cemeteries across the Paris Region. The biodiversity of gravel or landscaped cemeteries ranging from 2,700 m^2 to over 60 hectares was compared, as well as the influence of management practices ranging from the use of pesticides to ecological management. Initial results identify 421 species of plants, 175 groups of pollinators, 15 species of bats, and 4 groups of small terrestrial mammals (unpublished data, Paris Region Institute). The results indicate that cemeteries can be urban land uses with the potential to increase biodiversity and that the design and management principles can also have an influence. The gravel substrate used in cemeteries tends to support species adapted to hot and dry conditions.

Conclusions

Studies of many local initiatives reveal a growing shift in public acceptance of urban design and planning approaches that actively consider biodiversity and climate change issues. Over the last several decades, municipalities in the Paris Region are slowly transitioning from traditional engineering to an NBS approach to planning. They rely on NBS to adapt cities to heat and flooding, improve the health and well-being of inhabitants, and increase the available access to green spaces. Progressively, these NBS assets are replacing gray infrastructure, wetlands and flood expansion areas are restored for flood management, forests and urban woodlands projects are imagined to fight against summer heat, concrete banks are restored, and there is a growing interest for removing paving from urban areas. However, these local actions must multiply and be embraced by policymakers to be consistent across the region. Although interest in urban ecology has increased in recent years, responses to urban ecology are still insufficient to halt biodiversity decline. Beyond opportunistic actions, the Paris Region needs to build a clear road map for climate change adaptation and biodiversity governance, with more regulations and incentives that consider solutions at local scales and how they fit into a regional network. Research in urban ecology can provide a scientific framework and principles to plan, design, and manage for biodiversity (Clergeau 2020). However, this new paradigm of urban planning cannot be achieved without collaboration between ecologists and urban policymakers, designers, and managers.

Key companion papers

Barot, S., et al. 2019. Urban ecology, stakeholders and the future of ecology. *Science of the Total Environment* 667:475–484.

Clergeau, P. 2020. *Urbanisme et biodiversité, vers un paysage vivant structurant le projet urbain.* Éditions Apogée, Rennes, France.

Muratet, A., F. Chiron, M. Muratet. 2021. *Manual of Urban Ecology.* les presses du réel, Dijon, France.

Spotswood, E., et al. 2019. *Making Nature's City.* SFEI Contribution No. 947. San Francisco Estuary Institute, Richmond, CA.

References

Ahn, C., S. Schmidt. 2019. Designing wetlands as an essential infrastructural element for urban development in the era of climate change. *Sustainability* 11:1920.

Aronson, M.F., C.A. Lepczyk, K.L. Evans, M.A. Goddard, S.B. Lerman, J.S. MacIvor, C.H. Nilon, T. Vargo. 2017. Biodiversity in the city: Key challenges for urban green space management. *Frontiers in Ecology and the Environment* 15:189–196.

Barra, M. 2020. *Gestion des eaux pluviales et biodiversité: Revue bibliographique et préconisations.* ARB IdF, Paris, France.

Barra, M., H. Johan. 2021. *Ecology of Green Roof: Summary of the GROOVES (Green Roofs Verified Ecosystem Services) Study.* Paris Region Institute, Paris, France.

Baude, M., A. Muratet, C. Fontaine, M. Pellaton. 2011. *Plantes et pollinisateurs observés dans les terrains vagues de Seine-Saint-Denis.* Carnet de l'Observatoire Départemental de la Biodiversité Urbaine (ODBU), Seine-Saint-Denis Partner Resource Center, Seine-Saint-Denis, France.

Belliard, J., G. Gorges, C. Le Pichon, E. Tales. 2009. *Le peuplement de poissons du bassin de la Seine. Comprendre son évolution sous l'impact des changements générés par l'homme ou la nature*. Agence de l'eau Seine-Normandie, Paris, France.

Beninde, J., M. Veith, A. Hochkirch. 2015. Biodiversity in cities needs space: A meta-analysis of factors determining intra-urban biodiversity variation. *Ecology Letters* 18:581–592.

Bonthoux, S., M. Brun, F. Di Pietro, S. Greulich, S. Bouché-Pillon. 2014. How can wastelands promote biodiversity in cities? A review. *Landscape and Urban Planning* 132:79–88.

Clergeau, P. 2020. *Urbanisme et Biodiversité, vers un Paysage Vivant Structurant le Projet Urbain*. Éditions Apogée. Rennes, France.

Clergeau, P., J.P.L. Savard, G. Mennechez, G. Falardeau. 1998. Bird abundance and diversity along an urban-rural gradient: A comparative study between two cities on different continents. *The Condor* 100:413–425.

Connop, S. 2018. *Blandscaping: A Form of Ecological Gentrification*. Connecting Nature. https://connectingnature.eu/blog/blandscaping-form-ecological-gentrification

Cordeau, E. 2017. *Adapter l'Ile-de-France à la chaleur urbaine*. IAU île-de-France, Paris, France.

Cornet, N., A. Coquière. 2020. *La pleine terre: Nécessité d'une définition partagée dans les PLU. Note rapide Environnement, n° 884*. Paris Region Institute, Paris, France.

Deguines, N., R. Julliard, M. De Flores, C. Fontaine. 2016. Functional homogenization of flower visitor communities with urbanization. *Ecology and Evolution* 6:1967–1976.

Delaville, D., N. Laruelle, X. Opigez. 2021. *Requalifier les friches, un enjeu majeur de l'aménagement durable francilien, N°929*. Paris Region Institute, Paris, France.

Flégeau, M., H. Soubelet, S. Carré, S. Barot, X. Lagurgue, S. Bognon, P. Clergeau. 2021. What evidence exists on the possible effects of urban forms on terrestrial biodiversity in western cities? A systematic map protocol. *Environmental Evidence* 10:1–11.

Fosse, J., J. Belaunde, M. Dégremont, A. Grémillet. 2019. *Objectif "Zéro artificialisation nette": Quels leviers pour protéger les sols*. France Stratégie, Paris, France.

Gueymard, S., N. Laruelle. 2017. *Inégalités environnementales et sociales sont étroitement liées en Île-de-France, Note rapide Environnement, n° 749*. Paris Region Institute, Paris, France.

Hanford, J.K., C.E. Webb, D.F. Hochuli. 2020. Management of urban wetlands for conservation can reduce aquatic biodiversity and increase mosquito risk. *Journal of Applied Ecology* 57:94–805.

Kowarik, I. 2005. Wild urban woodlands: Towards a conceptual framework. Pages 1–32 in I. Kowarik, S. Körner, editors. *Wild Urban Woodlands*. Springer, Berlin, Germany.

Laruelle, N., M. Barra. 2020. *The Rise of Nature-Based Solutions*. Paris Region Institute, Paris, France.

Lososová, Z., M. Horsák, M. Chytrý, T. Čejka, J. Danihelka, K. Fajmon, O. Hájek, L. Juřičková, K. Kintrová, D. Láníková, Z. Otýpková. 2011. Diversity of Central European urban biota: Effects of human-made habitat types on plants and land snails. *Journal of Biogeography* 38:1152–1163.

Monberg, R.J., A.G. Howe, H.P. Ravn, M.B. Jensen. 2018. Exploring structural habitat heterogeneity in sustainable urban drainage systems (SUDS) for urban biodiversity support. *Urban Ecosystems* 21:1159–1170.

Muratet, A. 2006. *Diversité végétale en milieu urbain: L'exemple des Hauts-de-Seine*. Doctoral Dissertation, Université Pierre et Marie Curie, Muséum National d'Histoire Naturelle, Paris, France.

Muratet, A. 2016. *Etat de santé de la biodiversité en Île-de-France. Apport du programme de sciences participatives Vigie-Nature*. Natureparif, Paris, France.

Oertli, B., K.M. Parris. 2019. Review: Toward management of urban ponds for freshwater biodiversity. *Ecosphere* 10:e02810.

Pascal, M., S. Goria, V. Wagner, M. Sabastia, A. Guillet, E. Cordeau, C. Mauclair, S. Host. 2021. Greening is a promising but likely insufficient adaptation strategy to limit the health impacts of extreme heat. *Environment International* 151:106441.

Pellissier, V., M. Cohen, A. Boulay, P. Clergeau. 2012. Birds are also sensitive to landscape composition and configuration within the city centre. *Landscape and Urban Planning* 104:181–188.

Radford, J.Q., A.F. Bennett, G.J. Cheers. 2005. Landscape-level thresholds of habitat cover for woodland-dependent birds. *Biological Conservation* 124:317–337.

Rivallin, P. 2021. *Suivi de la Biodiversite du Parc Urbain du Vignois a Gonesse*. Paris Region Institute, Paris, France.

Rudd, H., J. Vala, V. Schaefer. 2002. Importance of backyard habitat in a comprehensive biodiversity conservation strategy: A connectivity analysis of urban green spaces. *Restoration Ecology* 10:368–375.

Scapino, J. 2016. *De la friche urbaine à la biodiversité: Ethnologie d'une reconquête: (La petite ceinture de Paris)*. Doctoral Dissertation. Muséum National d'Histoire Naturelle, Paris, France.

Schwoertzig, E., A. Hector, S. Kaempf, M. Trémolières, S. Brolly. 2015. *Comment concevoir des continuités écologiques en milieu urbain.* Sciences Eaux et Territoires, hors série, 7. Irstea, Antony, France.

Seddon, N., A. Smith, P. Smith, I. Key, A. Chausson, C. Girardin, J. House, S. Srivastava, B. Turner. 2021. Getting the message right on nature-based solutions to climate change. *Global Change Biology* 27:1518–1546.

Shanahan, D.F., C. Miller, H.P. Possingham, R.A. Fuller. 2011. The influence of patch area and connectivity on avian communities in urban revegetation. *Biological Conservation* 144:722–729.

Shwartz, A., A. Turbé, R. Julliard, L. Simon, A.C. Prévot. 2014. Outstanding challenges for urban conservation research and action. *Global Environmental Change* 28:39–49.

Soga, M., Y. Yamaura, S. Koike, K.J. Gaston. 2014. Land sharing vs. land sparing: Does the compact city reconcile urban development and biodiversity conservation? *Journal of Applied Ecology* 51:1378–1386.

Spotswood, E., R. Grossinger, S. Hagerty, M. Bazo, M. Benjamin, E. Beller, L. Grenier, R. Askevold. 2019. *Making Nature's City: A Science-Based Framework for Building Urban Biodiversity.* San Francisco Estuary Institute, Richmond, CA.

Szulczewska, B., R. Giedych, J. Borowski, M. Kuchcik, P. Sikorski, A. Mazurkiewicz, T. Stańczyk. 2014. How much green is needed for a vital neighbourhood? In search for empirical evidence. *Land Use Policy* 38:330–345.

Trentanovi, G., T. Campagnaro, I. Kowarik, M. Munafò, P. Semenzato, T. Sitzia. 2021. Integrating spontaneous urban woodlands into the green infrastructure: Unexploited opportunities for urban regeneration. *Land Use Policy* 102:105221.

28
REGENERATION OF URBAN NATURE
From planning to action

Diana Wiesner

Introduction

The Latin American and Caribbean region is one of the most urbanized in the world, as almost 80% of its population lives in cities. Likewise, this region has the greatest biodiversity on the planet since it contains about 60% of species of terrestrial, marine, and freshwater flora and fauna (UNEP 2016; Ávila and Rojas 2021). Although the challenges facing inequality, poverty, and socio-environmental problems are challenging, the presence of this generous biodiversity makes the development and planning of the region a top priority (Carrizosa Umaña 2014).

According to the United Nations (UNEP 2016), we are in the decade of ecosystem restoration, which means that the role of cities is fundamental to trigger changes not only in the forms of governance and local action plans that help conserve biodiversity. Cities in the Latin American and Caribbean region are in a position to rethink their paradigms of growth and human quality, recognizing the influence and benefits of nature and its ecosystem services and migrating towards the so-called ecological urbanism. In this chapter, I review the status of biodiversity in Colombia and Bogotá; describe how key concepts from landscape planning, green infrastructure, and nature-based solutions are incorporated into planning and design practice; and present a case study of Bogotá illustrating how planning and design are relevant in its specific socio-ecological and economic context.

Biodiversity in Colombia and Bogotá

Colombia is a megadiverse country: 52% of its land area is made up of forests, and it has 299,000 hectares of mangroves. It is an urban country with 77.1% of its population living in municipal capitals, and 20 of its main cities are supplied by the ecosystem services associated with the country's high mountain peaks (páramos). Seventy percent of the water for human consumption comes from these ecosystems. The country's population is projected to increase by 13 million inhabitants by 2050. The loss of biodiversity in Colombia is alarming: in the last year more than 80,000 hectares of forests have been deforested in Colombian Amazon (Amazonas, Caquetá, Guainía, Guaviare, Putumayo, and Vaupés), which shows that we are a society that does not know how to coexist with wildlife, which equates development to construction of hard infrastructure, and that does not conceive the balance between this and conservation (Ministerio de Ambiente y Desarrollo Sostenible de Colombia 2012).

DOI: 10.4324/9781003016120-32

The capital of the country, Bogotá, is a tropical mountain megacity, located 2,600 m above sea level, a peculiarity unique among the world's cities. With a privileged location, surrounded by a majestic mountain range in a territory with the world's largest moorland ecosystems, rich in biodiversity, multiple waterways descend from its mountains towards the savanna with the most fertile soils in the country. This savanna was covered by a large wetland and a lake system in which the ancient Muisca society existed. Today, urban pressure has reconfigured the savanna landscape with the appearance of developments on the urban-rural interface that ignore its rich geography. Planning has prioritized the economic and the functional, focusing on road infrastructure, and ignored the broader socio-ecological system (Figure 28.1). Approximately 4% of the ancient forests remain, and of the 80 wetlands inventoried by civil society, there are 15 legally registered for protection (Escobar n.d.).

Understanding the complexity of the current status of biodiversity in Bogotá lies in understanding the governance of a territory in which multiple and intricate forms of appropriation and planning converge and understanding the relationships and synergies that occur in that rich and diverse territory in which multiple views converge and there is a diversity of living beings, apart from humans. Historically, and especially in the case of the Eastern Hills, protected areas that cross the city, the governance of ecosystems has been diluted by the intermingling of public agencies with different responsibilities and the pressures of urban development.

Bogotá's more than 8 million inhabitants have diverse cognitive traditions, ways of life, and ancestral experiences that create social conflicts amid the abundance of a country rich in physical, biological, and cultural conditions. Additionally, as in many other Colombian cities, a segregated society results in unequal access to biodiversity and public green areas.

Figure 28.1 Aguanoso mountain, one of the Eastern Hills in Bogotá, Colombia. Picture Credit: Fernando Cruz – Fotografia Colombiana

Figure 28.2 Development in mountains surrounding Bogotá. Picture Credit: Fernando Cruz – Fotografia Colombiana.

Public access to biodiversity and land use conflicts in the buffer zones of the páramos are areas of critical need for action. The distribution of vegetation is uneven in different areas of the city and mainly affects socioeconomically vulnerable populations. Lands in the páramos have traditionally been used for subsistence agriculture by people living in informal settlements in the urban-rural interface. Subsistence agriculture is often in conflict with the broader conservation and climate mitigation goals for the páramos (Alcaldía Mayor de Bogotá 2021: Figure 28.2).

Public space in Bogotá can be classified as deficit since historically the effective occupancy indicator has been 3.4 m^2/inhabitant, compared to goals of 10 and 15 m^2/inhabitant (Alcaldía Mayor de Bogotá 2022). The distribution of the availability, accessibility, and ecological quality of effective public space in the city is uneven. Increased amounts of impervious surface has aggravated this situation in some sections of the city, limiting opportunities for planting trees and expanding green areas. Both are cost-effective strategies in the fight against the climate crisis.

The Bogotá government has developed a strategic plan, the Main Ecological Structure, to address unequal access to the social and cultural benefits and ecosystem services of public spaces (Alcaldía Mayor de Bogotá 2022). The goals of the Main Ecological Structure are to develop 20,000 m^2 of greening, which has led to prioritizing macro strategies over the issue. Bogotá has set itself various goals for 2024 regarding biodiversity. Among them is greening of vegetated infrastructure. Regarding urban tree planting, although there has been satisfactory progress, the goals of planting 80,000 trees and maintaining the 400,000 existing in urban public space have not yet been met. The objective of planting gardens in public space shows important advances. The Main Ecological Structure also addresses agroecology, waste management, and climate variability.

Lastly, the signing of the Bogotá 2020–2050 Climate Action Plan and the declaration of the climate emergency are actions aimed at specifying the necessary actions so that the goals regarding the protection of biodiversity in the city transcend the scope of political discourse (Alcaldía Mayor de Bogotá 2021).

Additional challenges emerge in using planning and design to address biodiversity conservation in Bogotá. The following can be highlighted:

- Promote the effectiveness of actions for the conservation of biodiversity through preservation, restoration, and connectivity of regional ecosystems and of green infrastructure.
- Incorporate the role of this biodiversity in enhancing the quality of citizens' lives, into decision-making and in closing the gap in access and in the benefits of citizens to it.
- Establish border pacts as forms of management (heritage, cultural) and integration.
- Productivity and technology that contributes to biodiversity and biological interconnectivity (water, soil, air, climate). The look cannot be reduced only to the indicators; it must transcend the operability and orient them towards the quality of life of the people.
- Manage the land, since a very high percentage of protected areas is privately owned. In the specific case of Bogotá, a very low percentage, close to 30%, is public property.
- Integrate citizenship with state programs. Although a program of caring for the other was started in Bogotá, the issues of self-care, caring for life and nature are still a challenge.
- Measure the variables that make it possible to analyze the quality and use of green areas and public spaces, with indicators that include other measurements: activity and type of users, frequency and time of use, informal occupation, participation of target populations, perception of quality and safety.
- Unite designs and actions that consider the complexity of each place in its cultural, geographical, and social particularities, among others, anchored to different logics. Consideration must be given to how to adapt to this complex and flexible framework.
- Increase public access to strategic green areas on the urban fringe such as the Eastern Hills, the Bogotá River, and the wetlands so that they can be enjoyed in a sustainable way. Consider possible restrictions on hours, uses, and capacity limits according to their capacity to load and the protection and management needs of each ecosystem.

Addressing the ecological and social issues concerning biodiversity in Bogotá requires an understanding of key principles in planning and design and how they incorporate social and ecological aspects of biodiversity conservation and management.

Key principles for landscape planning

Joan Nogué describes place and landscape as the two key concepts of landscape planning. Landscape is the area as it is perceived by the population, which is defined by the dynamic interaction of natural (topography, hydrology, flora, fauna) and human (activities, history, culture) factors. The landscape, defined by the Landscape Observatory of Catalonia as the area as it is perceived by the population, whose character is the result of the dynamic interaction of natural factors (such as relief, hydrology, flora, or fauna) and humans (activities or historical heritage) (Nogué and Wilbrand 2018).

Planning and design principles based on nature-based solutions can be incorporated into broader planning that addresses economic growth, create employment opportunities, and provide well-being (Eggermont et al. 2015). They should also be framed in what are known as projective ecologies or, better still, projective socio-ecologies, where key concepts from ecology are linked to design practice (Reed and Lister 2014). Planning and design principles bring together these ideas as and apply them, often through green infrastructure projects, to specific issues in cities as described in the next section (European Commission 2010).

Honor the soil. Landscape designers work with the soil, one of the resources that contains 1% of the species, and with relationships that seem invisible to many. Soil biodiversity is so fundamental because everything we eat and drink passes through the soil biota; that is, the health of the soil has implications for public health. Designers can give a voice to living soil and claim the importance of microbial decomposition and its contribution to climate change.

Despite being a renewable resource, soil has a very low recovery rate with very high costs. This vital resource, from the perspective of the natural or built space, acts as a support for vital processes and eco-systemic functions. Public and green spaces cushion the effects of urbanization on

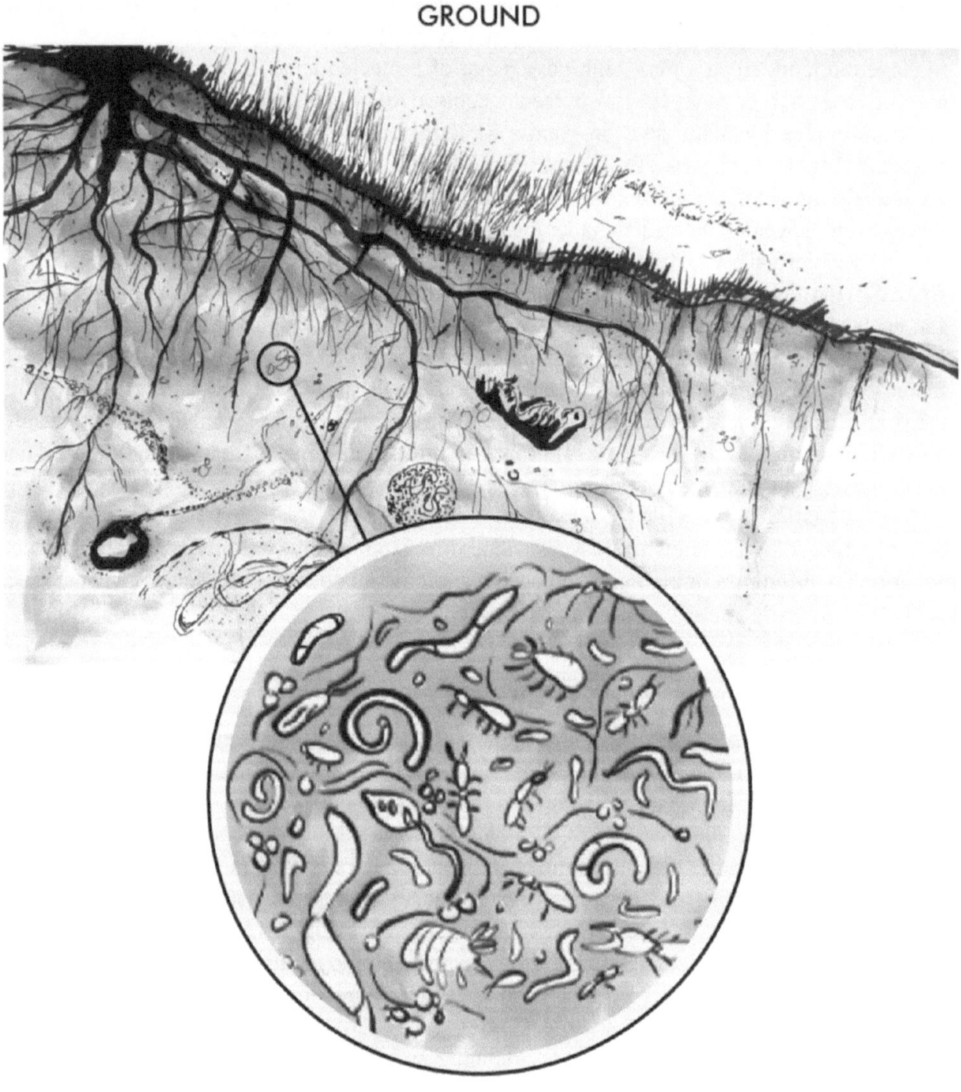

Figure 28.3 Soil and soil biota. Picture Credit: Diana Wiesner.

the ground. The valuation of the soil goes through parameters such as the control of contamination by liquid or solid waste, the stability and control of erosion, permeability, vegetation cover, and the management of runoff, among others.

Appropriate management of soil in building and construction, greening and green infrastructure, lands stabilization, and surface water and stormwater projects are some of the recommended approaches to minimizing soil loss and guaranteeing. It is equally important to generate among the public a deep ethic for understanding of the role of soil in the environment and its fragility. Soil is an increasingly scarce resource in cities, however, the need to reserve it for public use is not sufficiently appreciated. Having agricultural soils associated with a healthy life, urban agriculture and food culture, and the construction of collective work habits is an alternative that has shown its importance in many cities around the world.

Honor water. In cities with high rainfall, such as Bogotá, water can be an inspiring element for urban creation by making it visible and accessible. The designs or interventions must use a watershed approach that focuses on natural water bodies, water quality, and use of rainwater (Figure 28.4).

Comprehensive water management makes it possible to regulate rainy periods, and in public spaces it contributes, through appropriate applications of green infrastructure, to mitigating floods and reducing their speed. Sustainable water management strategies are being implemented that require the use of permeable materials in public spaces, directing of urban runoff through permeable pavements for infiltration into the land, and the creation of retention zones or the implementation of green areas that mimic natural infiltration processes.

Water should not be seen only as a resource. Rivers and streams should not be considered only as sources of energy or simple channels of liquid. It is necessary to delve into their deep and original meaning, that is, to see them as a source of life and recognize their importance as integrators and connectors of both sides.

Honor air. Human activity has generated various effects on air quality at the local, regional, and global levels. Therefore, the actions taken to improve it address external factors related to human behaviors and activities. Impacts such as the discharge of wastewater, the emissions of gases, vapors, and particles, as well as the release of aggressive or unpleasant odors are the origin of the main annoyances for public health.

Honor and manage biodiversity. Biodiversity is the fundamental basis for human well-being and the sustainable development of a nation. Its management is achieved through the provision of environmental services obtained through the conservation, preservation, and management of biological diversity. Resilience and sustainability are two essential keys in the management of biodiversity, and therefore, the evaluation of variables and criteria that generate a correct analysis of this relationship will allow evaluating appropriate environmental management policies for the proper functioning of ecosystems.

Public space constitutes an important element within urban systems, since it encourages habitability and setting the limits of use. In the same way, the inclusion of green areas in the urban environment determines the biological diversity of the area, generating decent habitat conditions for flora and fauna, thus enhancing the supply of ecosystem goods and services.

Although the benefits of green and public spaces are recognized, the goals of the local plans continue to limit compliance to quantitative factors, without thinking about quality. In addition to providing universal access, quality public spaces must be close to neighborhoods, within walking distances of local residents, and provide diverse and inclusive opportunities for recreation. New concepts such as "eco-urbanism" or "green infrastructure" are incorporated to contextualize urban environmental management based on a holistic and comprehensive view and assessment (European Commission 2010; Figure 28.5).

Figure 28.4 Design interventions to manage high rainfall on city streets. Picture Credit: Diana Wiesner and Fernanda Sanabria.

Vegetation. Good management of trees and other urban vegetation and urban trees is one of the most important components in the contribution to biodiversity. Globally, cities have focused on developing urban forestry and arboriculture programs. In recent years researchers and managers have documented the positive impact of management programs that include all strata of vegetation on birds, other native wildlife species, and human health. Given the recognized environmental,

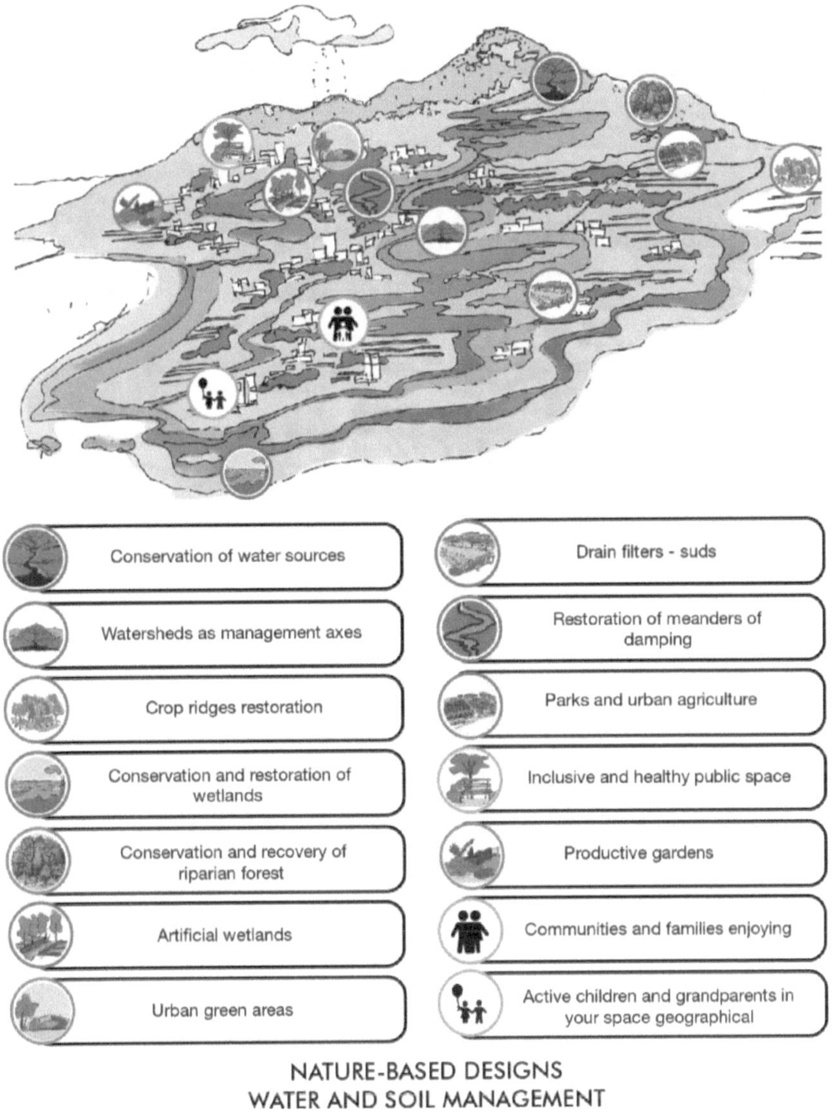

Figure 28.5 Nature-based solutions for water and soil in cities are part of biodiversity conservation. Picture Credit: Diana Wiesner and Fernanda Sanabria.

economic, and sociocultural functions of urban vegetation, prioritizing a long-term plan is imperative (Figure 28.6).

Voice for the communities. Within the planning process, give prominence to the actions of communities and citizens in favor of biodiversity. In addition to using knowledge generated by researchers, planners should listen to residents' local stories and act as spokespersons to

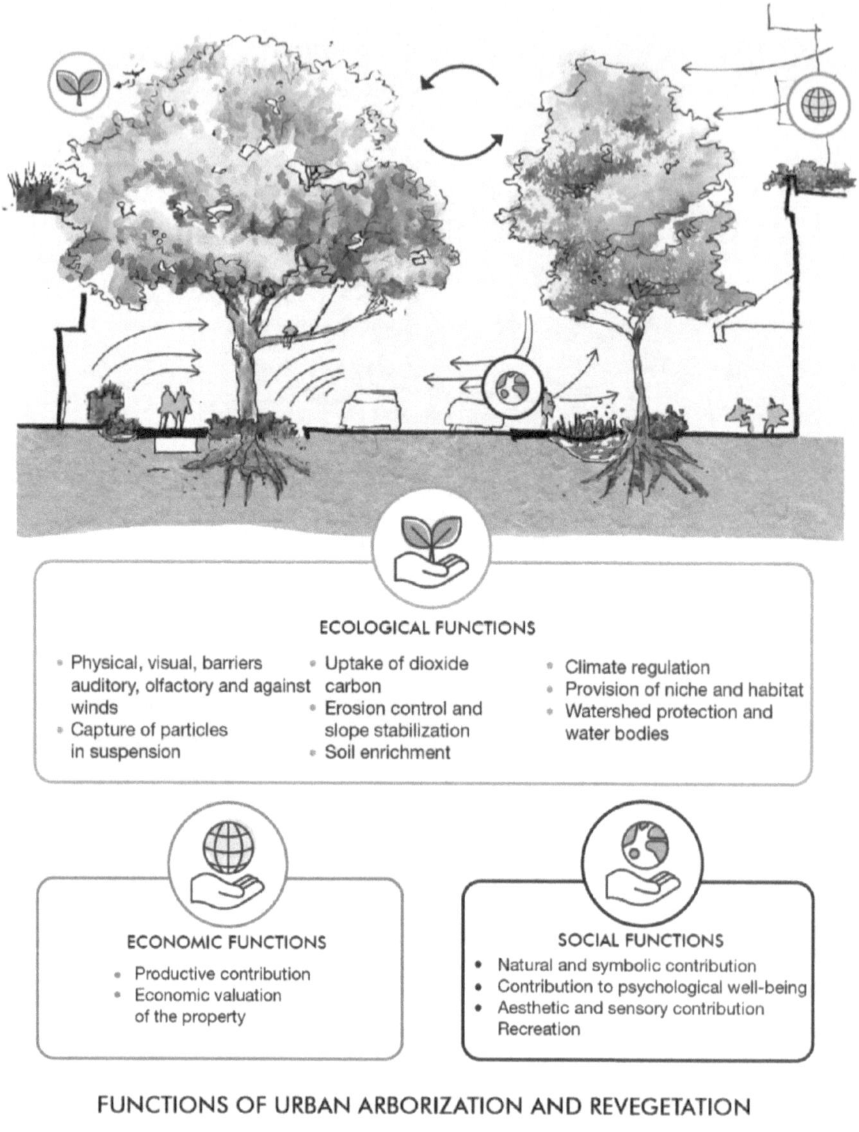

Figure 28.6 Ecological, social, and economic functions of urban greening projects. Picture Credit: Diana Wiesner and Fernanda Sanabria.

highlight residents' popular values to those who produce scientific knowledge. Planners should also give priority to local and ancestral knowledge as a way of challenging dominant and homogenizing global narratives about people and biodiversity. In this way, local residents and communities understand that citizenship includes co-responsibility for nature as well as for people (Figure 28.7).

Figure 28.7 Community engagement at Horizontes Reserve, Bogotá. Picture Credit: Juan Sebastian Pabón.

Planning and design for biodiversity

Planning and design have a fundamental role in understanding cities as urban ecosystems. All aspects of urban development are considered as part of the metabolism of the region where provision and disposal of materials are part of ecosystem processes. Cities have a dynamic associated with their larger context. They are not static. Population's migration, in the case of Colombia and Bogotá, affects ecosystems since they are the first places that are occupied informally (green areas, protected areas, border territories, among others).

In this sense, both planning and urban and landscape design must offer possibilities for the flora, fauna, and people to be better connected; they must promote a multifunctional urban-rural balance and generate a connected and permeable urban landscape with distribution and access to nature that reduce inequality gaps. Likewise, the strategies for expansion of green areas in the city must be aimed at balancing these patterns of inequity between the different areas and neighborhoods.

It is essential that the urban matrix that grows above geographies and ecosystems without adapting to them – at a time when development is extensive and has affected and fragmented said ecosystems – offers opportunities for biodiversity. In planning and design, goals, strategies, and programs can be managed that include solutions based on nature, aligned with these sustainable development goals (SDGs): (3) Well-being and health, (10) Reduction of inequalities, (11) Sustainable cities and communities, (13) Action for the climate, and (15) Life of terrestrial ecosystems (United Nations 2022).

Urban nature must be protected and managed not only by planners but by a "symphony" of actors, including citizens, who must be recognized in their processes and narratives of mobilization around the protection of biodiversity. These solutions provided by the communities should be integrated into institutional practices in a way that fosters multi-sectoral governance.

Landscape design and planning can specifically affect aspects such as measurement and monitoring of the sensitive quality of public space and green areas. And not only measure quantitative aspects but systematically evaluate others as important as proximity and accessibility, as well as establish ecological and physical quality indexes of the different types of spaces with the presence of nature.

It is important to integrate the multiple functions that green areas and public spaces perform into design. These spaces are valued for social and cultural reasons, as sites for economic and civic activities, for recreation, and community resilience, but also as places that connect the city with regional ecosystems and that support nature and biodiversity.

A Bogotá case study: design and citizen actions

Eleven years ago, in my dual role as designer and activist, I made a proposal to develop a plan rural-urban interface adjacent to the Eastern Hills of Bogotá (Wiesner Ceballos et al. 2016). I created a citizen organization, Fundación Cerros de Bogotá, with local residents to develop the plan because we see how it is essential to arrange models of mixed and innovative management not only from planning and design but from citizen action. Examples of hypotheses and practices that we test and constantly learn from are the following:

1 All the inhabitants of the city-region have the right to find a sustainable territory in the Eastern Hills of Bogotá, with the possibility for enjoyment and respectful public use.
2 Ecological knowledge and culture are fundamental for the citizen appropriation of nature.
3 Territories need to be structured and planned around water.
4 The restoration and care of nature generate civic transformations.
5 Art is a tool to generate landscape and channel pedagogical and playful processes.

The integration of the natural environment and the community is fundamental in restoration. In this sense, one of the foundation's strategies is to generate knowledge, but also to offer platforms for sharing and discussion information. The Cerros Bogotá Chair (lecture series), "From the Mountains to the People," is an example of this strategy. The lecture takes place every Friday at the Umbral Cultural Horizontes reserve and is led by experts on topics such as urban agriculture, sustainable mobility, green waste management, circular economy, ecological restoration, and water management, who kindly donate their time and share their knowledge with the foundation community. The chair, created at the end of 2014, and originally called "Frutalcerro," is open to all citizens and in the last six years has had 120 sessions and 3,600 participants within the territory of the hills of the Sabana de Bogotá (Figure 28.8).

Ecological restoration is fostering reconciliation and peace at the urban-rural interface in many places in Colombia. The restoration of ecosystems fosters the development of skills and knowledge that contribute to recovery from trauma associated with the armed conflict in Colombia (Correa Assmus 2015), which is why in 2021 the foundation named its campaign Me Restauro (I restore myself) (#MeRestauro), pointing out the importance of highlighting the vital role that plays in the regeneration of the relations of society with nature. By participating in the restoration, relations with other living organisms are also transformed.

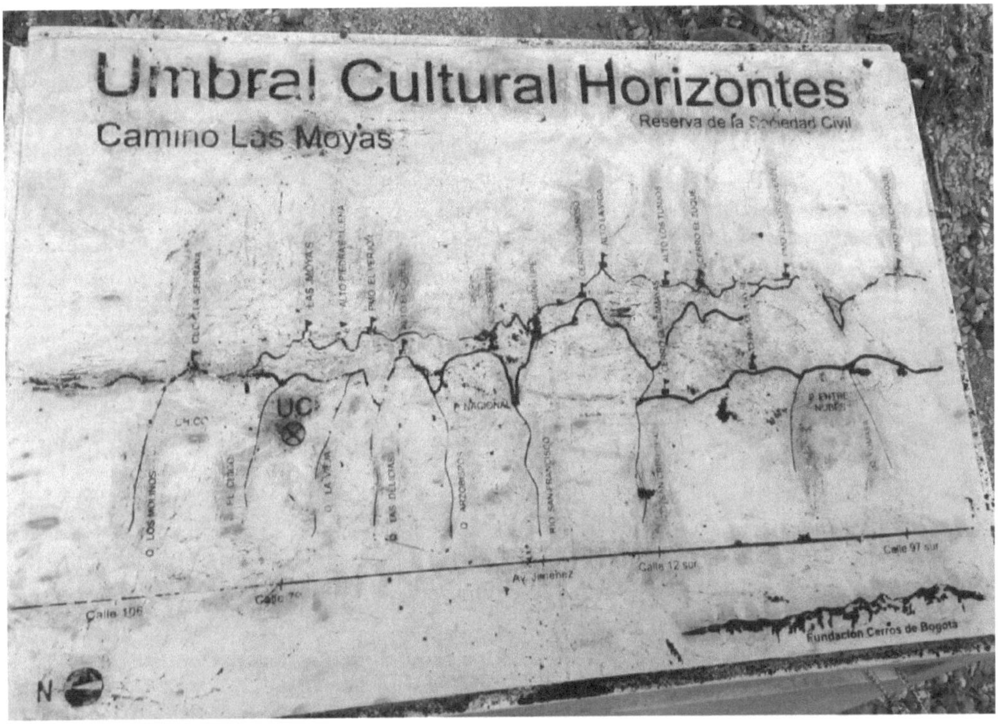

Figure 28.8 Horizontes Reserve, Bogotá, where the mountain talks to the citizens. Photo Credit: Diana Wiesner.

In the example that I present later as a citizen activist around biodiversity, we see that the role of the community is fundamental in restoration since restoration must be integral in the relationship of communities with their natural environment.

Citizenship in action for the regeneration of a society that coexists with a mountain range of life. The more than 13,000 hectares that make up the Protective Forest Reserve Bosque Oriental de Bogotá, commonly known as Cerros Orientales, located in the towns of Usaquén, Chapinero, Santa Fe, Candelaria, San Cristóbal, and Usme, seem imperturbable by what is happening in the city, in the country, in the world (Mesa and Ramírez 2016). The Eastern Hills of Bogotá constitute a set of environmental, economic, and aesthetic values linked to their unique context as relics that provide a natural landscape and environmental services to the main city of the country, which has made them central elements of the ecological heritage of the region.

The boundary area or border corridor between the city and the forest reserve had the greatest diversity of the high Andean Forest. Today, because it has been altered, it is one of the most degraded areas of the entire ecosystem. This intermediate area forms a corridor with a diverse population and an occupation with enormous potentials, but also with great weaknesses. Ninety-six percent of the native vegetation cover has been replaced with infrastructure related to urban development such as quarries, roads, and human settlements that block ecological connections between the mountains and the city, the hills with the city. Around 30 species of native mammals, 140 species of birds, six species of reptiles, eight species of amphibians, and three species of fish (one of them introduced) make these mountains their home.

In 2006 after an interdisciplinary team that toured the study area to determine its potential and weaknesses, we proposed a plan for public use of the largest ecological and recreational corridor in the city, and to establish a special management area within the Priority Public Occupation Area. Three strategies defined the plan: (1) an environmental and biophysical strategy, which sought to increase ecological connectivity and progressively reestablish the ecosystem at the foot of the slope that has to be altered by development and human activity; (2) a sociocultural strategy that sought social development, territorial appropriation, participatory planning, sustainability, and containment of expansion through the city's largest Border Pact; and (3) a spatial strategy to physically delineate the city limit within the preserve and make it available for public use for passive recreation, education, culture and conservation tourism projects. Consolidated as the most relevant Border Pact of the city, it was proposed from the beginning as a living regenerative plan that sought, in an inclusive way, to restore biodiversity as a strategy for social development and territorial appropriation by neighboring communities.

Environmental and social benefits. The socio-ecological corridor seeks to guarantee the preservation of the Cerros Orientales Forest Reserve with the generation of a corridor that definitively establishes the delimitation of urban uses. The benefits of the proposal include the improvement of ecological quality, the increase of biodiversity, and the restoration of the native forest, actions that will support recreational activities of contemplation, environmental education and research. Other benefits have to do with the promotion of social development, security, and coexistence linked to the corridor, the improvement of the quality of life (increase in the surface of the public space from 3.7 m^2/inhabitant to 4.4 m^2/inhabitant), public health, visual and air quality, as well as the urban articulation and containment of urban development, equity, the reduction of natural risks, and the creation of opportunities in the area of influence. Once the plan was proposed, we decided to create an organization of those who knew and love the hills to follow up on this plan, which was awarded and widely published and welcomed by the communities and citizens.

Since then, 11 years have passed. I am part of that group of citizens represented by the Cerros de Bogotá Foundation, which supports the dream of consolidating this corridor for public use and integration of social and ecological processes in the border area between the city and the great forest reserve of the Eastern Hills of Bogotá. This plan, although it has advanced from the citizenship by actions and initiatives of various sectors of the city, has not been executed by the mayor's office as political priority has not been given to the hills. Progress has been made in some aspects of prioritization and reactivation of some trails, but the comprehensive and integrating vision of governance has not been put in place (Robson et al. 2019).

It is precisely this scenario that has motivated us to insist on the possibility of building a shared vision and integral management for the Eastern Hills of the city. The foundation dreams of creating that socio-ecological corridor in the Adequacy Strip that restores existing trails with native species and integrates new public spaces. The dream also involves the neighbors and leaders of the hills in the care and management of this new urban ecosystem (what the foundation calls neighborhood pacts), who will be trained guides and will share stories about ethnobotany, geology, and environmental history in viewpoints, among cedars, tíbars, tunos, and amargosos, through visible and clean streams, spotting mountain turkeys, hummingbirds, weasels, and frogs at the same time.

The Horizontes Cultural Threshold Reserve: collaborative transformation laboratory. The foundation strengthened its network of collaborators by working with communities, research centers and universities, and with the local authority. We emphasized the importance of consolidating the socio-ecological corridor. One of the strategies used was to adopt a fragment of the hills and turn it into a participatory restoration project to build community.

In 2013, the State Council established that the Protective Forest Reserve Bosque Oriental de Bogotá, specifically its Adequacy Strip, should become a "priority public occupation area, in order to project a large ecological use area for the city's inhabitants." The Horizontes Cultural Threshold Reserve was declared a Civil Society Natural Reserve through Resolution 101 of 30 July 2015, intended for the enjoyment of the community and linked as a pilot project for the management and handling of the Eastern Hills. The reserve implemented a management model for the public use of a privately owned property, evidencing the possibility of managing and conserving territories belonging to civil society in an urban area. This reserve constitutes a laboratory for collaborative transformation and a space for the development of artistic, contemplative, civic and ecological education, and restoration and hiking activities (Figure 28.9).

The Umbral Cultural Horizontes reserve is one of the scenarios through which the foundation tests hypotheses and management practices, experimenting with the public policies that it promotes and which it follows up (commitments derived from the ruling of the State Council mentioned previously). These practices have to do with the possibility of developing a sustainable territory and a public space for enjoyment, generate knowledge and an ecological culture as a

Figure 28.9 Information sign for Horizontes Reserve showing information on Eastern Hills of Bogotá. Picture Credit: Diana Wiesner.

form of citizen appropriation of nature, planning the territories around water and using art as a tool to generate landscape and channel pedagogical and playful processes.

In the last five years, the foundation and its friends have planted more than 1,300 individuals of 33 species native to the hills. Through this project we have learned that involving the community is an essential component of the restoration. This led us to involve leaders from border neighborhoods to participate in the project and train them to be producers of native plant material from the hills (Figure 28.10).

Figure 28.10 Native plant nursery where local residents learn to be producers of native plant material to be used in restoration project. Picture Credit: Liseth Ramírez.

This affectionate and fruitful coexistence of the citizen with the imposing Eastern Hills, utopian or risky for some, has already been becoming a reality through a small living laboratory in the three hectares of the Umbral Cultural Horizontes. There, with the collaboration of volunteers and experts, talks, collaborative restoration with native species, orchards, and composting practices have been carried out, forming a replicable pilot project (Figure 28.11).

The Nature Reserve acts as a collaborative transformation laboratory where volunteers have been involved in participatory restoration for five years. We also generate collective works such as the creation of barriers and filters to retain plant material, thus mitigating the impact of the rains on the area. We produce works of art from the land (LandArt), and we offer weekly talks on urban ecology called Cátedra Cerros Bogotá in which the mountains speak (Figure 18.12).

Figure 28.11 Citizen actions in the mountain in the Umbral Cultural Horizontes living laboratory. Picture Credit: Diana Wiesner.

There are more than 500 plant species in the Eastern Hills, but our citizen nursery produces only 50 native species that can be used in the restoration project. However, this is the most diverse citizen-led native plant nursery in the region. The restoration of biodiversity is done in a pedogogical and participatory way. We continue to raise funds, lead walks in the hills, and collect seeds for the nursery.

This work is long, and it has not been easy. Sometimes we meet walkers who do not understand our work, we suffer troubles due to the lack of resources, or we are frustrated by the low impact that our activities sometimes have. We are also overwhelmed by the reality that we live in our country, where there is violence, attacks against environmental leaders, and increased poverty and widespread pain in a country that refuses to change, although it tries hard. Sometimes we are saddened by the lack of will and political actions to return the Eastern Hills to its role within the community. However, our hope lives on, and we continue to be motivated to contribute our little piece of peace to the life of this beloved corner of the hills of Bogotá.

Figure 28.12 Living laboratory before (2017) and after (2021) participatory restoration project. Picture Credit: Diana Wiesner.

Key companion papers

Ayos Figueroa, N. 2021. El conflicto armado, los acuerdo de paz, y la restauracion social con personas victimas y desplazadas en el Caribe Colombiano. *Revista de Investigaciones Universidad del Quindio* 33:256–270.

Gunderson, L.H., C.S. Holling, S. Light. 1995. *Barriers and Bridges to Renewal of Ecosystems and Institutions*. Columbia University Press, New York.

References

Alcaldía Mayor de Bogotá. 2021. *Plan de Acción Climática (PAC) Bogotá 2020–2050*. Bogotá, Colombia.

Alcaldía Mayor de Bogotá. 2022. *Estructura Ecológica Principal*. www.ambientebogota.gov.co/estructura-ecologica-principal

Ávila, J., F. Rojas. 2021. *Ciudades y Naturaleza, construyendo una nueva narrative hacia el urbanismo ecológico*. Hablemos de Sostenibilidad y Cambio Climatico. https://blogs.iadb.org/sostenibilidad/es/lo-que-pasa-con-la naturaleza-es-determinante-para-las-ciudades/

Carrizosa Umaña, J. 2014. *Colombia Compleja. Jardín Botánico de Bogotá José Celestino Mutis*. Instituto de Investigación de Recursos Biológicos Alexander von Humboldt. Bogotá.

Correa Assmus, G. 2015. Restauración ambiental y posconflicto. *Revista de la Universidad de LaSalle* 66:33–144.

Eggermont, H., E. Balian, J.M.N. Azevedo, V. Beumer, T. Brodin, J. Claudet, B. Fady, M. Grube, H. Keune, P. Lamarque, K. Reuter, M. Smith, C. van Ham, W.W. Weisser, X. Le Roux. 2015. Nature-based solutions: New influence for environmental management and research in Europe. *GAIA – Ecological Perspectives for Science and Society* 24:243–248.

Escobar, J.E. n.d. *Mapa de los Humedales de Bogotá*. Fundación Humedales Bogotá. http://humedalesbogata.com/mapa-humedales-bogata

European Commission. 2010. *Green Infrastructure Implementation*. https://ec.europa.eu/environment/nature/ecosystems/docs/GI%20programme.pdf

Mesa, C., A. Ramírez. 2016. Gobernanza a gran escala: La franja de adecuación de los Cerros Orientales de Bogotá. Pages 86–89 in M.A. Mejía, editor. *Naturaleza Urbana*. Instituto de Investigación de Recursos Biológicos Alexander von Humboldt. Bogotá DC, Colombia.

Ministerio de Ambiente y Desarrollo Sostenible de Colombia. 2012. *Política para la Gestión Integralde la Biodiversidad y sus Servicios Ecosistémicos*. Bogotá, Colombia.

Nogué, J., S.M. Wilbrand. 2018. Landscape identities in Catalonia. *Landscape Research* 43:443–454.

Reed, C., N.-M. Lister, editors. 2014. *Projective Ecologies*. HAR/CHRT edition. ACTAR, Harvard Graduate School of Design, Cambridge, MA.

Robson, E., L. van Kerkhoffm, S. Cork. 2019. Understanding citizen perceptions of the Eastern Hills of Bogotá: A participatory place-based ecosystem service assessment. *Urban Ecosystems* 22:19–35.

UNEP-WCMC. 2016. *The State of Biodiversity in Latin America and the Caribbean: A Mid-term Review of Progress Towards the Aichi Biodiversity Targets*. UNEP-WCMC, Cambridge.

United Nations. 2022. *The 17 Goals: Sustainable Development*. Department of Economic and Social Affairs. https://sdgs.un.org/goals

Wiesner Ceballos, D., L.F. González, L.M.W. Hoyos. 2016, Conectividad ecológica y social en el borde oriental de Bogotá. Pages 44–49 in M.A. Mejía, editor. *Naturaleza Urbana*. Instituto de Investigación de Recursos Biológicos Alexander von Humboldt. Bogotá DC, Colombia.

29

DESIGN FOR URBAN BIODIVERSITY

Applying research on biodiversity of urban lawns into landscape design practice

Maria Ignatieva

Introduction

Urban ecosystems are complex and diverse. Constant disturbances are a distinguishing feature of urban landscapes, often because urbanisation is always associated with the disappearance or degradation of native plant communities. There are almost always some remnants of native vegetation that can survive within urban boundaries. However, these remnants are subject to constant recreation pressure, changes of structure and composition, and the invasion of non-native species.

When original vegetation is cleared, two main processes shape the urban landscape: (1) spontaneous colonisation of disturbed areas and/or (2) deliberate planting of desirable plant assemblages.

Natural, semi-natural remnants, and spontaneously growing urban biotopes (i.e. habitats) were the subject of extensive research by European and North American urban ecologists for at least 50 years (Müller 1990; Rupprecht et al. 2015). Comparatively, designed plant communities were more commonly studied by landscape architects and horticulturalists with the emphasis on the design and aesthetical qualities, plant resistance to harsh urban environments, and maintenance and management. Such plant communities are seen more as a design material that should be staying in the static fixed stage like a non-living material such as stones or asphalt (Ignatieva 2021a).

Landscape architects look at the plants as living spatial elements that should create a three-dimensional outdoor-space environment that fulfils several functions (e.g. recreation, protection from sun and wind exposure). In urban areas, designed plant communities are prevalent and represented by park groves and shrubberies, hedges, street tree plantings, flowerbeds, and lawns. This aim of creating a special visual experience frequently results in designing assemblages where plants are chosen because of their aesthetical visual and structural qualities (e.g. height, form, texture, colour). Landscape architects combine these plants following the rules of visual composition (contrast, scale, sequence) and use a range of plants from different geographic regions and natural biomes. Once designed plant communities are planted, they follow the force of nature since they consist of living plants (Ignatieva 2021a).

However, designed biotopes are still ecological systems. Weeds emerge from soil seed banks or arrive with wind or animals from surrounding biotopes. Ants, microbes, fungi, and other organisms contribute to the complicated processes of functioning of urban living systems. Thus, urban ecologists are interested in studying how designed spaces can become new habitats for birds,

DOI: 10.4324/9781003016120-33

insects, invertebrates, or microbes and how these new biotopes are different from the natural (indigenous) ecosystems. Understanding the dual character of designed plant communities – that are shaped by design and ecological processes – is essential for successful landscape design.

Among all types of designed biotopes, lawns have a special position. They dominate in all urban green spaces such as public parks and private gardens, open sports grounds, road verges, and cemeteries. The lawn is also the most influential element in landscape design history (Ignatieva et al. 2020a). However, despite its dominance in the landscape, the lawn only became a subject of urban ecological studies in the last 30 years.

Definitions and approaches to lawn research

The lawn acts as a design element that visually "glues" together trees, shrub groups, and flowerbeds and highlights pavilions and architectural structures. There is an established lawn industry that supplies material on a large scale, as lawns are created by direct seeding on the ground or by planting prefabricated turf. Lawns are purely designed plant communities that originated in Europe by modifying natural grasslands and pasturelands and by selecting appropriate grass species and experimenting with maintenance schemes. The lawn was always a symbolic garden element invented for human pleasure such as sport and recreation (Ignatieva et al. 2018).

From an ecological point of view, lawns are the manmade plant communities created from selected grass species that are sowed or planted by vegetated parts of grasses (stolons) in a very high density that allows a dense durable surface to be created. Constant disturbance, mowing, provokes the growth of new shoots and makes this surface layer denser. The morphological and ecological responses of grasses to given environment conditions (e.g. soils, heat, precipitation) and to constant mowing (or grazing) resulted in forming a unique phenomenon – *turf* – which is a complex entity consisting of several parts: above the soil surface covered by dense intertwined grasses and some non-grasses shoots and some soil, and below the ground surface that consists of the mesh of densely bounded living roots and stolons which are in symbiosis with soil fauna (Harvey 2002; Ignatieva 2017). Thus, turf is the essential feature of urban lawns.

Compared to the lawn's prototypes, European meadows, the plant density in turf is much higher (up to 10,000 plant shoots compared to 3000–7000 per m^2; Goryshina and Ignatieva 2000), and the biodiversity of plants is much lower. For example, European meadows have up to 89 species (Wilson et al. 2012), compared to 1 to 4 species used for common turf mixtures. Meadow-like lawns also support more biodiverse invertebrate and soil microbial communities, compared to traditional mown lawns (Norton et al. 2019).

Once the turf is installed on the ground, it is transformed into a greater and more complicated organism – the *lawn* – that exists over a certain period. It becomes the specific plant community, the issue of different natural processes such as climate, growth of other species from existing seed banks, competition between species, development of the microbial communities, and changing of soil structure and its microbial populations. However, the most influential factor in the lawn's life is the maintenance operations such as mowing, fertilising, and aeration.

The turf growing industry has a long history. Horticulturists are looking for new grass species and advanced techniques for growing durable and resilient turfs. In the 20th century and 21st century in countries such as the USA, Australia, and New Zealand, the words "turf" and "lawns" are often used interchangeably by horticulturists, turf industry specialists, and common people without thinking about their differences ecologically. In countries such as Australia, all lawns are created from prefabricated turf grown on turf farms. When turf is installed in public or private space, it keeps the name "turf". Even in Australian scientific literature and in professional landscape architecture and garden industry, "turf" is also a prevalent term. For example, Grose (2010) noticed that turf is a dominant feature of Perth's open spaces and even has been synonymous with it.

In European countries, lawns are divided into different types such as sport, residential, public lawns, high grass, and meadow-lawn. This classification combines the use and ecological structural and compositional parameters of lawns (Ignatieva 2017). In Australia, the typology is referred to as turf and is based mostly on concrete use or ownership (e.g. sport turf, turf for private gardens, golf turf, turf for public parks). We hypothesise that such a difference in the use of "lawn" versus "turf" is also connected to sociocultural (a reinterpretation of British lawns in colonies) and ecological peculiarities (lawns as man-made plant communities with a limited grass species that aim to create mostly durable turf surface).

An important part of parks, gardens, and pleasure grounds, European lawns came through several stages of their development, from small medieval grass seats and grass plots in castle and monastery gardens (often embedded with wild flowering plants), to huge, uniformed parterres in the French-style gardens, to the vast green carpets in English parks of the 18th and 19th century (Ignatieva 2018; Figure 29.1). The lawn together with other European landscape planning and design heritage became the most influential type of green public open space in cities around the globe. Recent climate change, rapid urbanisation, and loss of biodiversity forced researchers and landscape professionals to search for more sustainable solutions in the organisation of open green spaces, including alternative solutions to traditional conventional shortcut grassy lawns.

Research on lawns

The dual nature of lawns – the designed element aimed to fulfil human needs and ecosystem services and the plant community, which is susceptible to natural forces that are beyond human control – has resulted in different approaches to lawn research.

Lawn as a horticultural crop and a provider of ecosystem services

Most of the existing literature on lawns concentrates on the successful growing of turf for creating different types of lawns. For example, turf researchers search for resistant grass species and their cultivars, effective soil mixtures, and techniques of site preparation and maintenance regime

Figure 29.1 Left: Grass seats with wild flowering plants, *The Virgin and Child in a Garden* by Martin Schongauer (1469–1491). Middle: Grass dominated parterre at the Chateau de Vaux-le-Vicomte, France. Right: English-style park in Blenheim, UK. Photos: M. Ignatieva

(i.e. irrigation and water use, applying fertilisers and pesticides, mowing frequencies). Australian lawn research is dominated by growing and maintaining turf (Barton and Colmer 2006; Cumming 2018). Research thus far is aimed at the turf industry itself and has lacked distribution to a wider interdisciplinary community.

This knowledge of how to grow, implement, and manage the turf is what landscape architects should know when designing parks and gardens. This requires understanding the basic needs of grass species, their availability from the turf farms, and understanding of soils and climatic peculiarities of a particular urban site. In typical landscape architecture, there is almost no understanding of lawns as an ecosystem that will be invaded by other undesirable species of plants and animals. Likewise, there are very few studies of lawn ecological succession under natural processes. At the same time, the knowledge of urban ecological basic processes would help urban planners, landscape architects, and park managers to maintain their urban green spaces more effectively.

Western Europe was the "giver" of lawns and the leader in the development of the industry and related research and practices. However, in the 20th and 21st centuries, the lawn establishment bloomed in the USA, Australia, New Zealand, and later in China, Iran, and other countries. Turf industries have strong research groups of biologists and soil scientists who dedicate their efforts to growing turf grasses in different climatic conditions (e.g. cold climates with heavy snowfall and arid zones with plus 40°–50°C; Khoshkholghsima and Rohollahi 2015). In all countries where the lawn has arrived as a legacy of European colonial culture or as a powerful symbol of market economy and globalisation, the research on lawns as a successful urban crop is particularly prevalent. For example, in the US, lawns are the biggest irrigated crop and cover almost 2% of the country's terrestrial area.

Another distinctive type of research related to urban lawns is on psychological, health wellbeing, and ecosystem services provided by lawns in urban green spaces (Monteiro 2017; Dickinson 2018; Ignatieva et al. 2020b). For example, research on living lawns as a mitigator of urban heat has been performed in Australia. Siebentritt (2020) showed that in Sydney on a day with a temperature of 28°C, a lawn area surface temperature was recorded to be between 28–33°C, while synthetic lawn areas and barren land adjacent to the site were recorded at 54–55°C. In Perth, the difference between living and artificial lawns was 30°C (Loveday 2019).

In the last 20 years, there is an increasing interest in lawns from an environmental history perspective, as well as interest in their sociocultural meaning, symbolism, and the history of landscape design. The European researchers were among the first who explored this theme (Ignatieva et al. 2018; Seiler 2018). However, in the last two decades, scholars from the US and Australia have actively explored this topic (Gaynor 2020). *The American Lawn* (1999) is one of the best examples of research on the United States' lawn social phenomenon, historic patterns, perceptions, status, and symbolism. There are also a group of related studies on the sociocultural benefits of urban green spaces where lawns play the leading role and, thus, received the attention of researchers (Dickinson 2018).

Lawns as urban ecosystems/biotopes

If landscape architects explore lawns from planning and design (green infrastructure and green spaces) and an environmental point of view (creating comfortable living space), ecologists look at lawns differently. Their interest is in the lawn as the result of installed turf which was exposed for some time to natural processes. The research on lawn biodiversity was first developed in Germany, other Central European countries, and the UK, but lately is an increasing topic of study in the USA, China, South Africa, Russia, and New Zealand (Yang et al. 2019; Stewart et al.

2009). The studies revealed the hypothesis of biological homogenisation of lawns, as well as the tendency of decreasing biodiversity (Müller et al. 2013). Plant biodiversity is still the dominant theme of the research with fewer works related to lawn ecology and lawn fauna (Ignatieva et al. 2020b).

Such studies on lawn biodiversity and compositional structure are also driven by the necessity to revisit the existing practice of lawns with high input of resources (e.g. water, fertilisers, herbicides) and associated costs. Because of the unification and globalisation of landscape architecture and planning styles, using the same grass species, soil mixtures, and similar routine maintenance lawns have become homogenised and low in biodiversity.

All ecological studies on lawns in the northern hemisphere logically conclude by suggesting alternative solutions for lawns, such as meadows, as a way to increase urban biodiversity. Publications on lawns and their alternatives have been growing over the last five years (Lampinen et al. 2020). However, the data obtained from biodiversity research on lawns is usually not accessible to or known by horticultural practitioners and park managers responsible for the implementation and management of urban vegetation.

Lawn as an ecological and cultural phenomenon: towards transdisciplinary research

The dual character of lawns resulted in parallel studies on growing and keeping turfs healthy, lawns as living ecosystems-biotopes, and lawns as a design material and a part of green infrastructure that provides human wellbeing. Recent climatic changes and pandemics call for new thinking and approaches to the research on lawns. There is a need for a transdisciplinary approach that addresses all of the ecological, sociocultural, and design perspectives of lawns. Once the turf has been established, it is not an agricultural crop anymore, but a new ecological entity as well as a social amenity, a design feature, and an important planning tool.

Another component of the transdisciplinary approach is the implementation of research results into practice. The novelty and importance of performing interdisciplinarity are in the transparency of science to the public and in involving the community in the research. For example, biodiversity research can be obtained by researching lawns in established openly accessible demonstration sites on traditional and alternative lawns. Then people can be asked about their preferences and thoughts on different design scenarios. Urban ecologists can also collaborate with social scientists and landscape architects on the composition, structure, maintenance, and location of different types of lawns in public or private urban spaces.

Recently, urban biodiversity scholars have pointed out the importance of integrating landscape design principles in the design of urban meadows (Wang et al. 2017). And vice versa: landscape architects should consider the results of ecological studies of different types of lawns and learn about the plant strategies and potential of lawns as wildlife biotopes if they want to create truly functional biodiverse lawns.

Most successful transdisciplinary studies on lawns were those where research results were applied and realised in the form of sustainable waterwise and alternative biodiverse solutions on the city's street verges and public parks or in the form of special educational demonstration sites and Living Labs. However, transdisciplinary research on lawns and the implementation of the results are still a rarity, especially in Australia, where lawns are seen mostly as a horticultural crop (turf) and not as a complex urban biotope. The main goal of this chapter is to share, discuss, and compare experiences and implementations of two transdisciplinary projects on alternative lawns in Sweden and Western Australia using experimental Living Labs.

Methodology

This chapter argues the importance of experimental Living Labs as an effective platform for applying the research on the biodiversity of urban lawns into planning and design practice and as an important educational facility.

The author used the results of two projects where she was the leader. The first project was *Lawn as an ecological and cultural phenomenon: Search for sustainable lawns in Sweden* (2013–2016) funded by the Swedish Research Council. From the very beginning, the project adopted a transdisciplinary approach (Ignatieva et al. 2020a). The team consisted of 11 researchers of different backgrounds (plant, pollinator and earthworm biodiversity, carbon sequestration, biomass production, energy use, social science, history, planning and design) and stakeholders (landscape architects, managers, and planners) from three Swedish municipalities.

This project went beyond the perspective of classical biology (biodiversity), geography, conservation ecology, or landscape architecture and "view lawns as a field of research shaped by environmental factors, human activities and management factors, and as a complex subsystem of a great urban ecosystem" (Ignatieva et al. 2020a, p. 2). One of the crucial points was to receive stakeholder feedback and updates about ongoing policies, upcoming changes of urban master plans, and how research results can be applied to the city's strategical planning and management documents. Native plant nurseries and park managers were crucial players in the project, helping in the design and installation of demonstration sites at the Ultuna Campus of the Swedish University of Agricultural Sciences (SLU) in Uppsala and later in other Swedish municipalities.

The vision for lawn alternatives for demonstration sites was based on the LAWN project results, including data on biodiversity, historic prototypes, the analysis of existing solutions in Sweden, the UK, and Germany, and consultations with the international advisory group (Ignatieva 2017). There were four different types of alternative lawns implemented in 2014 and 2016.

The second ongoing project, *Lawn as an ecological and cultural phenomenon in time of drying climate* (started in 2021), is supported by nine industrial partners (Western Australian turf industry, local plant nurseries, city councils, and government departments) and researchers from the University of Western Australia (UWA) and Murdoch University, Perth, Western Australia. This project investigates the phenomenon of lawns in Perth from environmental (biodiversity, soil quality, heat island effect, and water use), social (perception and attitudes), and design and planning perspectives to test alternative sustainable design solutions for urban lawns in a drying climate. The first Living Lab established in UWA Campus in 2021 aimed to monitor and test a series of traditional lawns and lawn alternatives that introduce new native species, designs, and management strategies.

The concept of Living Labs

According to Lupp et al. (2021), the concept of Living Labs ("living laboratories") originated in the 1990s in the USA as a tool for describing students' experimentation in one of Philadelphia's neighbourhoods. Later in the early 2000s, this concept was accepted in Europe as a progressive experimental way of doing projects and implementing results. In the 2020s, Europe went towards democratisation and innovation of science by involving stakeholders in the design and implementation of different research projects. Lupp et al. (2021) suggested investigating the opportunities of using the Living Lab concept for nature-based solutions ("Solutions that are inspired and supported by nature, which are cost-effective, simultaneously provide environmental, social, and economic benefits, and help to build resilience" European Commission – Directorate-General for Research and Innovation 2015. p. 70).

The participatory aspect – involving public and private actors in the process of designing and establishing experimental sites – makes Living Labs particularly attractive in transdisciplinary research.

Today there is no formal definition and understanding of Living Labs' scope in scientific and professional landscape design literature. In this chapter, Living Labs are understood as open and accessible transdisciplinary experimental research amenities aimed at encouraging joint scientific, communal, and educational activities concerning urban ecology and sustainable design. Compared to conventional plots of studying urban plant biodiversity (traditional quadrats), Living Labs involve design principles of space organisation (e.g. overall configuration of the demonstration area, particular details such as information signs, benches or sitting places, and art elements).

Living Labs are commonly established on university campuses and botanic gardens because of their strong emphasis on research and education. After testing in demonstration sites, new solutions become accessible and visible to the public and professionals (landscape designers, park managers) and policymakers. Living Labs discussed in this chapter were created with the participation of university students and staff as well as private and public actors. They are outcomes of the transdisciplinary research on urban lawns and allow observation of urban plant communities' dynamics and sociocultural feedback on alternatives to continue.

Living Labs sites

Four demonstration sites – Living Labs – displaying alternative lawns are in the Knowledge Garden and the central (Ultuna Plaza) and south parts of the Ultuna Campus of SLU. In this chapter, two sites are discussed in detail: a grass-free tapestry lawn and a perennial meadow established in May 2016. Ultuna is a municipality within Uppsala, the fourth biggest city in Sweden, with a population of 219,914 people, and has a Baltic climate, moderately continental climate with significant rainfall throughout the year.

Lawn as a Living Lab site is at the Nedlands Campus of the School of Design, the University of Western Australia (UWA) in Perth, Western Australia. The site was established in 2021. Perth is the capital of Western Australia with a population of 2,093,000 and has a Mediterranean climate with hot, dry summers and cool, wet winters.

Living Labs – alternative lawns in the SLU campus, Uppsala

The choice of alternatives to traditional lawns for Living Labs in Ultuna shares the European vision. European postglacial landscapes typically consisted of a variety of forests, wetlands, and meadows. Swedish lawns have the favourable precondition of pre-existing pastures and meadows, which were used by our ancestors for collecting seeds and cutting turfs for ornamental lawns. Pieces of cut meadow-like turfs were used for green roofs in countryside houses. In modern Swedish municipalities, there are four types of lawns: ornamental, conventional, high grass, and meadow-like. The existence of high grass and meadow-like categories among the others indicates the progressive environmental thinking and recognition of the importance of diverse plant communities for local biodiversity (Ignatieva 2017). Although the percentage of such types is quite small compared to conventional regularly mowed lawns.

The landscape architecture concept of establishing alternative lawns in Ultuna was mainly inspired by an acceptance and appreciation of Nordic "meagre nature" and the use of only native species (Ignatieva 2018). This vision corresponds with the character of Swedish historic landscapes and the modern environmental movement toward saving native grasslands and creating a sustainable and biodiverse urban environment. The implementation of this idea was easy since all types of lawns in Swedish cities are dominated by native species. The project researched "old" lawns (at least 10 years old). There were several native perennial and annual herbaceous species that embedded the original grasses. These plants germinated from seedbanks or were delivered by wind or animals. Some of the species from the conventional lawns such as *Prunella vulgaris, Bellis perennis, Potentilla reptans, Lotus corniculatus*, and *Achillea millefolium* were recommended and used in tapestry (grass-free lawns). These species are very beneficial for biodiversity.

Tapestry lawn

Ultuna's grass-free (tapestry) lawn was based on 29 native perennial flowering species that could potentially provide a similar level of dense plant cover as a conventional grass lawn. The original idea of a tapestry lawn was inspired by the medieval castle's flower-rich lawns, which were widely used throughout Europe and the recent works by Lionel Smith in the UK (Smith 2014), who used a wide range of low-growing herbaceous native as well exotic species. Because of its highly decorative appearance and flowering patterns during the spring and summer seasons, we were permitted to locate our experimental tapestry lawn in the heart of the campus, next to the main administrative buildings (Figure 29.2).

Figure 29.2 Tapestry lawn at the Swedish University of Agricultural Sciences (SLU), Ultuna Campus, Uppsala, Sweden, June 2019. Photo: M. Ignatieva.

The name "tapestry lawn" reflected the character of the designed plant community. It is a mosaic of attractive flowering plants that can live, compete, and create a dense, walkable, and at the same time, biodiverse ground surface. The inspiration for plant composition came from the existing natural prototypes: the Swedish open edges of forests and semi-natural grasslands. However, this type is not a "true" lawn since it is grass-free, and plants are not able to create the turf. Ultuna's tapestry lawn was created by planting pre-grown plug plantings that formed desirable aesthetical and biodiversity effects straight away just within one month. However, this method is more expensive compared to sowing seeds (labour efforts and cost of plant material). Grass-free lawns were cut only two times per season and were not irrigated.

The natural processes were already visible in the first season. The weeds from the seed bank appeared. However, the dense plantings still allowed planted species to compete with them. The development of a new plant community was influenced by climatic (dry versus wet summers) and anthropogenic factors (trampling, invasion of some undesirable species from grassy margins of the site and dumping snow from the surrounding roads in winter). The successional pattern was easy to follow since the plants were arranged according to the planting plan. The most significant change in the second and following years of the experiment was the invasion by two legumes: *Trifolium repens* and *Medicago lupulina*. These two species are particularly successful in most Swedish lawns (Figure 29.3).

Figure 29.3 White clover (*Trifolium repens*) invaded the tapestry lawn. SLU Ultuna Campus, August 2018. Photo: M. Ignatieva.

During the last five years, the successional pattern has changed and some plants have not survived or have decreased in numbers (for example, *Thymus serpyllum* (thyme) which struggled in the hot and dry periods of 2020–2021). The surviving pattern indicated the advantage of plants spreading by stolons, such as *Fragaria vesca*, *Potentilla anserina*, and *Pilosella officinarum*, and mat-like growth forms such as *Potentilla tabernaemontani* (Figure 29.4).

The Swedish growing season is quite short; thus, the most attractive period is from the middle of May to the end of September. However, the lawns are usually free of snow from March onward. After the first two seasons, the tapestry lawn was beautified by planting spring exotic bulbs (*Crocus* spp., *Scilla siberica* and *Chionodoxa* spp.) that prolonged the aesthetical season and provided an additional resource for spring pollinators (Figure 29.5).

For reinforcing the process of covering the gaps on the ground, the extra sowing of biennial and annual seeds (*Viola tricolor* and *Bellis perennis*) proved to be effective. In the first season, people were cautious to walk on the tapestry lawn because of a psychological awareness of the possibility of crushing beautiful delicate flowers, but in the second, third, and fourth seasons, there were more explorations and the use of grass-free lawn (Figure 29.2).

Figure 29.4 Species with stolons and mat-like growth forms are the best for grass-free lawns.SLU Ultuna Campus, May 2020. Photo: M. Ignatieva.

Figure 29.5 Crocuses in the tapestry lawn. SLU Ultuna Campus, April 2019. Photo: M. Ignatieva.

Perennial meadow

The second type of demonstration site was the "*Swedish meadow*": a perennial meadow community made from a mixture of native grasses and perennial herbaceous species that flower from summer – early autumn. This type of alternative lawn is one of the most common in Germany, England, Finland, and France, among other countries. Representatives from the Sheffield School of Planting Design, James Hitchmough and Nigel Dunnett, developed naturalistic herbaceous plantings-meadow-like communities which were made from native and attractive exotic species. Similarly, perennial prairie-like meadows are popular for alternative lawn solutions in the USA (Ignatieva 2017).

Perennial meadow mixtures usually contain 20% wildflowers and 80% grass species. By manipulating the mixture composition, it is possible to achieve a desirable dominant aspect of a particular species in a certain growing period. For example, by adding seeds of *Leucanthemum vulgare* in larger proportions into the seed mixtures, the landscape designer can create the "white" meadow. Existing soil conditions, hydrology, and microclimate are the most influential parameters for the choice of meadow seed mixtures. Most successful biodiverse meadows in Europe are established on nutrient-poor soils (Andren 2008).

Sweden still has a significant number of native natural and semi-natural grasslands. However, their number has dramatically declined in the last 20 years. Swedish meadows at the Ultuna Campus used the seed mixtures from Swedish company, Pratensis AB, who specialises in producing wild-flower seeds from native plant communities. The desirable ornamental and ecological effects from designed perennial meadows can be predicted by researching existing meadow plant communities or by monitoring experimental plantings. Implemented perennial meadows can exist for up to 15–20 years. The development of successional patterns in meadow plant communities is quite difficult to predict because of unpredictable natural forces in combination with weather patterns, soil moisture, the quality of original seeds, the competition direction, and the role of soil and fauna populations.

The Swedish Meadow was inspired by the traditional Swedish lövängar or grazed meadows between fruit trees. The original site in the Ultuna Campus had cherry trees planted on the conventional lowed lawn. The seed mixture for the meadow consisted of 30 species (17.5% of forbs, 82.5% of grasses). The best ornamental outcome was in the second summer season (2017) when three forbs *Echium vulgare* (blue flowers), *Leucanthemum vulgare* (white), and *Lotus corniculatus* and *Anthemis tinctoria* (yellow) were in full bloom and dominated the site. In the third season, *Echium vulgare* (biennial species) had almost disappeared, and the *Leucanthemum* and *Anthemis* decreased in numbers, but legumes, such as *Lotus corniculatus*, became more dominant. In the next season, the species that bloom later in the summer season (July–August), such as *Centaurea*, became more dominant in the plant community (Figure 29.6).

The central position of experimental sites within the Ultuna Campus provides a good visibility and educational opportunity for students, researchers, staff, and visitors. They had a chance to see the development of different types of meadows. This project was influential to local (Uppsala), national (Sundbyberg, Örebro, and Norrtälje Municipalities), and international communities (Finland, Estonia, and Russia) who successfully established alternative lawns.

Lawn as a Living Lab at the Nedlands Campus of the University of Western Australia, Perth

Compared to Europe, lawns in Western Australia have different social and ecological histories. Lawns arrived in Australia together with the first European settlers as a part of the colonial aesthetic heritage of England, together with other planning, landscape, and architectural styles and

Figure 29.6 Perennial meadows (Swedish Meadow) successional pattern from 2017 to 2019. (A) In July 2017 (second season) dominant plants were *Echium*, *Leucanthemum*, and *Anthemis*. (B) In August of 2018, *Lotus* was dominant. (C) In July 2019 (fourth season), *Centaurea* and *Cichorium* were most visible. Photos: Maria Ignatieva and John Lööf Green.

practices. Perth was created in 1829 on the territory of the Aboriginal Noongar people. The natural landscapes of Western Australia (WA) are very diverse and are part of the unique southwest floristic region of WA, a biodiversity hotspot (as one of the world's 36 biodiversity hotspots). It is one of the richest floras on Earth, with 11,713 native higher vascular plants (Florabase 2021). Natural plant communities such as Banksia and Eucalypt woodland, kwongan shrublands, and wetlands looked very different from the English emerald, green parklike landscape with deciduous trees and grasslands (Figure 29.7).

There are some grass species in local Western Australian flora, but none of them are suitable for creating the durable dense sod similar to traditional grassy turf. Settlers found native vegetation unattractive and alienating. Then they observed that local soils were very poor. Lawns became an important tool in "civilising" new cities and "allies in a war against heat and dirt" (Gaynor 2020). Native vegetation was cleared and replaced by private gardens and public parks where lawns play an essential role. Today Perth metropolitan areas contain a mean average of 17.3% urban green spaces. It was quite a challenging process of finding and trying appropriate grass species which could survive the heat. Tremendous efforts were put into arranging a proper irrigation system for watering lawns as well as into developing a lawn-care industry in Western Australia.

From the very beginning, lawns in Perth became an important attribute of the urban landscape and included a very strong social and moral message: well-kept lawns indicated a good citizen. It was the tool of mitigating the heat and dust and the condition of a comfortable lifestyle. These historical and social peculiarities of lawns in Australia resulted in seeing lawn mostly as a planning and designed element and not as an urban plant community also influenced by natural processes. The biggest difference from European and North American visions of lawns is that in Western Australia, lawns are not seen as a part of urban nature. Lawns in Australia are completely designed urban plant communities, consisting of exotic grass cultivars capable to create dense turf. Turf is grown on farms, mostly from stolons, and is sold in the mats ready to be installed on sites. Prepared urban soil mixtures for lawns are very different from natural soils.

In the 1970s, Australia experienced a strong environmental movement that aimed to research and protect unique native biodiversity that dramatically declined because of urbanisation (suburbanisation) and human-related disturbances. Since that time in Australia, the term "biodiversity"

Figure 29.7 Left: Eucalyptus woodland with grass trees (*Xanthorrohoea* spp.) in Western Australia. Right: English countryside with parklike landscapes with patches of deciduous forest and pastures. Photos: Maria Ignatieva.

is used to reflect only native biodiversity. The wilderness, otherwise referred to as "the bush", had become a subject of admiration and thorough research. Urban landscapes where lawns and exotic decorative plants are dominated were delineated from native bushlands and became the subject of public open spaces (POS), researched mostly from the angle as an important amenity for human wellbeing and creating a comfortable living urban environment (Figure 29.8).

This dichotomy of the existing native bush, remnant vegetation in the reserves, and national parks and urban public open spaces with designed and managed ecosystems were also fixed by the governance. POS is taken care of by local government bodies, such as the City of Perth. Native bushes are the responsibility of the Department of Biodiversity, Conservation, and Attraction, with the help of the local governments if a native remnant is located on their territories. The biodiversity of public open spaces in Australian cities is included in the sphere of scientific research only from the angle of the habitat provision for native fauna.

Perth public open green spaces are made using standard design and planning structure. They are lawns based where trees (native and exotic) are scattered within the grass areas (Grose 2010). In all governmental documents (master plans, urban development strategies) and landscape architecture publications, the term "turf" but not "lawn" dominates. This indicates misunderstanding lawns as living, complicated ecosystems with specific relationships between plants, soil, wildlife, and humans and as a part of urban nature.

There are numerous publications dedicated to questions about the growth and care of turfs in Western Australia. Research initiated by the Australian Turf Industry has focussed on the success

Figure 29.8 Public open spaces consisting of lawns and exotic trees.Kings Park, Perth, Western Australia. Photo: Maria Ignatieva.

of different species in the Australian climate, irrigation and water use, water repellency of sandy soils (Barton and Colmer 2011), and maintenance practices such as mowing height and frequency (Turf WA 2022).

However, there is no empirical research on the biodiversity of lawns, their structure and typology, or their potential as urban biotopes (habitats). The recent heatwaves and coming restrictions on using bore water for irrigation require researching lawns from ecological perspectives. There is also limited research on possible alternative solutions for traditional short-cut lawns. The only related study in Perth is on the rewilding of street verges and replacing lawns with native plantings (Ligtermoet et al. 2022).

One of the most recent "Woody Meadow" projects was inspired by the English naturalistic meadows of James Hitchmough. Several woody meadows appeared in Melbourne and Perth. However, these meadows look very similar to usual low height native species plantings in Perth and other Australian cities (Figure 29.9). Such plantings never could provide walkable surfaces similar to classic lawns created from turfs.

Lawn as a Living Lab was established in May 2021 at the UWA Nedlands Campus. The foundation of this project is an interdisciplinary collaboration between ecologists, sociologists, landscape architects, and plant nurseries, WA turf industry, and local and state governments. This Living Lab tests different lawn typologies and possible sustainable solutions for lawns. It also aims to educate students, the community, and campus visitors about different types of lawns and management practices. We explore lawn structure and composition, irrigation and maintenance variations, and the potentials of native ground cover vegetation to mitigate the local climate.

The Lawn as a Living Lab exhibits the three most common lawn species used in Perth with conventional irrigation and mowing (*Cynodon dactylon, Stenotaphrum secundatum*, and *Cenchrus clandestinus*) and native species – potential ground covers – *Dichondra repens*, *Eremophila, Myoporum*, and *Scaevola* species (Figure 29.10). One of the plots is "spontaneous lawn": an empty spot that was left alone for natural processes. Another interesting plot is the "old" lawn, the existing turf that was not mown. Thus, this is an equivalent of "high grass" typology in European

Figure 29.9 Design composition with native plants in recycled pipes and turf seats in Shifting Sands-Demonstration site in the School of Design, UWA, Perth, WA. Lawns provide sitting useable space to enjoy native plants. Photos: Maria Ignatieva.

Figure 29.10 Lawns as Living Labs in the end of May 2021 (Left: one month after the installation) and in October 2021 (Right: six months after installation). The "old lawn" and "spontaneous lawn" are the most dynamic among the ten plots. Photos: Maria Ignatieva.

classification. One plot was covered by a plastic lawn. The measure of surface temperature allows us to see the influence of an urban heat island on different ground surfaces. We also regularly measure the percent of coverage of each plant species, their heights and phenological phases (Figure 29.11).

Lawns in Perth are irrigated three times a week for nine months (from September to May). Perth's lawns are completely dominated by exotic species. The dominant species are grasses from the original turf. There are also spontaneously appeared plants from the existing seed banks or from newly used soils such as *Medicago polymorpha* and different species of clover (*Trifolium repens* and *Trifolium tomentosum* are the most common). *Dichondra repens* is the only native species that was observed in Perth's lawns. *Dichondra* was used in one of the demonstration plots as a potential option for alternative lawns. Some annual weeds from Northern Hemisphere such as *Conyza* spp. Are very drought tolerant and are actively colonising urban lawns and wastelands (Figure 29.12; Ignatieva 2021b).

Western Australia is a unique part of the Australian continent with one of the oldest surfaces and endemic vegetation which existed without human disturbance for thousands of years. This led to the development of unique specialisation, sensitivity, and a very high level of plant biodiversity. During the clearing of native vegetation, the upper soil layer was removed and replaced by a new one. The native flora has no opportunity to regenerate and to compete with new arrivals: the pioneer plants such as *Conyza, Sonchus*, and other weeds. Western Australian flora is vulnerable to human disturbances and weed invasion. All urban soils in Perth are full of alien weeds which

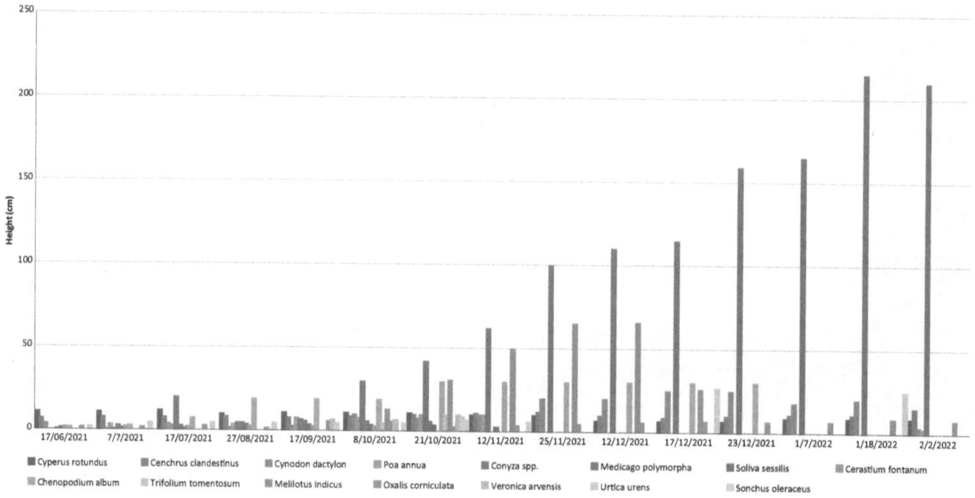

Figure 29.11 Height of 15 species found in the "spontaneous" lawns from June 2021–February 2022, UWA, Perth, Western Australia.

regenerate year-round. Warm temperatures combined with irrigation and constant human-induced disturbances created ideal conditions for growing weeds.

Discussion

The research on lawns and their alternative in Uppsala and Perth, the cities of a different climate, cultural, social, and planning history, confirmed the dominant position of lawns in urban green spaces and their complex character as urban biotopes. Lawns are designed plant communities that are created from specially selected grass species produced in globalised turf industries. Despite global distributions of lawns, their biodiversity and plant communities' essence are less researched. Australian lawns are designed and heavily maintained and are highly contrasting to surrounding native ecosystems (Figure 29.7). Without human input and control (i.e. watering, mowing, fertilising), lawns in Australia will die or turn into weedy plant communities.

Because of the unified technique of lawn establishment and maintenance, they share similar structure and composition and thus lead to homogenisation of urban ecosystems. Grass species are dominant, but other herbaceous species can appear in lawns and increase their biodiversity. In both cities, the group of herbaceous representatives of legumes, such as *Trifolium* and *Medicago*, were among the most successful species.

The major difference between Europe and Australia is the absolute domination of exotic species and a very small percentage of native plants that are capable of colonising Australian lawns. European flora and native vegetation are very "young" compared to "old" Australian Gondwana land. Glaciation in Europe resulted in developing plants with pioneer strategies (ruderals) that can be very receptive to different disturbances and capable of colonising new areas, including urban biotopes. Urban soils in European cities have many native species which are capable to be established in the "body" of grassy lawns (Ignatieva 2021b). Despite native fauna species that can use lawn biotopes in Australian cities (Ignatieva et al. 2020b), the urban lawns in Australia are not seen as an ecosystem and are not researched from a biodiversity point of view.

Figure 29.12 *Conyza* spp. Appeared in "spontaneous" plot in July 2021, and in six months it reached the height of two meters. Lawn as Living Lab, UWA Campus, Perth. Photos: Maria Ignatieva.

There are still many research questions that should be explored, such as possible successional dynamics of lawn plant communities and the ways of navigating and increasing the desirable flora and, thus, fauna biodiversity. Another important direction can be the search for native species that can provide a dual effect: the durability and resistance to trampling, similar to turf, and the value for biodiversity.

These Living Labs provide educational and research opportunities for long empirical research that can be also visible to the community and practitioners. Publicly accessible experimental sites are growing in European parks and university campuses. For example, recent interdisciplinary research (plant biodiversity, invertebrates, soil characteristics, and soil biological analysis) used nine experimental sites of perennial meadows as alternatives to short-cut grasslands in residential areas of Bedford, Luton, and Cranford University in England. The interdisciplinary method allowed us to follow some vegetation patterns and learn about the seed bank longevity and the role of soil disturbance in plot cultivations, which can be later used in practical landscape design applications (Norton et al. 2019).

Theoretical and empirical research on lawns and implementation of alternative solutions into demonstration sites reveals that successful implementation of biodiverse lawns requires careful use of landscape design principles (Ignatieva 2017; Wang et al. 2017). Biodiversity research is better understood by the public when experimental lawns employ appropriate design frameworks such as "cues to care" where cut grass strips or crushed natural stones for trails are combined with specially designed information signs and art objects.

The results of social surveys on alternative lawn solutions in Sweden (Ignatieva et al. 2017) helped to reveal not only preferable alternative biodiverse scenarios for the public green spaces but also a failure in planning and design practice. Most cities have a lot of unused conventional lawns which could easily be redesigned as alternative meadows. The manual "Lawn Alternatives from Theory to Practice" was based on the results of transdisciplinary research and the experiences from Living Labs in Ultuna and other European examples and suggested a series of retrofitting scenarios of turning short-cut lawns into different types of alternative lawns (Ignatieva 2017).

Both projects have shown the importance of addressing and respecting the local context and people's opinions, while suggesting new lawn typology and sustainable scenarios. For example, in Sweden, urban lawns already have the prerequisites to be more ecologically "proven" since they consist of native species. Here, lawns have no pressure to be irrigated because of climate or to use pesticides since they are already prohibited in Swedish public spaces since the 1990s. Many traditional short-cut lawns can simply adopt more relaxed mowing regimes that allow blooming herbaceous plants before cutting. In Sweden, there is also a network of specialised nurseries producing seed mixtures or prefabricated meadow mats of different heights that can be quickly installed in any local neighbourhood.

As an opposite, Western Australia is a country of "lawnscapes" that cool the city. People can play sports and enjoy recreation year-round. Australians are attached to their green lawns and do not mind investing a lot of resources to keep them healthy and alive. There is a clear division between native biodiversity (the bush) and urban public spaces represented mainly by lawns that have no analogs in local Australian native ecosystems. The approach to retrofitting such spaces into more sustainable and biodiverse is more complicated. There is also no knowledge of existing urban lawn biodiversity which could help to reveal possible sustainable solutions.

Compared to Europe, where lawns are an important part of urban nature, in Australia, all spontaneously appeared plants in lawns are only considered to be weeds that are harmful to lawns. Lawns themselves are seen as a horticultural element – turf, the main design element of public open spaces. European models of alternative lawns such as high uncut grass or the "go wild" approach do not work in Australia. The "woody meadows" created from low-growing shrubs are seen as alternatives for the specific site such as road verges or some plantings in public and private gardens.

There are only a few researchers and practitioners trying to rethink lawnscapes in Western Australia. Grose (2010) proposed a new design concept for POS, which she named "turf that works". These are lawns needed for a specific purpose, such as sports or dedicated recreation spaces. Other lawns should be replaced by native ground covers, understorey planting, and introducing more native trees, thus increasing biodiversity and decreasing water use and ongoing maintenance.

This call for a new approach in planning, design, and management of POS in Perth, and particularly for lawns, is supported in many countries. However, suggested theoretical concepts might be implemented into practice only after conducting transdisciplinary research that involves social, ecological, and design studies. For quite a long time, Perth was luckier than other Australian cities and has not experienced heavy water restrictions. However, starting in 2022, the restriction on private garden irrigation applies. This requires urgent actions and revisiting policies related to public open spaces, and thus, lawns.

In the case of Western Australia and other countries with Mediterranean and arid climates, the issues of mitigating urban heat are directly connected to irrigated lawns that are an important tool for cooling cities. Recent research in Santiago, Chile, revealed that irrigation has positive implications for the quality of urban life and positive ecosystem services in cities with a Mediterranean climate. On the other hand, climate change and increasing water shortage require urgent studies on the reduction of the amount of water for irrigation of urban vegetation without jeopardising the survival of urban plants (Reyes-Paecke et al. 2019).

The question of searching for alternative solutions for Perth requires a complex approach. It includes finding more sustainable solutions for watering parks and gardens, changing the approach to planning and designing "leftover places" usually covered by lawns and finding appropriate urban vegetation. Another option is to try new draught tolerant mixtures of grasses in combination with some non-grass species such as *Dichondra*, or even to think about a more relaxed approach to lawn's composition and allow some non-native flowering species (that are already living in lawns and attract wildlife) to stay on lawns. However, these new lawn scenarios for Perth need empirical biological and social confirmation and testing in experimental trials. Addressing lawns as not only a public open space element but as an ecological system and potential valuable urban habitat can cover the missing gap between native nature and urban nature in Australian cities.

Future education and research on urban biodiversity

Landscape architects and horticulturists are responsible for creating and managing designed plant communities in cities. These two professions should logically become the main contributors in changes and reinforcing the value of urban biodiversity. However, in the current situation, university and college curricula don't include an integrated approach to urban green spaces which acknowledges the dual character of urban green spaces as a design element and as a dynamic urban ecosystem. In most landscape architecture and horticulture curricula, urban ecology (including urban biodiversity components) is not studied. Ecology students can learn some aspects of urban ecosystems (e.g. wildlife, plants) but are not able to assess and explain why such habitats have appeared in cities in the first place.

Landscape architecture students study lawns as a spatial element, and they know the essentials of how to create and manage the grassed area. On the other hand, landscape architecture students have difficulties understanding the ecological traits of grass-dominated communities and trouble decoding their ecological essence. Occasionally, landscape architecture curricula can include general botany and ecology courses. However, this is not enough, since urban ecosystems are very different from native biomes.

In Sweden and other European countries, the majority of landscape architecture programmes fall under the umbrella of agricultural or forestry schools. The plant and ecology related disciplines have more "rights" and teaching hours (Ignatieva 2019). In Australia, all landscape architecture university programmes are included in design, architectural, or built environment schools. Here the plant and ecological subjects are quite limited. There is a clear emphasis on the design, where plants are just the material for creating multifunctional spaces. This also explains a limited number of urban biodiversity research on non-native biotopes in Australian cities.

Lawns have already become a universal model for the research of urban biodiversity and a place for experimental design and monitoring ecological processes in urban environments. Now and in the future, the task of designing biodiverse landscape composition is very much dependent on creating a new generation of landscape architecture and horticultural programmes where urban ecology and urban biodiversity would have an important place.

Key companion papers

Ignatieva, M. 2017. *Lawn Alternatives in Sweden: From Theory to Practice*. Department of Urban and Rural Development, Swedish University of Agricultural Sciences, Uppsala, Sweden.

Ignatieva, M., D. Haase, D. Dushkova, A. Haase. 2020. Lawns in cities: From a globalised urban green SPACE phenomenon to sustainable nature-based solutions. *Land* 9(3):73.

Ligtermoet, E., C.E. Ramalho, J. Foellmer, N. Pauli. 2022. Greening urban road verges highlights diverse views of multiple stakeholders on ecosystem service provision, challenges and preferred form. *Urban Forestry & Urban Greening* 74:127625.

Müller, N. 1990. Lawns in German cities. A phytosociological comparison. Pages 209–222 in H. Sukopp, S. Hejný, I. Kowarik, editors. *Urban Ecology*. SPB Academic Publishing, The Hague, The Netherlands.

Müller, N., et al. 2013. Patterns and trends in urban biodiversity and landscape design. Pages 123–174 in T. Elmqvist et al., editors. *Urbanization, Biodiversity and Ecosystem Services: Challenges and Opportunities*. Springer Netherlands, Dordrecht, The Netherlands.

References

Andren, H. 2008. *Utemiljö*. Svensk Byggtijänst, Stockholm, Sweden.

Barton, L., T.D. Colmer. 2006. Irrigation and fertiliser strategies for minimising nitrogen leaching from turfgrass. *Agricultural Water Management* 80(1–3):160–175.

Barton, L., T.D. Colmer. 2011. Granular wetting agents ameliorate water repellency in turfgrass of contrasting soil organic matter content. *Plant and Soil* 358:411–424.

Cumming, J. 2018. *Environmental Assessment of the Australian Turf Industry: Horticulture Innovation Project No. TU16000 Benchmarking Report 2018*. Infotech Research, Hort Innovation, North Sydney, NSW.

Dickinson, D.C. 2018. *GREENspace Perth: A Social-Ecological Study of Urban Green Space in Perth, Western Australia*. Thesis, The University of Western Australia, Perth, Australia.

European Commission – Directorate-General for Research and Innovation. 2015. *Towards an EU Research and Innovation Policy Agenda for Nature-Based Solutions & Re-Naturing Cities*. European Commission, Brussel, Belgium.

Florabase. 2021. *Florabase*. Accessed 08/12/2022 https://florabase.dpaw.wa.gov.au/

Gaynor, A. 2020. Lawnscaping Perth: Water supply, gardens, and scarcity, 1890–1925. *Journal of Urban History* 46(1). https://doi.org/10.1177/0096144217692991

Goryshina, T., M. Ignatieva. 2000. *Botanical Excursions Around the City*. Chimisdat, St. Petersburg, Russian Federation.

Grose, M. 2010. *'Turf that Works': Changing Turf in Perth's Public Open Spaces*. Green Fields, Brown Fields, New Fields: Proceedings of the 10th Australasian Urban History, Planning History Conference, University of Melbourne, Melbourne.

Harvey, G. 2002. *The Forgiveness of Nature. The Story of Grass*. Vintage, London.

Ignatieva, M. 2017. *Lawn Alternatives in Sweden: From Theory to Practice*. Swedish University of Agricultural Sciences, Uppsala, Sweden.

Ignatieva, M. 2018. Biodiversity-friendly designs in cities and towns: Towards a global biodiversinesque style. Pages 216–235 in A. Ossola, J. Niemela, editors. *Urban Biodiversity: From Research to Practice*. Routledge, Oxon, Oxfordshire.

Ignatieva, M. 2019. Notes on a global experience of landscape architecture education from Sweden, Russia, USA, New Zealand and Australia. Pages 204–205 in L. Gao, S. Egoz, editors. *Lessons from the Past, Visions for the Future: Celebrating One Hundred Years of Landscape Architecture Education in Europe*. ECLAS and UNISCAPE Annual Conference 2019, Norwegian University of Life Sciences, Ås, Norway.

Ignatieva, M. 2021a. Evolution of the approaches to planting design of parks and gardens as main greenspaces of green infrastructure. Pages 1–25 in C. Catalano, M.B. Andreucci, R. Bretzel, M. Leone, S. Pasta, editors. *Urban Services to Ecosystems: Green Infrastructure Benefits from the Landscape to the Urban Scale. Future City*, Vol. 17. Springer, Cham, Switzerland.

Ignatieva, M. 2021b. Spontaneous urban nature: Opportunities for planting design in Europe and Australia. *Kerb* 29:32–35.

Ignatieva, M., F. Eriksson, T. Eriksson, P. Berg, M. Hedblom. 2017. The lawn as a social and cultural phenomenon in Sweden. *Urban Forestry & Urban Greening* 21:213–223.

Ignatieva, M., F. Eriksson, T. Eriksson, T. Kätterer, P. Tidåker, J. Wissman, K. Ahrné, J. Bengtsson, M. Hedblom. 2020a. Pros and cons of transdisciplinary research: A case study of Swedish lawns and their sustainable alternatives. *Urban Forestry & Urban Greening* 56:126799.

Ignatieva, M., C. Florgård, K. Lundin. 2018. Lawns in Sweden: History and etymological roots, European parallels and future alternative pathways. *Bebyggelsehistorisk tidskrift* 75:26–47.

Ignatieva, M., D. Haase, D. Dushkova, A. Haase. 2020b. Lawns in cities: From a globalised urban green SPACE phenomenon to sustainable nature-based solutions. *Land* 9(3):73.

Khoshkholghsima, N., I. Rohollahi. 2015. Evaluating biochemical response of some selected perennial grasses under drought stress in Iran. *Horticulture, Environment, and Biotechnology* 56(3):383–390.

Lampinen, J., M. Tuomi, L.K. Fischer, L. Neuenkamp, J.G. Alday, A. Bucharova, L. Cancellieri, I. Casado-Arzuaga, N. Čeplová, L. Cerveró, B. Deák, O. Eriksson, M.D.E. Fellowes, B.F. de Manuel, G. Filibeck, A. González-Guzmán, M.B. Hinojosa, I. Kowarik, B. Lumbierres, A. Miguel, R. Pardo, X. Pons, E. Rodríguez-García, R. Schröder, M.G. Sperandii, P. Unterweger, O. Valkó, V. Vázquez, V.H. Klaus. 2020. Acceptance of near-natural greenspace management relates to ecological and socio-cultural assigned values among European urbanites. *Basic and Applied Ecology* 50:119–131.

Ligtermoet, E., C.E. Ramalho, J. Foellmer, N. Pauli. 2022. Greening urban road verges highlights diverse views of multiple stakeholders on ecosystem service provision, challenges and preferred form. *Urban Forestry & Urban Greening* 74:127625.

Loveday, J., G. Loveday, J.J. Byrne, B. Ong. 2019. Seasonal and diurnal surface temperatures of urban landscape elements. *Sustainability* 11(19):5280.

Lupp, G., A. Zingraff-Hamed, J.J. Huang, A. Oen, S. Pauleit. 2021. Living Labs – A concept for co-designing nature-base solutions. *Sustainability* 13:188.

Monteiro, J.A. 2017. Ecosystem services from turfgrass landscapes. *Urban Forestry & Urban Greening* 26:151–157.

Müller, N. 1990. Lawns in German cities. A phytosociological comparison. Pages 209–222 in H. Sukopp, S. Hejný, I. Kowarik, editors. *Urban Ecology*. SPB Academic Publishing, The Hague, The Netherlands.

Müller, N., M. Ignatieva, C.H. Nilon, P. Werner, W.C. Zipperer. 2013. Patterns and trends in urban biodiversity and landscape design. Pages 123–174 in T. Elmqvist, M. Fragkias, J. Goodness, B. Güneralp, P.J. Marcotullio, R.I. McDonald, S. Parnell, M. Schewenius, M. Sendstad, K.C. Seto, C. Wilkinson, editors. *Urbanization, Biodiversity and Ecosystem Services: Challenges and Opportunities*. Springer, Dordrecht, The Netherlands.

Norton, B.A., G.D. Bending, R. Clark, R. Corstanje, N. Dunnett, K.L. Evans, D.R. Grafius, E. Gravestock, S.M. Grice, J.A. Harris, S. Hilton, H. Hoyle, E. Lim, T.G. Mercer, M. Pawlett, O.L. Pescott, J.P. Richards, G.E. Southon, P.H. Warren. 2019. Urban meadows as an alternative to short mown grassland: Effects of composition and height on biodiversity. *Ecological Applications* 29(6):e01946.

Reyes-Paecke, S., J. Gironás, O. Melo, S. Vicuña, J. Herrera. 2019. Irrigation of green spaces and Residential Gardens in a Mediterranean Metropolis: Gaps and opportunities for climate change adaptation. *Landscape and Urban Planning* 182:34–43.

Rupprecht, C.D.R., J.A. Byrne, J.G. Garden, J.-M. Hero. 2015. Informal green spaces: A trilingual systematic review of its role for biodiversity and trends in the literature. *Urban Forestry & Urban Greening* 14(4):883–908.

Seiler, J. 2018. Historic lawn management regimes. The skill of the gardeners at Gunnebo House. *Bebyggelsehistorisk tidskrift* 75:8–25.

Siebentritt, M. 2020. *Conveying the Benefits of Living Turf – Mitigation of the Urban Heat Island Effect.* Report. Horticulture Innovation Australia, North Sydney, Australia.

Smith, L. 2014. Developing grass-free lawns. *Plantsman* 13(3):182–187.

Stewart, G.H., M.E. Ignatieva, C.D. Meurk, H. Buckley, B. Horne, T. Braddick. 2009. Urban biotopes of Aotearoa New Zealand (URBANZ) (I): Composition and diversity of temperate urban lawns in Christchurch. *Urban Ecosystems* 12:233–248.

Teysott, G. 1999. *The American Lawn*, 1st ed. Princeton Architectural Press, New York.

Turf, W.A. 2022. *Turf Growers Association of WA*. Accessed 8/12/2022 https://tgawa.com.au

Wang, J.W., C.H. Poh, C.Y.T. Tan, V.N. Lee, A. Jain, E.L. Webb. 2017. Building biodiversity: Drivers of bird and butterfly diversity on tropical urban roof gardens. *Ecosphere* 8:e01905.

Wilson, J.B., R.K. Peet, J. Dengler, M. Partel. 2012. Plant species richness: The world records. *Journal of Vegetation Science* 23(4):796–802.

Yang, F., M. Ignatieva, A. Larsson, N. Xiu, S. Zhang. 2019. Historical development and practices of lawns in China. *Environment and History* 25(1):23–54.

INDEX

Note: Page numbers in *italic* indicate a figure and page numbers in **bold** indicate a table on the corresponding page